MRI
MUSCULOSKELETAL SYSTEM

JAVIER BELTRAN, MD

ASSOCIATE PROFESSOR OF RADIOLOGY
CHIEF, MUSCULOSKELETAL RADIOLOGY DIVISION
DEPARTMENT OF RADIOLOGY
THE OHIO STATE UNIVERSITY HOSPITALS
COLUMBUS, OHIO

FOREWORD BY

WILLIAM F. ENNEKING

Distinguished Service Professor of Orthopaedics
W. Thaxton Springfield Center for Orthopaedic Study and Research
Department of Orthopaedics
College of Medicine, University of Florida
Gainesville, Florida

J.B. LIPPINCOTT COMPANY • PHILADELPHIA
GOWER MEDICAL PUBLISHING • NEW YORK • LONDON

Distributed in USA and Canada by:
J.B. Lippincott Company
East Washington Square
Philadelphia, PA 19105
USA

Distributed in UK and Continental Europe by:
Harper & Row Ltd.
Middlesex House
34-42 Cleveland Street
London W1P 5FB
UK

Distributed in Australia and New Zealand by:
Harper & Row (Australasia) Pty Ltd.
P.O. Box 226
Artarmon, N.S.W. 2064
Australia

Distributed in Southeast Asia, Hong Kong, India and Pakistan by:
Harper & Row Publishers (Asia) Pte Ltd.
37 Jalan Pemimpin 02-01
Singapore 2057

Distributed in Japan by:
Igaku Shoin Ltd.
Tokyo International
P.O. Box 5063
Tokyo
Japan

Library of Congress Cataloging-in-Publication Data

Beltran, Javier.
MRI : musculoskeletal system / Javier Beltran ; foreword by William F. Enneking.
p. cm.
Includes bibliographical references.
ISBN 0-397-44644-6
1. Musculoskeletal system—Magnetic resonance imaging. I. Title.
[DNLM: 1. Bone Diseases—diagnosis. 2. Joints—anatomy & histology. 3. Magnetic Resonance Imaging—methods. 4. Muscular Diseases—diagnosis. 5. Musculoskeletal System—anatomy & histology. WE 141 B453m]
RC925.7.B45 1990
616.7'07548—dc20
DNLM/DLC
for Library of Congress 89-71484
CIP

British Library Cataloguing in Publication Data
Beltran, Javier
MRI : musculoskeletal system.
1. Man. Musculoskeletal system. Diagnosis. Magnetic resonance imaging
I. Title
616.707575

ISBN 0–397–44644–6

ISBN 0-397-44644-6

Editors: Nancy Berliner, William B. Millard
Art Director: Jill Feltham
Illustrators: Laura Pardi Duprey, Sue Ann Fung, Seward Hung, Ronald S. Futral
Designer: Maura Burke

10 9 8 7 6 5 4 3 2 1

Printed in Hong Kong

To my wife Andrea

To Chuck McConnell

10-10-90

CONTRIBUTORS

ROBERT V. BULAS, MD
Clinical Instructor
Department of Radiology
Medical College of Virginia
Richmond, Virginia

DONALD W. CHAKERES, MD
Associate Professor of
Radiology
Chief, Neuroradiology Division
Clinical Director of Magnetic Resonance Imaging
Department of Radiology
The Ohio State University Hospitals
Columbus, Ohio

VIJAY CHANDNANI, MD
Clinical Instructor, Diagnostic Radiology
Department of Radiology
The Ohio State University Hospitals
Columbus, Ohio

ANDREW J. CURTIN, MD
Fellow, Neuroradiology Division
Department of Radiology
The Ohio State University Hospitals
Columbus, Ohio

JAMES C. MOSURE, MD
Fellow, Musculoskeletal Radiology
Department of Radiology
The Ohio State University Hospitals
Columbus, Ohio

PETRA SCHMALBROCK, PhD
Consultant Physicist for General Electric Co.
Department of Radiology
The Ohio State University Hospitals
Columbus, Ohio

PREFACE

Clinical magnetic resonance imaging (MRI) was first applied to the central nervous system in the early 1980s. Initial attempts to study other areas of the anatomy were discouraging, primarily because of motion artifacts occurring in the chest and abdomen. In addition, it was assumed that bone imaging with MRI would not be feasible because of the lack of signal from cortical bone. However, it soon became apparent that the musculoskeletal system was well suited for MRI. Respiratory motion was not a consideration in musculoskeletal imaging, and careful analysis of images showed that, despite the lack of cortical bone signal, the signal from surrounding soft tissues was sufficient to provide exquisite delineation of the complex anatomy of the joints and spine. The potential for application of surface coil imaging in the extremities appeared promising.

The knee was among the first areas to be evaluated by MRI. Encouraging results were obtained in the detection of meniscal tears in humans and laboratory animals. Improved software and surface coil design, as well as the availability of medium- and high-strength magnets, led to the highly accurate technique for MRI knee imaging known today. The specificity and sensitivity of this technique is over 90 percent, not only for evaluation of meniscal lesions but also for diagnosis of lesions of the ligaments, tendons, and cartilage.

Bone marrow lesions, particularly avascular necrosis of the femoral head, were of interest to early researchers. Well-conducted studies comparing MRI with other modalities demonstrated the high accuracy of MRI in the assessment of avascular necrosis and other bone marrow diseases.

MRI of musculoskeletal tumors was another subject of early investigations. Large series studies have now demonstrated the superiority of MRI in the staging of bone and especially of neoplasms of soft tissue. However, no significant progress has yet been achieved in tissue characterization with relaxometry.

Barely five years after its first clinical application, MRI is a routine modality for the evaluation of musculoskeletal disorders. It has been applied to the evaluation of articular cartilage injuries, infections of the musculoskeletal system, and bone trauma, and the potential for its use continues to grow. Recent technical advances include off-center imaging, oblique planes of section, fast imaging techniques, refined techniques of spectroscopy, updated work stations, and dedicated pulse sequences for bone marrow imaging. Such developments have enhanced MRI's role in evaluation of musculoskeletal disorders. The increased sensitivity and decreased imaging time have improved patient throughput and lowered cost.

The intent of this book is to acquaint the reader with the common MRI manifestations of musculoskeletal disorders. Chapters 2 through 8 open with a description of the pertinent imaging techniques. Following each introductory section is a discussion of the normal anatomy of the particular joints, with extensive illustration. The third section of each chapter is devoted to the pathologic processes that affect these areas. A detailed discussion of bone marrow diseases is found in Chapter 6, *The Pelvis and Hip*, since most diffuse marrow abnormalities can be seen in the pelvis. Avascular necrosis is covered as a separate entity in the chapters dedicated to the joints. Chapter 9 is dedicated to the spine. Chapters 10 and 11 summarize the current application of MRI in musculoskeletal infections and tumors.

Most of the images presented in the book were obtained with a 1.5 Tesla superconductive magnet (Signa, General Electric). The technical considerations discussed in each chapter, therefore, are primarily based on the technical parameters used at this field strength.

It is beyond the scope of this book to discuss in depth the physical principles upon which magnetic resonance imaging is based, but Chapter 1 presents a brief summary of the more important physical concepts pertinent to MRI of the musculoskeletal system. Readers interested in more detailed discussion of the physics and imaging techniques of MRI are referred to widely available texts on the subject.

Javier Beltran, MD

ACKNOWLEDGEMENTS

I must thank Mr. John Croyle for his invaluable photographic work, which was performed meticulously and promptly; Melinda Vasila, Chief Technologist of the Magnetic Resonance Facility at OSU, for her cooperation and technical help in performing many of the MRI studies in normal volunteers and patients presented in this book; Paul J. Keller, Ph.D., from the Barrow Neurological Institute, for the helpful suggestions in the physics chapter; and Nancy Wilburn for her secretarial assistance.

Special acknowledgement is made to Dr. A.J. Christoforidis, Chairman of the Department of Radiology at OSU, for his continued support and encouragement, and to the medical staff of the Division of Orthopedic Surgery at OSU, who understood very early the potential of MRI in the investigation of musculoskeletal disorders.

FOREWORD

Magnetic Resonance Imaging (MRI) has become one of the most important diagnostic tools in the armamentarium of both the orthopaedic surgeon and the radiologist. From its recent inception to the present day, there has been an explosive development of technology with an almost constant improvement in the acquisition and interpretation of the images. The technique provides a noninvasive look at the entire anatomy and pathology of the musculoskeletal system, with a resolution of millimeters. As knowledge has been acquired through correlative studies, the accuracy of the interpretation has kept pace with the expanding technology. Dr. Beltran has provided us with the benefits of his extensive experience and produced a state-of-the-art text on the use and limitations of MRI in the musculoskeletal system. He has painstakingly matched images of striking clarity with diagrams documenting their anatomical correlation and pathologic findings that allow readers of all levels of expertise access to his diagnostic acumen. He has provided significant insight into the theory, technology, interpretation, and usefulness of Magnetic Resonance Imaging for every physician who deals with afflictions of the musculoskeletal system. For this effort we are all in his debt.

William F. Enneking, MD

CONTENTS

CHAPTER FOUR
THE ELBOW

CHAPTER FIVE
THE HAND AND WRIST

CHAPTER SIX
THE PELVIS AND HIP

CHAPTER SEVEN
THE KNEE

CHAPTER EIGHT
THE ANKLE AND FOOT

CHAPTER NINE
THE SPINE

CHAPTER TEN
MUSCULOSKELETAL TUMORS

CHAPTER ELEVEN
MUSCULOSKELETAL INFECTIONS

CHAPTER ONE

Petra Schmalbrock
Javier Beltran

Principles of Magnetic Resonance Imaging

THE INGREDIENTS

Magnetic resonance images are composed of signals produced by interactions of homogeneous external magnetic fields, magnetic field gradients, radiowaves, and the nuclei of body tissues.

The MRI unit can receive these signals from a particular location for a specific tissue type.

A homogeneous external magnetic field is generated by a magnet with a field strength of 0.2–1.5 Tesla. Hydrogen proton nuclei are used to generate images because they are the most abundant nuclei in tissue and the most sensitive of all nuclei to magnetic fields and radiowaves. These nuclei, or spins as they are also called, orient themselves with

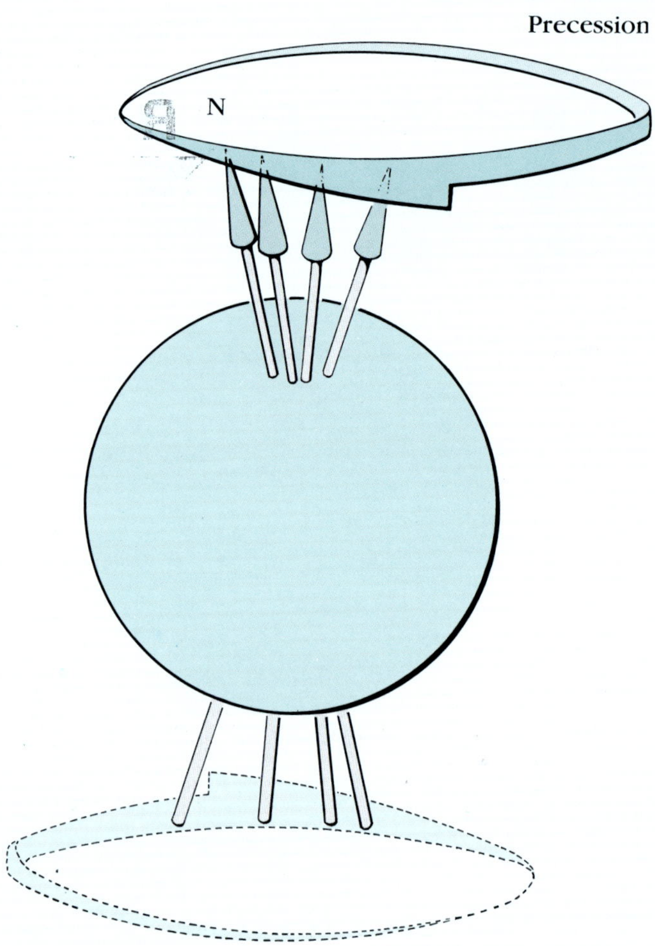

FIGURE 1.1. Hydrogen proton nucleus or spin. Note the rotational motion or precession about the magnetic field.

the external magnetic field. In addition, spins rotate (precess) about the magnetic field with a specific Larmor frequency that is determined by the strength of the external magnetic field and the internal magnetic properties of the nuclei (Fig. 1.1). The Larmor frequency for hydrogen nuclei is approximately 64 MHz (1).

In addition to the MHz variations in Larmor frequency, spins have smaller frequency differences, called chemical shift, which depend on the chemical composition of the environment. For example, water and lipid protons have a 220-Hz frequency difference at 1.5 Tesla.

Without an external magnetic field, the spins point randomly in all directions with a net magnetization of zero. Once the patient is placed in the magnet, the spins align with or against the direction of the external field to form a net longitudinal magnetization (Fig. 1.2).

A signal is generated from the tissue by application of a short radiofrequency (rf) pulse equal to the Larmor frequency of the precessing spins. This causes the net longitudinal magnetization to rotate away from its alignment by an angle (the excitation flip angle) that depends on the amplitude and/or duration of the rf pulse. The resulting vector there-

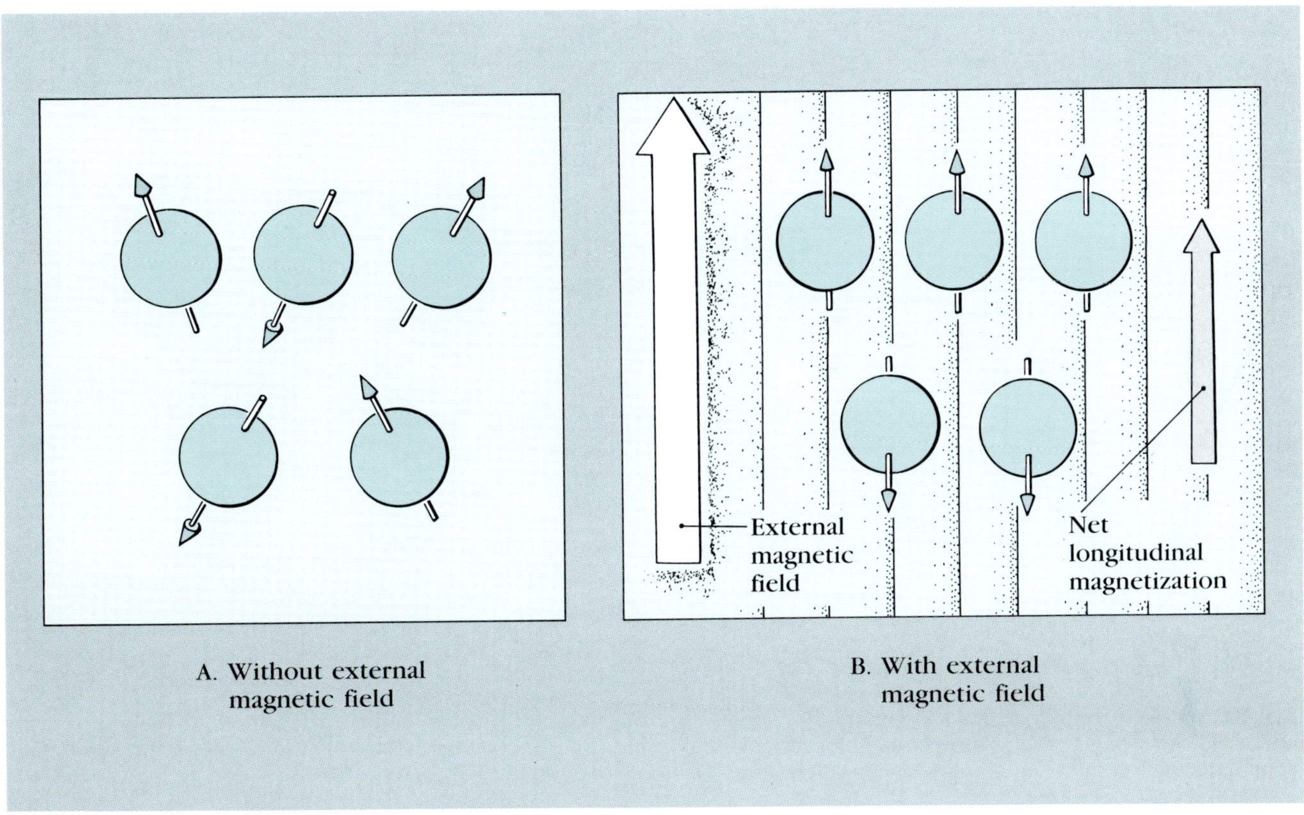

FIGURE 1.2. Spin orientation before (A) and after (B) application of an external magnetic field. Following application of an external magnetic field, spins orient themselves with the direction of the field or against it. The net longitudinal magnetization is oriented in the direction of the external field.

fore has a longitudinal component (longitudinal magnetization) and a horizontal component (transverse magnetization) (Fig. 1.3).

A receiver coil connected to an amperemeter detects the sinusoidal alternating current (the MR signal) originated by the spins precessing perpendicular to the plane of the receiver coil. This MR signal is then amplified and fed to the computer, where the time information of the signals is converted to frequency information by means of mathematical analysis (Fast Fourier Transformation or FFT).

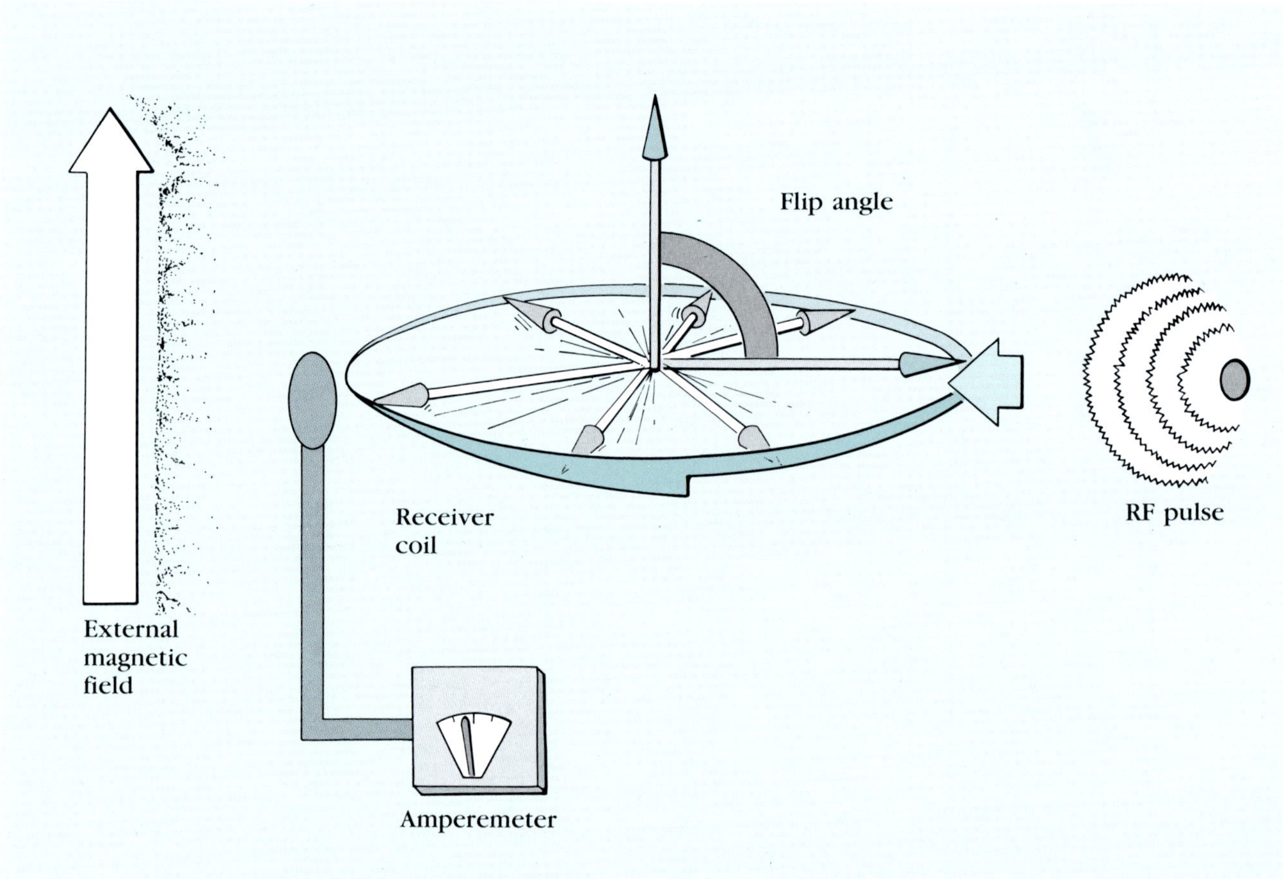

FIGURE 1.3. Generation of the MR signal. The applied rf pulse causes the spins to rotate away from their previous alignment, inducing a component of transverse magnetization. The transverse magnetization originates an alternating current in the receiver coil (the MR signal), which can be measured with an amperemeter.

The MR signal is localized by use of the gradient coils (1,3) which change the strength of the magnetic field at different positions. Since the Larmor frequency depends on the strength of the magnetic field, the position-dependent field changes imposed by the gradient coils translate into position-dependent resonance frequencies (Fig. 1.4). During the short time that the gradient coils are on, the transverse magnetization at different locations precesses with different frequencies. When the coils are turned off, spins at different locations will have precessed at different angles and acquired different phase (Fig. 1.4).

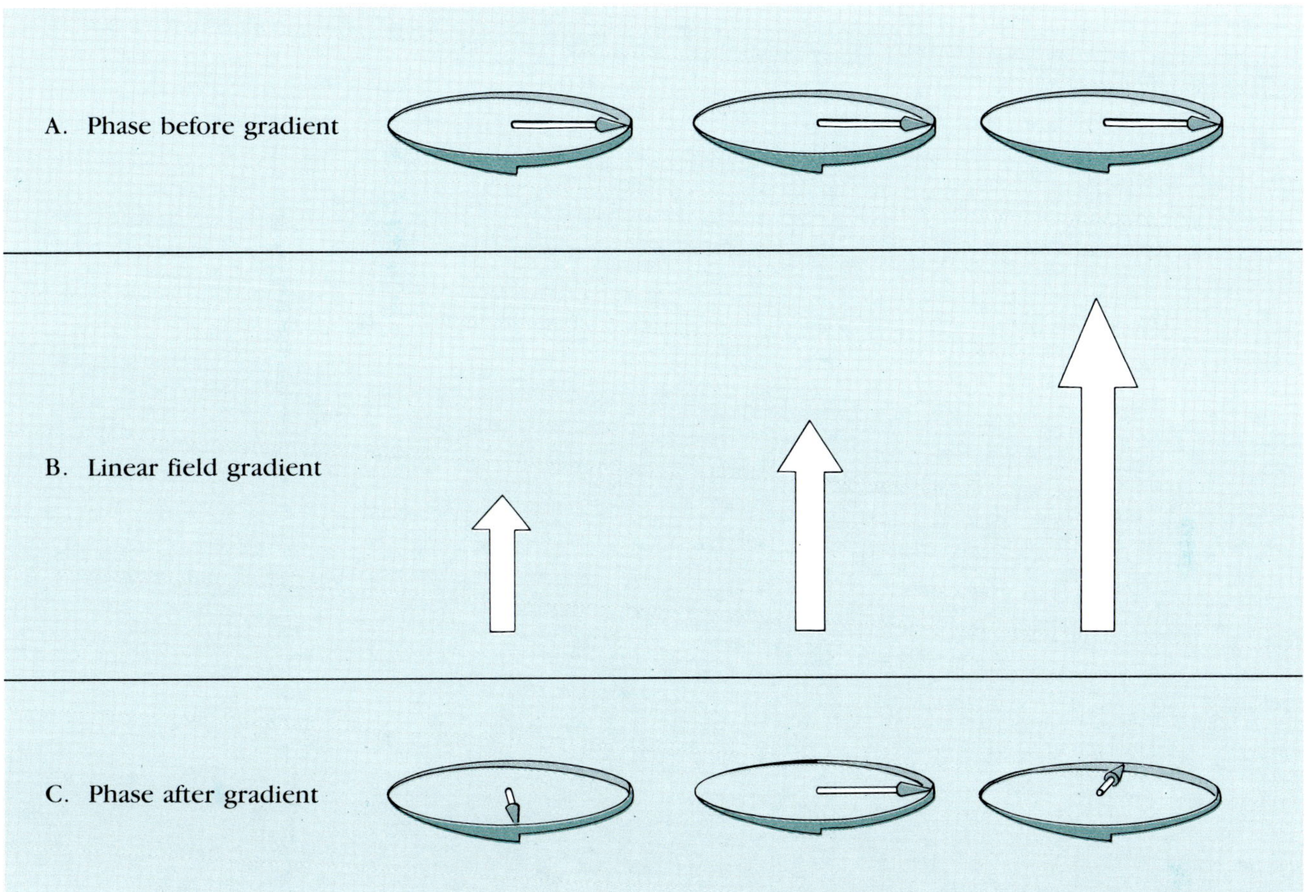

FIGURE 1.4. Effects of linear field gradients. (A) Without a field gradient, all spins precess with the same Larmor frequency and point in the same direction (phase). **(B)** A linear field gradient changes the strength (not the direction) of the main field. **(C)** When the gradient is turned on, the field is changed depending on the location, and spins therefore precess with different frequencies. Consequently, when the gradient is turned off again, spins from different positions will point in different directions, having acquired different phase.

IMAGE FORMATION

Different techniques for image formation have been described (4,5). The most commonly used methods are 2-dimensional (2D) spin warp, and 3-dimensional (3D) Fourier imaging.

With the 2D spin warp technique, sectional images are generated by manipulating the gradient fields and/or rf pulses. For slice selection, a nearly rectangular rf pulse is emitted while the slice selection gradient is turned on in the slice selection direction. This causes the spins to precess with different frequencies at different locations. The rf pulse, which covers only a given range of frequencies, excites spins in a specific slice (Fig. 1.5).

A second gradient (readout or frequency encoding gradient) is turned on while the signal is measured. Spins at different locations will then yield signals with different frequencies which can be encoded using FFT (Fig. 1.6). Finally, a third gradient (phase encoding gradient), turned on between excitation and acquisition, causes the excited spins to rotate with a position-specific frequency which, when the gradient is turned off (see Fig. 1.4), results in a different phase for different locations.

In 3D Fast Fourier Transform (3DFFT) the signal is received from the entire volume imaged for each acquisition step. Instead of using slice excitation in one dimension, and frequency encoding and phase encoding in the other two, 3DFFT imaging phase encoding is done in the first two dimensions and frequency encoding in the last. To keep imaging time within a reasonable range, 3DFFT must be

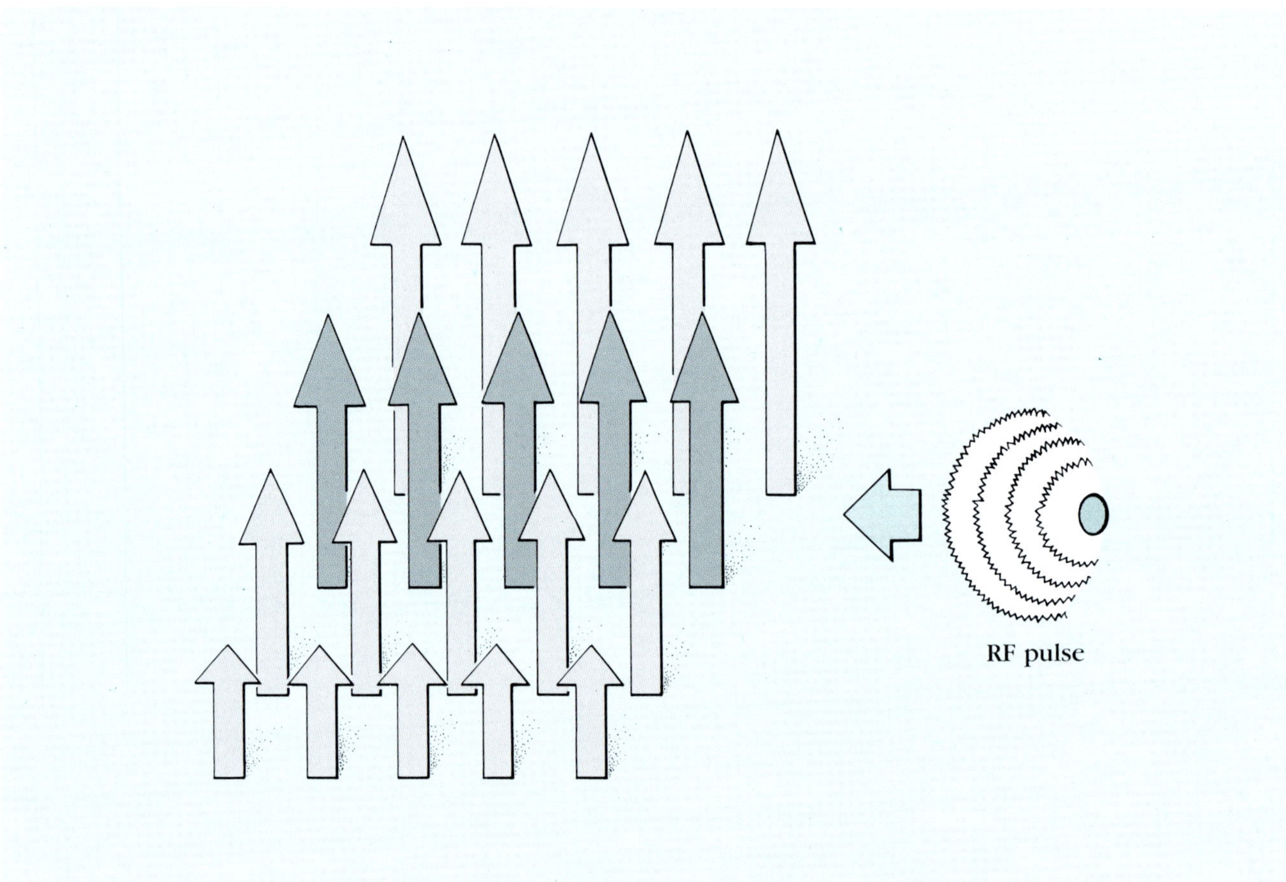

FIGURE 1.5. Slice selection. Spins are excited only if the frequency of the radiowave is equal to the Larmor frequency, which in turn is proportional to the magnetic field strength. If the rf pulse has a specific frequency range and is turned on simultaneously with a linear field gradient, only spins of a determined region are excited. This is utilized to define the first image plane, the sectional slice.

used in combination with fast scan techniques.

Three-dimensional imaging has two major advantages over 2D imaging. First, since the signal originates from the entire volume, the overall signal is much stronger than the background noise. Second, 3DFFT allows thinner slice sections to be viewed. The routine application of 3DFFT imaging is hampered only by the relative unavailability of powerful computers and computer programs to efficiently handle the flood of data.

Three prominent artifacts originate in MR imaging and should be recognized: aliasing, truncation, and chemical shift artifacts. The most frequent artifact is the aliasing, or wrap-around ar-

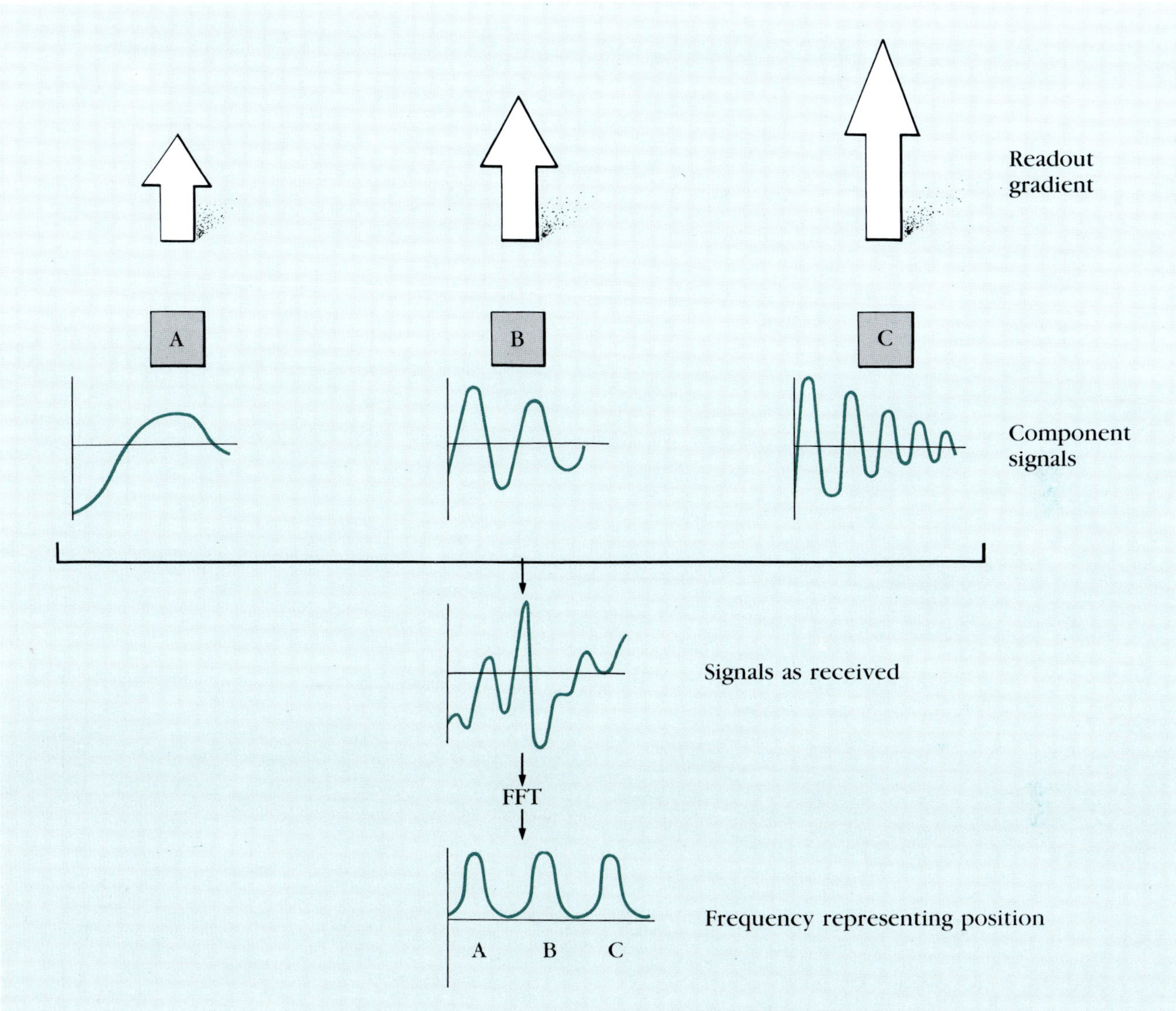

FIGURE 1.6. Frequency encoding. A linear field gradient is applied along the line of three regions **(A,B,C)** in a slice. Since the spins in region A "see" a lower field strength than in regions B or C, they precess slower, thereby producing a signal of lower frequency. Signal from all three regions is simultaneously induced in the receiver coil. The total measured signal, therefore encompasses components from all three regions. The frequency information contained in the time signal can be extracted using FFT.

tifact, and occurs when the sampled signal contains time fluctuations that are faster than the sampling rate (2). The aliasing artifact appears as a "ripple" or "fold" in the image. Aliasing is dominant in the phase encoding direction, but can occur in all imaging directions (Fig. 1.7).

Truncation artifacts result when there are discontinuities in the sampled data (2). These artifacts appear as low intensity, parallel striations, are found adjacent to fairly straight tissue interfaces (Fig. 1.8), and are more pronounced in the phase encoding direction. The use of higher resolution matrix will decrease or eliminate this artifact.

Chemical shift artifacts result from the frequency differences of tissue chemical composition. They appear as bright and dark boundaries at interfaces between fatty tissue and tissue containing predominantly water (Fig. 1.9), and are observed in the frequency encoding direction. Changing phase and frequency encoding directions can be helpful in preventing these interpretation errors.

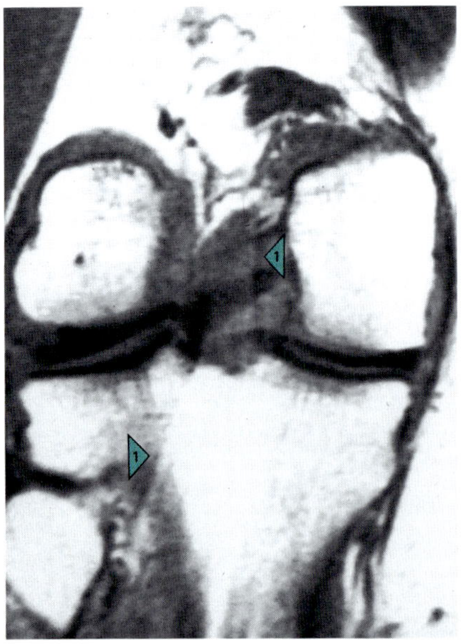

FIGURE 1.7. Wrap-around or aliasing artifact. Note the oblique lines crossing the popliteal space in this coronal knee image. Arrows **1** Wrap-around artifact.

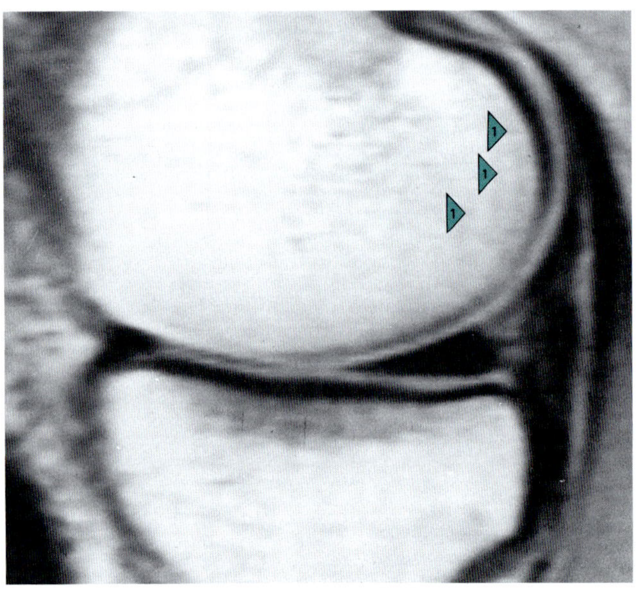

FIGURE 1.8. Truncation artifact. Note the low SI lines parallel to the posterior cortex of the medial femoral condyle. Arrows **1** Truncation artifact.

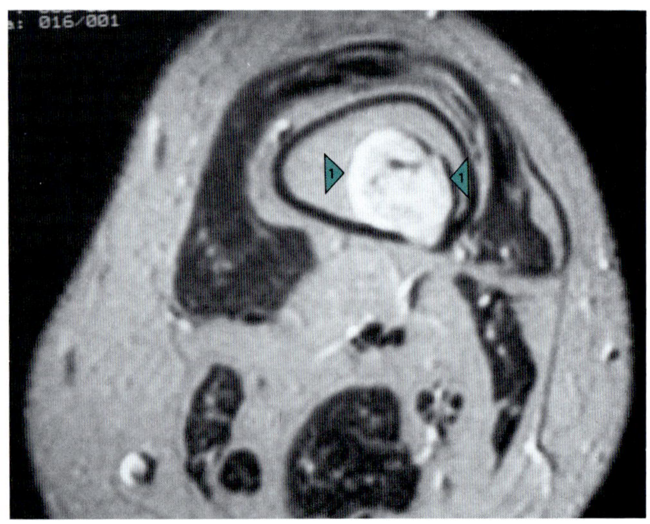

FIGURE 1.9. Chemical shift artifact. Axial section (SE 2000/80) through a chondrosarcoma of the distal femur. Note the low SI line between the tumor and the fatty marrow in one side and the hyperintense line in the other side, following the direction of the frequency encoding gradient. Arrows **1** Chemical shift artifact.

SIGNAL CHARACTERISTICS

Anatomic images are generated by (a) spatially localizing the signal and (b) receiving signals with different intensities from different types of tissue. These differences in signal intensity (image contrast) are dependent on the behavior of the spins after manipulation by rf pulses.

The spins do not maintain the excited state, which is induced by an rf pulse, but eventually will realign with the external magnetic field. This process is called longitudinal or spin-lattice relaxation, because the interaction between the spins and the surrounding lattice accelerates the realignment. The time it takes for this realignment to occur is called T1; T1 is tissue dependent and, in human tissues, varies between 300 and 1500 msec (Fig. 1.10). Since T1 relaxation cannot be observed directly, special pulse sequences have to be designed

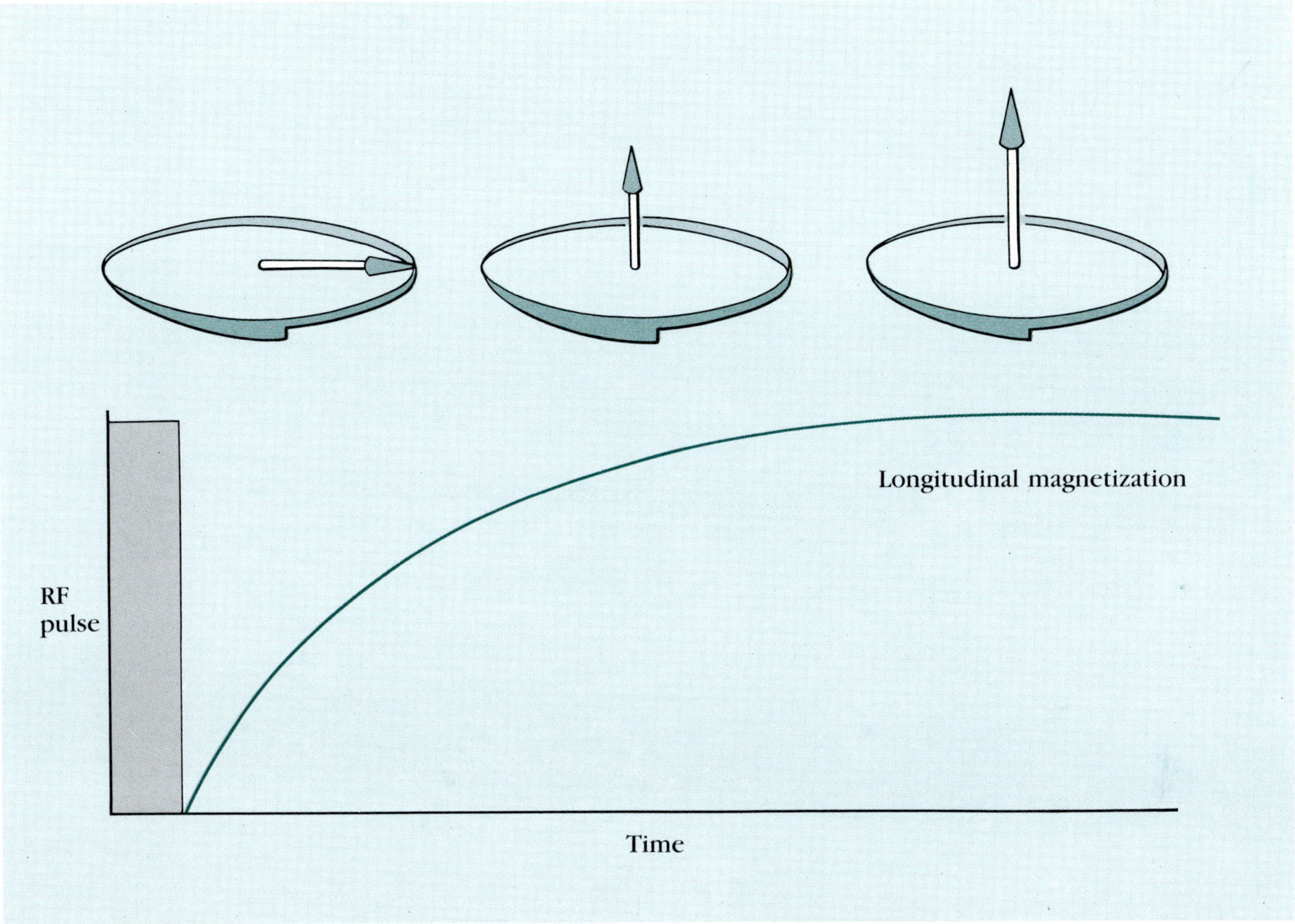

FIGURE 1.10. T1 relaxation. After excitation with a 90-degree rf pulse, the spins realign with the field and the net longitudinal magnetization is zero. The process occurs exponentially until the original net magnetization is reached; the time constant for this process is the T1 relaxation time.

to impose longitudinal T1 information on the observable transverse magnetization.

A second phenomenon occurs following rf excitation. Initially, all individual spins within the excited volume will rotate with the same frequency and the same phase angle. Consequently, both the net transverse magnetization and the observed signal are large. After some time, however, the spins start to dephase due to local changes in the magnetic field, and the net transverse magnetization decreases, leading to an exponential signal loss called free induction decay (FID) (Fig. 1.11). Local changes in the magnetic field are due to linear field gradients, inhomogeneities of the external magnetic field, and, more importantly, the interaction between adjacent spins—hence the name "spin-spin relaxation" given to the decay time (T2*) of the transverse magnetization.

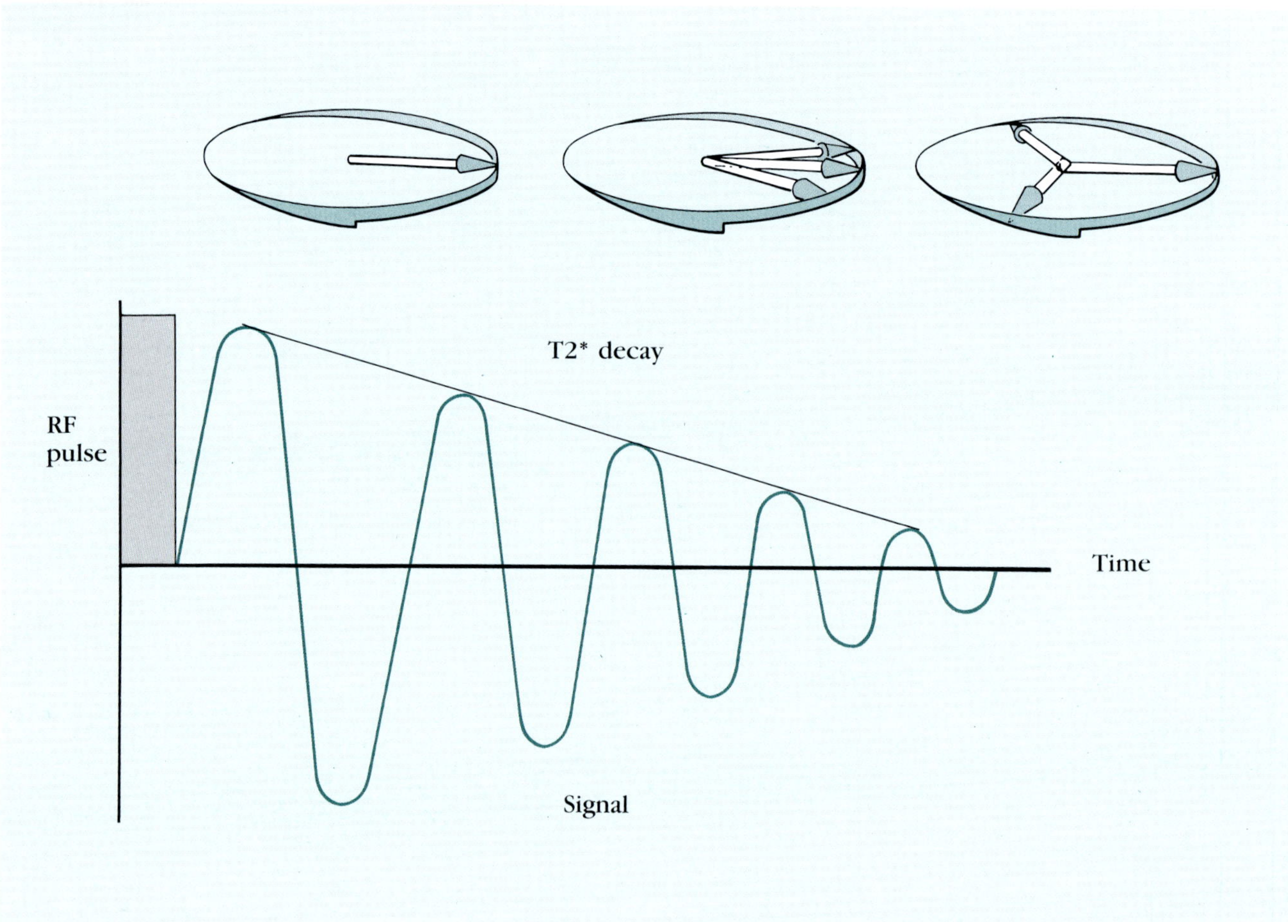

FIGURE 1.11. T2* decay free induction decay (FID). Immediately after excitation with rf pulse, all spins rotate with the same frequency and phase. The net magnetization (and therefore the measured signal) is large. After some time, the spins dephase, reducing the net transverse magnetization and leading to T2* decay or exponential signal loss. The time constant for this decay depends on effects of linear field gradients, main field inhomogeneity, and T2 relaxation.

PULSE SEQUENCES

Pulse sequences are designed to enhance the differences in tissue relaxation times (T1 and T2) and thus provide the necessary image contrast. Three pulse sequences are frequently used in musculoskeletal MRI: spin echo, inversion recovery, and fast scan techniques.

In spin echo pulse sequences, the spins are initially excited by a 90 degree rf pulse (Fig. 1.12). After a short time, the spins start to dephase. A 180 degree rf inversion pulse is then applied, causing inversion of the rotation direction. After a time equaling the interval between the 90 and the 180 degree pulse, all spins are in phase again. The echo time (TE) is the time from the initial 90 degree excitation to the echo signal.

If 180 degree rephasing pulses are repeated, echo signal will form repeatedly. The strength of

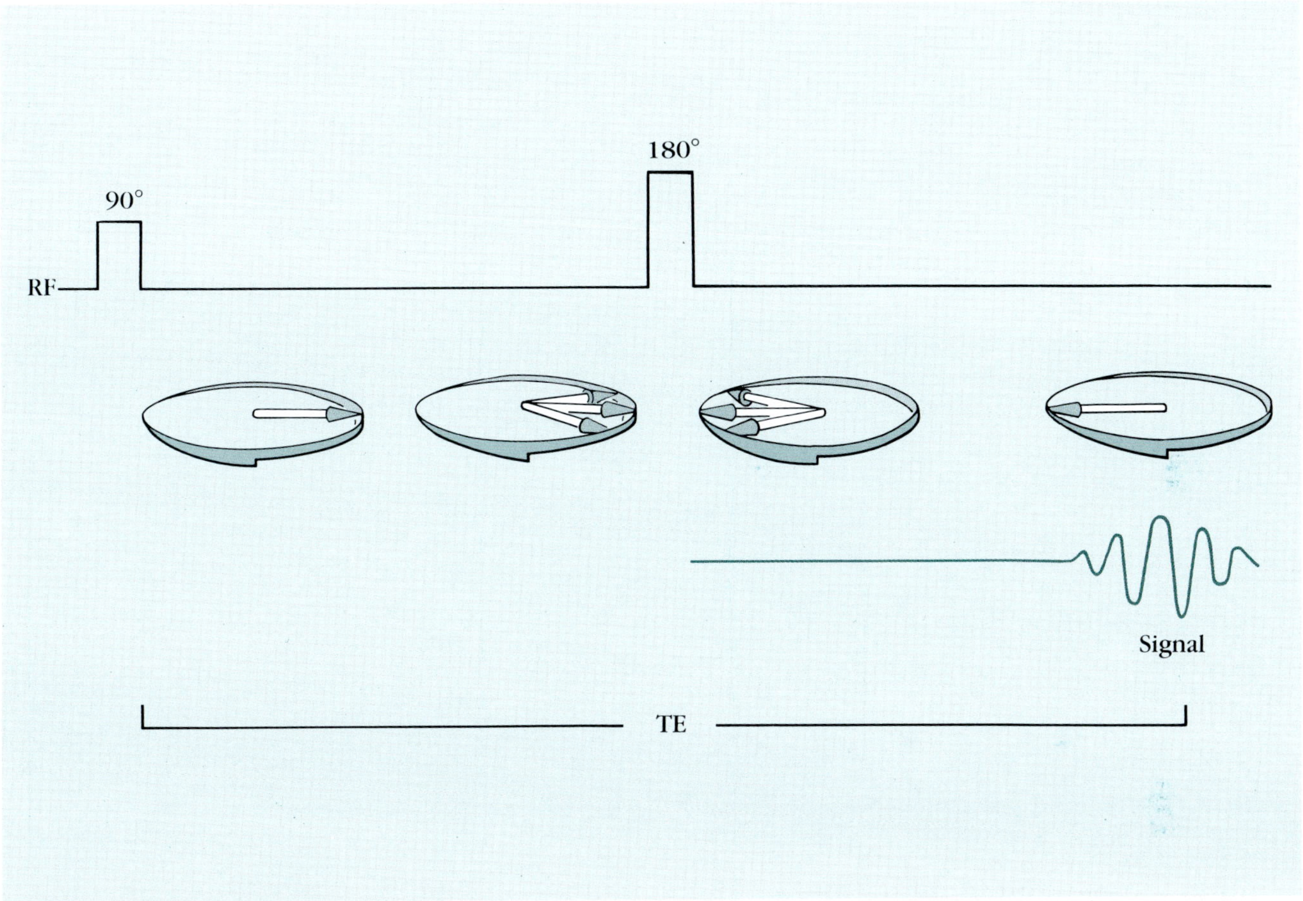

FIGURE 1.12. Spin echo pulse sequence. Spins initially excited by a 90-degree rf pulse, after a short time, dephase. A subsequently applied 180-degree inversion pulse causes inversion of the rotation direction. Spins with previous phase advantage lose this advantage, and slower precessing spins catch up. After a period equal to the duration between the 90- and 180-degree pulses, all spins are in phase again, resulting in an echo signal.

each consecutive echo then reflects T2 relaxation (Fig. 1.13). Such a multiecho pulse sequence can be used to simultaneously obtain images containing proton density as well as T2 information. Repetition time (TR) between 90 degree excitations is long compared with the T1 relaxation time; when used, spins completely relax between excitations. Consequently, the signal intensity of the first echo in a multiecho sequence depends primarily on the number of protons contributing to the signal (pro-

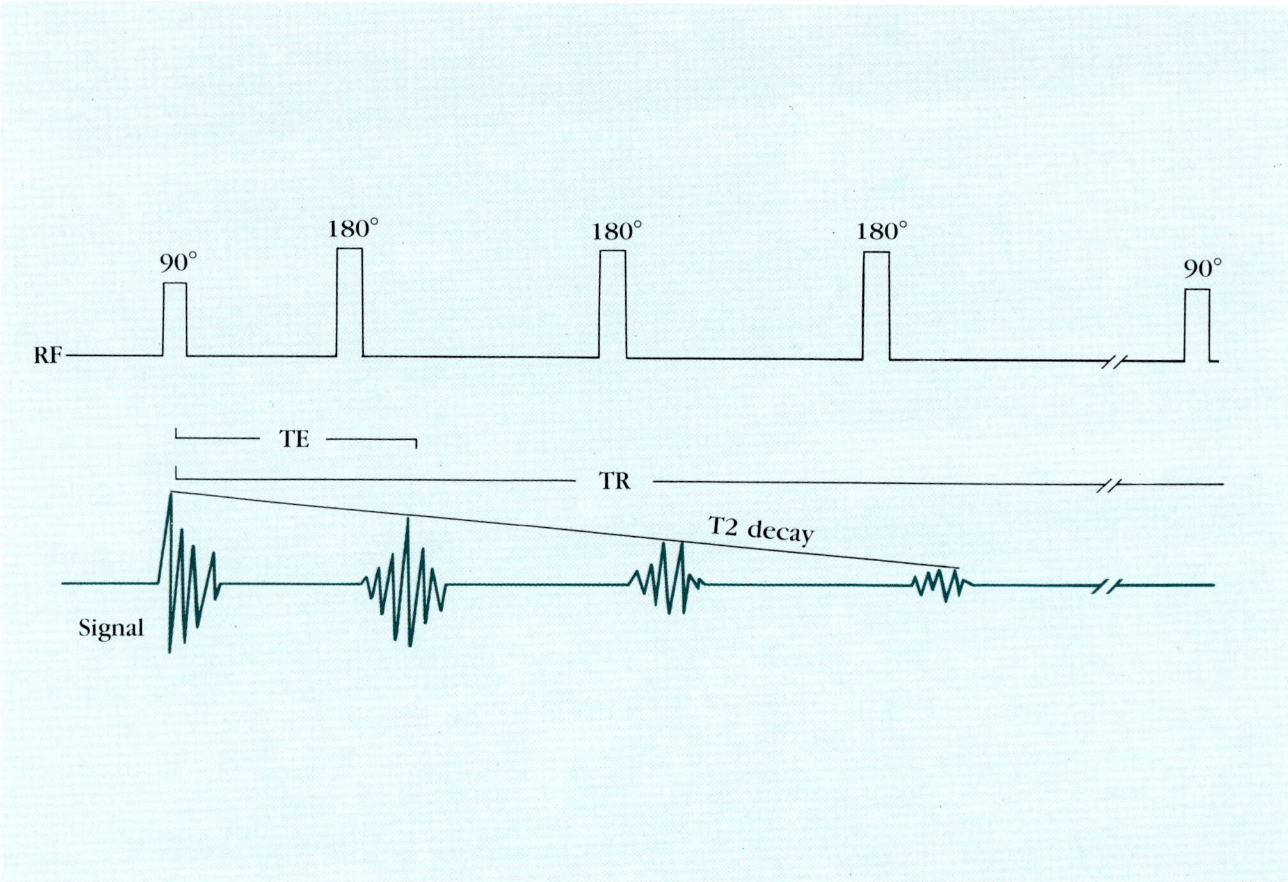

FIGURE 1.13. Multiecho pulse sequence. The initial 90-degree pulse is followed by several 180-degree inversion pulses, leading to repeated echo information. The signal intensity of subsequent echoes is reduced, since the microscopic T2 relaxation effects are not recovered in the spin echo experiment.

ton density weighted image). An echo acquired later will reflect T2 relaxation (T2-weighted image).

Signal intensity is a function of TR and T1. If a spin echo sequence with a TR shorter than the T1 of the tissues is used, T1 information is obtained (T1-weighted image) (Fig. 1.14). During the short TR, the longitudinal magnetization does not have enough time to fully recover and is, therefore, smaller than initially. The next 90 degree pulse

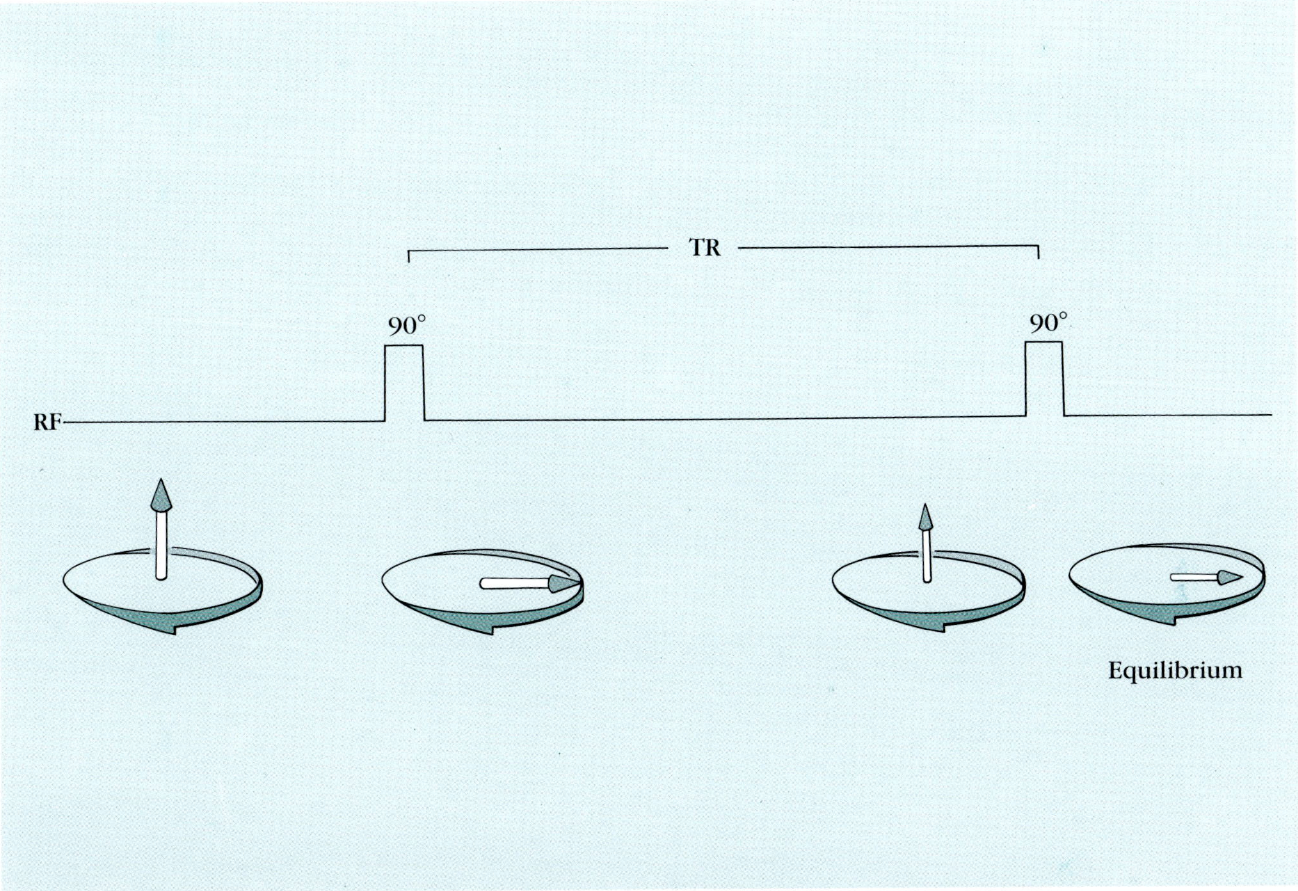

FIGURE 1.14. T1 contrast using short TR. If consecutive repetitions of 90-degree excitation pulses occur faster than the T1 relaxation time, the longitudinal magnetization does not have enough time to recover during the TR interval. Consequently, the second excitation pulse can only convert a smaller longitudinal magnetization component into transverse magnetization. Successive rf excitations lead to equilibrium. The observable transverse magnetization yields a signal amplitude corresponding to the T1 relaxation time.

converts this reduced longitudinal magnetization into the transverse plane. The now observable transverse magnetization yields a signal amplitude corresponding to the amount of relaxation that occurred during the TR interval. Successive rf excitations cause the spins to attain equilibrium.

In inversion recovery sequences, spins are initially excited by a 180 degree rf pulse resulting in inversion of the longitudinal magnetization (Fig 1.15). After a pause (TI inversion time), partial recovery of the longitudinal magnetization takes place. Spins from tissue with short T1 relaxation time may recover fully. The 90 degree pulse following the TI converts the recovered longitudinal magnetization into observable transverse magnetization. The signal strength now directly reflects T1 relaxation. The sequence is repeated after a TR longer than the T1. Inversion recovery can be combined with multiplanar imaging to shorten scan time. With a short TI, in the range of 100–150 msec, the effects of prolonged T1 and T2 relaxation times are cumulative and the signal from fat is nulled. This technique, called short thau inversion recovery (STIR), has been useful for evaluation of bone tumors (Fig. 1.16).

In fast scan techniques, the longitudinal magnetization is not completely transformed into transverse magnetization (limited flip angle excitation)

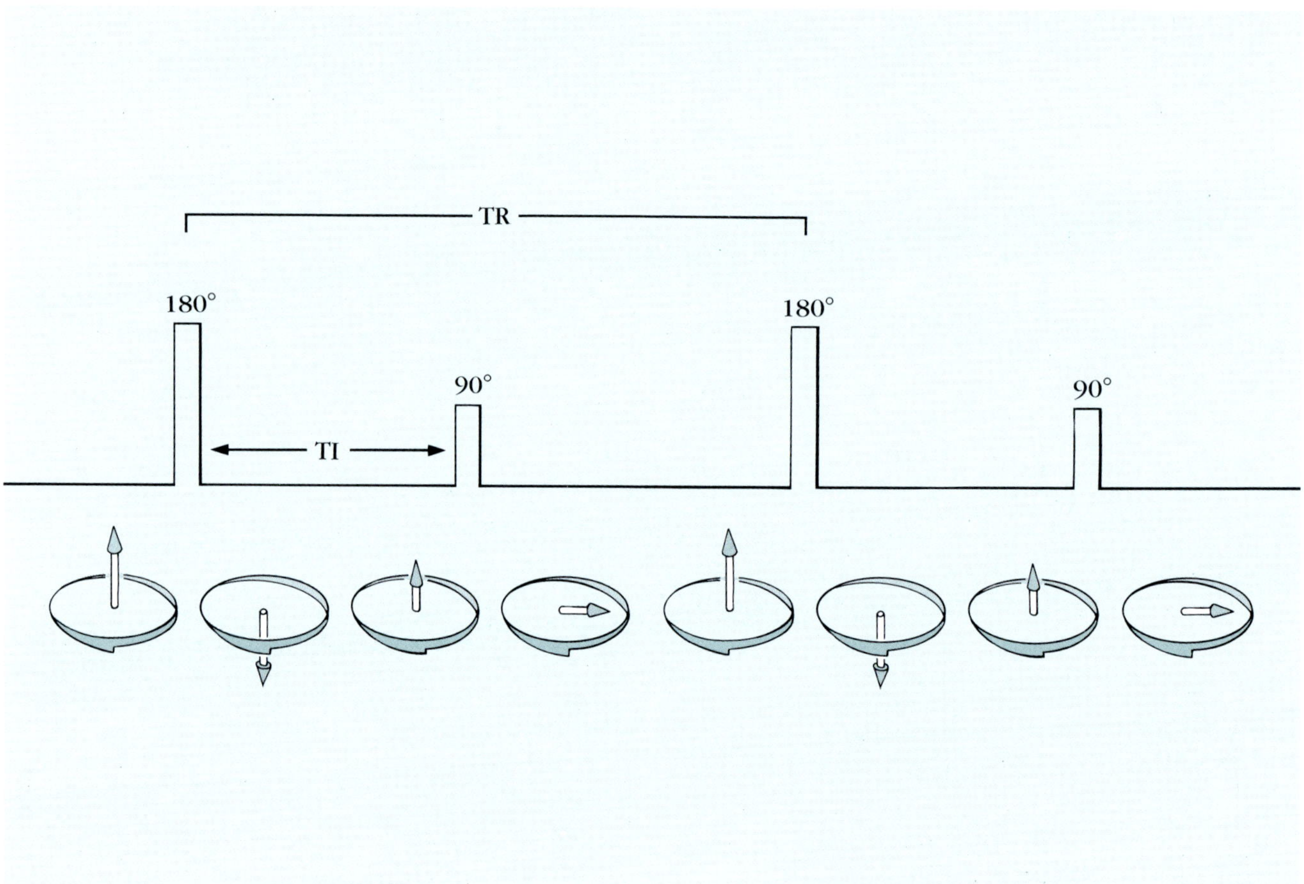

FIGURE 1.15. Inversion recovery. Spins are initially excited by a 180-degree inversion pulse, producing inversion of the longitudinal magnetization. After the time TI, the inversion pulse is followed by a 90-degree pulse. Since relaxation occurs during the time TI, the observable transverse magnetization is a direct measure for T1 relaxation. After the time TR, which is typically much longer than T1, the sequence is repeated.

(6,7). Successive rf excitations bring the magnetization to equilibrium. For flip angles of 30–60 degrees, the resulting signal amplitude is determined predominantly by T1. For smaller flip angles, T1 relaxation effects play a less significant role and the signal is dependent on the number of protons. If TR comparable to the T2 is used (TR <100 msec), a T2-weighted image is obtained.

Low flip angle techniques are highly sensitive to inhomogeneities of the external and local magnetic fields (magnetic susceptibility). Magnetic susceptibility is particularly large between interfaces of tissue and air, and between soft tissue and dense bone, leading to artifacts in the interface region.

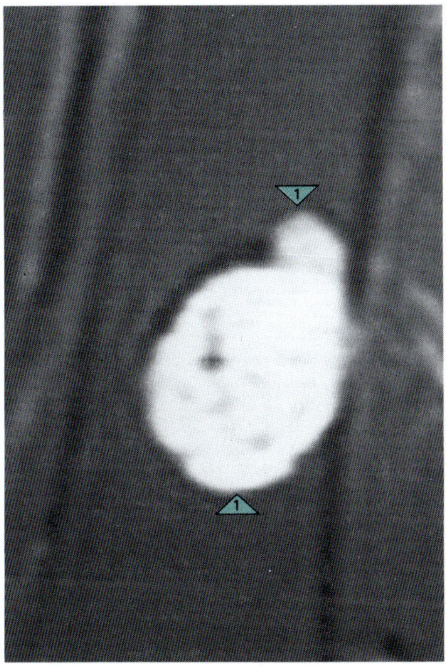

FIGURE 1.16. STIR pulse sequence (1500 TR/ 145 TI/ 30 TE), through the distal femur, in the sagittal plane. This pulse sequence effectively suppresses the signal from the fatty marrow and subcutaneous fat, providing excellent contrast between the tumor and the surrounding tissues. Arrows **1** Chondrosarcoma.

REFERENCES

1. Curry TS, Dowdey JE, Murry RC: Introduction to the physics of diagnostic radiology. Philadelphia, Lea & Febiger, 1984.
2. Henkelman RM, Bronshill MJ: Artifacts in magnetic resonance. 1987; 2:1.
3. Keller PJ: Basic principles of MR imaging. General Electric Company Publication, 1988.
4. Mansfield P, Morris PG: NMR imaging in biomedicine. London, Academic Press, 1982.
5. Partain CL, James AE, Rollo RD, Price, RR: Nuclear magnetic resonance imaging. Philadelphia, WB Saunders, 1983.
6. Wehrli FW: Introduction to fast-scan magnetic resonance. General Electric Company Publication, 1986.
7. Wehrli FW, Shaw D, Kneeland JB: Biomedical magnetic resonance imaging. New York, VCH, 1988.

CHAPTER TWO

Javier Beltran

The Temporomandibular Joint

TECHNICAL CONSIDERATIONS

IMAGING PLANES AND PULSE SEQUENCES

The sagittal plane is ideal for imaging the temporomandibular joint (TMJ). The long axis of the mandibular condyle is angled internally at about 20 degrees from the horizontal line that transects the condyles on an axial projection. Theoretically, oblique sagittal sections perpendicular to the long axes of the condyles would provide a better depiction of the TMJ (Fig. 2.1). However, straight sagittal sections can also be used in visualization of the joint anatomy without significant compromise. Both closed- and open-mouth positions should be obtained to assess meniscal dynamics (Fig. 2.2).

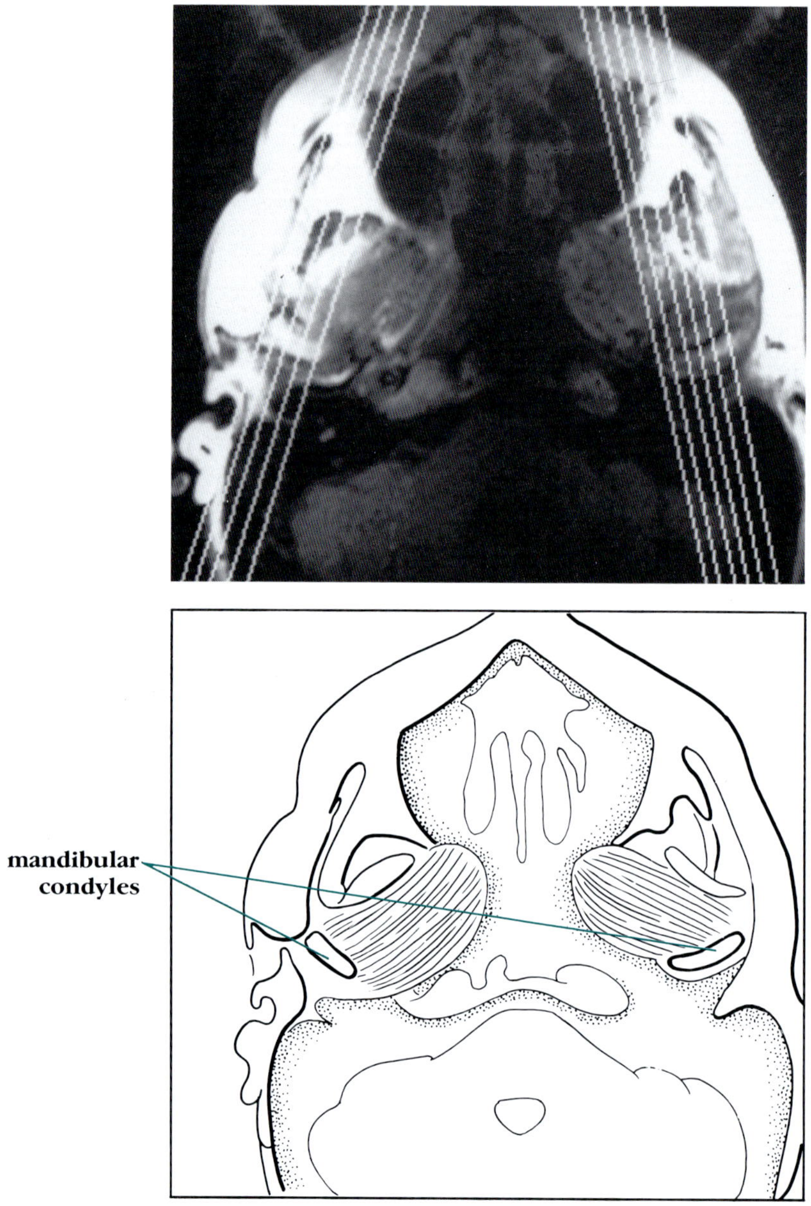

FIGURE 2.1. Orientation of sections through mandibular condyles. Axial section [SE 500/20]. Oblique lines perpendicular to the long axes of the mandibular condyles indicate the orientation of the sagittal sections. The condyles are rotated 15–20 degrees from the horizontal plane.

The axial plane is useful for localization of the TMJ and for determination of the oblique sagittal planes of sections. Coronal plane images can be used to assess medial or lateral meniscal displacement.

The most widely used acquisition technique is 2DFT multislice. For axial localizer images, SE pulse sequences with short TR and short TE can be used. T1WI with short TR/TE are also used with 2 NEX for straight or oblique sagittal sections in the closed-mouth position. If two surface coils are used for simultaneous imaging of both TMJ's, a longer TR may be necessary in order to provide enough sections through both joints. Straight or oblique sagittal sections can then be obtained in the open-mouth position using the same pulse sequences. A disposable mouth prop can be used to stabilize the mandible in the open-mouth position, thus preventing motion artifacts (Fig. 2.2). Occasionally, T2WI with long TR/TE may be necessary to evaluate inflammatory changes of the TMJ (see Fig. 2.8).

In order to decrease imaging time, low flip-angle, fast-scanning techniques with 2DFT, 3DFT GRASS, or FLASH can also be used (Fig. 2.3). Accurate dynamic evaluation of the TMJ can be obtained by using 2DFT partial flip-angle techniques during various stages of mouth opening. The degree of opening can be regulated with a Burnett TMJ device

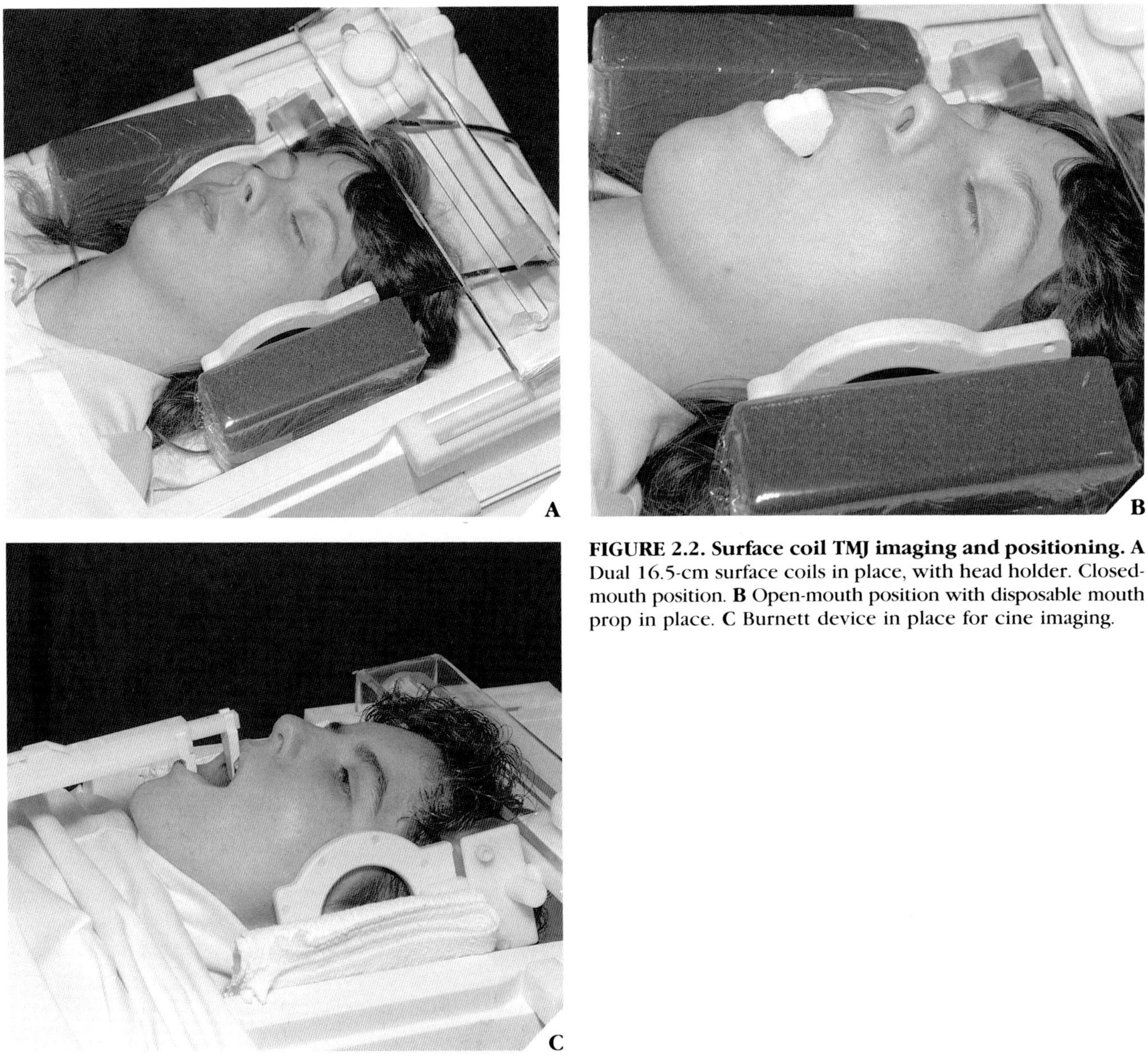

FIGURE 2.2. Surface coil TMJ imaging and positioning. A Dual 16.5-cm surface coils in place, with head holder. Closed-mouth position. **B** Open-mouth position with disposable mouth prop in place. **C** Burnett device in place for cine imaging.

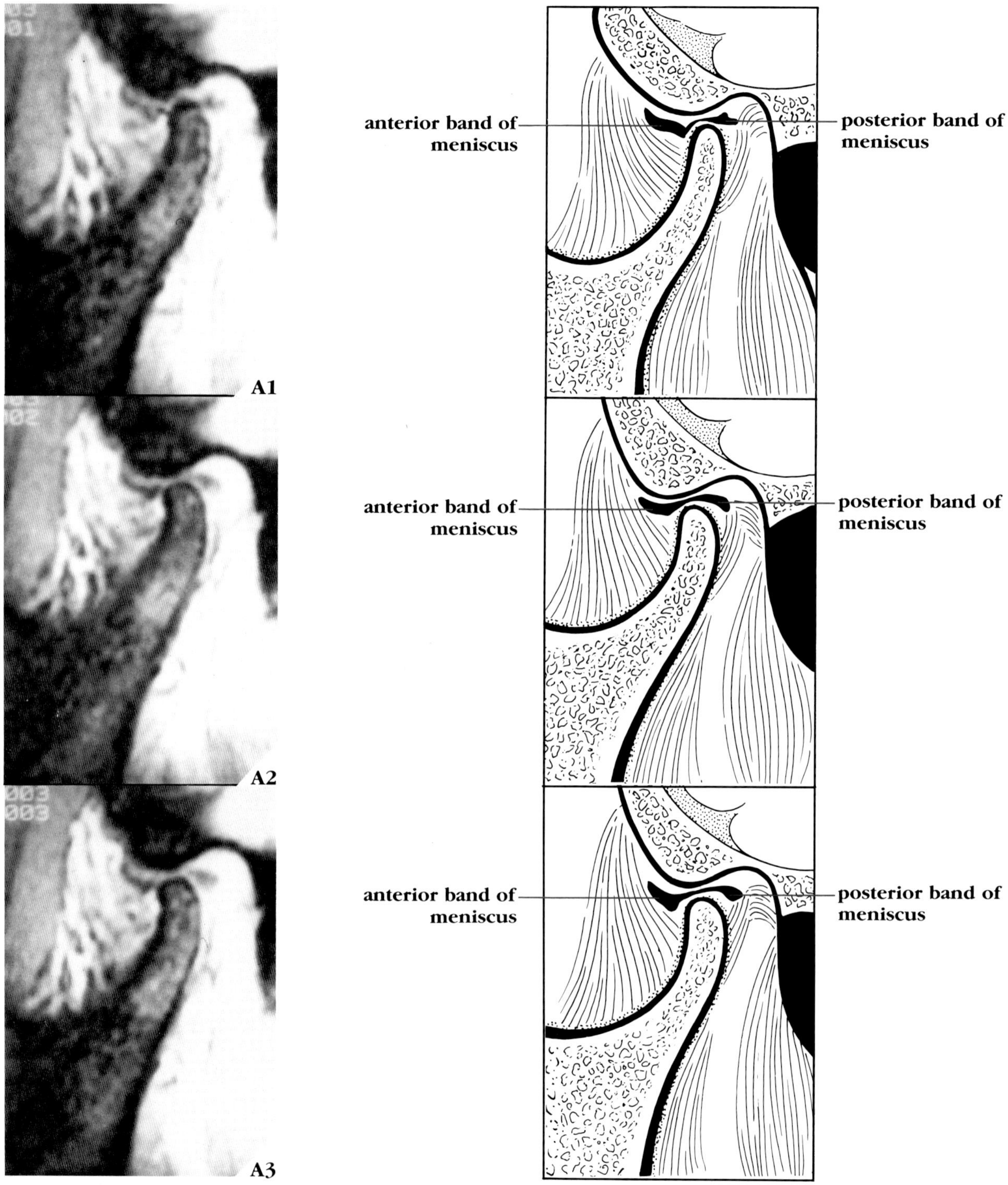

FIGURE 2.3A. Cine MRI during opening. Six sagittal sections (**A1–A6**) during various stages of opening [GRASS 100/13, flip angle 20 degrees]. The meniscus is in normal position.

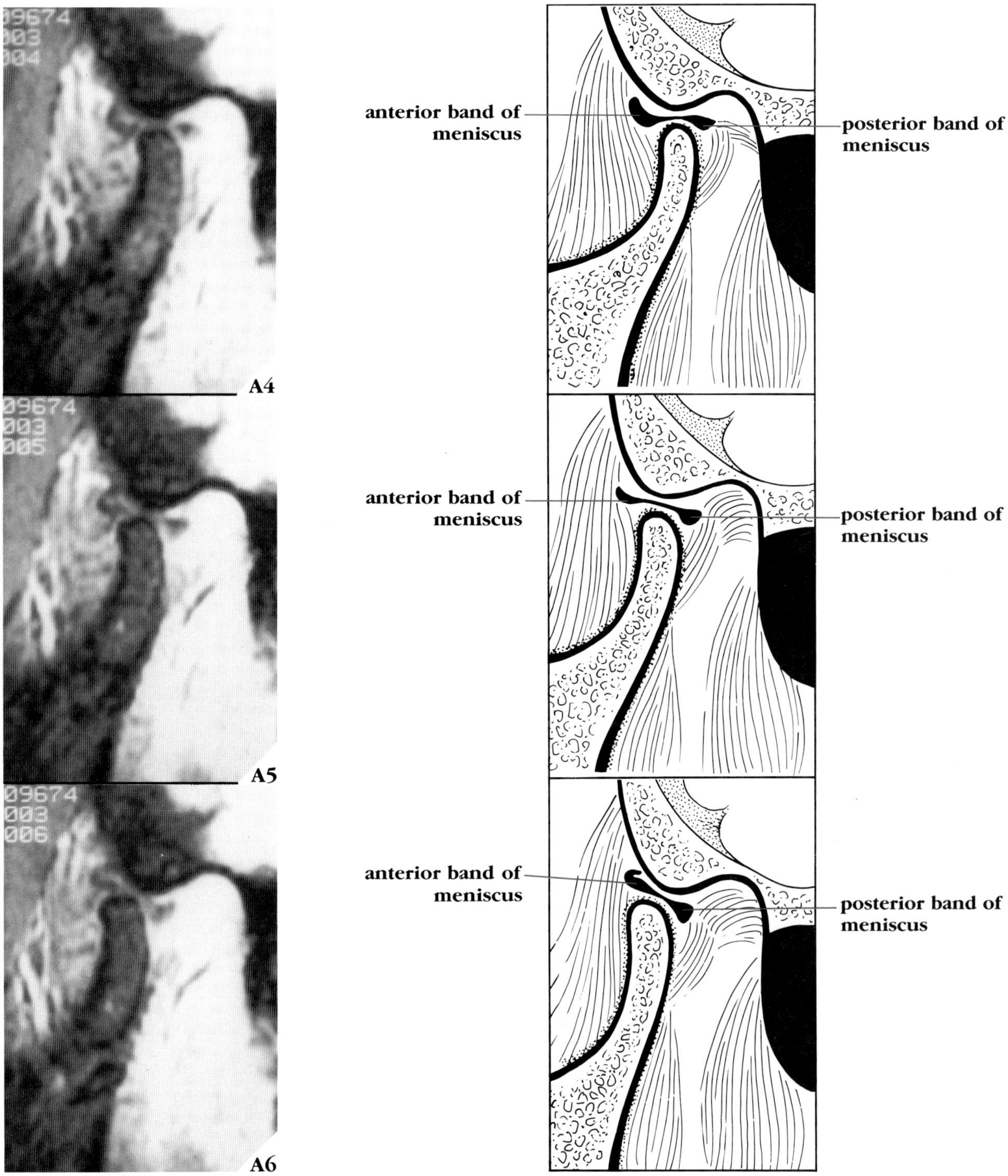

FIGURE 2.3A (*cont'd*)

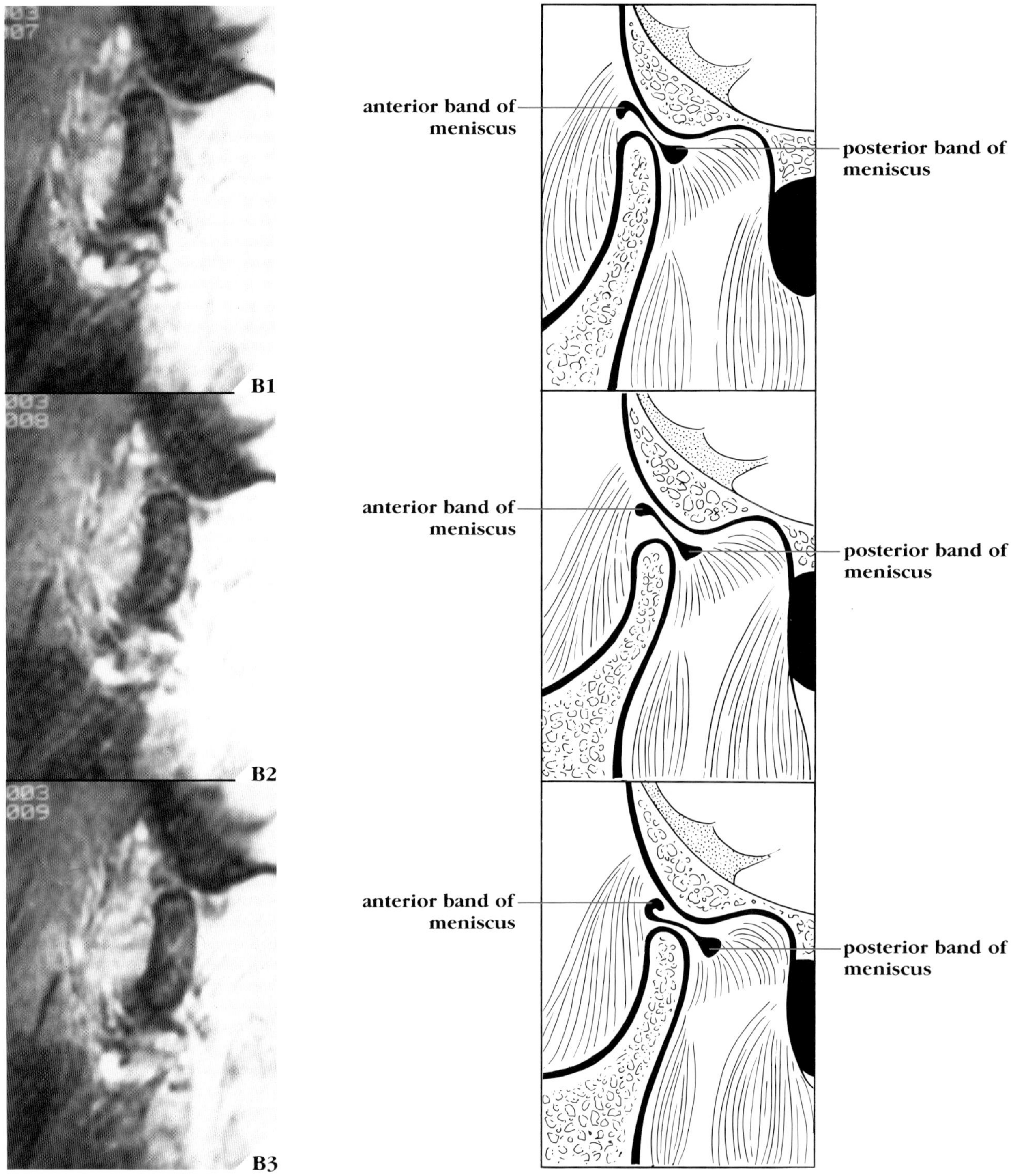

FIGURE 2.3B. Cine MRI during closing. Six sagittal sections (**B1–B6**) during various stages of closing [GRASS 100/13, flip angle 20 degrees]. Same patient as in Fig. 2.3A. Normal motion of the meniscus is also seen on closing.

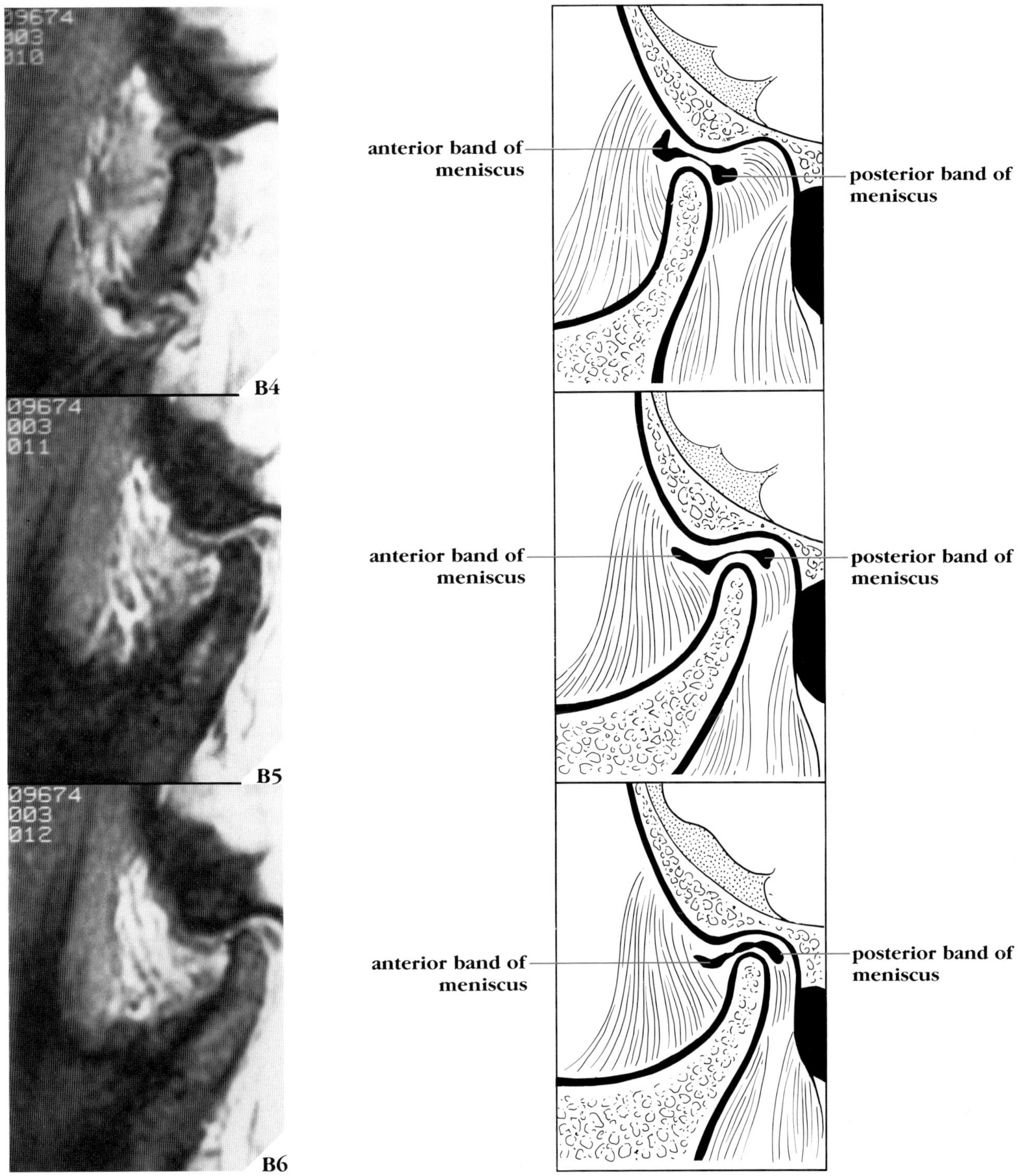

FIGURE 2.3B (*cont'd*)

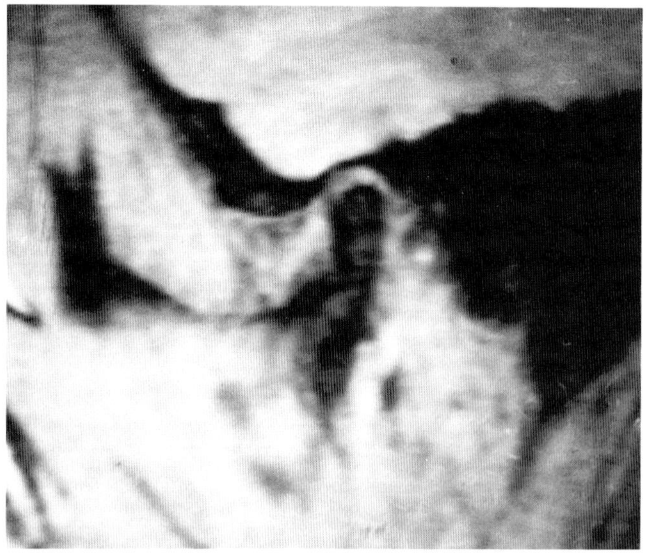

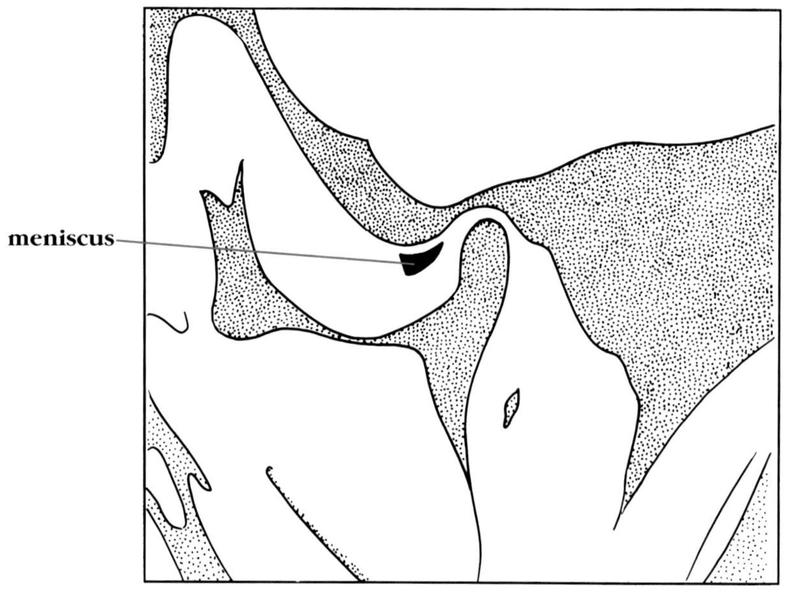

FIGURE 2.3C 3DFT TMJ imaging. Sagittal section, closed-mouth position [GRASS 50/11, flip angle 25 degrees]. This 1.5-mm sagittal image was reconstructed from a 3-D data set obtained with a matrix of 256 × 128 pixels and 32 gradient steps. NEX: 1. Imaging time: 3.25 min. The low SI meniscus is anteriorly displaced. Bone marrow signal is low due to magnetic susceptibility originated by fast scanning techniques.

(Medrad, Inc.) (Fig. 2.2). Eight to 12 images can be obtained during opening steps of 3 to 5 mm (Fig. 2.3) (10).

3DFT partial flip-angle technique provides an improved S/N and thin slices which are truly contiguous (Fig. 2.3) (2).

FIELD OF VIEW

In the axial plane, an FOV of 16 cm is used to include both TMJ's in the same image. A smaller FOV of about 12 cm is used for the sagittal planes.

SLICE THICKNESS

Three-millimeter slices with 0.5-mm interslice gaps provide enough contrast resolution and S/N for imaging the TMJ's.

POSITIONING

The patient is positioned supine in the gantry head first. If oblique sagittal sections are desired, and no oblique plane capability is available, the head can be turned about 20 degrees toward the side to be imaged. If oblique plane capability with dual surface coil imaging is available, the head is positioned straight in the gantry. Oblique or perpendicular sections through both TMJ's can be graphically assigned on the first axial localizer scan (Fig. 2.1). The patient is instructed not to move, especially during the open-mouth views.

SURFACE COILS

A small diameter (16.5 cm) receive-only, circular coil is positioned over the region of the TMJ (uni- or bilaterally), with a head-holder to support the coils (Fig. 2.2) (3). The improved S/N achieved with surface coils allows better visualization of the small anatomical structures that compose the TMJ.

NORMAL ANATOMY

The TMJ is a synovial joint located between the mandibular condyle and the temporal fossa. There are two synovial spaces (superior and inferior), separated by a fibrocartilaginous structure (meniscus). The superior and inferior synovial spaces do not normally communicate. The meniscus has an elliptical shape with a central thin portion and peripheral thickening. The anterior and posterior thickenings are called the anterior and posterior bands.

The meniscus is attached peripherally to the articular capsule. Anteriorly, it is attached to the tendon of the lateral pterygoid muscle. The meniscus is secured posteriorly by the bilaminar attachment with fibrous connections to both the temporal bone and the posterior aspect of the mandibular condyle (Fig. 2.4). The bilaminar zone is composed of two layers of elastic tissue separated by a highly vascular zone that contains the endings of the auriculotemporal, masseteric, and posterior, deep temporal nerves (1). On mandibular opening,

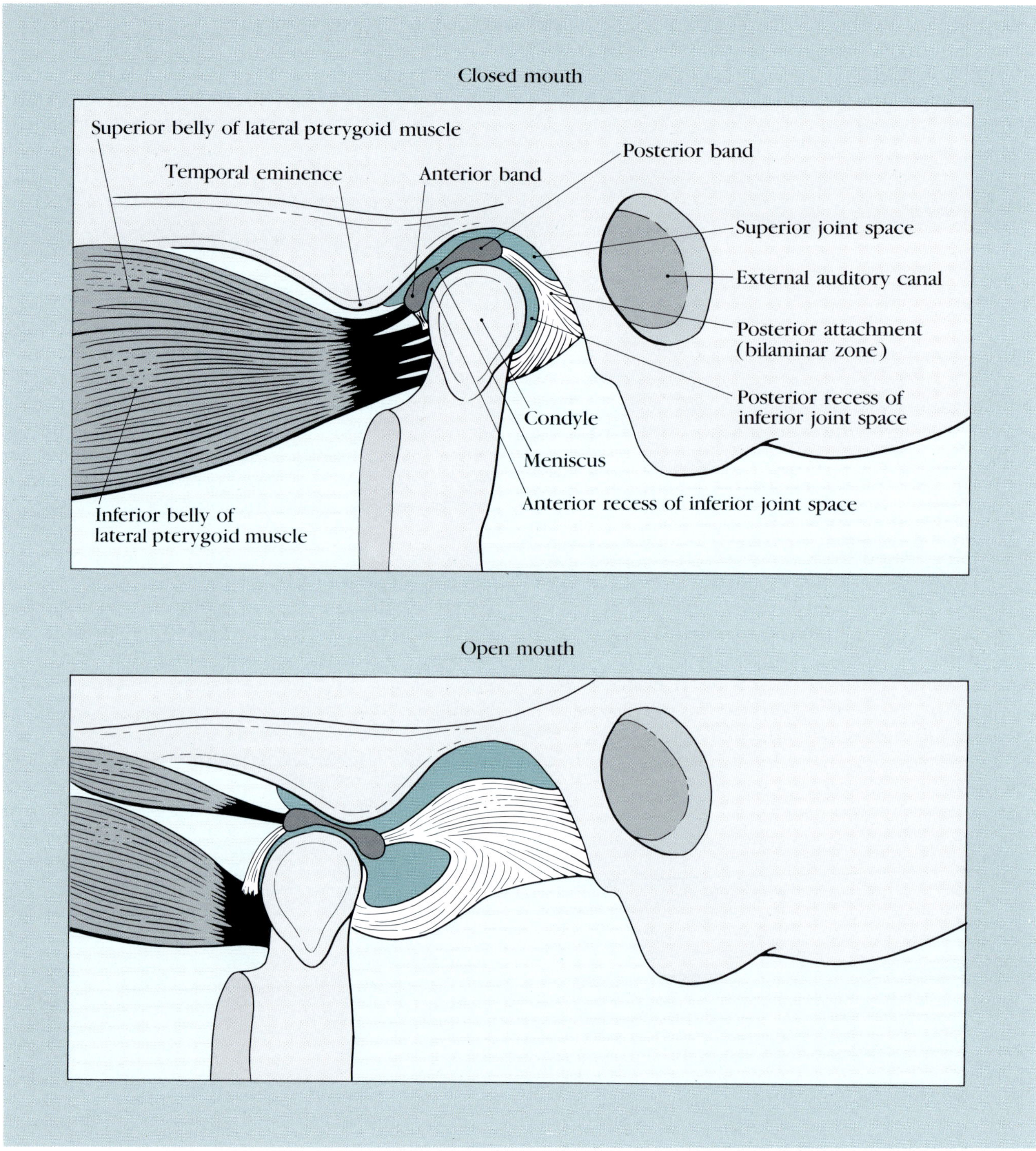

FIGURE 2.4. Normal anatomy of the TMJ in the (A) closed-, and (B) open-mouth positions. The posterior meniscal attachment becomes markedly stretched in the open-mouth position. In the closed-mouth position, the posterior band of the meniscus is located adjacent to the superior aspect of the condyle. In the open-mouth position, the posterior band is located posterior to the condyle. The anterior recess of the inferior joint space becomes collapsed on opening, whereas the posterior recess enlarges.

the neurovascular zone engorges with blood, partially filling the space vacated by the mandibular condyle.

▼▼▼ *SAGITTAL PLANE*

In the closed-mouth position, the mandibular condyle articulates with the temporal fossa. The bone marrow space of the condyle and the temporal eminence are identified on T1WI by the hyperintense fatty marrow (Fig. 2.5) (5). The bones are outlined by the low SI cortex. The low SI meniscus is located between the condyle and the temporal fossa, with its anterior band only slightly anterior, and its posterior band located on top of the condyle. An area of high SI, believed to represent myxoid degeneration, can occasionally be seen within the posterior band (11) (see Fig. 2.13).

The lateral pterygoid muscle with its superior and inferior bellies is seen as the intermediate SI structures, anterior to the condyle. The tendon of the inferior belly of the lateral pterygoid muscle is frequently seen as a low SI structure, and should not be confused with the meniscus.

In the open-mouth position the condyle translates anteriorly over the temporal eminence, and the meniscus moves with the condyle in such a way that when the mouth is fully opened, the posterior band is located behind the superior portion of the condyle, and the anterior band is located anterior to it. Occasionally the posterior ligament can be seen in the open-mouth projection as a thin low SI line extending from the posterior band of the meniscus to the temporal bone.

▼▼▼ *AXIAL PLANE*

The mandibular condyle forms an angle of 15–20 degrees with the horizontal plane (Fig. 2.6). The medial and lateral pterygoid muscles extend from the pterygoid plate to the medial aspect of the mandible. The masseter muscle can be identified in the lateral aspect of the mandible. The coronoid process often appears as a low SI line, anterior to the mandibular condyle. The temporalis muscle is adjacent to the medial aspect of the coronoid process. The external auditory canal is located immediately posterior to the TMJ. The eustachian tube, the carotid artery, the petrous portion of the temporal bone, and the jugular fossa are located medially to the TMJ.

▼▼▼ *CORONAL PLANE*

The temporal fossa, mandibular condyle, and the intervening meniscus can be identified in the coronal plane using surface coil imaging (Fig. 2.7). The lateral pterygoid muscle insertion and the masseter muscle are seen in the medial and lateral aspects of the mandible, respectively. The medial pterygoid muscle is identified inferiorly and medially. The parotid gland is located in a position lateral to the TMJ.

INTERNAL DERANGEMENT

TMJ dysfunction due to internal derangement is an extremely frequent abnormality which is believed to affect large segments of the population (9). The clinical syndrome consists of pain, headache, earache, limitation of opening, joint clicking, and emotional stress. The etiology of internal derangement of the TMJ involves a spectrum of traumatic conditions, including microtrauma (intubation, prolonged dental procedure, bruxism, masticatory muscle spasm, and malocclusion), macrotrauma (fracture, dislocation, hemarthrosis), and inflammatory joint disease (osteoarthritis, rheumatoid arthritis, psoriasis). Emotional stress may also play an important etiological role.

The pathophysiology of internal derangement of the TMJ is believed to be related to loss of the ability of the posterior meniscal attachment to counteract the pulling forces of the lateral pterygoid muscle. When this phenomenon occurs, the meniscus remains anterior to the condyle during closure, and the condyle impinges directly upon the posterior neurovascular zone (8).

Previously, arthrography was the classic procedure used to evaluate internal derangements of the TMJ. With the advent of MRI, however, arthrography is fast becoming a secondary procedure due mainly to its invasive nature.

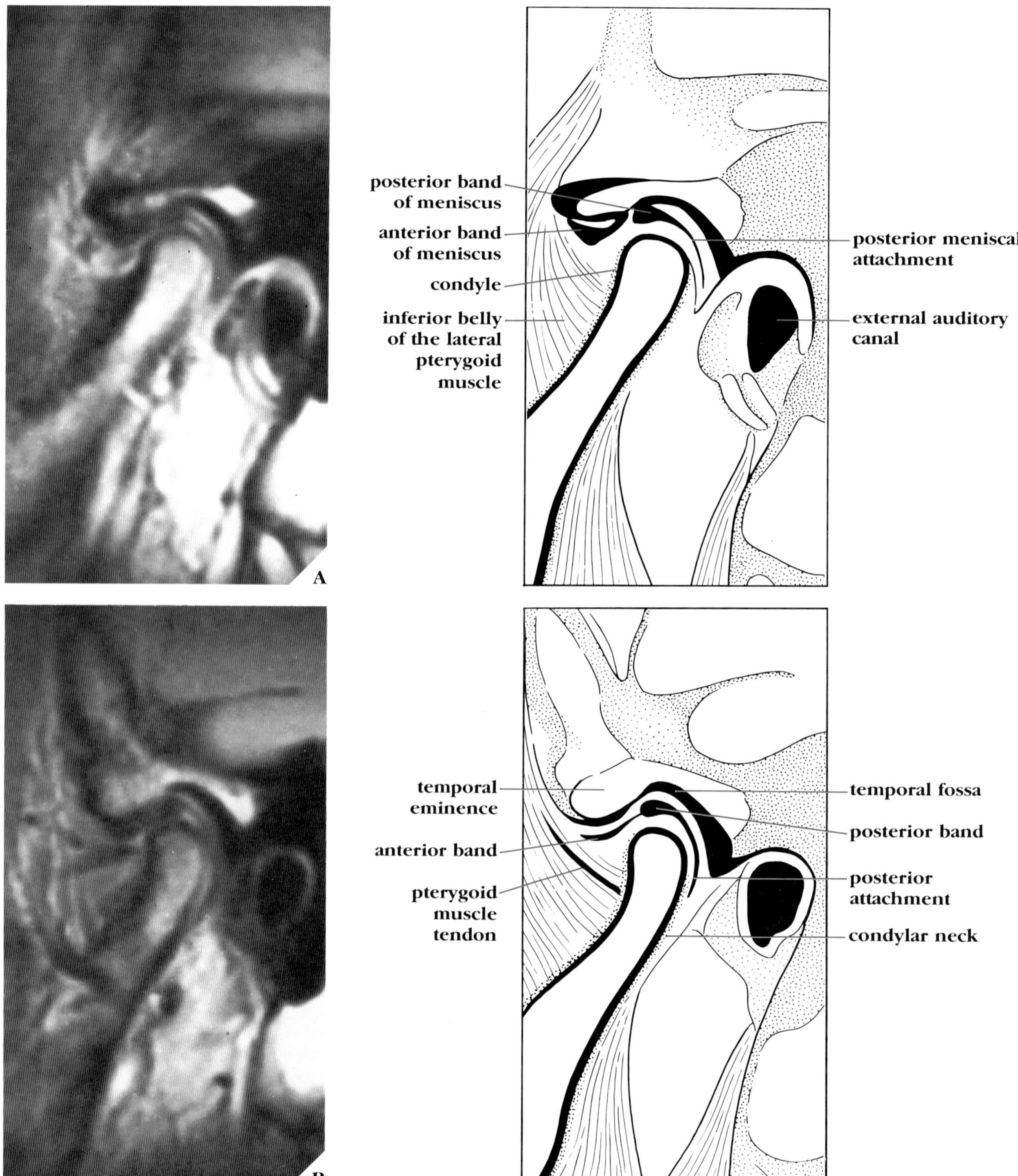

FIGURE 2.5. Normal anatomy. A, B Two consecutive sagittal sections, closed-mouth position [SE 1200/20]. The anterior and posterior meniscal bands are identified in their normal position. The posterior meniscal attachment is well-demonstrated.

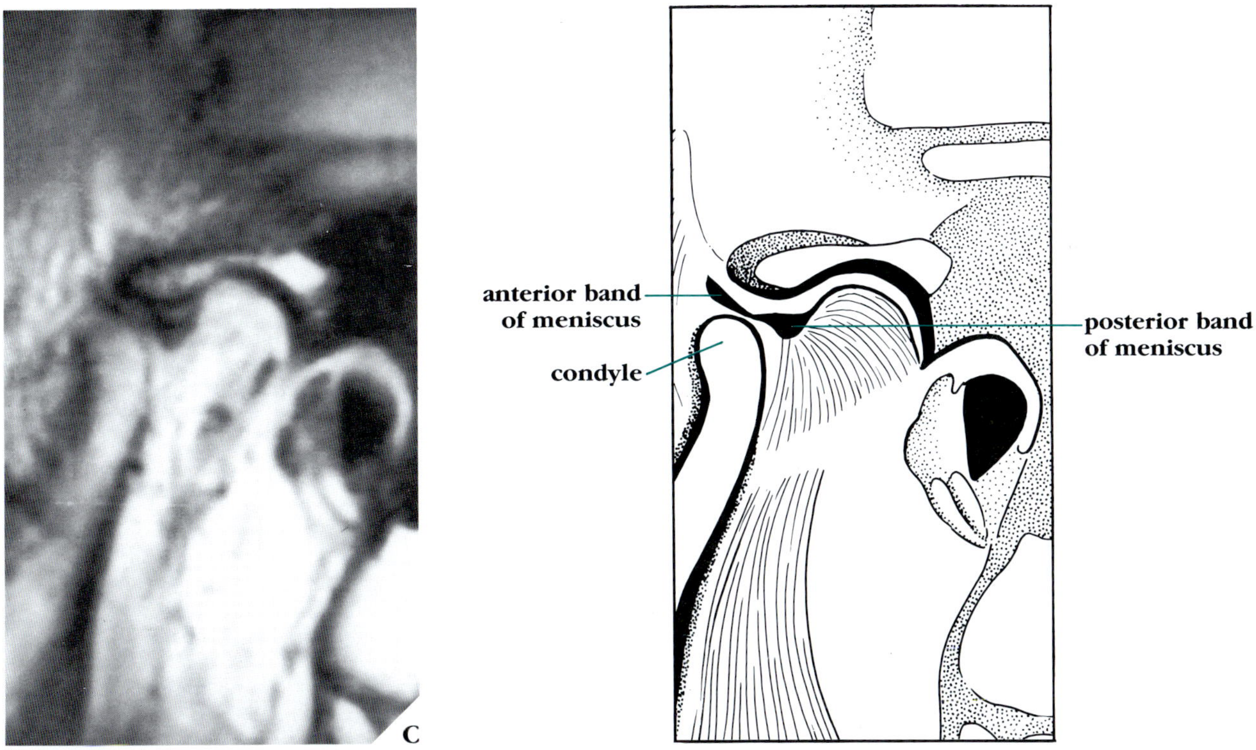

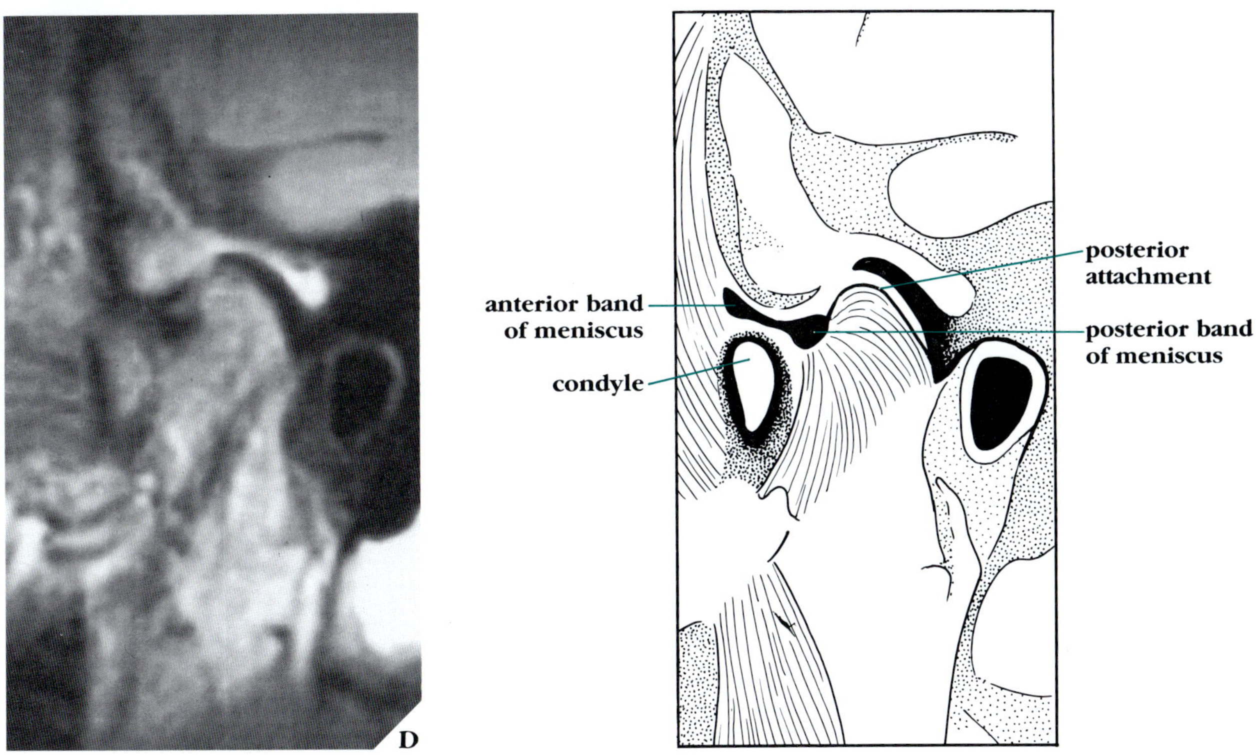

C, D Two consecutive sagittal sections, open-mouth position [SE 1200/20]. The condyle is at the level of the temporal eminence. The posterior band is posterior to the condyle.

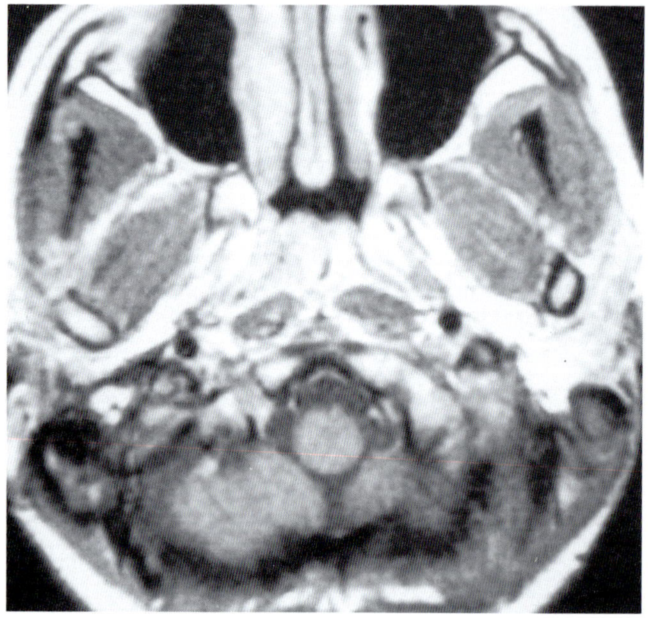

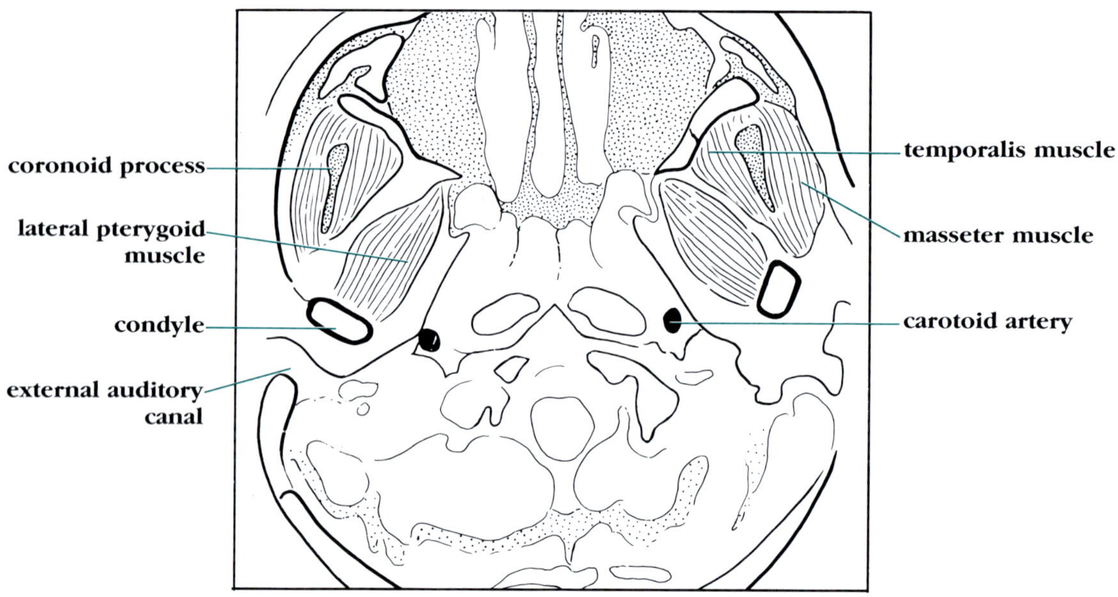

FIGURE 2.6. Normal anatomy. Axial section [SE 500/20]. The mandibular condyles are tilted 15–20 degrees from the horizontal plane. The condyles show high SI fatty marrow on this T1WI. Muscles are seen as intermediate SI structures. The coronoid process is seen as a low SI band between the masseter and temporalis muscles. The external auditory canals are located immediately posterior to the TMJ's.

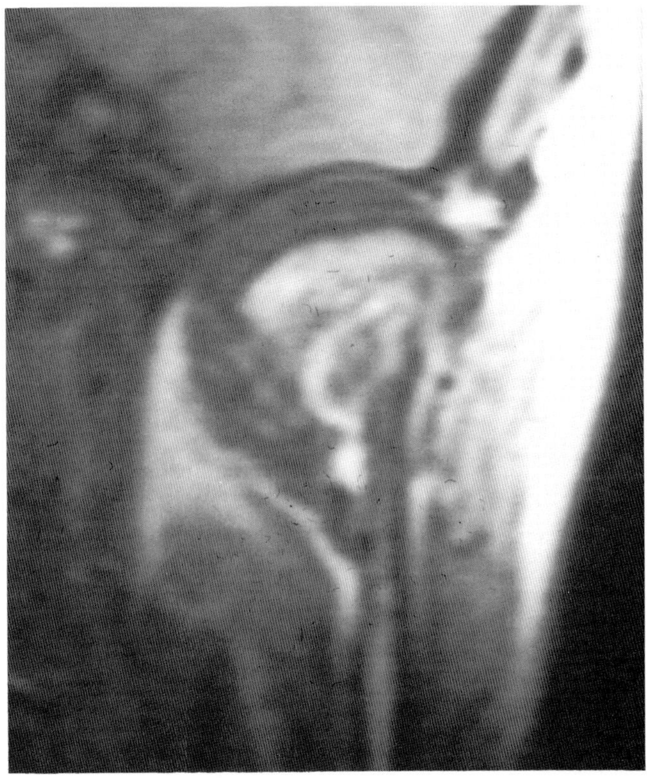

FIGURE 2.7. Normal anatomy. Coronal section [SE 500/20]. The low SI meniscus is interposed between the temporal fossa and the mandibular condyle.

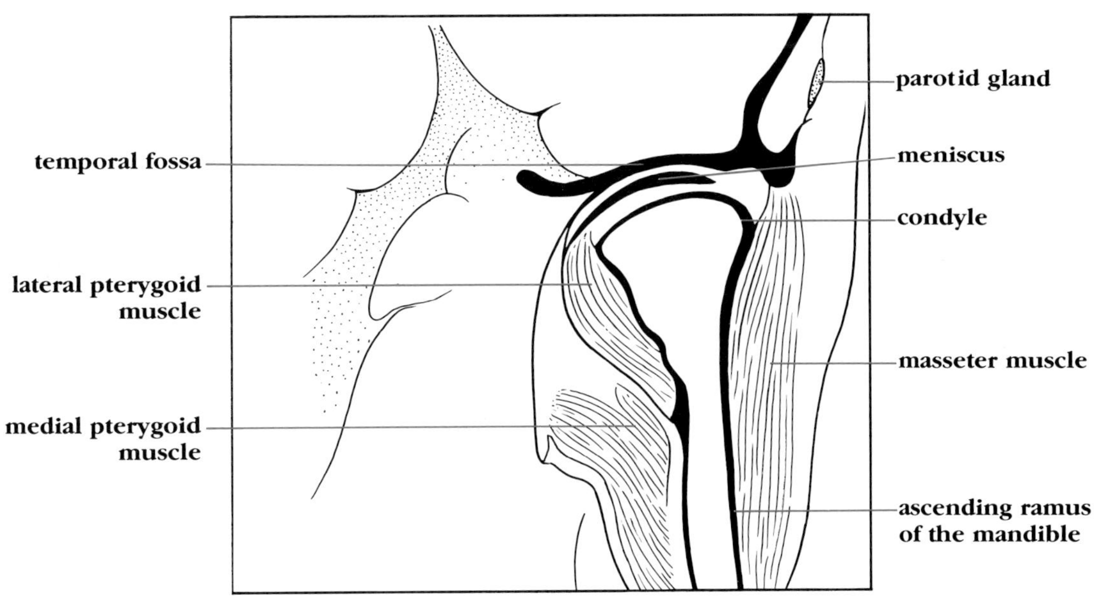

Meniscal Subluxation

Anterior meniscal displacement with reduction on opening (subluxation) is often accompanied clinically by clicking noises (Fig. 2.8). Subluxation can be diagnosed with MRI when the low SI meniscus moves anterior to the mandibular condyle in the closed-mouth view, but returns to its normal position (on top of the condyle) in the open-mouth view. Medial or lateral rotation of the meniscus frequently occurs in association with subluxation or dislocation. Careful analysis of contiguous sagittal sections or coronal sections can be used to assess medial or lateral displacement (6).

Meniscal Dislocation

Anterior meniscal displacement without reduction on opening (dislocation) represents a more severe stage of internal derangement of the TMJ, and is

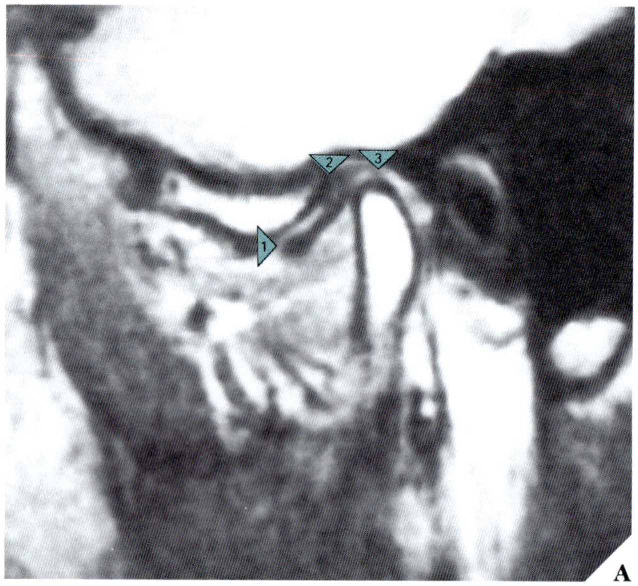

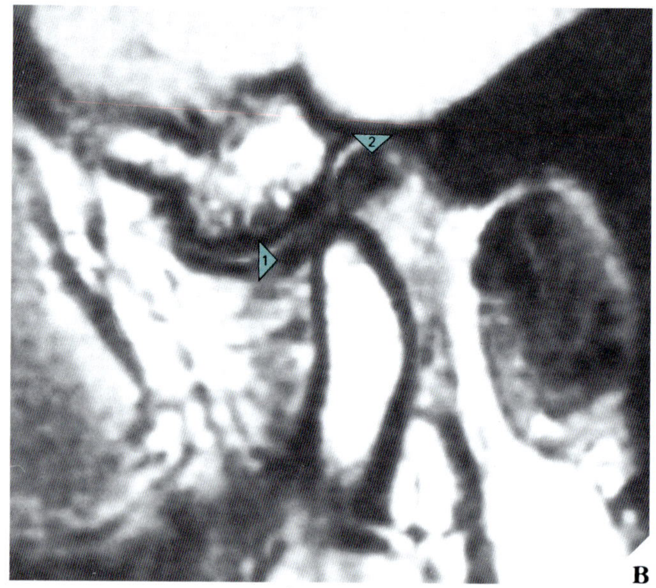

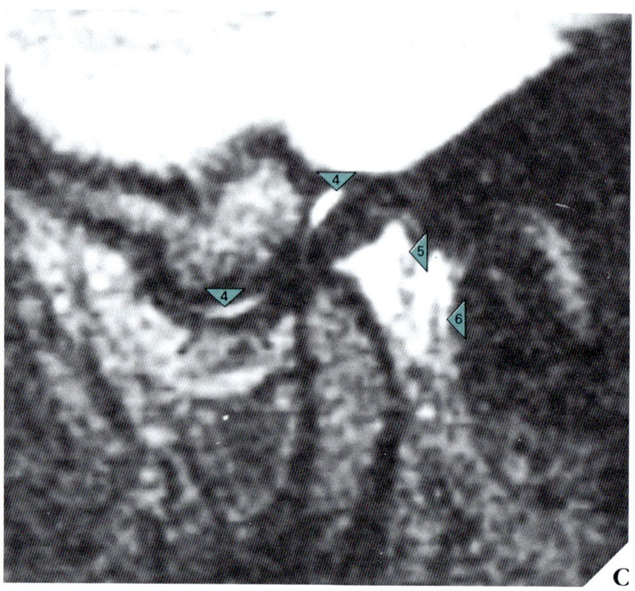

FIGURE 2.8. Subluxation of the meniscus with early reduction. A Closed-mouth sagittal section [SE 1200/20]. The meniscus is anteriorly displaced. The posterior band can be identified by the small area of increased SI. The posterior attachment is stretched over the condyle. **B** Early opening mouth, sagittal section [SE 1200/20]. The condyle has moved forward, but is not yet at the level of the temporal eminence. The meniscus is reduced. The posterior band is located posterior to the condyle. **C** Open-mouth sagittal section [SE 2000/80]. A small amount of high SI fluid in the superior and inferior joint spaces and inflammatory changes in the region of the posterior attachment can be seen on this T2WI. Arrow **1** Anterior band of meniscus. Arrow **2** Posterior band of meniscus. Arrow **3** Posterior attachment. Arrows **4** Fluid in the superior joint space. Arrow **5** Fluid in the posterior recess of the inferior joint space. Arrow **6** inflammatory changes of the posterior attachment.

often accompanied by chronic pain and limitation of opening (closed lock) (Fig. 2.9). Dislocation can be diagnosed with MRI when the meniscus is located anterior to the condyle in both closed- and open-mouth views (2). Chronically displaced meniscus may show fragmentation and distortion of its normal "bow tie" configuration.

Joint Effusion

Joint effusion can be present in cases of chronic anterior meniscal displacement, due to the inflammatory changes induced by chronic joint dysfunction. T2-weighted images can demonstrate the presence of the effusion in the superior or inferior spaces (Fig. 2.8).

Meniscal Perforation

Perforation of the meniscus or its attachments is frequently present in cases of chronic and severe derangement of the TMJ, and is often associated with anterior dislocations. MRI diagnosis of perforation is difficult because of the small size of the

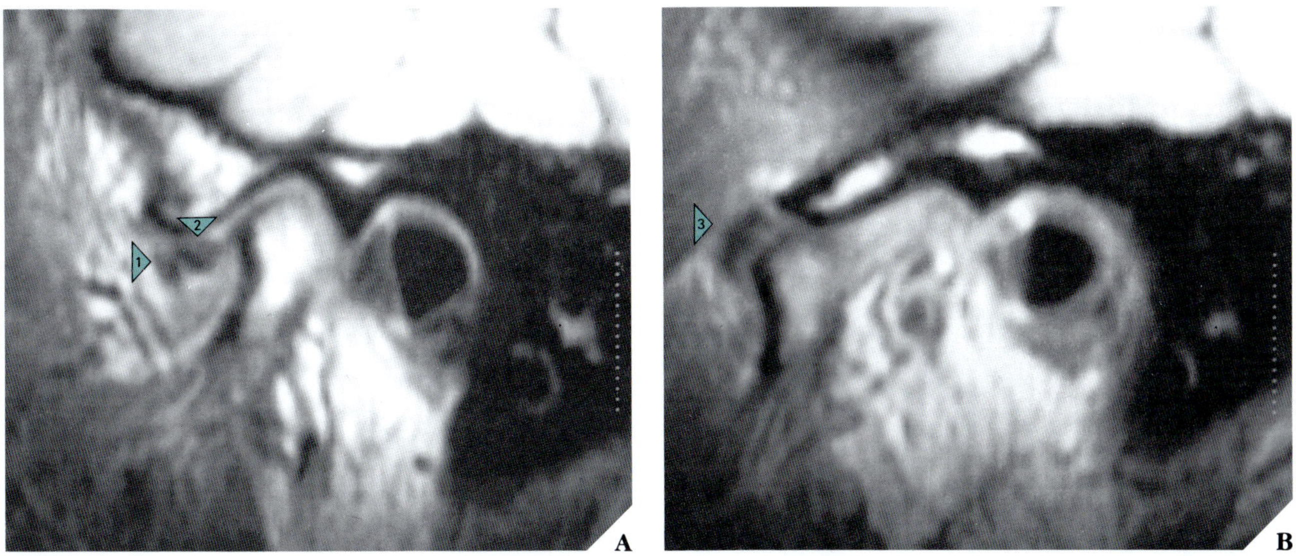

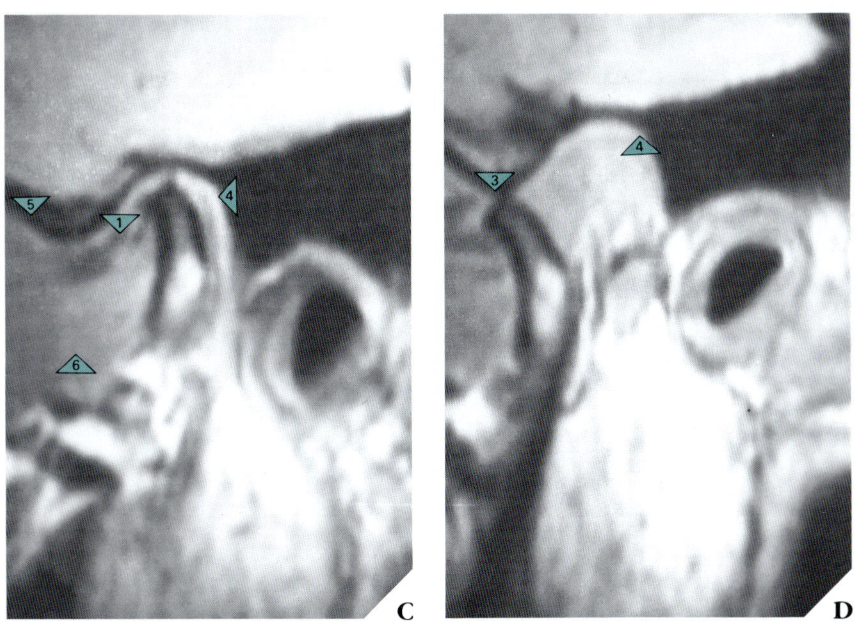

FIGURE 2.9. Dislocation of the meniscus. A Closed-mouth sagittal section [SE 1200/20]. The meniscus is anteriorly displaced and folded. **B** Open-mouth sagittal section [SE 1200/20]. The meniscus is still anteriorly displaced. **C** Closed-mouth sagittal section [SE 1200/20]. Anterior dislocation of the meniscus in another patient. **D** Open-mouth sagittal section [SE 1200/20]. The meniscus remains anteriorly displaced. The posterior attachment is stretched over the condyle. Arrow **1** Anterior band of meniscus. Arrow **2** Posterior band of meniscus. Arrow **3** Meniscus. Arrow **4** Posterior attachment. Arrow **5** Superior belly of the lateral pterygoid muscle. Arrow **6** Inferior belly of the lateral pterygoid muscle.

structures involved. Perforation is suspected when there are erosive changes of the articular eminence, discontinuity of the meniscus, anterior dislocation of the meniscus, osteoarthritis with remodeling of the mandibular condyle, or loss of the joint space, with bone-to-bone contact (Figs. 2.10 and 2.11).

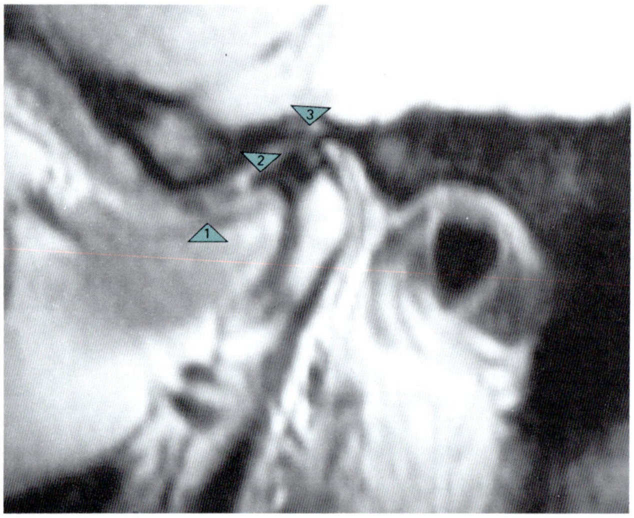

FIGURE 2.10. Osteoarthritis, meniscal dislocation, and perforation. Closed-mouth sagittal section [SE 1200/20] There is decreased SI of the condyle, an anterior osteophyte, and superior narrowing of the joint space, with bone-to-bone contact. The meniscus is anteriorly displaced. Perforation of the posterior attachment was found at surgery. Arrow **1** Meniscus. Arrow **2** Osteophyte. Arrow **3** Bone-to-bone contact with small subchondral cyst in the temporal fossa.

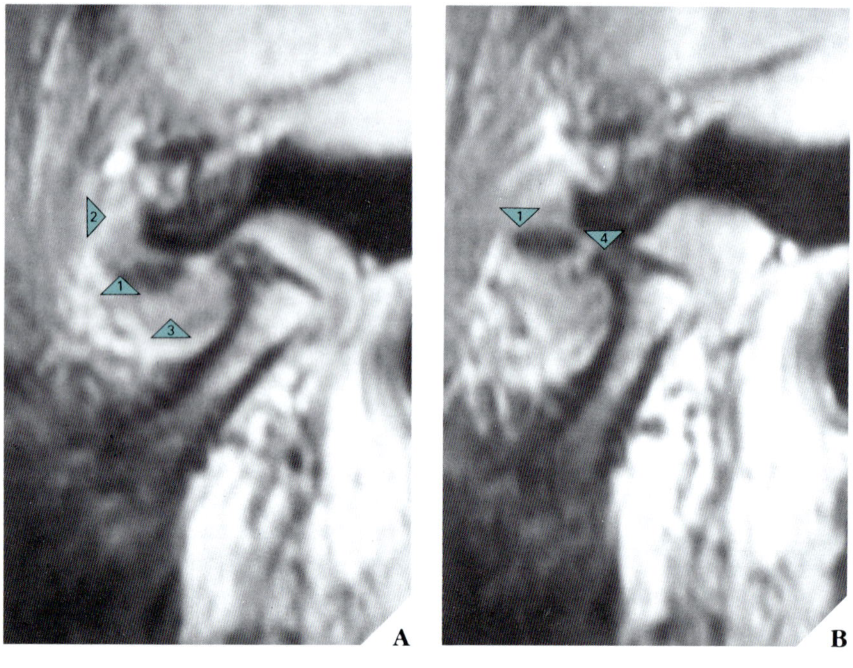

FIGURE 2.11. Osteoarthritis, meniscal dislocation, and perforation. A Closed-mouth sagittal section [SE 1200/20]. There is decreased SI of the mandibular condyle and temporal eminence (associated with anterior osteophyte formation), narrowing of the anterior joint space, and anterior displacement of the meniscus which has lost its normal configuration. A small amount of intermediate SI fluid is present in the superior and inferior joint spaces. **B** Open-mouth sagittal section [SE 1200/20]. The meniscus remains anteriorly displaced. There is bone-to-bone contact between the irregular mandibular condyle and the temporal eminence. Perforation of the posterior attachment was found at surgery. Arrow **1** Mensicus. Arrow **2** Fluid in superior joint space. Arrow **3** Fluid in anterior recess of inferior joint space. Arrow **4** Osteophyte.

Fibrosis of the Lateral Pterygoid Muscle

Fibrosis and fatty infiltration of the superior belly of the lateral pterygoid muscle will be observed in cases of chronic meniscal dislocation with perforation, and also after meniscectomy (Fig. 2.12).

Myxoid Degeneration

Myxoid degeneration of the meniscus has been described in patients with or without meniscal displacement (10, 11). High SI within the posterior band can be identified on MR imaging examinations (Fig. 2.13).

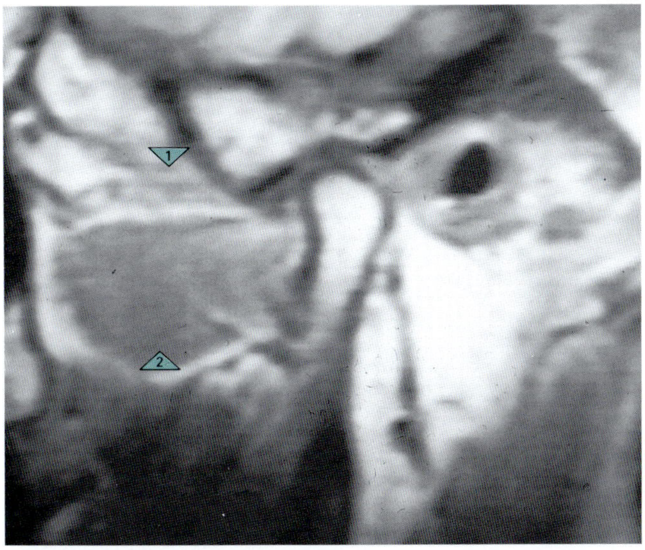

FIGURE 2.12. Fatty infiltration of the superior belly of the lateral pterygoid muscle. Closed-mouth sagittal section [SE 1200/20]. The superior belly of the lateral pterygoid muscle is ill-defined and shows increased SI due to atrophy and fatty infiltration. Arrow **1** Superior belly of the lateral pterygoid muscle. Arrow **2** Inferior belly of the lateral pterygoid muscle.

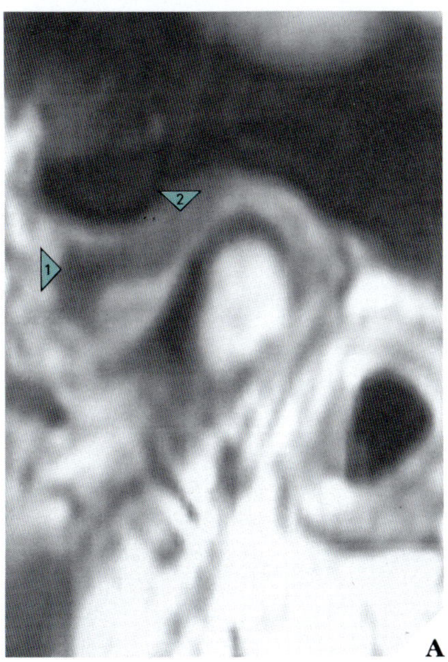

A

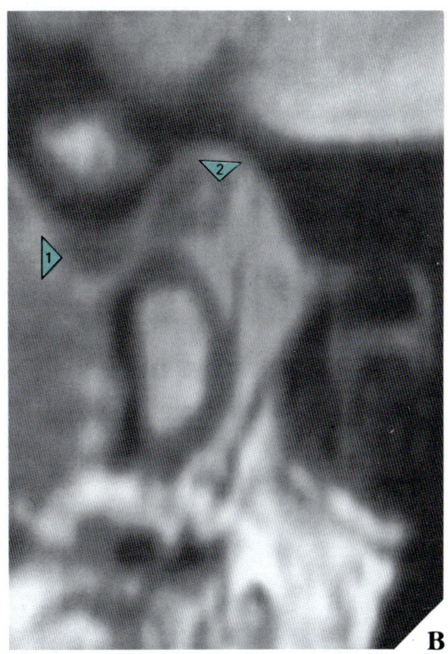

B

FIGURE 2.13. Myxoid degeneration of the meniscus with subluxation. A Closed-mouth sagittal section [SE 1200/20]. The meniscus is anteriorly displaced. The posterior band shows higher SI than the anterior band. **B** Open-mouth sagittal section [SE 1200/20]. The meniscus is reduced to its normal position. The posterior band shows a central portion of increased SI, probably due to myxoid degeneration. Arrow **1** Anterior band of meniscus. Arrow **2** Posterior band of meniscus.

DEGENERATIVE AND INFLAMMATORY DISORDERS

Osteoarthritis

Osteoarthritis may involve the TMJ as it does any other joint in the body. In general, osteoarthritis is secondary to malfunction of the joint due to internal derangement and meniscal displacement. The classical manifestations of osteoarthritis include narrowing of the joint space, sclerosis of the articular surfaces, subchondral cysts, and osteophyte formation. Additional findings in osteoarthritis of the TMJ are anterior meniscal displacement (often with fragmentation) and loss of the normal configuration (Figs. 2.10 and 2.11).

Inflammatory Arthritis

Inflammatory arthritis, including rheumatoid arthritis and seronegative spondyloarthropathies (psoriasis, Reiter's syndrome, ankylosing spondylitis), can involve the TMJ. Destructive changes with condylar erosions and meniscal displacement can be seen on MRI (Fig. 2.14).

POSTTRAUMATIC AND POSTTHERAPEUTIC CHANGES

Condylar Fracture

Posttraumatic derangement of the TMJ may result in meniscal dislocation or subluxation, accompanied by pain as well as decreased motion. Condylar

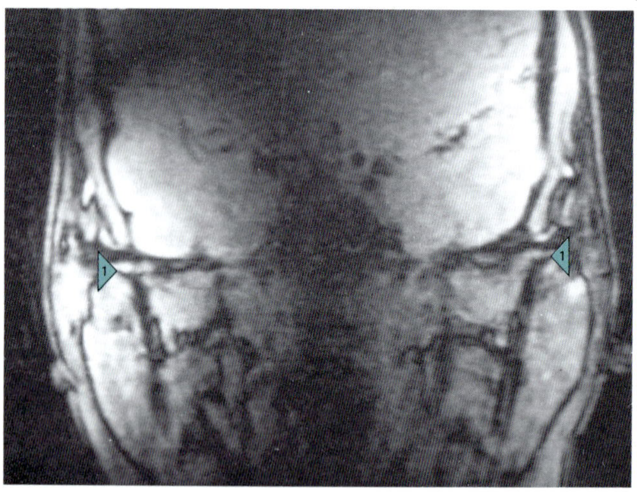

FIGURE 2.14. Rheumatoid arthritis. Coronal section through both TMJ's [GRASS 1000/12, flip angle 30 degrees]. There is flattening and erosive changes of both TMJ's. The left joint space is narrowed. Arrows **1** Condyles.

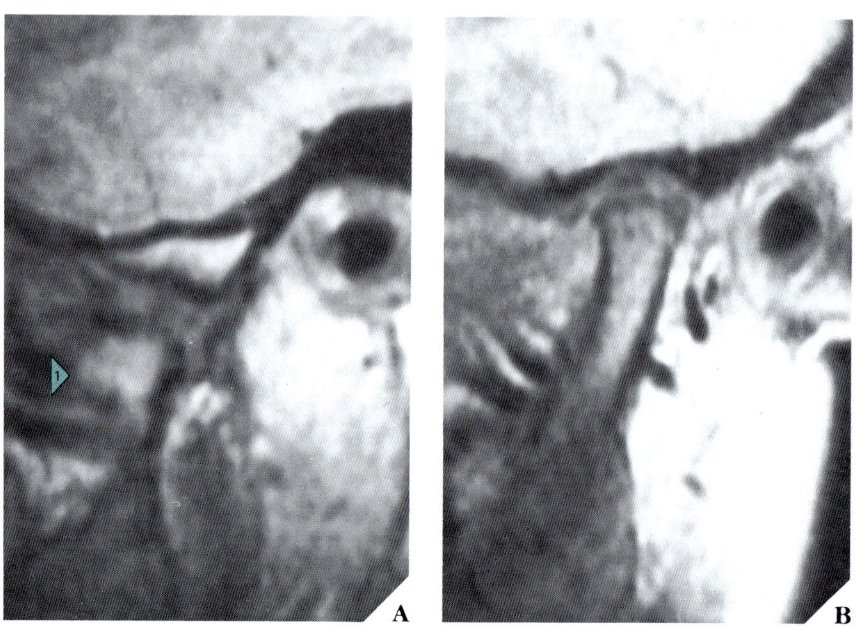

FIGURE 2.15. Condylar fracture. A Right TMJ, closed-mouth sagittal section [SE 1200/20]. The condyle is rotated and anteriorly displaced on the outside of the temporal fossa. The patient sustained a condylar neck fracture one year prior to this MRI examination. **B** Left TMJ, closed-mouth sagittal section [SE 1200/20]. There are early osteoarthritic changes with flattening of the condyle, due to dysfunction of the contralateral joint. The meniscus is normally positioned. Arrow **1** Condyle.

fractures can produce adhesive capsulitis and hematomas but they can also induce early degenerative changes in the contralateral TMJ (Fig. 2.15) (10).

Avascular Necrosis

Avascular necrosis of the mandibular condyle can occur following mandibular fracture or orthognathic and surgical manipulation of the mandible (10). Low SI of the condyle and irregularity of the articular surface may be seen on MRI (Fig. 2.16).

Osteochondritis Dissecans

Osteochondral fractures as osteochondritis dissecans can produce a sharp indentation on the articular surface of the condyle, with the osteocartilaginous fragment becoming detached as a loose body (Fig. 2.17).

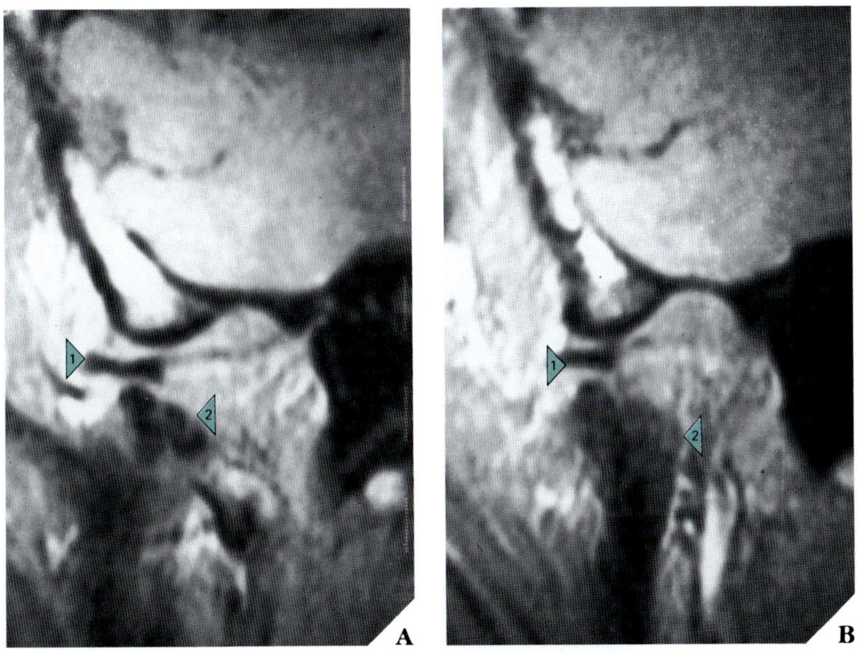

FIGURE 2.16. Avascular necrosis. Open-mouth sagittal consecutive sections **(A,B)** [SE 1200/20]. There is low SI in the condyle, with irregularity of the articular surface. The meniscus is in normal position. Arrows **1** Meniscus. Arrows **2** Condyle.

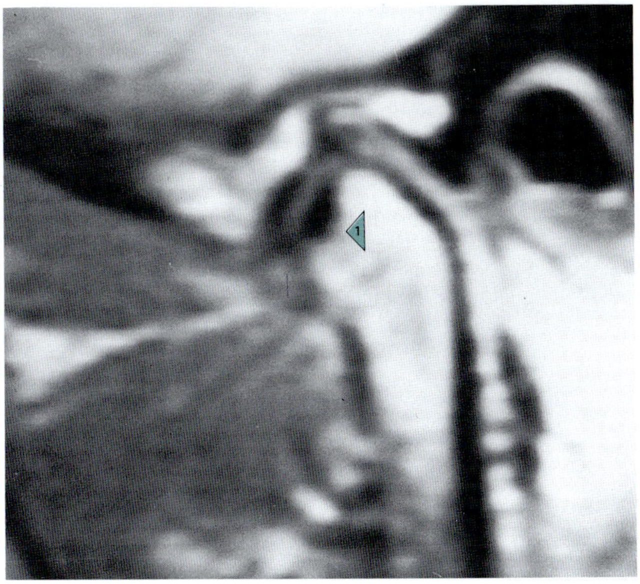

FIGURE 2.17. Ostechondritis dissecans. Sagittal sections [SE 1500/20]. There is a sharp indentation of the articular surface of the condyle representing an osteochondral lesion. The SI of the bone marrow space is otherwise normal. Arrow **1**: osteochondritis dissecans.

Lack of recognition of avascular necrosis and osteochondral fractures as important causes of internal derangement can lead to failed orthognatic procedures.

Splint Therapy

Anterior mandibular reposition splint is one of the techniques used for conservative treatment of meniscal subluxations. In an attempt to recapture the meniscus, an occlusal appliance is positioned between the teeth to produce a forward displacement of the condyle in the closed-mouth position. MRI can be used to assess the effectiveness of the occlusive appliance (Fig. 2.18).

Meniscectomy

Simple meniscectomy may be performed in cases of chronic dislocation with degeneration of the meniscal tissue. MRI will show the absence of the meniscus (Fig. 2.19) (10).

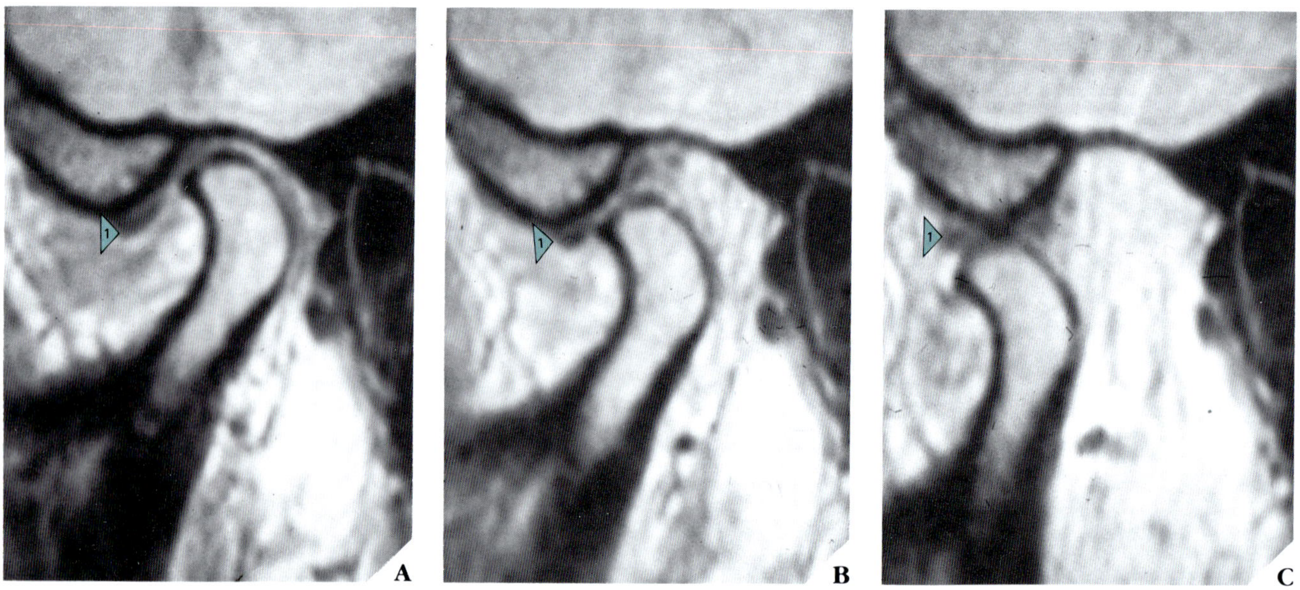

FIGURE 2.18. Meniscal subluxation with reduction following splint therapy. A Closed-mouth sagittal section without splint [SE 1200/20]. The meniscus is anteriorly displaced. **B** Closed-mouth sagittal section with splint]SE 1200/20]. The condyle is more anteriorly positioned. The meniscus is reduced to its normal position. **C** Open-mouth sagittal section]SE 1200/20]. The meniscus is normally positioned on top of the condyle. Arrows **1** Meniscus.

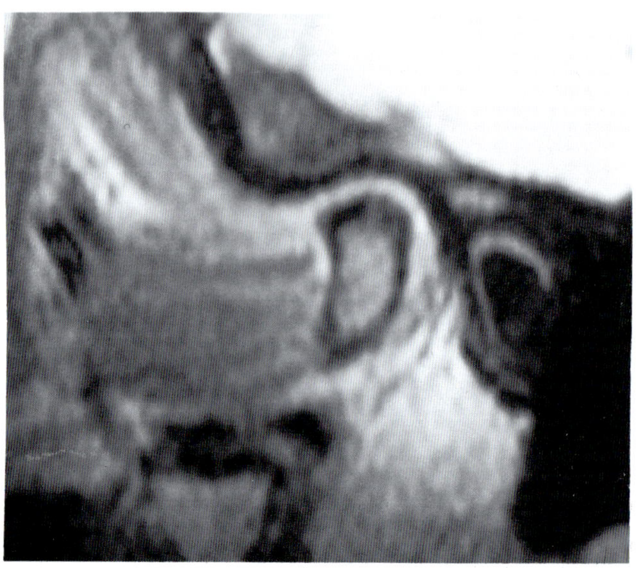

FIGURE 2.19. Status post meniscectomy. Closed-mouth sagittal section [SE 1200/20]. The meniscus has been removed. There is minimal narrowing of the joint space.

Implant Therapy

Teflon–Proplast or Silastic implant interposition has been used in recent years in cases of advanced TMJ derangement. However, complications due to foreign body reaction with bone destruction, and perforation with or without dislocation of the implant, are relatively frequent (4,7,12). MRI is used to assess the integrity and location of the implant (Fig. 2.20).

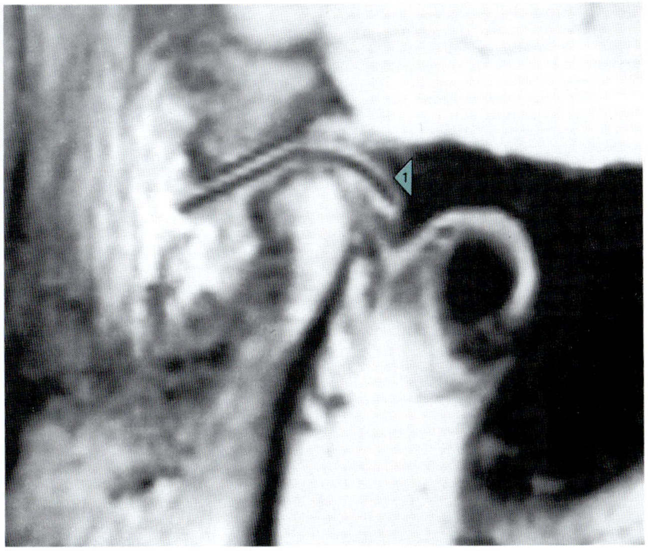

FIGURE 2.20. Status post-teflon implant. Closed-mouth sagittal section [SE 1200/20]. The low SI Teflon implant is interposed between the condyle and the temporal fossa. No fragmentation or displacement is seen. Arrow **1** Teflon implant.

REFERENCES

1. Griffin DJ, Hawthorn R, Harris R: Anatomy and histology of the human temporomandibular joint. Monogr Oral Sci 1975;4:1–26.
2. Harms SE, Wilk RM: Magnetic resonance imaging of the temporomandibular joint. Radiographics 1987;7:541–542.
3. Harms SE, Wilk RM, Wolford LM, et al: The temporomandibular joint: Magnetic resonance imaging using surface coils. Radiology 1985;157:133–136.
4. Kaplan PA, Ruskin JD, Tu HK, et al: Erosive arthritis of the temporomandibular joint caused by Teflon–Proplast implants: plain film features. AJR 1988;151:337–339.
5. Katzberg RW, Bessette RW, Tallents RH, et al: Normal and abnormal temporomandibular joint: MR imaging with surface coil. Radiology 1986;158:183–189.
6. Katzberg RW, Westesson PL, Tallents RH, et al: Temporomandibular joint: MR assessment of rotational and sideways disk displacements. Radiology 1988;169:741–748.
7. Kneeland JB, Ryan DE, Carrera GF, et al: Failed temporomandibular joint prostheses: MR imaging. Radiology 1987;165:179–181.
8. Mercury LG: Intra-articular meniscus dysfunction and the spectrum of TMJ problems. Compend Contin Educ Dent 1985;6:107–115.
9. Murphy WA: Arthrography of the temporomandibular joint. Radiol Clin North Am 1981;19:365–378.
10. Schellhas KP, Wilkes CH, Fritts HM, et al: Temporomandibular joint: MR imaging of internal derangements and postoperative changes. AJR 1988;150:381–389.
11. Schellhas KP, Wilkes CH, Omlie MR, et al: The diagnosis of temporomandibular joint disease: two-compartment arthrography and MR. AJR 1988;151:341–350.
12. Tallents RH, Katzber RW, Miller TL, et al: Arthrographically assisted splint therapy. J Prosthet Dent 1985;53:235–238.

CHAPTER THREE

Javier Beltran

The Shoulder

TECHNICAL CONSIDERATIONS

IMAGING PLANES AND PULSE SEQUENCES

Axial, sagittal, and coronal plane images should be obtained for a complete MRI examination of the shoulder (Fig. 3.1). The axial plane is ideal for evaluation of the glenoid labrum, bicipital groove, biceps tendon, and subscapularis tendon. Imaging in the coronal and sagittal planes is obtained in an oblique direction, following the longitudinal and transverse axes of the supraspinatus muscle (4,16,19). This is achieved by angling the plane of the section parallel and perpendicular to the plane of the glenoid fossa (Figs. 3.2 and 3.3). The oblique sagittal and coronal planes more clearly demonstrate the muscles and tendons of the rotator cuff.

A short TR/TE (T1WI) pulse sequence sufficiently demonstrates the structural anatomy in the axial plane. Using this pulse sequence, the labrum is seen as a low SI structure, contrasting with the intermediate SI of the articular cartilage of the glenohumeral joint. This contrast represents the difference in structural composition of the fibrocartilage (labrum) and hyaline articular cartilage, and is similar to that which occurs in the structural anatomy of the knee joint (see Chapter 7). In cases of suspected labral tear or capsular detachment, T2W1 in the axial plane may provide further information. Proton density and T2WI are used in the oblique coronal and sagittal planes. This pulse sequence provides the necessary information for the evaluation of pathology in the rotator cuff, joint space, subacromial region, and the bones. Normal muscle exhibits intermediate SI, and the tendons exhibit low SI in all SE pulse sequences. Any area of high SI on T2W1 in the subacromial region is abnormal.

The use of respiratory compensation bellows helps to diminish respiratory motion artifacts.

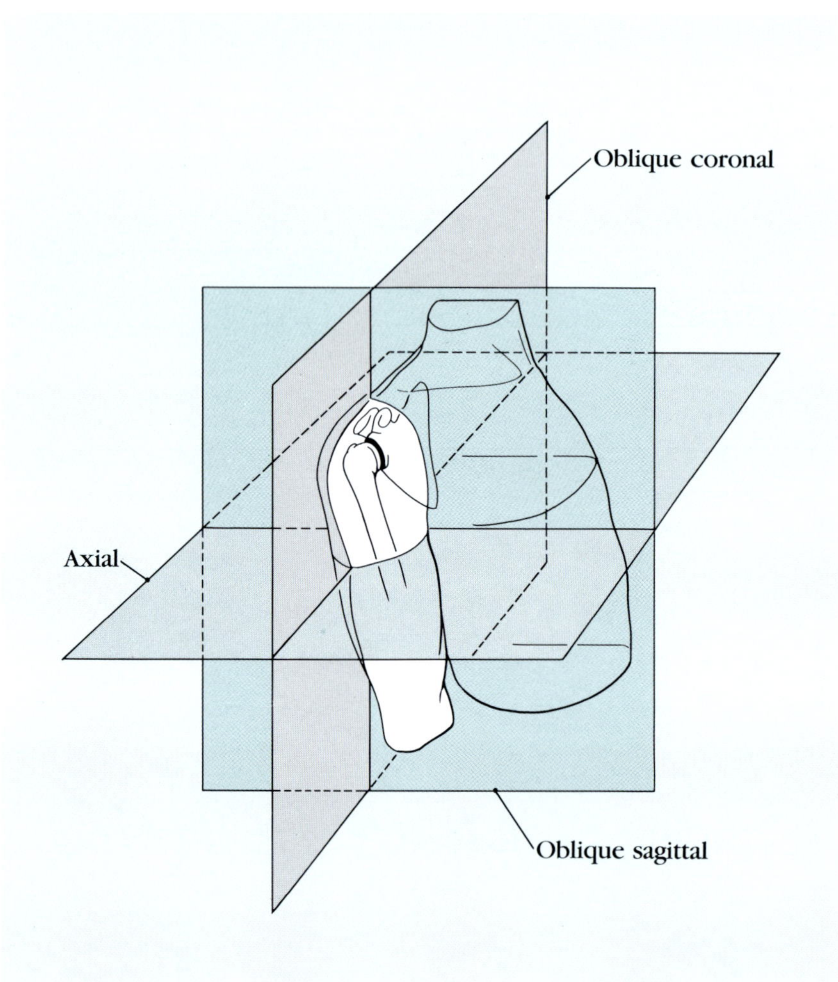

FIGURE 3.1. Routine planes of section for shoulder imaging.

FIELD OF VIEW

In the majority of patients, an off-center FOV of 16 cm sufficiently covers the entire shoulder and axillary regions, with adequate S/N. The area between the acromioclavicular joint and the inferior margin of the glenoid is covered in the axial plane. The anterior and posterior glenoid margins are the limits for coverage in the oblique coronal plane. The area between the glenoid fossa and the most lateral aspect of the humerus is covered in the oblique sagittal plane (Figs. 3.2 and 3.3).

POSITIONING

The patient is placed in a head-first, supine position with the arm in a neutral position, alongside of the body. Large patients may have to place their hands over their chests in order to be accommodated within the gantry.

ACQUISITION MATRIX

A high resolution matrix of 256 × 256 pixels is used in the axial plane. A matrix of 256 × 128 pixels is used in the oblique coronal and sagittal planes to keep imaging time to a minimum.

SLICE THICKNESS

A 5-mm slice thickness is used in the axial plane, with a 3-mm thickness in the oblique coronal and sagittal planes. An interslice gap of 1 mm is used in all three planes.

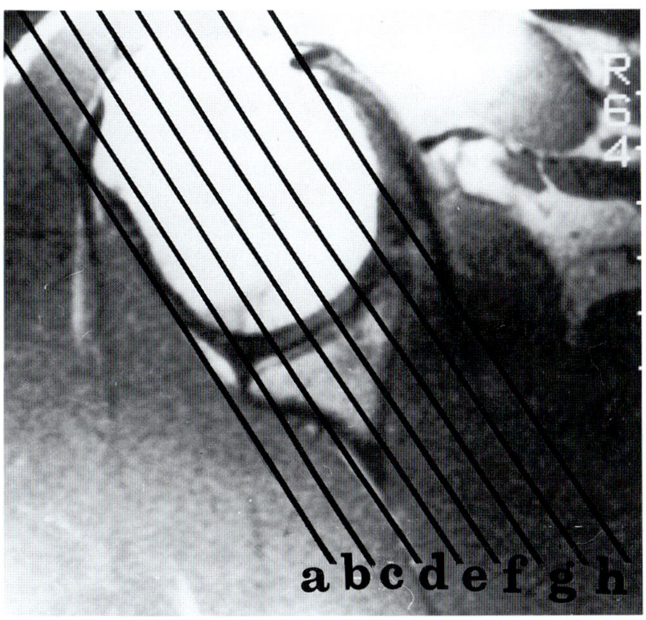

FIGURE 3.2. Right shoulder. Axial MRI. *Lines a–h* perpendicular to the glenoid fossa illustrate the orientation of the oblique coronal sections.

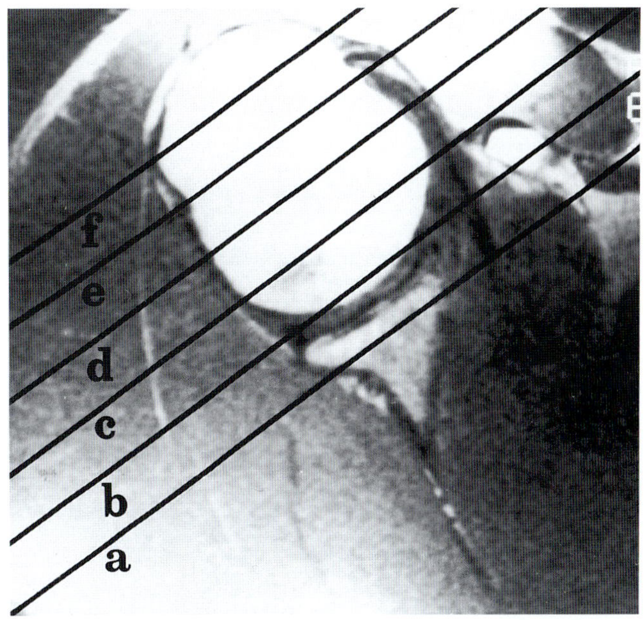

FIGURE 3.3. Right shoulder. Axial MRI. *Lines a–f* parallel to the glenoid fossa illustrate the orientation of the oblique sagittal sections.

SURFACE COILS

Surface coils must be used for shoulder imaging. Coupled pair coils, such as receive-only loop-gap resonator coils (angled pair, 4-inch diameter), are used most frequently (Fig. 3.4) (18, 22). These coils provide relatively uniform signal distribution with good S/N. The coils should be centered over the humeral head, avoiding a placement that is either too medial or too inferior. The coils can be secured with angled sponges and compression bands. With large patients, a signal drop-off can occur in the central area of the shoulder. In this case, anatomically shaped, flexible wraparound coils, Hemholtz coils (15,17), and planar coils (36) can be used instead. However, planar coils produce excessive signal drop-off in the region away from the coil and around the glenohumeral joint and are therefore not recommended for routine shoulder imaging. Alternately, a coupled pair of 5-inch receive-only circular coils placed anteriorly and posteriorly can be used.

NORMAL ANATOMY

The glenohumeral joint has the greatest range of motion (over 180 degrees in several planes) of any joint in the human body. The relatively large articular surface of the humeral head compared with the small articular surface of the glenoid cavity causes the extended mobility of this joint. Because of its wide range of motion, the glenohumeral joint is more susceptible to dislocations, subluxations, and lesions related to chronic stress in the surrounding soft tissues.

Internal and external rotation and some degree of abduction are provided by the rotator cuff muscles (supraspinatus, infraspinatus, subscapularis, and teres minor) (Fig. 3.5).

The joint capsule is reinforced anteriorly by the anterior glenohumeral ligaments. The capsular mechanism provides stabilization of the glenohumeral joint. The anterior capsular mechanism (26,31,34) includes the fibrous capsule, the anterior glenohumeral ligaments, the synovial membrane and its recesses, the fibrous glenoid labrum, the subscapularis muscle and tendon, and the scapular periosteum. There are three types of anterior capsular insertion determined by the proximity of insertion to the glenoid margin (Fig. 3.6) (40). Type I is the closest; type III the furthest. The further the anterior capsular insertion from the glenoid margin, the more unstable the glenohumeral joint will be.

The posterior capsular mechanism (31) is formed by the posterior capsule, the synovial membrane, the glenoid labrum and scapular periosteum, and the posterosuperior tendinous cuff and associated muscles (supraspinatus, infraspinatus, and teres minor). The long head of the biceps tendon inserting in the superior aspect of the labrum, and the triceps inserting in the infraglenoid tubercle inferiorly, provide additional supportive structures to the glenohumeral joint.

In practically all SE pulse sequences, the tendons, ligaments, fibrocartilage (labrum), and cortical

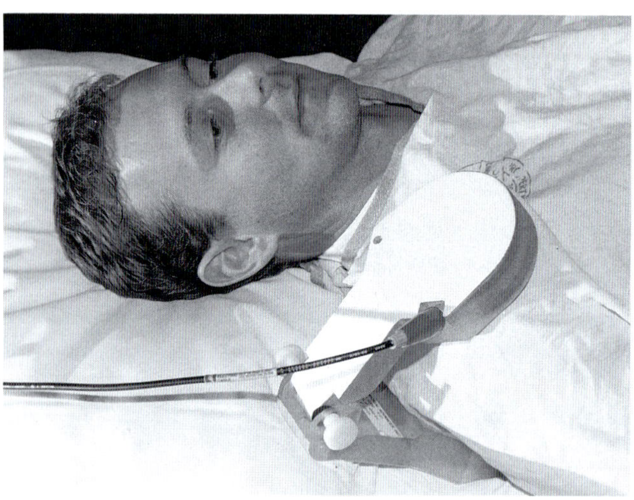

FIGURE 3.4. A loop-gap resonator coil positioned for MR imaging of the shoulder.

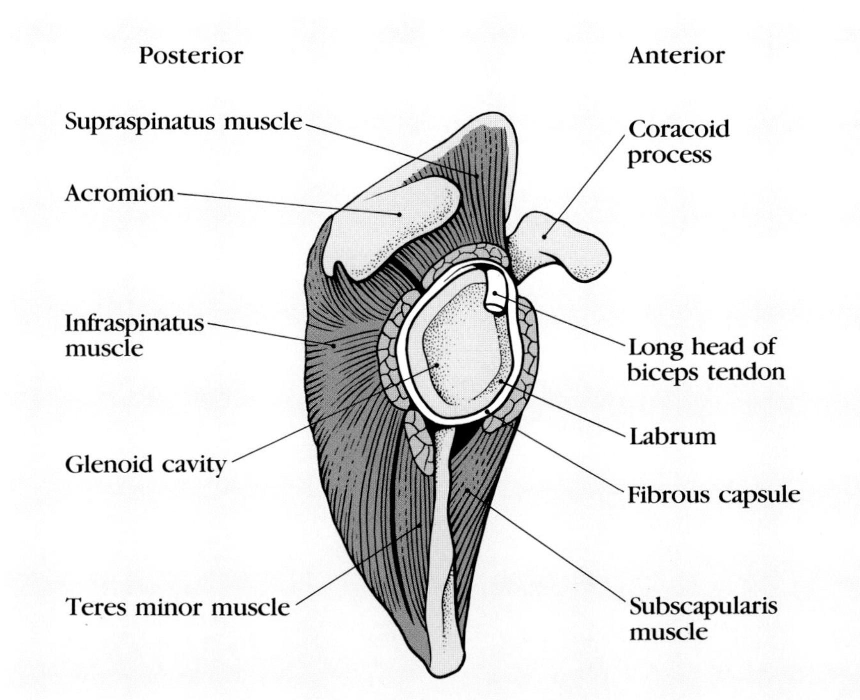

FIGURE 3.5. This schematic of the glenoid fossa (with the humerus removed) shows the distribution of the muscles of the rotator cuff and the intracapsular portion of the long head of the biceps tendon.

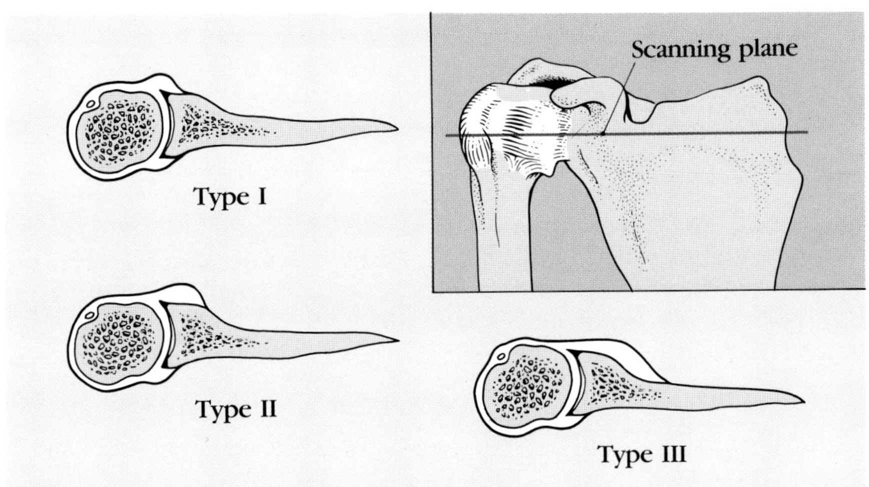

FIGURE 3.6. This schematic shows the three types of anterior capsular insertion. Type III predisposes the patient to anterior shoulder instability.

bone are seen on MRI as low SI structures. This is consistent with imaging in other joints. Muscles exhibit intermediate-to-low SI in all pulse sequences. Subcutaneous fat and fatty bone marrow exhibit high SI on T1WI, and low SI on T2WI.

AXIAL PLANE

The labrum is seen on axial sections as two small black triangles located anterior and posterior to the glenoid margin (Fig. 3.7) (13,22,35). The articular cartilage of the glenoid fossa and humeral head is seen as a thin, intermediate SI band contrasting with the low SI of the labrum. Since the labrum represents an infolding of the capsule, it may be seen separated from the cortex of the glenoid fossa by a thin layer of articular cartilage.

Superior sections through the subacromial arch show the supraspinatus muscle. Inferior sections through the midglenohumeral joint more clearly demonstrate the long head of the biceps tendon in the bicipital groove, and the subscapularis tendons and muscle. The subscapularis tendons can be followed to their insertion in the lesser tuberosity of the humerus. Inferior sections can demonstrate the anterior capsular insertion in the glenoid (Fig. 3.7).

OBLIQUE CORONAL PLANE

The coronal plane is ideal for evaluation of the subacromial region, the acromioclavicular joint, and the rotator cuff (Figs. 3.8 and 3.9). Normally, the subscapularis and infraspinatus muscles each have four to five tendons, arising in the muscle bellies. These tendons converge peripherally with single tendons from both the supraspinatus muscle and the teres minor muscle, to form the rotator cuff (23). The insertion of the supraspinatus and infraspinatus tendons in the greater tuberosity can be identified in the anterior and posterior oblique coronal planes, respectively. The subacromial–subdeltoid bursa is not clearly identified on MRI, unless it is distended by fluid or thickened by inflammation, but the layer of fat surrounding the bursa is normally identified.

The superior and inferior labrum are well-identified in this plane. In some cases, the inferior glenoid labrum may be unusually prominent. This is due to a summation of the low SI inferior capsule with the axillary recess and the inferior labrum. The long head of the biceps tendon is partially seen in the more anterior oblique coronal planes and can normally be followed until it becomes intracapsular in the bicipital groove.

OBLIQUE SAGITTAL PLANE

The sagittal plane shows the anatomy of the acromial arch and the subacromial structures (Fig. 3.10). The acromioclavicular joint, coracoclavicular ligament, coracoacromial ligament, glenohumeral ligament, and rotator cuff tendons and muscles can be assessed in the sagittal plane.

The teres minor muscle and tendon are best seen in the sagittal plane in posterior location, attached to and below the infraspinatus muscle and tendons (Fig. 3.10) (22). Three tendons can be seen inserting in the coracoid process in the sagittal plane, from medial to lateral: the pectoralis minor, the coracobrachialis, and the short head of the biceps tendons.

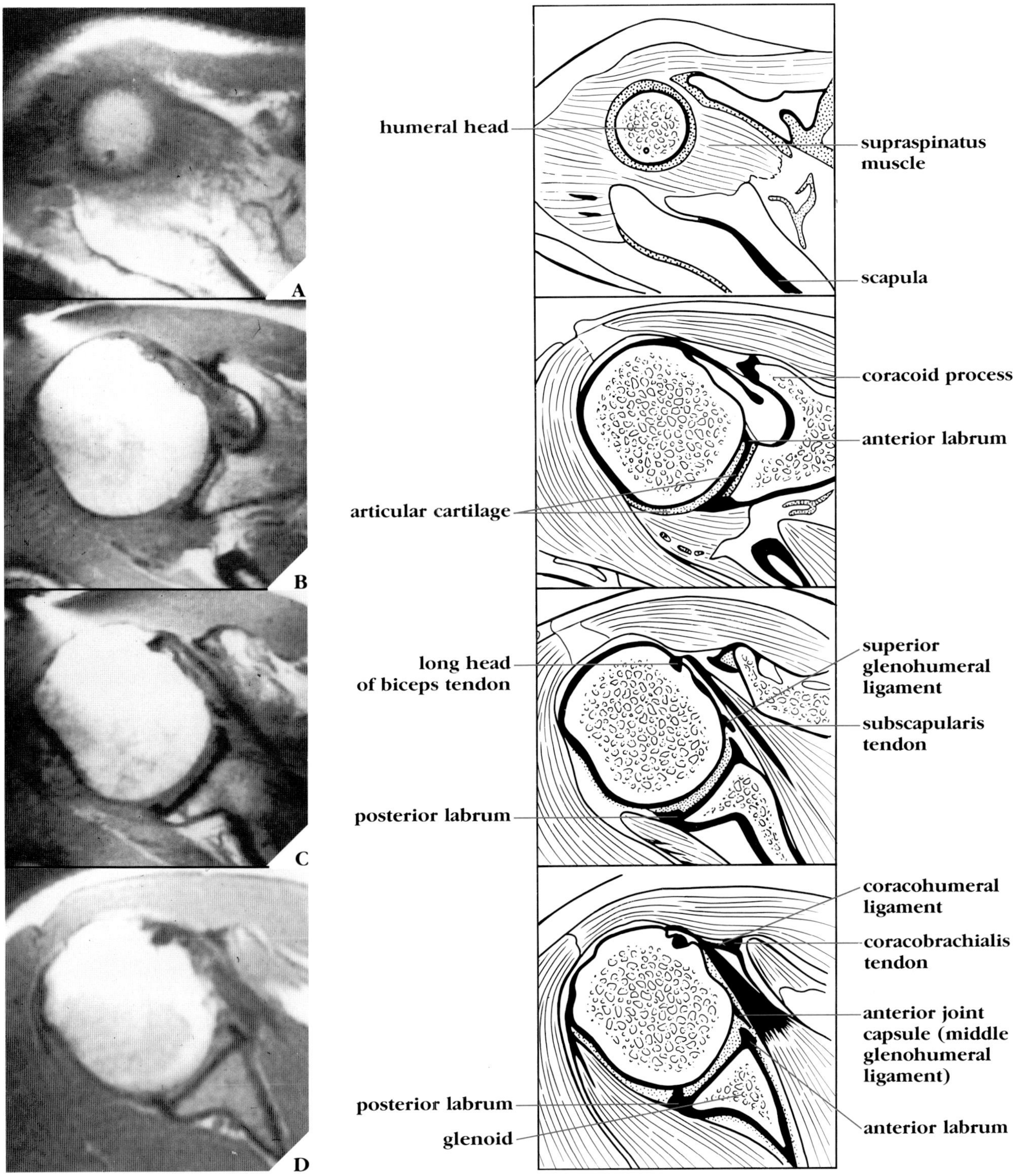

FIGURE 3.7. Normal glenohumeral joint. Consecutive axial sections from superior **(A)** to inferior **(D)** [SE 2000/20]. Note the small, low SI triangles anterior and posterior to the glenoid, representing the fibrous labrum. Occasionally, the anterior capsular insertion can be identified **(D)**.

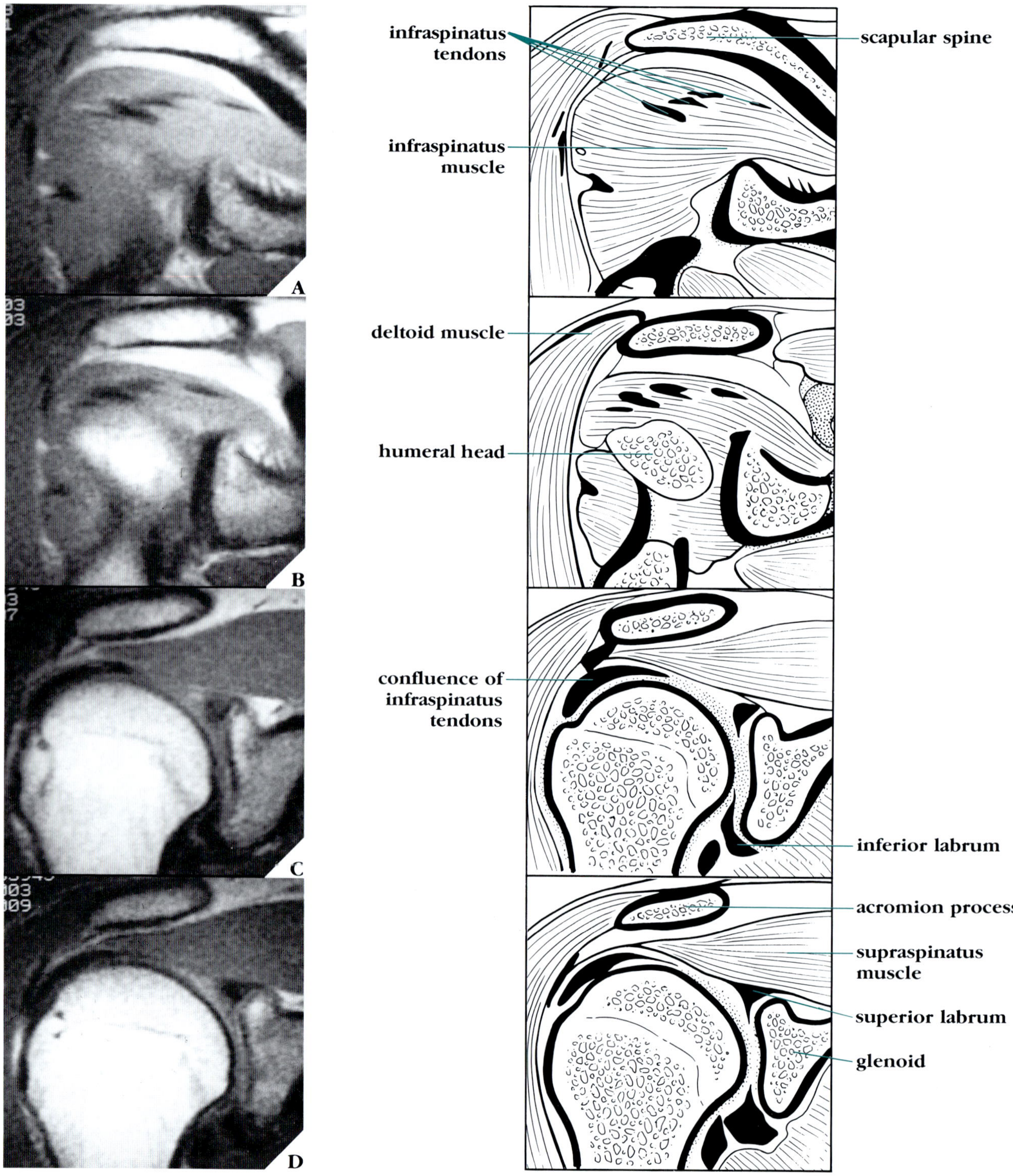

FIGURE 3.8. Normal posterior glenohumeral joint. Consecutive oblique coronal sections [SE 2000/20]. Note the multiple tendons of the infraspinatus muscle **(A)**.

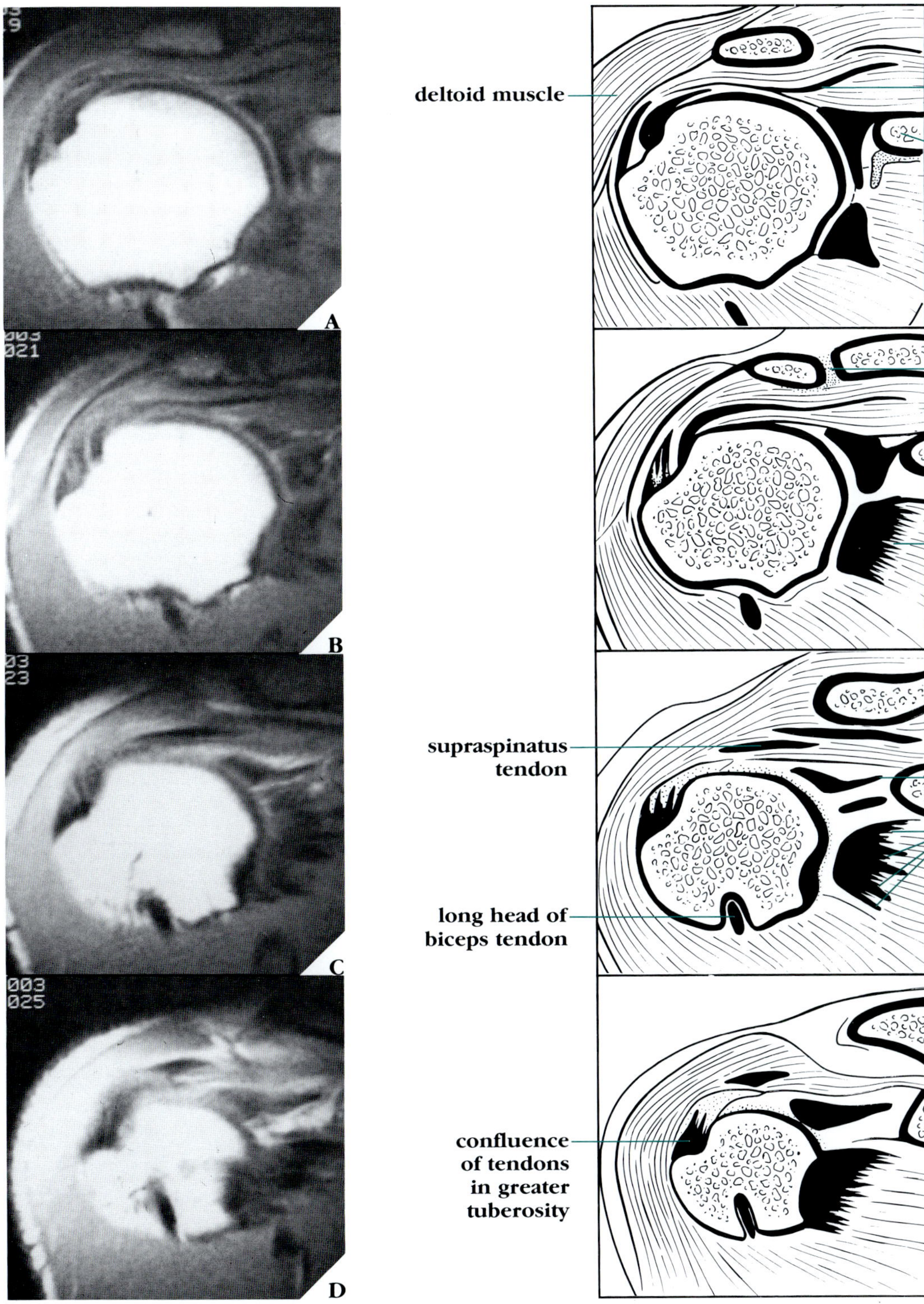

FIGURE 3.9. Normal anterior glenohumeral joint. Consecutive oblique coronal sections [SE 2000/20]. Note the single tendon of the supraspinatus muscle **(A)** and the multiple tendons of the subscapularis muscle **(C)**. The long head of the biceps tendon is seen in the more anterior sections **(A–D)**, when it becomes intracapsular in the bicipital groove.

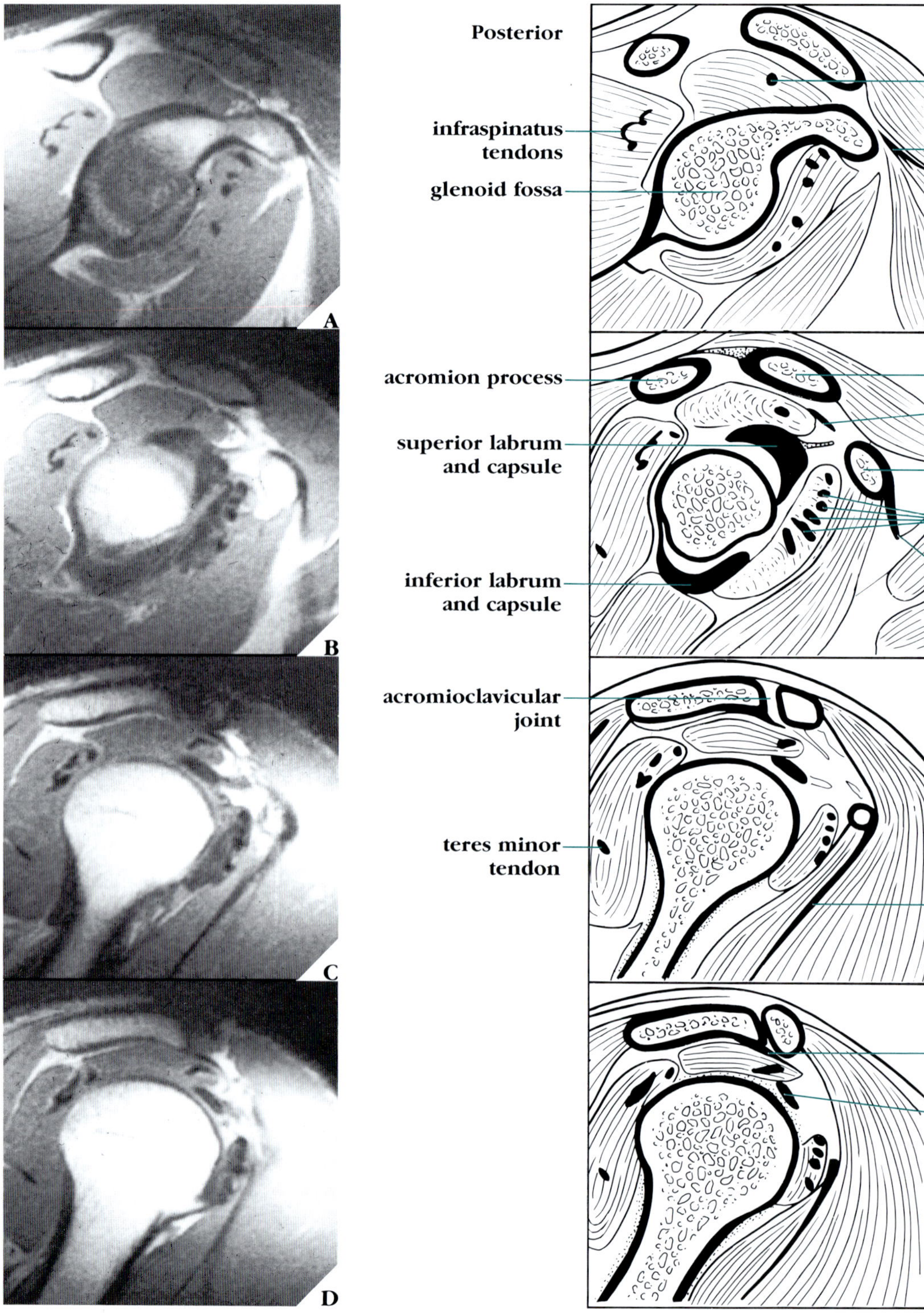

FIGURE 3.10. Normal glenohumeral joint. Consecutive oblique sagittal sections [SE 2000/20] from medial **(A)** to lateral **(D)**. Note the insertion of the three tendons in the coracoid process (pectoralis minor, coracobrachialis, and short head of the biceps tendons). The thickness of the supraspinatus muscle and tendon in the subacromial arch can be assessed.

SHOULDER IMPINGEMENT SYNDROME AND ROTATOR CUFF TEARS

Impingement syndrome is defined by Dr. C. S. Neer (27) as a spectrum of diseases resulting from an impingement upon the rotator cuff, the subacromial bursa, and the long head of the biceps tendon by the anterior edge of the acromial arch, acromioclavicular joint, coracoclavicular ligament, and occasionally the coracoid process (9).

These three stages of impingement have been defined (12,27,29): **Stage I** typically occurs in young individuals engaged in activities such as swimming, which require excessive use of the arm above the head. Pathologically, edema and hemorrhage of the subacromial bursa will be seen. If impingement persists, thickening and fibrosis of the bursa and tendinitis of the cuff take place and the syndrome progresses to stage II. **Stage II** is most often seen in patients 25 to 40 years old. In **Stage III,** usually seen in patients over 40 years old, there is further degeneration of the cuff with complete or incomplete tears. Rotator cuff tears occur frequently in the supraspinatus tendon, 1 cm proximal to its insertion in the greater tuberosity of the humerus. This region is called the "critical area" because of its relative hypovascularity compared with the rest of the tendon.

Once the rotator cuff becomes inflamed, the synovium of the joint capsule and its extension over the long head of the biceps tendon also develop inflammatory changes, hence the frequent association of impingement syndrome with biceps tenosynovitis (28).

Another condition involving the rotator cuff is "Milwaukee shoulder," which is seen in elderly individuals who have severe osteoarthritis of the glenohumeral joint and superior migration of the humeral head. Over 90 per cent of these patients have rotator cuff tears (33). Acute trauma and rheumatoid arthritis can also produce rotator cuff tears (32,33).

Nonsurgical management techniques are indicated during the early stages of impingement syndrome (12). These include rehabilitation, changing the patient's activities, and administering antiinflammatory medication. In late stage II and early stage III impingement syndrome, anterior acromioplasty and coracoacromial ligament resection can provide decompression of the subacromial space (29). In late stage III, the rotator cuff tear must be surgically repaired.

Multiple techniques are available for the evaluation of patients with impingement syndrome or suspected rotator cuff tear. All have advantages and disadvantages:

Plain-film radiography findings are normal during the early stages of the impingement syndrome. In late stage II and in stage III, radiographs can demonstrate cystic changes around the greater tuberosity, sclerotic changes and osteophytes at the anterior and inferior aspect of the acromion, degenerative osteoarthritis of the acromioclavicular joint, and late narrowing of the subacromial space (5,11,14,35).

Single- or double-contrast arthrography in complete rotator cuff tears demonstrates leakage of contrast material from the glenohumeral joint into the subacromial space with almost 100 per cent accuracy (10, 25). Subacromial bursography can show incomplete tears of the superior aspect of the rotator cuff in patients with normal glenohumeral arthrography (8).

Ultrasonography (US) has been used as a screening test for patients suspected of having rotator cuff tears (6). US has several advantages over other imaging techniques. It is readily available, noninvasive, inexpensive, and quickly performed. (The entire examination can be completed in 10 to 15 minutes.) There are several disadvantages, however: (a) Differentiation between complete and incomplete tears and evaluation of the size of the tear may be difficult; (b) the rotator cuff below the acromion cannot be assessed; (c) osseous and deep soft-tissue components of the shoulder joint, such as the labrum and glenohumeral ligaments, cannot be evaluated (23); and (d) this technique requires significant operator and interpretative skills.

Direct sagittal CT arthrography (CTA) allows assessment of the size and the integrity of the cuff. Incomplete rotator cuff tears not detected by conventional arthrography can be diagnosed by this technique (2). There are three main disadvantages of direct sagittal CTA, however. First, the early stages of impingement syndrome may produce a negative study. Secondly, CTA is an invasive technique, and, thirdly, it requires additional CT room time.

MR imaging has recently been introduced as a new technique for the evaluation of patients with shoulder impingement syndrome and rotator cuff tear. The primary advantage of MRI over other imaging modalities is that it can demonstrate changes in all three stages of the impingement syndrome (17–19,35):

Stage I Bursal thickening and effusion (subacromial bursitis) can be seen during the early stages, with high SI of the fluid on T2WI. Normal SI is seen in the area of the rotator cuff (17,35) (Fig. 3.11).

Stage II When edema and inflammatory changes in the supraspinatus tendon develop, MRI can demonstrate areas of high SI on T2WI within the substance of the tendon (Fig. 3.12) and occasionally compression of the tendon by a hypertrophied acromioclavicular joint.

Stage III In cases of complete rotator cuff tear, defects in the tendon, with or without retraction of the supraspinatus muscle, can be seen (Figs. 3.13 and 3.14). In the region of the defect, extending into the subacromial bursa and joint space, the associated high SI fluid will appear on T2WI.

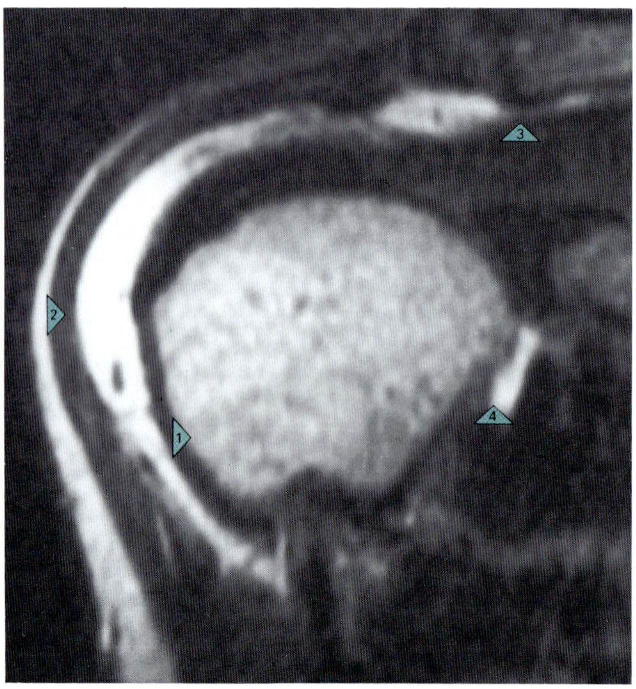

FIGURE 3.11. Subacromial bursitis. Oblique coronal MRI [SE 2000/80]. There is fluid in the subacromial bursa and in the joint space. Arrow **1** Humeral head. Arrow **2** Subacromial bursitis. Arrow **3** Clavicle. Arrow **4** Joint effusion.

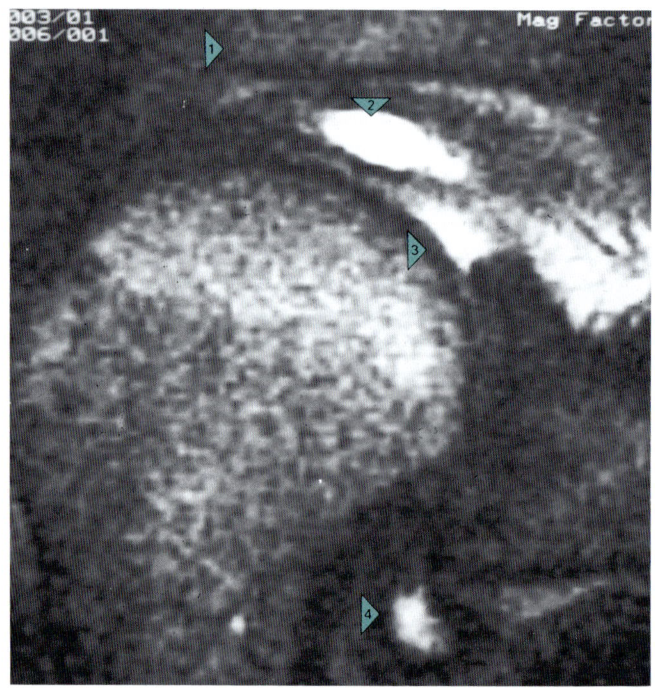

FIGURE 3.12. Stage II impingement syndrome. Oblique coronal MRI [SE 2000/80]. Forty-two-year-old male with shoulder pain and limitation of motion for the past few months. The area of high SI within the substance of the rotator cuff represents tendinitis. The thickness of the cuff is preserved. Differentiation between tendinitis and a partial rotator cuff tear may be difficult on MRI. A small amount of joint effusion is present. Arrow **1** Acromion process. Arrow **2** Rotator cuff tendinitis. Arrow **3** Joint effusion. Arrow **4** Fluid in axillary recess.

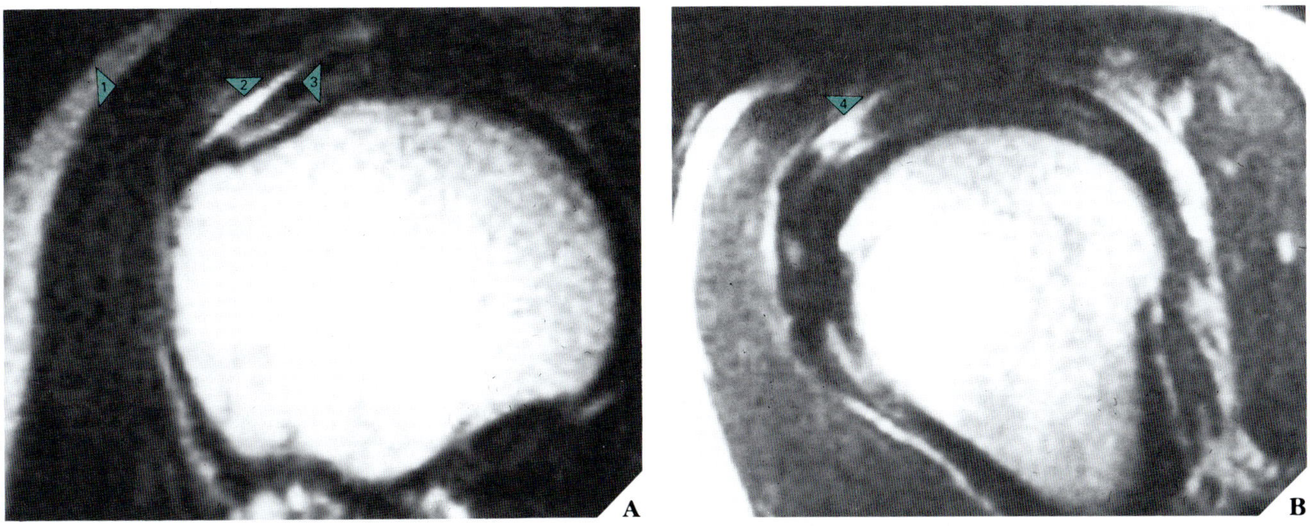

FIGURE 3.13. Stage III impingement syndrome with small, full-thickness rotator cuff tear. A Oblique coronal and **B** sagittal MRI [SE 2000/80]. This 53-year-old female had a long history of shoulder pain, becoming more severe in the few months preceding this visit. **A** MRI demonstrates thinning of the supraspinatus tendon near its insertion in the greater tuberosity, with fluid surrounding the tendon. **B** The oblique sagittal plane more clearly demonstrates a gap in the rotator cuff. Arrow **1** Deltoid muscle. Arrow **2** Fluid around supraspinatus tendon insertion. Arrow **3** Supraspinatus tendon. Arrow **4** Rotator cuff tear.

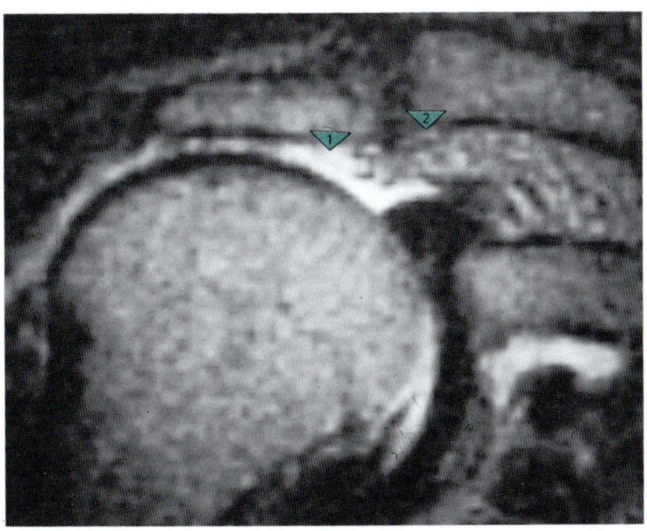

FIGURE 3.14. Stage III impingement syndrome with complete rotator cuff tear in a 60-year-old male. Oblique coronal MRI [SE 2000/80]. The supraspinatus muscle is retracted medially. No tendon tissue is present in the subacromial region. Joint fluid occupies the space of the rotator cuff tear. Arrow **1** Rotator cuff tear. Arrow **2** Retracted supraspinatus muscle.

However, differentiation between complete rotator cuff tear without muscle retraction, and incomplete tear or inflammatory changes in the supraspinatus tendon, may not be possible with MRI.

Rotator cuff tear occurring as a result of acute trauma, rheumatoid arthritis, or cuff arthropathy have the same MRI manifestations as rotator cuff tears occurring during the late stage of impingement syndrome.

GLENOHUMERAL JOINT INSTABILITY

Glenohumeral instability is manifested clinically by recurrent shoulder dislocations anteriorly, posteriorly, or multidirectionally. These occur as a result of direct anterior or posterior trauma, or of a fall on an outstretched arm (20,31).

As previously described, stability of the shoulder is maintained by the capsular mechanism. Lesions of the capsular mechanism causing or contributing to the production of shoulder instability include (a) partial or complete detachments of the capsule at its anterior or posterior insertions; (b) formation of a large anterior pouch; (c) labral lesions and glenoid fractures (Bankart lesions); and (d) tendinous and ligamentous ruptures or laxity. Lesions of the capsular mechanism are often multiple. Preoperative knowledge of the extent and type of lesions will help to insure good postsurgical results. Therefore, there is a strong need to develop techniques that provide the surgeon with an accurate diagnosis.

Treatment of shoulder instability is focused on strengthening the capsular structures or on mechanically impeding the anterior or posterior displacement of the humeral head. Frequently performed surgical procedures include (a) subscapularis tendon shortening (Putti–Platt procedure); (b) shortening and transfer of the subscapularis tendon (Magnusen–Stack procedure); (c) capsular tightening; (d) staple capsulorrhaphy (Bankart procedure); (e) anterior bone block (Eden–Hybinette procedure); and (f) anterior bone block with coracoid tip transfer (Bristow procedure) (38).

Unless a fracture of the glenoid (Bankart lesion) or the humeral head (Hill–Sachs or trough lesion) occurs, plain films do not provide useful information regarding the extent of capsular damage. Double contrast arthrography followed by conventional tomography has been shown to be quite accurate in demonstrating labral lesions (7), but it does not provide information regarding the other components of the capsular mechanism such as capsular insertion, glenohumeral ligaments, and subscapularis tendon.

Axial CT following double-contrast arthrography of the glenohumeral joint is very accurate in demonstrating subtle lesions not only involving the glenoid labrum, but also the capsular insertions, ligaments, tendons, and bony structures (21,30,31,37,38).

MRI has recently emerged as an alternative noninvasive imaging technique for evaluation of glenohumeral instability. Preliminary experience with state-of-the-art technology (18) indicates that the findings seen with CTA can also be detected with MRI (16,34,40). However, further investigation is necessary to validate these early results.

Labral lesions can be diagnosed with MRI when the normally low SI triangular labrum is attenuated, avulsed, or fragmented (16) (Figs. 3.15 and 3.16). T2WI can demonstrate high SI fluid within the labral lesion (34).

Capsular lesions are difficult to evaluate with MRI unless fluid distends the joint space. In these cases, separation of the capsule from its glenoid insertion can be demonstrated (16) (Fig. 3.15).

Abnormal areas of high SI on T2WI in the region of the subscapularis tendon reflect tendon rupture seen in association with posterior labral tears in cases of posterior glenohumeral dislocation (Fig. 3.16). Hill–Sachs lesions and fractures of the bony glenoid are also easily detected with MRI (Fig. 3.17). Other signs of shoulder instability include high SI fluid on T2WI along the anterior or posterior margin of the scapula, indicating capsular detachment and atrophy with fatty infiltration of the subscapularis muscle (34).

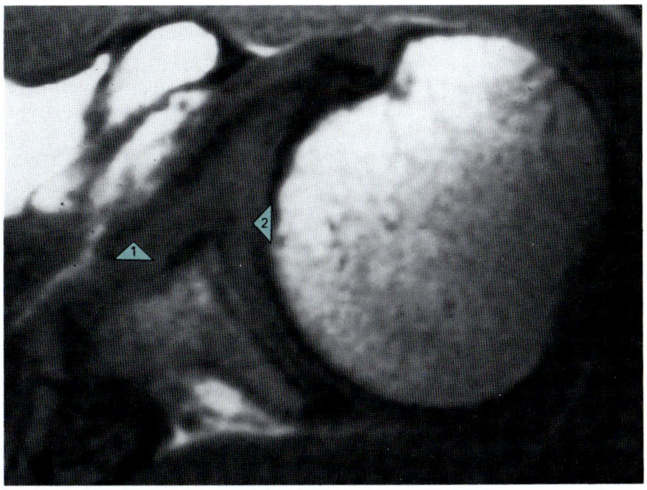

FIGURE 3.15. Anterior labral tear. Axial MRI [SE 2000/20]. The anterior labrum is small and irregular when compared with the normal posterior labrum. Arrow **1** Capsular detachment. Arrow **2** Anterior labrum.

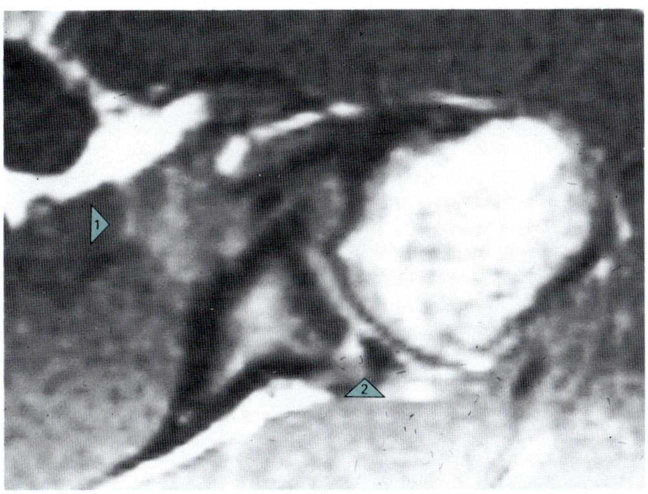

FIGURE 3.16. Posterior labral tear. Axial MRI [SE 2000/80]. The posterior labrum is avulsed from the glenoid margin. This young football player suffered anterior trauma to the shoulder which posteriorly dislocated the glenohumeral joint. The MRI was obtained after reduction of the posterior dislocation. There is residual posterior subluxation of the humeral head. The high SI in the region of the subscapularis tendon represents a tendon rupture. Arrow **1** Subscapularis tendon rupture. Arrow **2** Posterior labral tear.

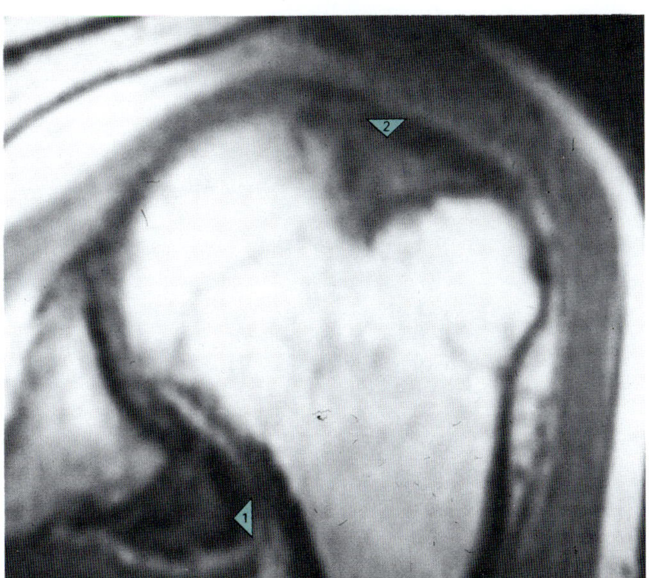

FIGURE 3.17. Anterior glenohumeral instability. Oblique coronal MRI [SE 2000/20]. There is a large Hill–Sachs fracture of the humeral head. The inferior labrum is hypertrophied and irregular. Arrow **1** Anterior–inferior labral tear. Arrow **2** Hill–Sachs lesion.

BICEPS TENDON LESIONS

Biceps Tenosynovitis

Biceps tenosynovitis is very frequently associated with the advanced stages of impingement syndrome, although it is also associated with infectious, inflammatory, and traumatic processes. In these cases, MRI shows high SI fluid on T2WI around the synovial sheath of the long head of the biceps tendon in the bicipital groove (Fig. 3.18) (24). Joint fluid can also be seen extending into the tendon sheath of the long head of the biceps tendon in cases of joint effusion without associated tendinitis. Theoretically, biceps tenosynovitis should be diagnosed only if high SI changes within the tendon are seen on T2WI.

Biceps Tendon Rupture

Acute rupture of the long head of the biceps tendon (Figs. 3.19 and 3.20) can be diagnosed clinically, when the patient presents with significant biceps weakness and obvious deformity of the anterior aspect of the arm. MRI can help determine the rupture site, which has important treatment implications. Ruptures at the musculotendinous interface are treated conservatively, while ruptures of the tendon substance are treated surgically.

SYNOVIAL LESIONS

Joint Effusion

Synovial effusions are associated with many different disorders of the glenohumeral joint, including acute or chronic trauma, inflammation, and infection among others. Effusions are well detected on T2WI because of the high SI of the fluid (3) distending the synovial recesses of the capsule (Figs. 3.18 and 3.21). Subacute hemorrhage, following trauma or in patients with bleeding disorders, may exhibit high SI on T1WI and T2WI (39). The capsular insertion in the glenoid can only be evaluated

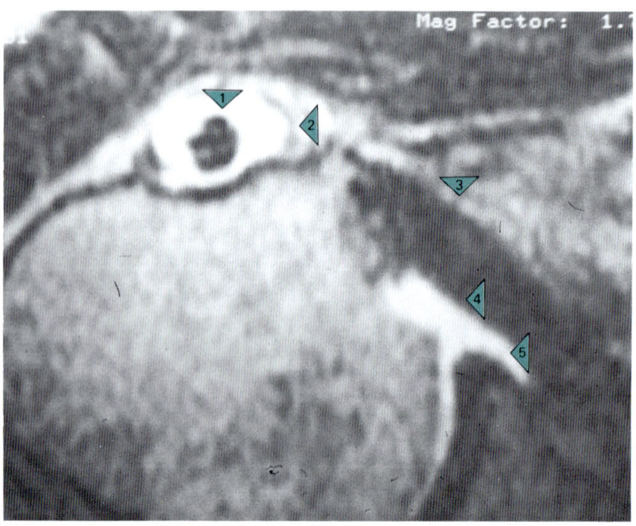

FIGURE 3.18. Biceps tenosynovitis. Axial MRI [SE 2000/80]. The long head of the biceps tendon is surrounded by fluid distending the synovial sheath. Note joint effusion outlining the anterior labrum. Arrow **1** Long head of biceps tendon. Arrow **2** Fluid distended synovial sheath. Arrow **3** Subscapularis tendon. Arrow **4** Joint effusion. Arrow **5** Anterior labrum.

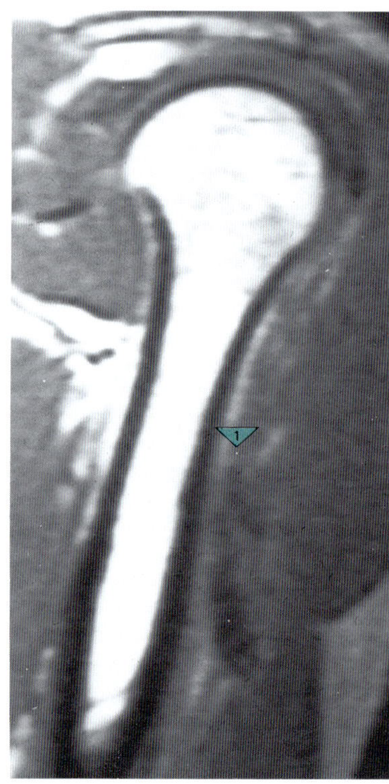

FIGURE 3.19. Rupture of the biceps tendon. Sagittal MRI [SE 2000/20]. The biceps tendon is seen in a low position and is markedly irregular. Arrows **1** Biceps tendon.

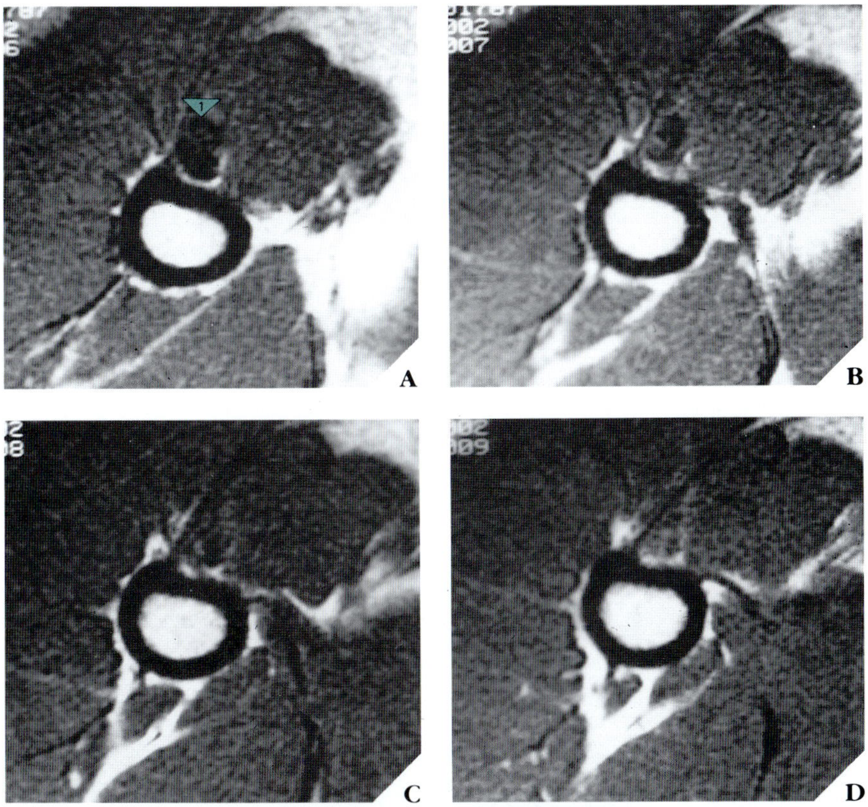

FIGURE 3.20. Rupture of the biceps tendon (same patient as in Fig. 3.19). Axial sections [SE 2000/20] from distal **(A)** to proximal **(D)**. Note the irregularity of the biceps tendon in the more distal sections, and its absence in the proximal sections, at the level of the bicipital groove. Arrow **1** Biceps tendon.

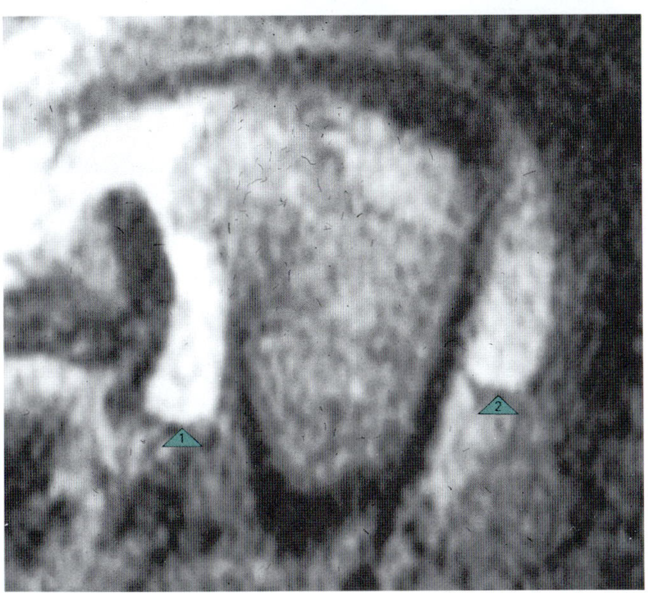

FIGURE 3.21. Joint effusion. Oblique coronal MRI [SE 2000/80]. High SI fluid is seen in the axillary recess of the joint space and laterally, around the humeral head. Arrow **1** Fluid in axillary recess. Arrow **2** Fluid around humeral head.

with MRI if fluid distends the synovial recesses. Joint fluid is often seen extending downward into the biceps tendon synovial sheath (Fig. 3.18).

Synovial Cyst

Synovial cysts are associated with traumatic, degenerative, or inflammatory processes of the joint, including ankylosing spondylitis and rheumatoid arthritis (33). A painful periarticular mass, with limitation of motion and the occasional compression of the adjacent neurovascular structures, can be seen clinically. Synovial cysts of the shoulder region are seen on MRI as fluid collections with low SI on T1WI and high SI on T2WI. They are adjacent to the glenohumeral or acromioclavicular joints (Fig. 3.22), and are often accompanied by joint effusion.

Bursitis

Subacromial–subdeltoid bursitis may occur as an isolated condition following trauma or it may occur in association with other inflammatory disorders, such as rheumatoid arthritis or impingement syndrome. The MRI characteristics of fluid within the bursa are no different than the MRI characteristics of fluid in other areas (Fig. 3.23).

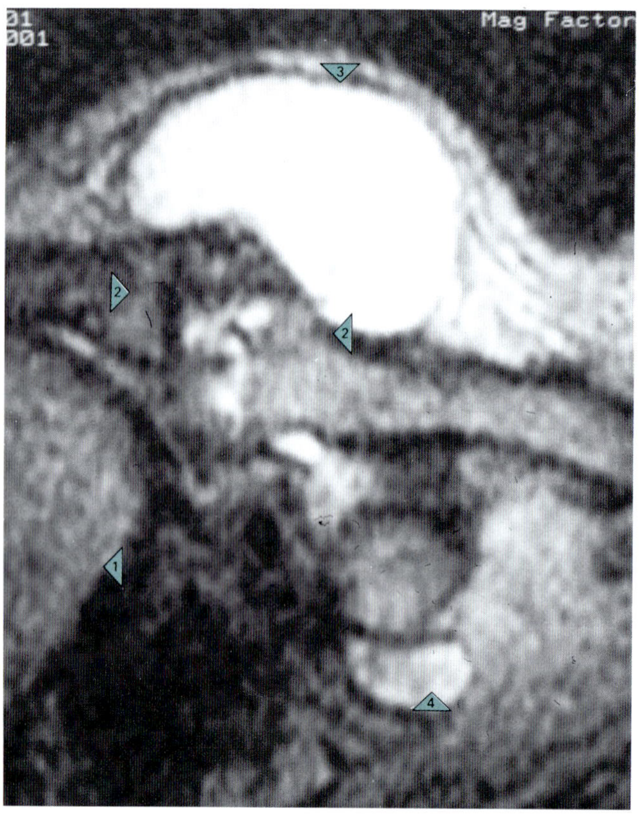

FIGURE 3.22. Acromioclavicular synovial cyst. Oblique coronal MRI [SE 2000/80]. There is a high SI, large synovial cyst in the superior aspect of the acromioclavicular joint. A small joint effusion is also present, with fluid in the subscapularis recess. Arrow **1** Humeral head. Arrows **2** Acromioclavicular joint. Arrow **3** Synovial cyst. Arrow **4** Fluid in subscapularis recess.

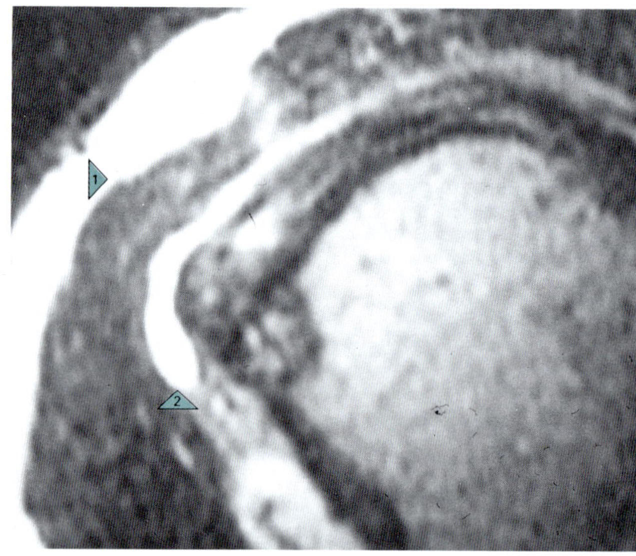

FIGURE 3.23. Subacromial bursitis. Oblique coronal MRI [SE 2000/80]. There is high SI fluid in the subacromial–subdeltoid bursa in this patient with stage III impingement syndrome and rotator cuff tear. Arrow **1** Deltoid muscle. Arrow **2** Fluid in subacromial-subdeltoid bursa.

Inflammatory Arthritis

Rheumatoid arthritis can involve the shoulder joints bilaterally. The glenohumeral joint and the acromioclavicular joint can be affected. Progressive articular cartilage destruction leads to narrowing of the joint space. Erosions can occur in the humeral head, glenoid fossa, acromioclavicular joint, acromion, and the distal undersurface of the clavicle (33). Synovial inflammation can involve the bursae of the shoulder as well as the rotator cuff. This may result in rotator cuff tear. In patients with rheumatoid arthritis, MRI of the shoulder can depict bone erosions that are sometimes not seen on plain films, with low SI lesions of the subchondral cortex. Cartilage destruction is seen on MRI as areas of thinning, irregularity, and low SI of the articular cartilage. Joint effusion, synovial cysts, and subacromial bursitis can also be demonstrated by MRI (Fig. 3.24) (1,3).

Hemophilic Arthropathy

Hemophilic arthropathy is a diffuse inflammatory reaction of the synovium, with fibrosis and hemosiderin-deposition due to repeated episodes of intra- and periarticular bleeding. The MRI pattern of hemophilic arthropathy is characterized by thickening of the joint capsule with the characteristically low SI on T1WI and T2WI, due to the paramagnetic effect of hemosiderin deposition. Joint space narrowing, subchondral cysts, bony erosions, and secondary degenerative changes with osteophytes are associated findings seen on MRI (Fig. 3.25).

Synovial Osteochondromatosis

Synovial osteochondromatosis also occurs in the shoulder joint, although less frequently than in the knee, hip, and elbow. The clinical manifestations of this condition include a chronic history of joint pain, a limitation of motion, and in some instances joint instability. Secondary osteoarthritis may be present in the later stages (33). The MRI manifestations of synovial osteochondromatosis in the shoulder are similar to those seen in other joints (intraarticular loose bodies of varying SI, depending on the amount of bone marrow formation or car-

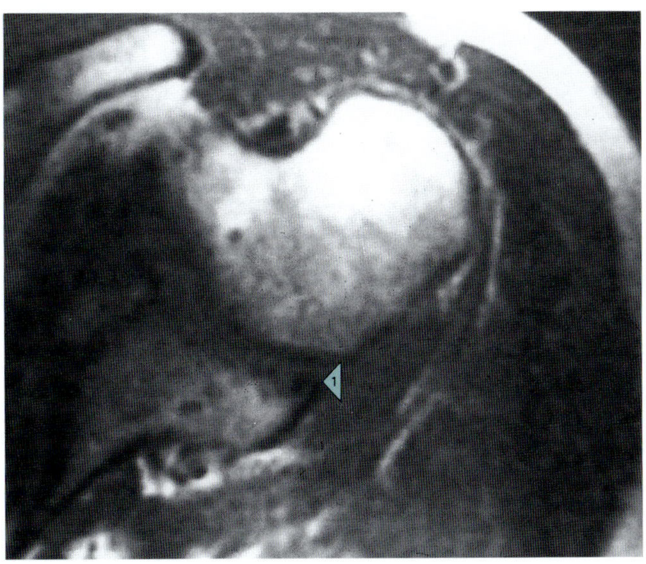

FIGURE 3.24. Rheumatoid arthritis. Axial MRI [SE 2000/20]. The joint space is narrowed due to cartilage destruction. The articular surfaces of both the humerus and the glenoid are eroded. Arrow **1** Irregularity and narrowing of glenohumeral joint.

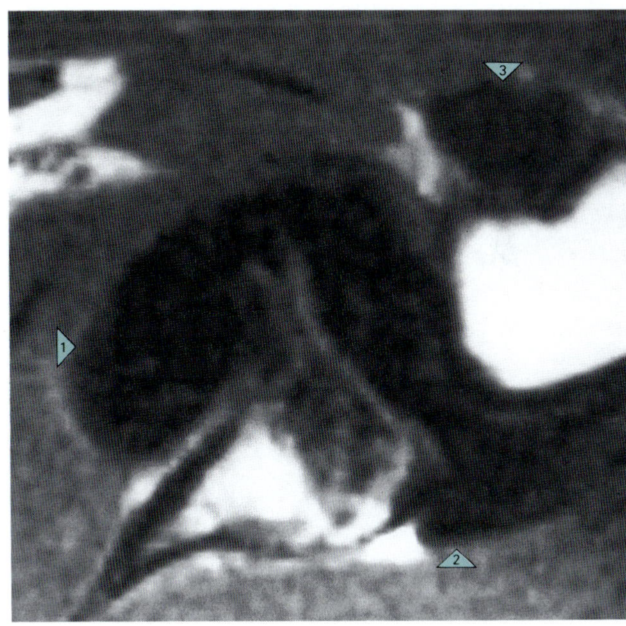

FIGURE 3.25. Hemophilic arthropathy. Axial MRI [SE 2000/20]. The synovium and capsule are markedly thickened and exhibit low SI due to periarticular fibrosis and hemosiderin deposition. The synovial sheath around the long head of the biceps tendon in the bicipital groove is also abnormally thickened. Arrow **1** Anterior synovial recess. Arrow **2** Posterior synovial recess. Arrow **3** Biceps tendon sheath.

tilaginous metaplasia) (Fig. 3.26). Joint effusion, sometimes bloody, can be associated with this condition.

AVASCULAR NECROSIS

Avascular necrosis of the humeral head can be related to fractures (33), or to a variety of other conditions including corticosteroid therapy, alcoholism, dysbaric disorders, radiation, pregnancy, and collagen vascular disorders, among others.

Most patients are referred to MRI for evaluation of shoulder pain, to rule out rotator cuff tear or impingement. Retrospectively, most cases will exhibit plain-film findings suggestive of avascular necrosis.

The MRI findings characteristic of avascular necrosis in other areas will also be seen in the humeral head. These include small or large areas of low SI in all SE pulse sequences sometimes associated with the "double-line sign" (see Chapters 6 and 7), or collapse of the articular surface of the humeral head with or without joint effusion (Fig. 3.27).

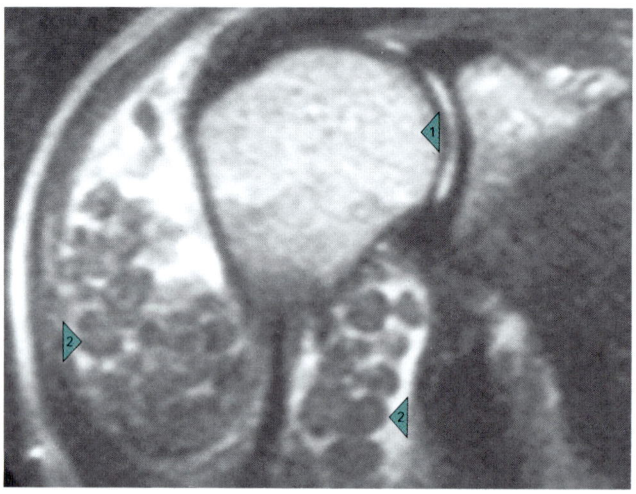

FIGURE 3.26. Synovial osteochondromatosis. Oblique coronal MRI [SE 2000/80]. There is a large joint effusion. Multiple low SI osteochondromas are seen within the joint space. Arrow **1** Humerus. Arrows **2** Synovial osteochondromas.

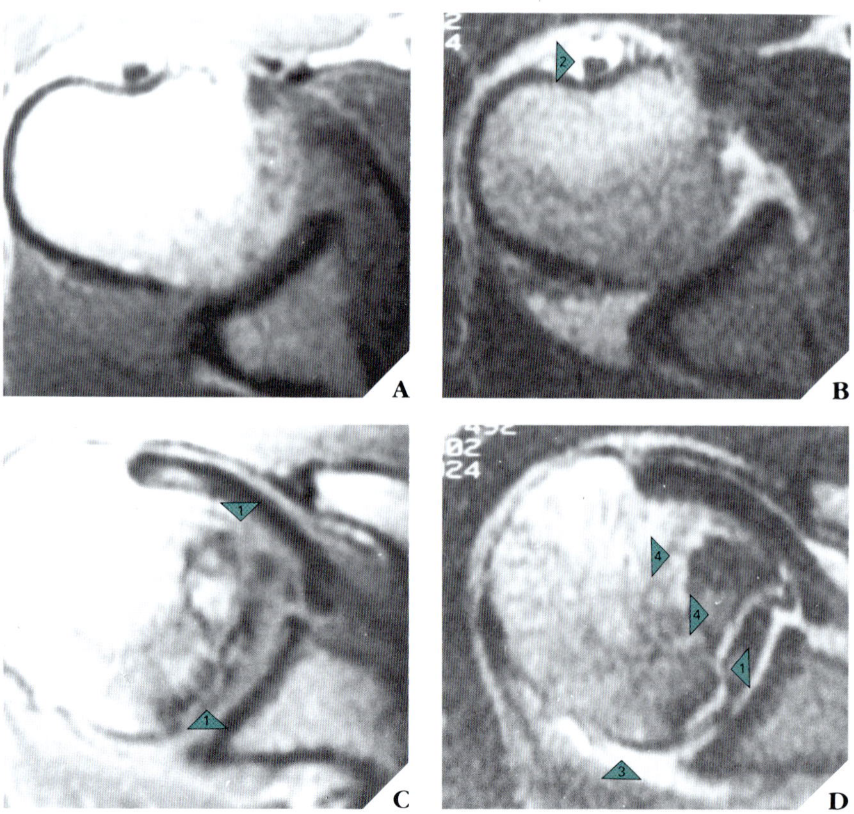

FIGURE 3.27. Large avascular necrosis with partial collapse of the humeral head. Axial MRI [SE 2000/80] from distal **(A)** to proximal **(D)**. There is a relatively large area of inhomogeneous SI in the humeral head, associated with a joint effusion. The area of avascular necrosis is surrounded by a band of high SI **(D)** inside another broader and less-defined band of low SI. This is the equivalent of the "double line" sign seen in avascular necrosis of the femoral head and femoral condyles, and represents the hypervascular area (high SI) and sclerotic, bone-forming area (low SI) of the reactive interface of avascular necrosis. Arrows **1** Avascular necrosis. Arrow **2** Fluid in biceps tendon sheath. Arrow **3** Joint effusion. Arrows **4** "Double line" sign.

REFERENCES

1. Beltran J, Caudill JL, Herman LJ, et al: Rheumatoid arthritis: MR imaging manifestations. Radiology 1987;165:153–157.
2. Beltran J, Gray LA, Bools JC, et al: Rotator cuff lesions of the shoulder; evaluation by direct sagittal CT arthrography. Radiology 1986; 160:161–165
3. Beltran J, Noto Am, Herman LJ, et al: Joint effusions: MR imaging. Radiology 1986;158:133–137.
4. Burk DL, Dalinka MK, Schiebler ML, et al: Strategies for musculoskeletal magnetic resonance imaging. Radiol Clin North Am 1988;26:653–672.
5. Cone RO, Resnick D, Danzing L: Shoulder impingement syndrome: radiographic evaluation. Radiology 1984;150:29–33.
6. Crass JR, Craig EV, Feinberg SB: Sonography of the postoperative rotator cuff. AJR 1986;146:561–564.
7. El-Khoury GY, Kathol MH, Chandler JB, Albright JP: Shoulder instability: impact of glenohumeral arthrotomography on treatment. Radiology 1986;160:669–673.
8. Fukuda H, Mikasa M, Yamanaka K: Incomplete thickness rotator cuff tears diagnosed by subacromial bursography. Clin Orthop 1987;223:51–58.
9. Gerber C, Terrier F, Ganz R: The role of the coracoid process in the chronic impingement syndrome. J Bone Joint Surg (Br) 1985; 67:703–708.
10. Hall FM, Goldberg R, Wyshak G, Kilcoyne RF: Shoulder arthrography: comparison of morbidity after use of various contrast media. Radiology 1985;154:339–341.
11. Hardy DC, Vogler JB, White RH: The shoulder impingement syndrome: prevalence of radiographic findings and correlation with response to therapy. AJR 1986;147:557–561.
12. Hawkins RJ, Abrams JS: Impingement syndrome in the absence of rotator cuff tear (stages 1 and 2). Orthop Clin North Am 1987;18:373–382.
13. Huber DJ, Sauter R, Mueller E, et al: MR imaging of the normal shoulder. Radiology 1986;158:405–408.
14. Kerr R, Resnick D, Pineda C, Haghighi P: Osteoarthritis of the glenohumeral joint: a radiologic–pathologic study. AJR 1985;144:967–972.
15. Kieft GJ, Bloem JL, Oberman WR, et al: Normal shoulder: MR imaging. Radiology 1986;159:741–745.
16. Kieft GJ, Bloem JL, Rozinf PM, Oberman WR: MR imaging of recurrent anterior dislocation of the shoulder: comparison with CT arthrography. AJR 1988;150:1083–1087.
17. Kieft GJ, Bloem JL, Rozinf PM, Oberman WR: Rotator cuff impingement syndrome: MR imaging. Radiology 1988;166:211–214.
18. Kneeland JB, Carrera GF, Middleton WD, et al: Rotator cuff tears: preliminary application of high-resolution MR imaging with counter rotating current loop–gap resonators. Radiology 1986;160:695–699.
19. Kneeland JB, Middleton WD, Carrera GF, et al: MR imaging of the shoulder: diagnosis of rotator cuff tears. AJR 1987;149:333–337.
20. Lanzer WL: Clinical aspects of shoulder injuries. Radiol Clin North Am 1988;26:157–160.
21. McNiesh LM, Callaghan JJ: CT arthroraphy of the shoulder: variations of the glenoid labrum. AJR 1987;149:963–966.
22. Middleton WD, Kneeland JB, Carrera GF, et al: High-resolution surface coil MR imaging of the normal rotator cuff. AJR 1987;148:559–564.
23. Middleton WD, Reinus WR, Melson GL, et al: Pitfalls of rotator cuff sonography. AJR 1986;146:555–560.
24. Middleton WD, Reinus WR, Totty W, et al: US of the biceps tendon apparatus. Radiology 1985;157:211–215.
25. Mink JH, Harris E, Rappaport M: Rotator cuff tears: evaluation using double-contrast shoulder arthrography. Radiology 1985;157:621–623.
26. Mosley HF, Overgaard B: The anterior capsular mechanism in recurrent anterior dislocations of the shoulder. J Bone Joint Surg (Am) 1962;44:913–927.
27. Neer CS: Impingement lesions. Clin Orthop 1983;173:70–77.
28. Neviaser TJ: The role of the biceps tendon in the impingement syndrome. Orthop Clin North Am 1987;18:383–386.
29. Post M, Cohen J: Impingement syndrome. A review of late stage II and early stage III lesions. J Bone Joint Surg (Am) 1986;207:126–132.
30. Rafii M, Firooznia H, Bonamo J, et al: Athlete shoulder injuries: CT arthrographic findings. Radiology 1987;162:559–564.
31. Rafii M, Firooznia H, Golimbu C, et al: CT arthrography of capsular structures of the shoulder. AJR 1986;146:361–367.
32. Resnick D: Shoulder pain. Orthop Clin North Am 1986;14:81–97.

33. Resnick D, Niwayama G: Diagnosis of Bone and Joint Diseases, ed 2. Philadelphia, WB Saunders, 1988.
34. Seeger LL, Gold RH, Bassett LW: Shoulder instability: evaluation with MR imaging. Radiology 1988;168:695–697.
35. Seeger LL, Gold RH, Basset LW, Ellman H: Shoulder impingement syndrome: MR findings in 53 shoulders. AJR 1988;150:343–347.
36. Seeger LL, Ruszkowski JT, Basset LW, et al: MR imaging of the normal shoulder: anatomic correlation. AJR 1987;148:83–91.
37. Singson RD, Feldman F, Bibliani L: CT arthrographic patterns in recurrent glenohumeral instability. AJR 1987;149:749–753.
38. Singson RD, Feldman F, Bibliani LU, Rosemberg ZS: Recurrent shoulder dislocation after surgical repair: double-contrast CT arthrography. Radiology 1987;164:425–428.
39. Unger EC, Glazer HS, Lee JKT, et al: MRI of extracranial hematomas: preliminary observations. AJR 1986;146:403–408.
40. Zlatkin MB, Bjorkengren AG, Gylys-Morin V, et al: Cross-sectional imaging of the capsular mechanism of the glenohumeral joint. AJR 1988;150:151–158.

FIGURE CREDITS

Figure 3.6. Adapted from Resnick D, Niwayama G: Diagnosis of Bone and Joint Diseases, ed 2. Philadelphia, WB Saunders, 1988.

Figures 3.8–3.10, 3.12, 3.14, 3.16–3.18, 3.20. From Beltran J: The shoulder: new imaging techniques. In: Radiology, edited by Taveras JM and Ferrucci JT. Philadelphia, JB Lippincott, 1989.

Figure 3.26. Courtesy of L Weiss, MD, and G Arbona, MD.

CHAPTER FOUR

Javier Beltran

The Elbow

TECHNICAL CONSIDERATIONS

▼▼▼ *IMAGING PLANES AND PULSE SEQUENCES*

Axial, sagittal, and coronal planes are routinely used for elbow imaging. The axial plane is best for demonstrating the proximal radioulnar joint, and the coronal and sagittal planes are best for the trochlear-ulnar and capitular-radial head articulations (3).

T1-weighted images with short TR/TE are obtained in the axial plane, and PDWI and T2WI with long TR/short and long TE, in the sagittal and/or coronal planes. Depending on the pathologic process under evaluation, the sagittal or coronal planes can be spared, if enough information has already been obtained in the other two planes. Alternatively, T1WI can be obtained in the sagittal or coronal plane to decrease imaging time.

▼▼▼ *FIELD OF VIEW, MATRIX, AND SLICE THICKNESS*

A small FOV of 16 cm with a matrix of 256 × 128, and 5 mm slice thickness with 1-mm interslice gaps provide adequate spatial resolution and S/N.

▼▼▼ *SURFACE COILS AND POSITIONING*

Two approaches can be used for elbow imaging. If the patient can position his or her arm overhead in an oblique prone position, a general purpose, receive-only extremity coil can be used. If mobilization of the elbow is painful or difficult, the arm can be anatomically positioned and a paired set of coils placed anteriorly and posteriorly to the elbow joint (Fig. 4.1). Off-center FOV acquisition is necessary in both situations.

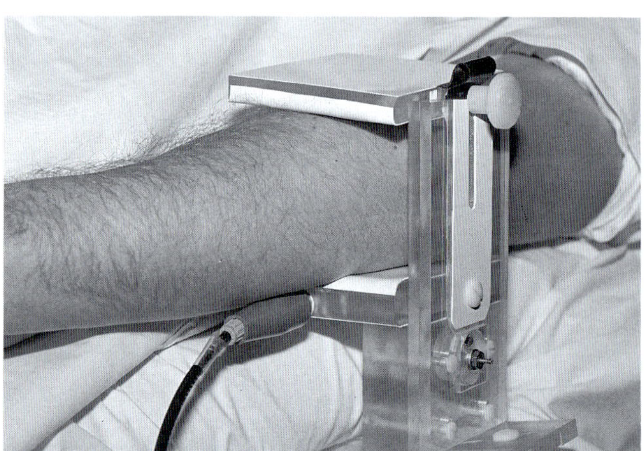

FIGURE 4.1. Positioning. Paired set of surface coils in position for elbow imaging. The arm is placed in a comfortable and anatomical position.

NORMAL ANATOMY

▼▼▼ *AXIAL PLANE*

The axial plane is ideal for visualization of the articulations between the capitulum and the head of the radius, and the trochlea and ulna (Fig. 4.2). The articular cartilage is seen as an intermediate SI band between the low SI cortical surfaces on T1WI and PDWI. The annular ligament is seen as a low SI line surrounding the radial head, and separated from it by the articular cartilage which extends around the outer radial head circumference (3).

In the posterior elbow compartment, the anconeus muscle and the triceps muscle and tendon are identified. The biceps brachii and tendon and brachialis muscles are seen in the anterior compartment. The lateral compartment muscles include the supinator, extensors of the wrist and digits, and the brachioradialis muscles. The medial compartment muscles include the pronator teres, the flexors of the hand and wrist, and the palmaris longus muscles (3).

The radial and ulnar arteries and corresponding veins are seen in the anterior aspect of the elbow. The radial artery runs between the brachioradialis and pronator teres, and the ulnar artery runs deep to the pronator teres.

The median nerve is seen anterior to the brachialis muscle proximally. The radial nerve is located between the brachialis and the brachioradialis muscles. The superficial and deep divisions of the radial nerve can often be identified. The ulnar nerve is seen posterior to the medial epicondyle.

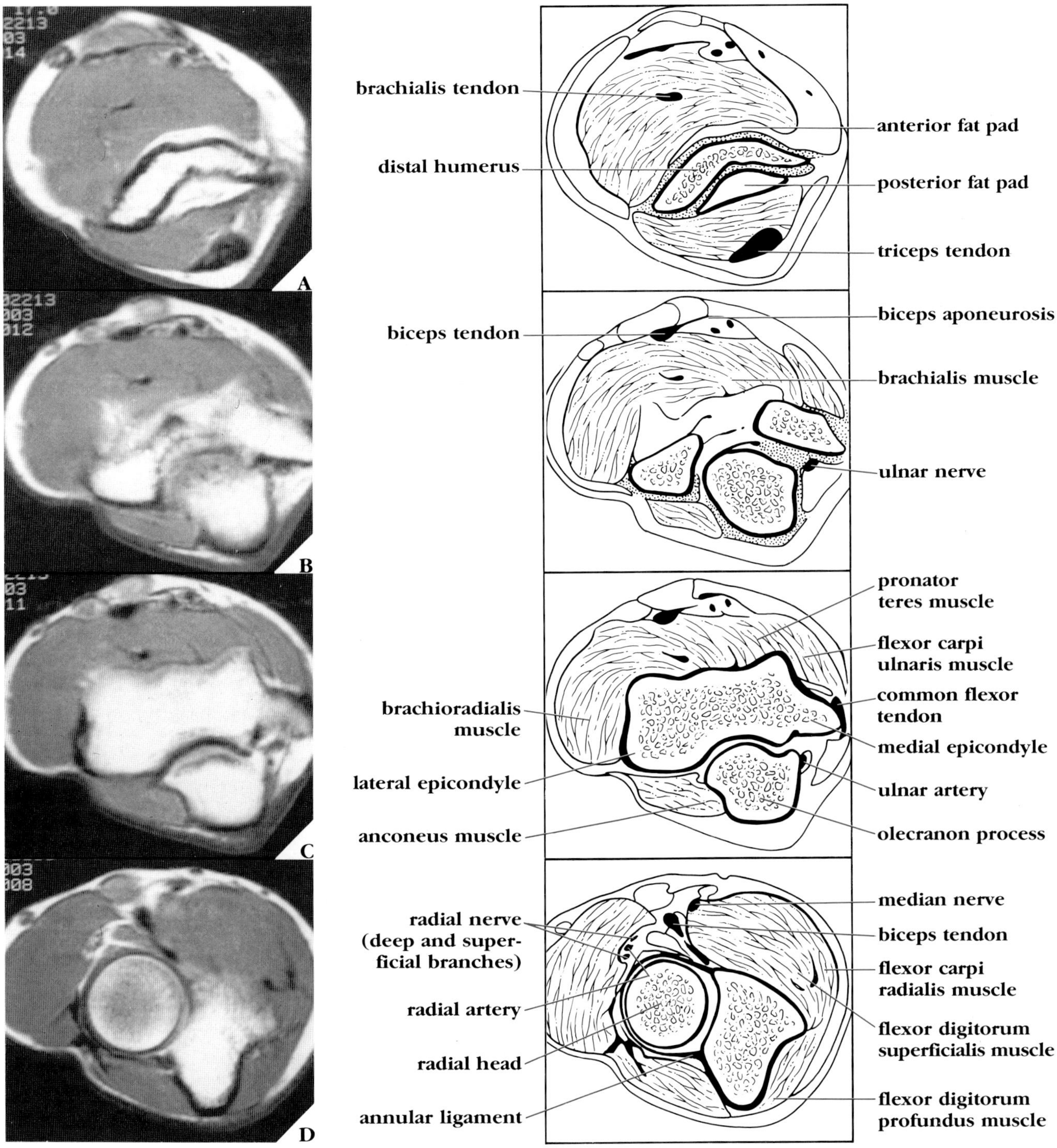

FIGURE 4.2. Normal anatomy. Axial plane [SE 1500/20] from proximal **(A)** to distal **(D)**. Muscles and tendons of the elbow region are identified. The annular ligament is the only ligament seen in axial images around the head of the radius. It inserts in the ulna, anteriorly and posteriorly. Note the ulnar nerve in the canal between the ulna and the medial epicondyle.

▼▼▼ *CORONAL PLANE*

The articular surfaces of the trochlea, capitellum, ulna, and radius are well seen in the coronal plane. The medial collateral ligament originates in the medial epicondyle and inserts distally in the trochlear notch of the ulna (Fig. 4.3). The lateral collateral ligament is thinner than the medial collateral ligament. It originates in the lateral epicondyle and inserts in the annular ligament (3) (Fig. 4.3).

▼▼▼ *SAGITTAL PLANE*

The radiocapitellum, and ulnar-trochlear articulations can be evaluated separately in the sagittal plane. A small amount of fat is present in the coronoid and olecranon fossae (Fig. 4.4). Posteriorly, the triceps tendon can be seen inserting in the olecranon. Anteriorly, the biceps tendon can be seen inserting in the radial tuberosity (3).

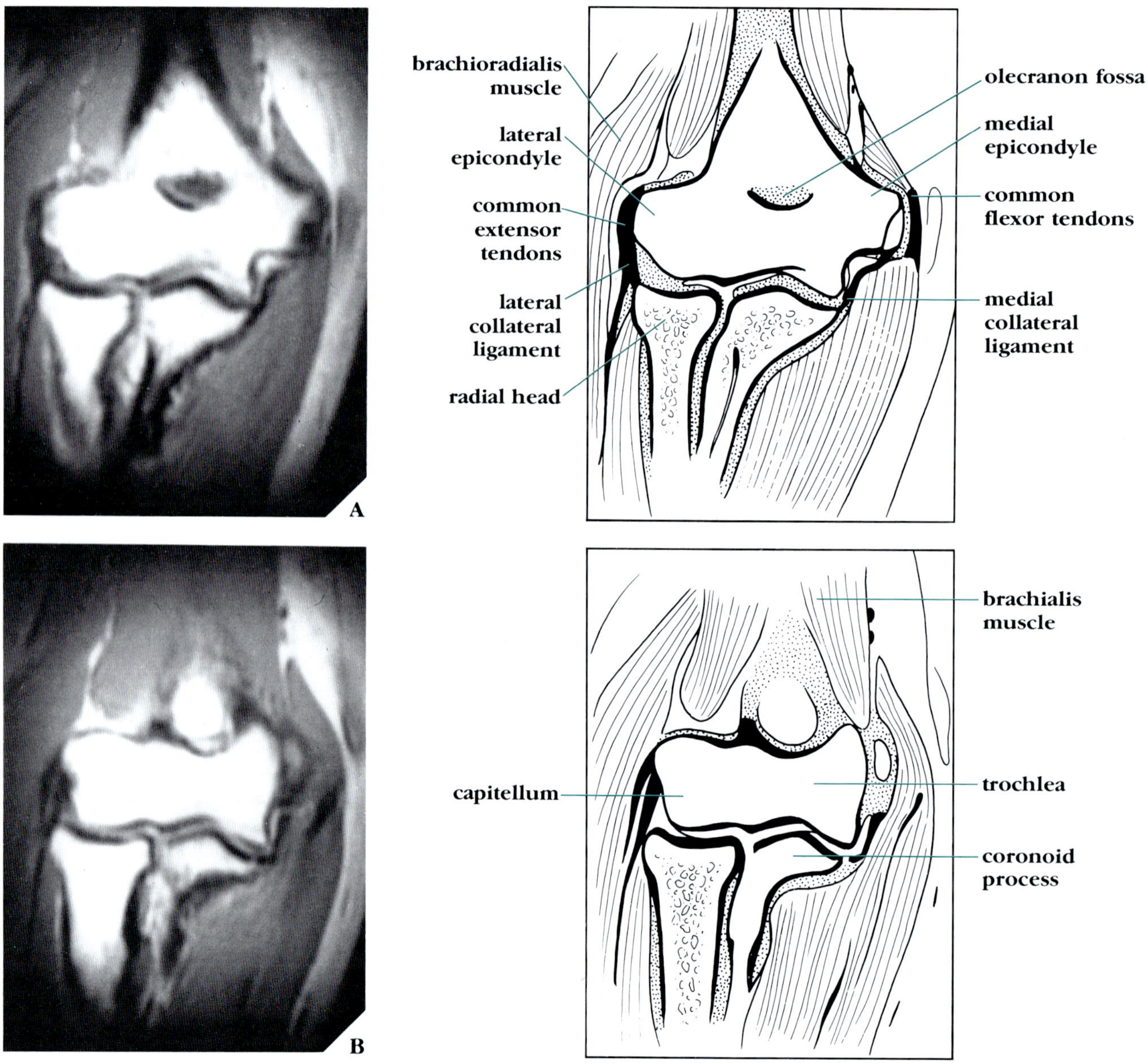

FIGURE 4.3. Normal anatomy. Coronal plane [SE 1500/20] from posterior **(A)** to anterior **(B)**. The integrity of the medial and collateral ligaments can be assessed in coronal images.

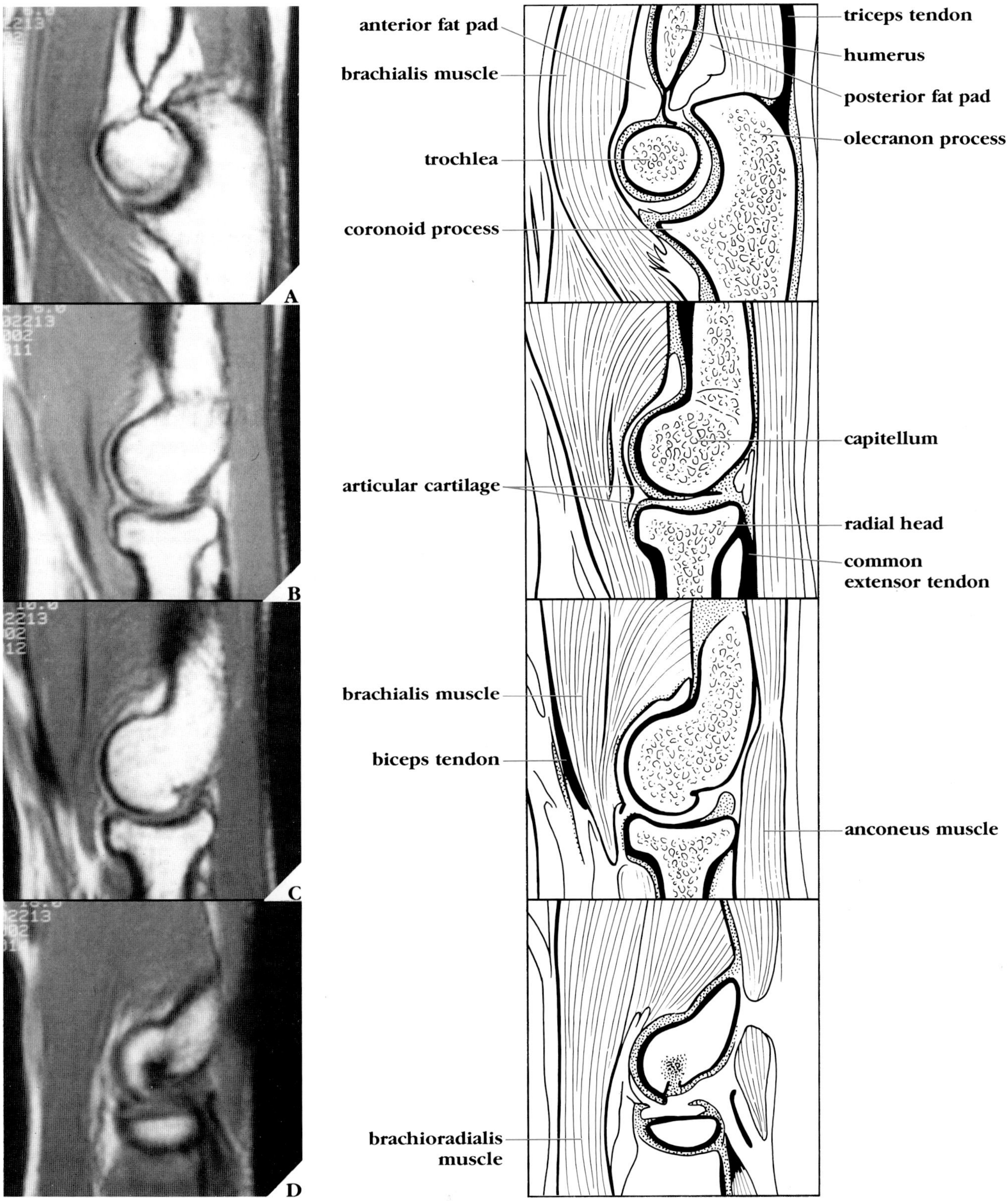

FIGURE 4.4. Normal anatomy. Saggital plane [SE 1500/20] from medial **(A)** to lateral **(D)**. The articulations between the radius, ulna, and distal humerus are well seen in sagittal images. Note that the superior and lateral aspects of the radial head are covered by the articular cartilage. The anterior and posterior fat pads are well demonstrated in the medial sections through the trochlea.

TRAUMATIC LESIONS

The vast majority of traumatic injuries to the elbow can be evaluated using conventional techniques. However, in some cases MRI may provide further information, particularly in soft tissues lesions.

Tendon Lesions

Avulsion, or rupture, of the biceps tendon occurs more frequently at the level of the long head of the biceps tendon (see Chapter 3), but occasionally it can involve the distal tendon at the level of its insertion in the radial tuberosity. Abnormally high SI on T2WI and irregularity of the tendon are seen on MRI (Fig. 4.5). Similar findings may be seen in the unusual case of triceps tendon rupture or avulsion.

Osteochondral Fracture

Osteochondral fractures or osteochondritis dissecans of the elbow joint are frequently related to chronic or acute sports trauma. The vast majority of these lesions occur in baseball players in the capitellum of the humerus (2). Most patients are males between 9 and 15 years of age, and the disease occurs in the dominant side (5). The injury results from the valgus position during the acceleration phase of pitching with subsequent compression of the radial side of the joint and distraction of the ulnar side (9). The osteochondral fragment can be found attached to the capitellum or as a free, loosened fragment floating within the joint space (4). MRI can provide information regarding the extent of the lesion and the localization of the fragment (Fig. 4.6). Curettage of the osteochondral fragment with or without drilling is the treatment of choice (8).

Osteochondrosis of the Capitellum (Panner's Disease)

Osteochondrosis of the capitellum is frequently confused with osteochondritis dissecans, as it occurs in the same age and gender group (5–16 year old boys) and is also linked to a history of trauma. The disorder is commonly attributed to pitching, and is thus occasionally referred to as "little-leaguer's elbow." The natural history and radiographic findings of this condition are reminiscent of Legg-Calvé-Perthes disease of the hip, and it is assumed to represent avascular necrosis of the capitellum (5,6). MRI can demonstrate the irregularity of the involved fragment of the capitellum with SI changes similar to those seen in avascular necrosis involving other epiphyses (Fig. 4.7).

Epiphyseal Fractures

Elbow epiphyseal injuries are common in children. Because of the multiple ossification centers existing in the elbow, radiographic diagnosis may be chal-

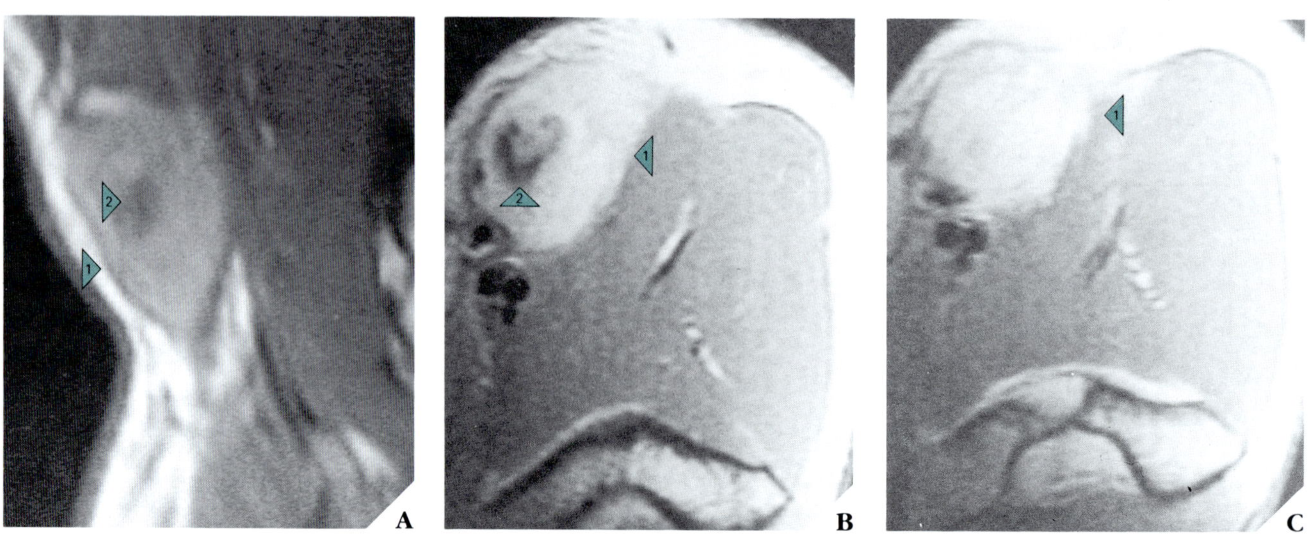

FIGURE 4.5. Avulsion of the distal insertion of the biceps tendon. Sagittal section [SE 2000/20] **(A)** and axial sections [SE 2000/20] **(B** and **C)** from proximal to distal. The ruptured biceps tendon is avulsed from its distal insertion in the tuberosity of the radius and is retracted proximally. The tendon is not seen in the distal axial section. There is a large hematoma at the site of the tendon rupture. Arrows **1** hematoma. Arrows **2** ruptured biceps tendon.

lenging (6). Assessment of the extension of the fracture lines is important for appropriate treatment (9). Due to the lack of ossification of the cartilaginous centers, radiographic evaluation may be misleading. In these cases MRI can provide further information (Fig. 4.8).

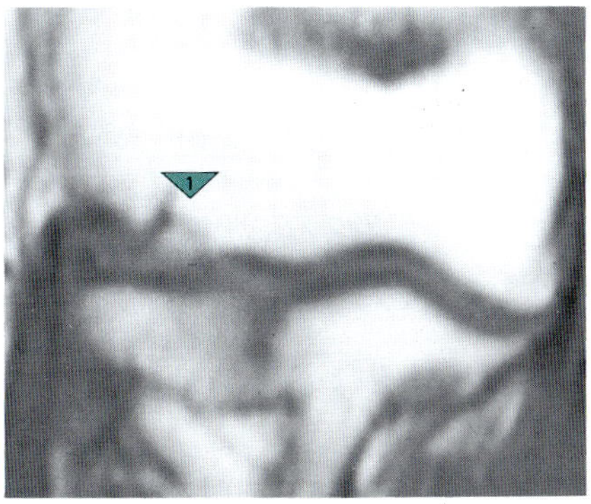

FIGURE 4.6. Osteochondral fracture of the capitellum. There is irregularity of the articular surface of the capitellum, with a small bone fragment in this 30-year-old male injured while pitching baseball. Arrow **1** osteochondral fracture.

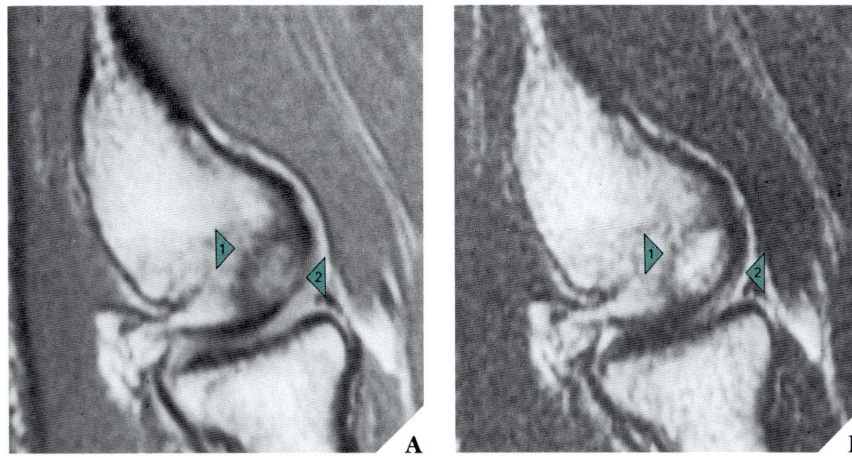

FIGURE 4.7. Osteochondrosis of the capitellum (Panner's disease). Sagittal sections [SE 2000/20] **(A)** and [SE 2000/80] **(B)**. There is an area of intermediate SI on PDWI **(A)** and high SI on T2WI **(B)**, surrounded by a low SI band. The articular cartilage is normal. The MRI findings resemble those of avascular necrosis of the femoral head. This patient was a 16-year-old baseball player. Arrows **1** area of osteochondrosis. Arrows **2** normal articular cartilage.

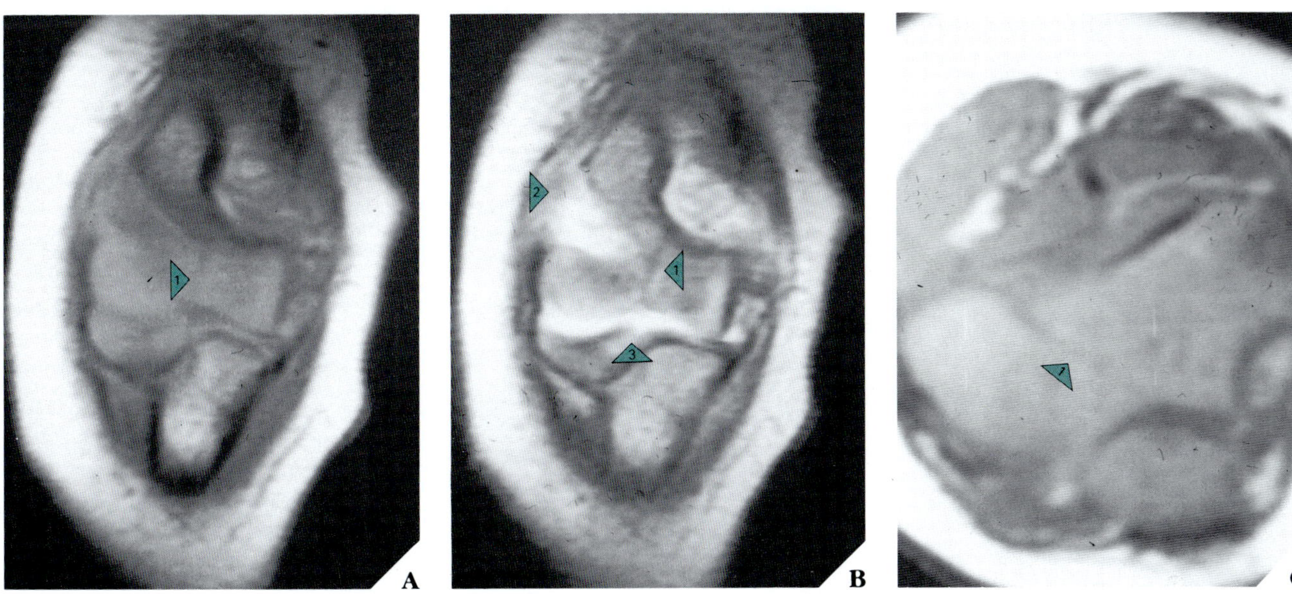

FIGURE 4.8. Intercondylar fracture. Coronal plane **(A** and **B)** [SE 2000/20—80] through the elbow joint, and axial plane **(C)** [SE 800/20] through the distal humeral epiphysis. A metaphyseal fracture of the distal humerus was detected on plain films (not shown) on this 6-month-old patient. Since the ossification centers of the elbow are not seen on plain films at this age, an MRI was obtained to assess the possible extension of the fracture line into the distal humeral epiphysis. The coronal and axial images demonstrate a low SI line through the trochlea representing the fracture through the cartilaginous epiphysis. The articular alignment is preserved. There is a triangular area of high SI in the metaphysis on T2WI **(B)**, probably due to hemorrhage in the region of the fracture seen on plain films. A small joint effusion is present. Arrows **1** fracture line. Arrow **2** metaphyseal area of hemorrhage. Arrow **3** joint effusion.

Elbow Dislocations

Dislocations of the elbow joint can involve the radius and/or ulna and are often associated with fractures. The radius and/or ulna can be displaced posteriorly, anteriorly, medially, or laterally. Posterior and posterolateral dislocations are by far the most frequent, usually resulting from a fall on the outstretched hand with the arm in extension and abduction (9). Extensive soft tissue damage is present, including capsular disruption, collateral and annular ligament rupture, triceps tendon lesions, and brachialis muscle compression. The median nerve can also be entrapped by fracture fragments. The presence of a dislocation and associated fractures can be confirmed with radiographs, but the extent of soft tissue damage is better evaluated with MRI (Fig. 4.9).

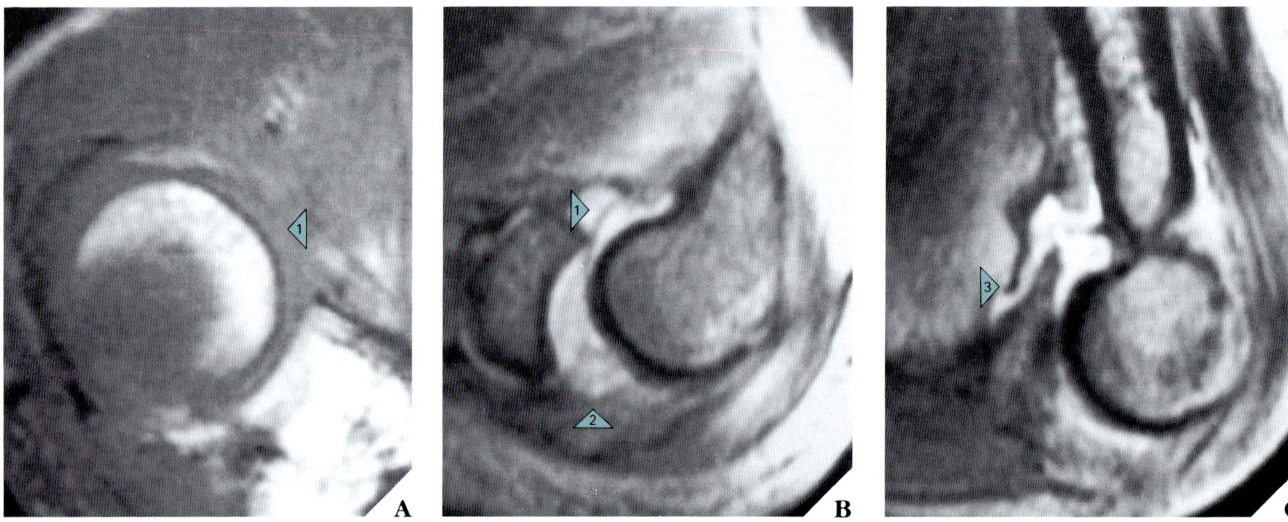

FIGURE 4.9. Elbow dislocation. Axial section [SE 800/20] **(A)** and sagittal sections [SE 2000/80] **(B** and **C)** through the proximal radioulnar joint **(A)**, capitellum **(B)**, and trochlea **(C)**. This 35-year-old male sustained a posterior elbow dislocation. The MRI examination was obtained postreduction to assess the extent of capsular damage. The axial image **(A)** demonstrates absent annular ligament due to rupture (compare with normal annular ligament in Fig. 4.2D). Sagittal images **(B** and **C)** show inferior subluxation of the head of the radius and a joint effusion. Torn ligamentous material is seen interposed between the radius and the capitellum in **B**. Anterior capsular detachment is seen at the level of the coronoid process in **C**. High SI soft tissue edema is seen posteriorly on T2WI **(B** and **C)**. Arrows **1** torn annular ligament. Arrow **2** joint effusion. Arrow **3** anterior capsular detachment.

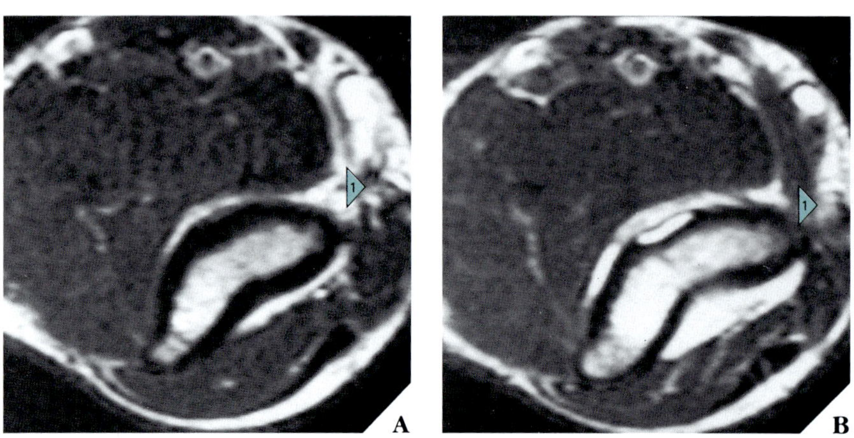

FIGURE 4.10. Medial epicondylitis. Axial plane [SE 2000/80]. Two consecutive T2WI **(A** and **B)** showing high SI at the level of the medial epicondyle in this baseball player. His clinical symptoms were compatible with partial avulsion of the common flexor tendons. Arrows **1** high SI adjacent to the medial epicondyle.

Epicondylitis

Repetitive movements of the wrist with pronation, or supination of the forearm can lead to tendinous lesions, including partial or complete tendon rupture, tenosynovitis, and peritendinitis. Clinically, the term epicondylitis is frequently used, and when the symptoms are referred to the lateral epicondyle it is called "tennis elbow." Due to the inflammatory changes occurring at the level of the insertion of the extensor tendons in the lateral epicondyle or flexor tendons in the medial epicondyle, high SI changes may be seen on MRI using T2WI (7) (Fig. 4.10).

SYNOVIAL AND BONE MARROW LESIONS

Joint Effusion

Distension of the joint capsule with displacement of the anterior and posterior fat pads are characteristic plain film findings of joint effusion. MRI can depict accurately the soft tissue changes associated with joint effusion of any etiology (Fig. 4.11).

Synovial Cysts

Synovial cysts can occur as an isolated entity or associated with inflammatory or traumatic conditions. They can occur in the region of the elbow joint as well as in any other joint. Well-demarcated fluid collections with low SI on T1WI, and high SI on T2WI are seen on MRI (Fig. 4.12).

Rheumatoid Arthritis, and Olecranon Bursitis

The elbow joint is affected by rheumatoid arthritis in over one third of the patients with this condition. Articular cartilage changes and bony erosions can be well depicted with MRI (1).

The olecranon bursa is frequently involved in rheumatoid arthritis, but gout, trauma, or infection

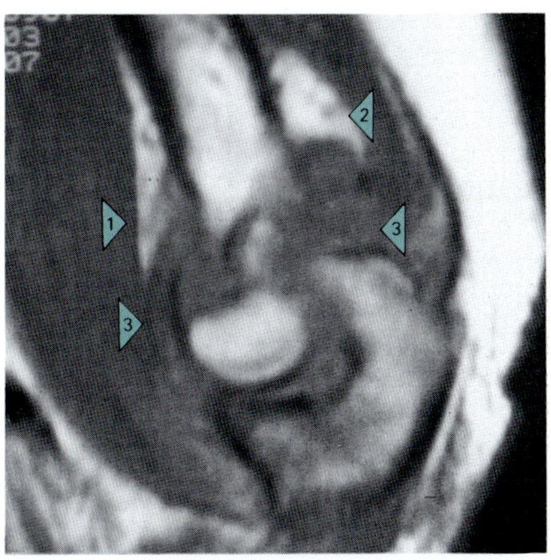

FIGURE 4.11. Chronic joint effusion. Sagittal section [SE 800/20] through the trochlea showing displacement of the anterior and posterior high SI fat pads on this T1WI, produced by a low SI chronic elbow joint effusion. Arrow **1** anterior fat pad. Arrow **2** posterior fat pad. Arrows **3** joint effusion.

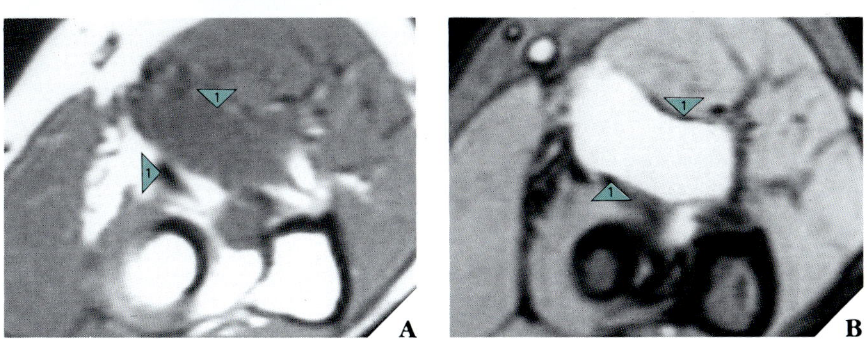

FIGURE 4.12. Synovial cyst. Axial plane [SE 800/20] **(A)** and (2DFT GRASS 1000/12, 20 degrees flip angle) **(B).** There is a fluid collection anterior to the elbow joint. Arrows **1** synovial cyst.

can also produce olecranon bursitis (6). A fluid collection in the region of the olecranon bursa can be identified with MRI (Fig. 4.13).

Avascular Necrosis

Until fairly recently, avascular necrosis around the elbow joint was considered to be a relatively rare occurrence. However, as a result of the more routine use of MRI for evaluation of joint symptoms, this entity is now being seen more often. The association of avascular necrosis with predisposing conditions such as trauma and steroid use is well known (see Chapter 6). The characteristic MRI findings of avascular necrosis are discussed in more depth in Chapter 6, but typically one may see low SI undulating lines on T1WI and T2WI in the bone marrow, with or without extension into the subchondral cortex (Fig. 4.14). High SI bands on T2WI may occasionally be seen ("double-line" sign). Associated joint effusion is a frequent finding.

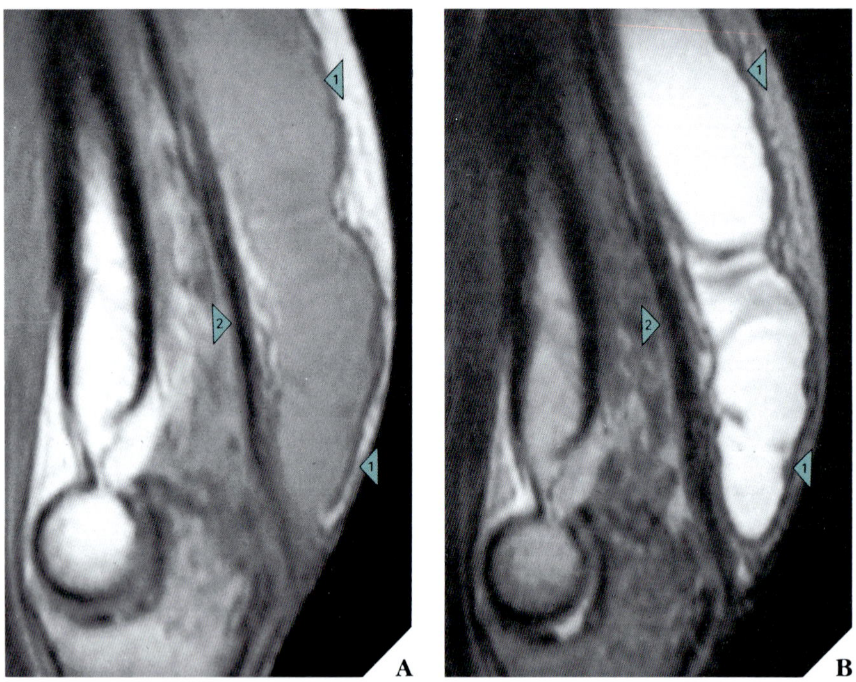

FIGURE 4.13. Rheumatoid arthritis with olecranon bursitis. Sagittal plane [SE 2000/20] **(A)** and [SE 2000/80] **(B)**. There is a large fluid collection posterior to the triceps tendon, representing olecranon bursitis in this patient with long-standing rheumatoid arthritis. Arrows **1** olecranon bursitis. Arrows **2** triceps tendon.

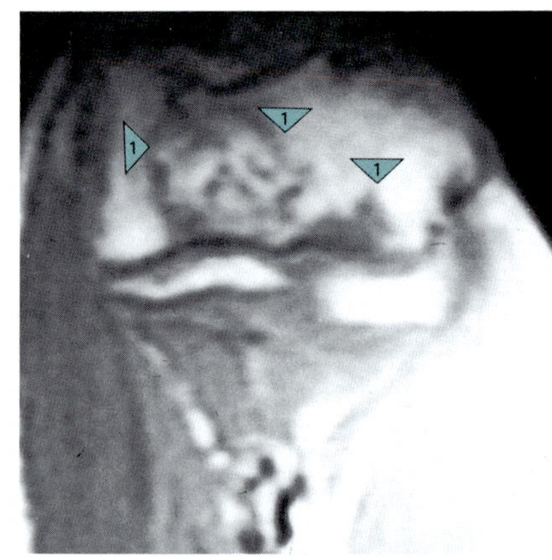

FIGURE 4.14. Avascular necrosis. Coronal section [SE 2000/20] through the elbow demonstrating low SI lines representing avascular necrosis in this patient on steroids following renal transplant. Arrows **1** avascular necrosis.

REFERENCES

1. Beltran J, Caudill JL, Herman LA, et al. Rheumatoid arthritis: MR imaging manifestations. Radiology 1987;165:153–157.
2. Gore RM, Rogers LF, Bowerman J, et al. Osseous manifestations of elbow stress associated with sports activities. AJR 1980;134:971–977.
3. Middleton WD, Macrander S, Kneeland JB, et al. MR imaging of the normal elbow: anatomic correlation. AJR 1987;149:543–547.
4. Mitsunaga MM, Adishian DA, Bianco AJ Jr. Osteochondritis dissecans of the capitellum. J Trauma 1982;22:53–55.
5. Ozonoff MB. Pediatric orthopedic radiology. W.B. Saunders, Philadelphia, 1st ed., 1979.
6. Resnick D, Niwayama G. Diagnosis of Bone and Joint Disorders. W.B. Saunders, Philadelphia, 2nd ed., 1988.
7. Stoller DW. Magnetic Resonance in Orthopedics and Rheumatology. J.B. Lippincott, Philadelphia, 1989.
8. Tivnon MC, Anzel SH, Waugh TR. Surgical management of osteochondritis dissecans of the capitellum. Am J Sports Med 1976;4:121–128.
9. Weissman BNW, Sledge CB. Orthopedic Radiology. W.B. Saunders, Philadelphia, 1st ed., 1986.

FIGURE CREDITS

Figure. 4.7. Courtesy of Dr. Lawrence Burk, Thomas Jefferson University.

CHAPTER FIVE

James C. Mosure
Javier Beltran

The Hand And Wrist

TECHNICAL CONSIDERATIONS

▼▼▼ *IMAGING PLANES AND PULSE SEQUENCES*

Images of all three orthogonal planes—axial, sagittal, and coronal—are essential for complete examination of the hand and wrist.

Authors have noted previously the value of the axial view in demonstrating the normal and pathologic anatomy of the carpal tunnel, as well as in delineating individual tendons, muscles, bones, and neurovascular bundles (4,6,24,36,37,54,55). The axial plane also demonstrates the important volar radial carpal, intercarpal, and dorsal radial carpal ligaments.

The coronal plane best demonstrates the course of the tendons, nerves, and blood vessels, together with their relationship to the osseous structures of the hand and wrist. Additionally, the interosseous ligaments, triangular fibrocartilage complex (TFCC), and radial and ulnar collateral ligaments are best seen in this plane.

The sagittal plane demonstrates the normal and pathologic anatomy of the digits and has recently been found to more clearly depict the volar ligaments of the wrist (23).

For the reasons noted above, imaging in all three planes has become routine for complete examination of the hand and wrist.

Normal hand and wrist anatomy is best demonstrated on T1W1 and PDW1, whereas their pathology is best demonstrated on T2W1. Imaging with fast gradient echo 3DFT offers the advantages of thin, contiguous sections with high resolution anatomic detail, short imaging times, multiplanar reconstruction capability, and, compared with SE multisection technique, improved S/N (52). Gradient echo imaging provides a positive contrast background within the joint spaces, resulting in finely detailed images of the wrist's ligamentous structures.

▼▼▼ *FIELD OF VIEW*

A 12-cm FOV is usually adequate for imaging the hand and wrist. This provides adequate S/N with 3-mm slice thickness and 2 NEX. Increased spatial resolution, for detailed work in the wrist or fingers, can be obtained with an 8-cm field of view. Four NEX can be used, if needed, to improve S/N.

▼▼▼ *POSITIONING*

The patient is positioned supine, head first in the gantry (a more oblique position can be used, depending on the patient's condition). The arm rests at the side with the wrist in the surface coil, as depicted in Figs. 5.1 and 5.2. For an off-centered FOV, which is sometimes difficult to obtain with 3DFT, it may be necessary to position the wrist centrally within the bore of the magnet.

▼▼▼ *ACQUISITION MATRIX*

A matrix of 256 × 128 or 256 × 192 is adequate for routine imaging of the hand and wrist. However, under certain conditions, such as use with a limited FOV, a 256 × 256 matrix could be preferable. For

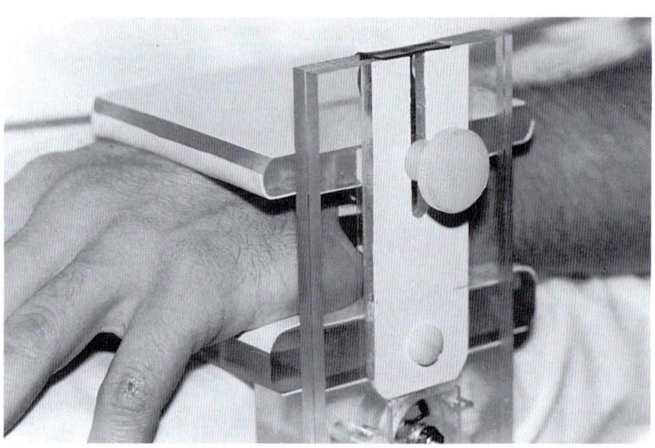

FIGURE 5.1. Helmholtz coil in place for hand and wrist imaging

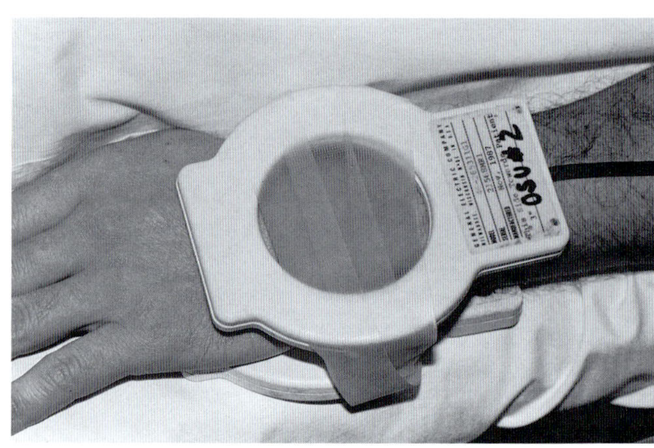

FIGURE 5.2. Two circular three-inch coils (General Electric) in place for hand and wrist imaging.

example, in the axial plane with T1W1, a 256 × 256 matrix and an 8-cm FOV could be used with a 4 NEX to increase S/N without significantly increasing imaging time. The 256 × 256 matrix also offers increased spatial resolution when visualization of the volar radial carpal and intercarpal ligaments is desirable.

▼▼▼ *SLICE THICKNESS*

A slice thickness of 3 mm with a 0.6-mm gap has been found optimal for studying the hand and wrist. With fast gradient echo 3DFT, 0.8-mm contiguous slice thickness can be achieved, resulting in the excellent depiction of detailed anatomy.

▼▼▼ *SURFACE COILS*

The use of surface coils is essential to obtain the high resolution necessary to demonstrate the intricate anatomy of the hand and wrist. Either a Helmholtz coil or two circular coils can be used (see Figs. 5.1 and 5.2). Both coils provide adequate and uniform signal through the hand and wrist.

NORMAL ANATOMY

The normal anatomy of the hand and wrist is illustrated in this chapter using SE 2DFT and 3DFT images. In general, the anatomy of tendons is well demonstrated with both techniques. The anatomy of ligamentous structures, however, is best demonstrated with 3DFT.

▼▼▼ *AXIAL PLANE*

Tendons

The anatomy of the hand and wrist as depicted by MRI has been well demonstrated in cadaver specimens and in volunteers (4,6,24,36,37,54,55). As with joints, tendons are individually seen on all pulse sequences as low SI tubular structures. The synovial sheaths do not appear as distinct entities unless distended by fluid. Figure 5.3 is an illustration

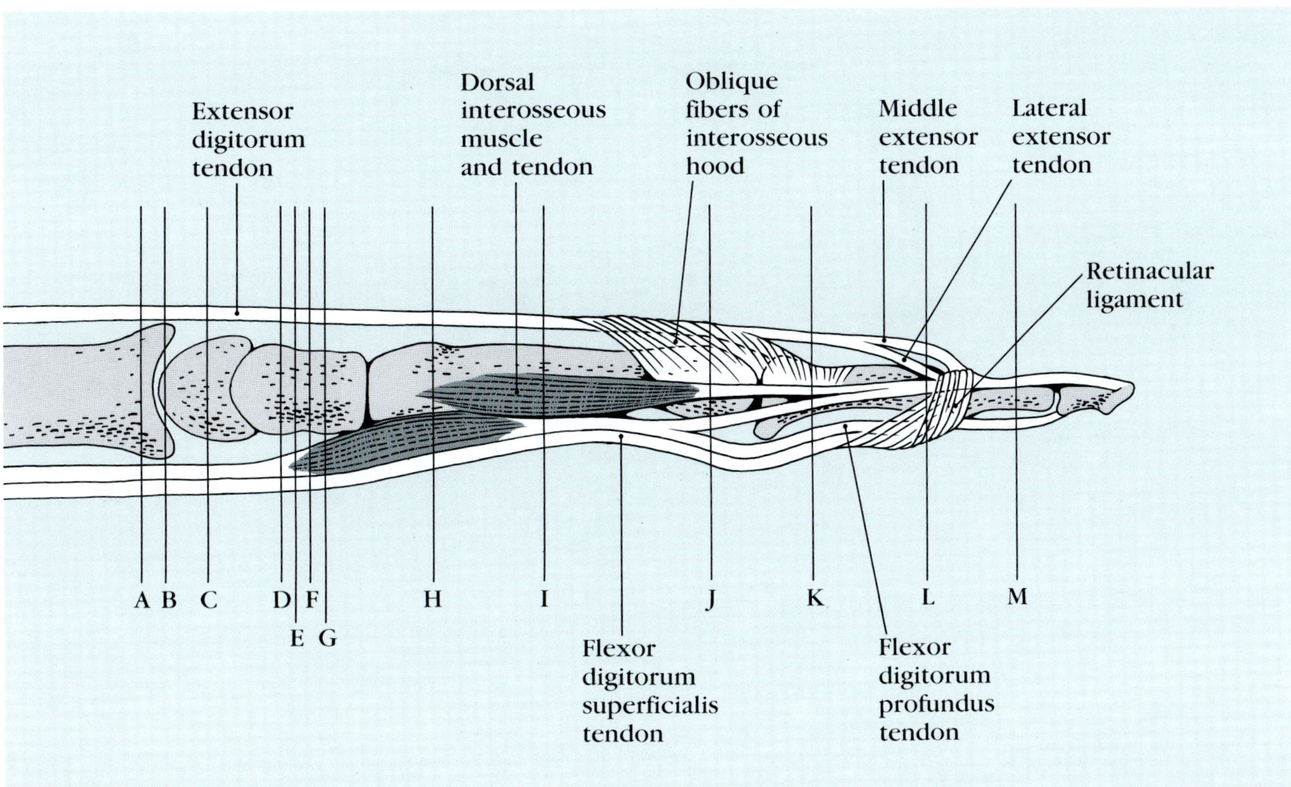

FIGURE 5.3. Lateral projection of the wrist and hand. *Vertical lines A–M* indicate the planes of section used to obtain the images in Fig. 5.4–5.16

of the lateral view of the hand and wrist. Each vertical line corresponds to the axial MR images shown in Figs. 5.4 through 5.16.

At the level of the wrist, the extensor tendons, invested by the low SI supratendinous and infratendinous dorsal retinaculum, are divided into six compartments by the retinacular septa (Figs. 5.4–5.6). The first compartment, lateral to the radius, contains the abductor pollicis longus and extensor pollicis brevis. The second compartment, lateral to

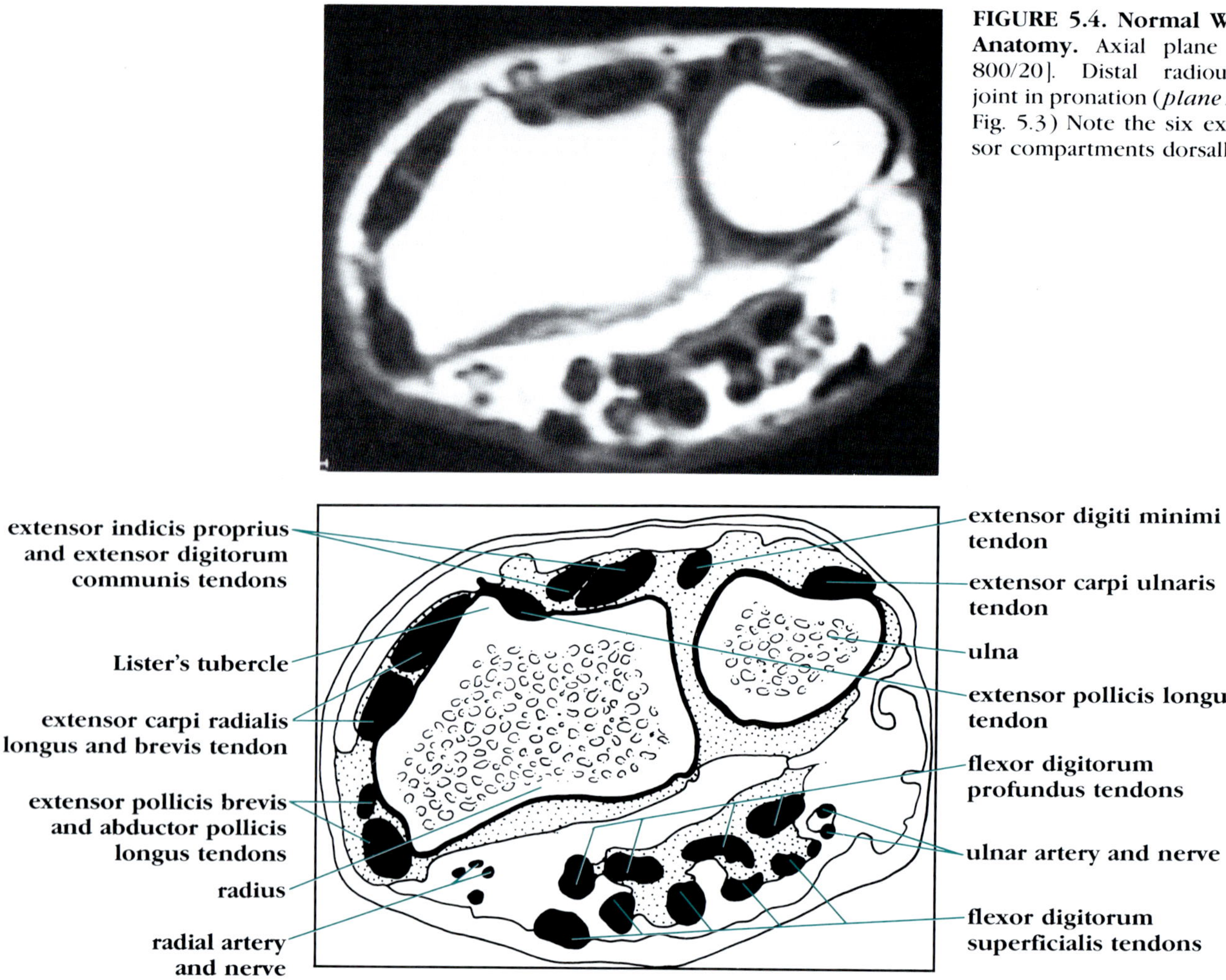

FIGURE 5.4. Normal Wrist Anatomy. Axial plane [SE 800/20]. Distal radioulnar joint in pronation (*plane A* in Fig. 5.3) Note the six extensor compartments dorsally.

Lister's tubercle, contains the extensor carpi ardialis brevis and longus. Continuing ulnarly, the third compartment contains the extensor pollicis longus; the fourth, the extensor indicis proprius and extensor digitorum communis; the fifth, the extensor digiti minimi; and, last, the sixth compartment, as seen medially, contains the extensor carpi ulnaris tendon. Each of these tendons is surrounded by a synovial sheath.

Volar to the wrist carpal bones are the flexor tendons, with their investing ulnar and radial bursae. These tendons are individually delineated as

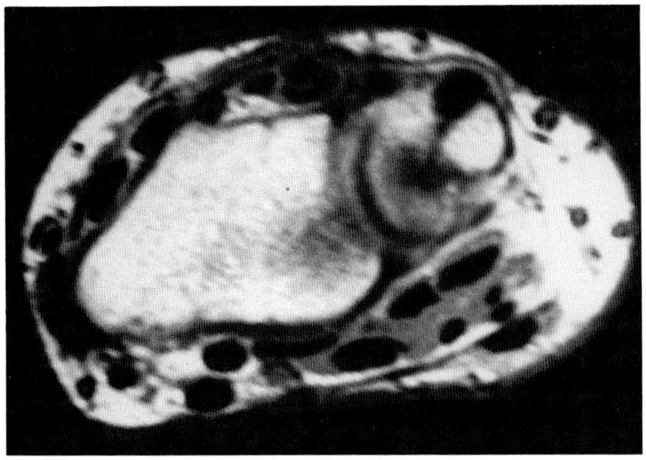

FIGURE 5.5. Normal Wrist Anatomy. Axial plane [SE 2000/20]. Distal radius and ulna (*plane B* in Fig. 5.3). The dorsal and volar distal radioulnar ligaments are seen as thin bands that bridge the ulna and radius anteriorly and posteriorly.

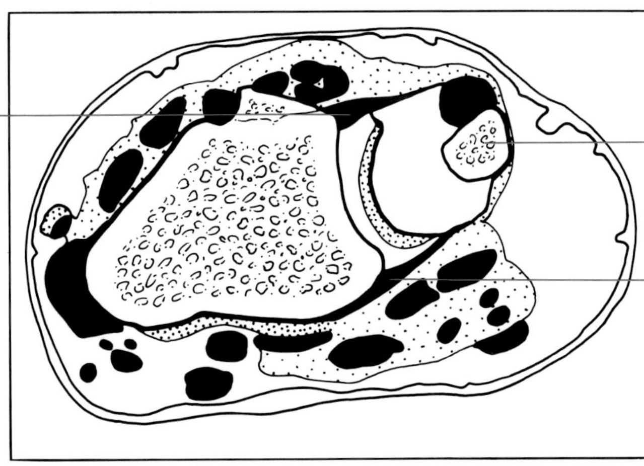

black oval structures within the carpal tunnel and extend distally over their respective metacarpals to their phalangeal insertions. The flexor digitorum profundus tendons lie deep to the volar, medial, and lateral borders of the carpal tunnel, which is formed by the concave carpal bones with their low

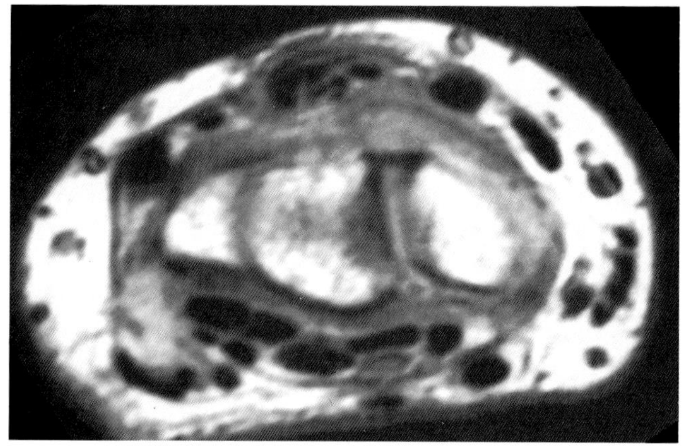

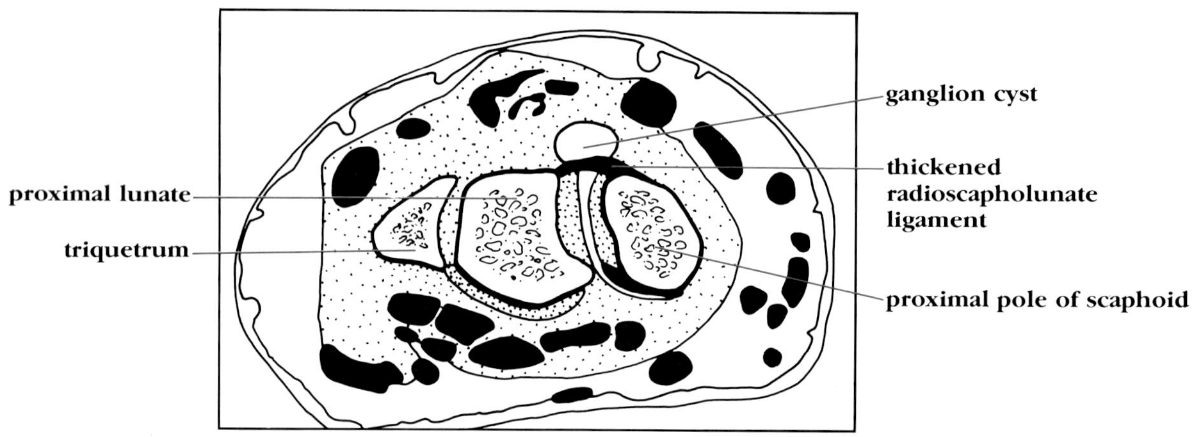

FIGURE 5.6. Normal Wrist Anatomy. Axial plane [SE 2000/20] through the proximal wrist (*plane C* in Fig. 5.3). The dorsal radioscapholunate ligament is demonstrated. The ligament is thickened in this patient, probably due to the presence of a ganglion cyst.

SI interosseous, intercarpal, and volar radial carpal ligaments (Figs. 5.6–5.9). An intervening layer of fat may be seen dorsal to the deep flexors at the level of the carpal tunnel (Figs. 5.7 and 5.10).

The flexor digitorum superficialis tendons are more volar, adjacent to the slightly bowed linear low SI flexor retinaculum (transverse carpal ligament) forming the volar boundary of the carpal

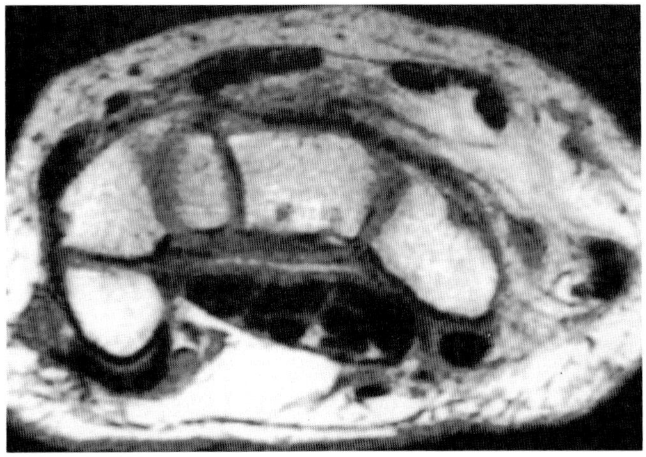

FIGURE 5.7. Normal Wrist Anatomy. Axial plane [SE 800/20]. Proximal wrist in pronation (*plane D* in Fig. 5.3). Note the flexor tendons within the carpal tunnel bounded volarly by the flexor retinaculum. Dorsally, a fat strip is seen interposed between the deep flexor tendons and the radiolunatotriquetral ligament, bridging the respective carpal bones.

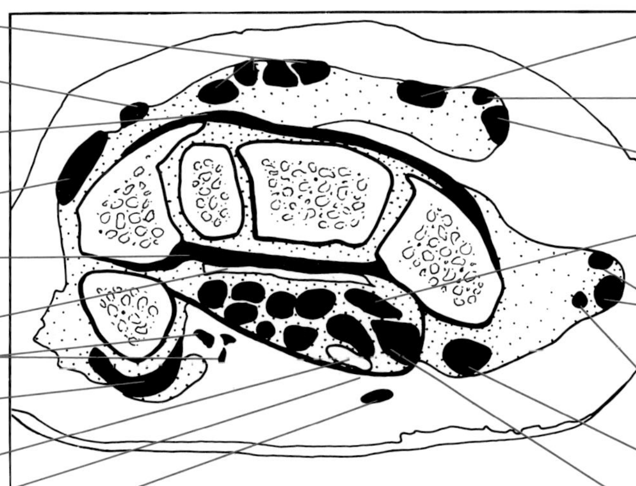

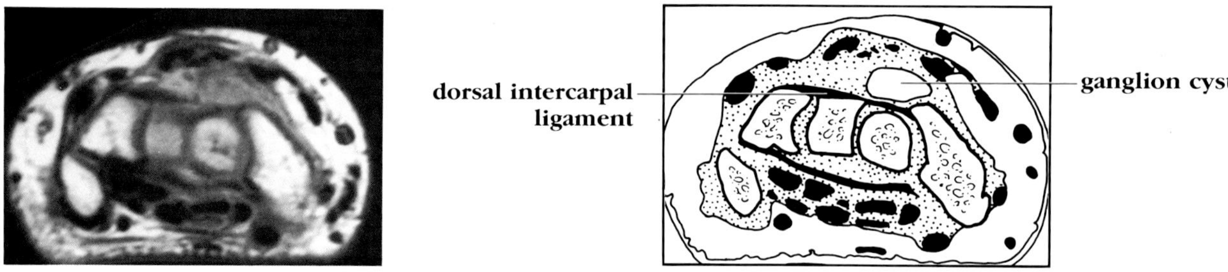

FIGURE 5.8. Normal Wrist Anatomy. Axial plane [SE 2000/20]. Proximal midwrist in pronation (*plane E* in Fig. 5.3; same patient as in Fig. 5.6). The linear low SI of the dorsal intercarpal (scaphotriquetral) ligament extends from the scaphoid to the triquetrum at the dorsum of the wrist. There is an incidental ganglion cyst dorsally, which is most likely responsible for the thickening of the dorsal intercarpal ligament evident in Fig. 5.6.

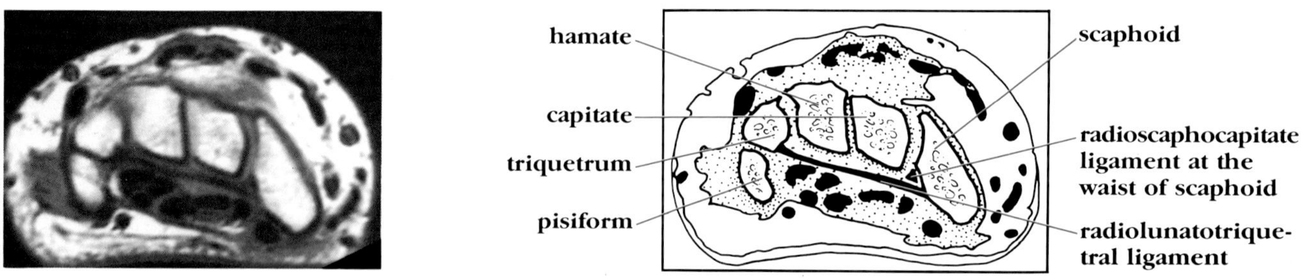

FIGURE 5.9. Normal Wrist Anatomy. Axial plane [SE 2000/20] through the midwrist (*plane F* in Fig. 5.3). The linear low SI radiolunatotriquetral ligament is seen interposed between the carpal tunnel and the proximal carpal row. The triangular-shaped low SI radioscaphocapitate ligament is seen at the volar waist of the scaphoid bone.

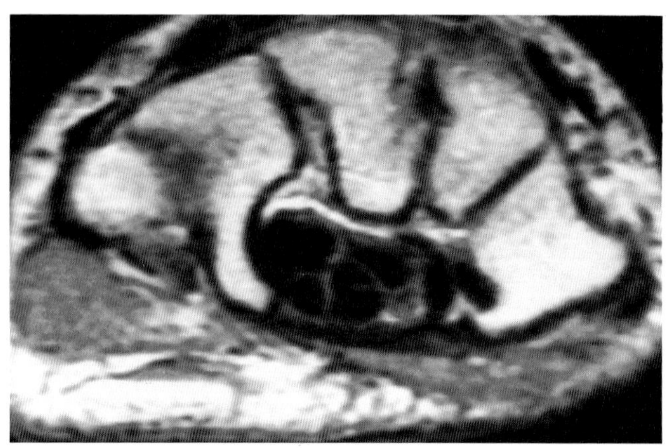

FIGURE 5.10. Normal Wrist Anatomy. Axial plane [SE 2000/20] through the carpal tunnel, in pronation (*plane G* in Fig. 5.3). Note the intermediate SI of the median nerve in the carpal tunnel.

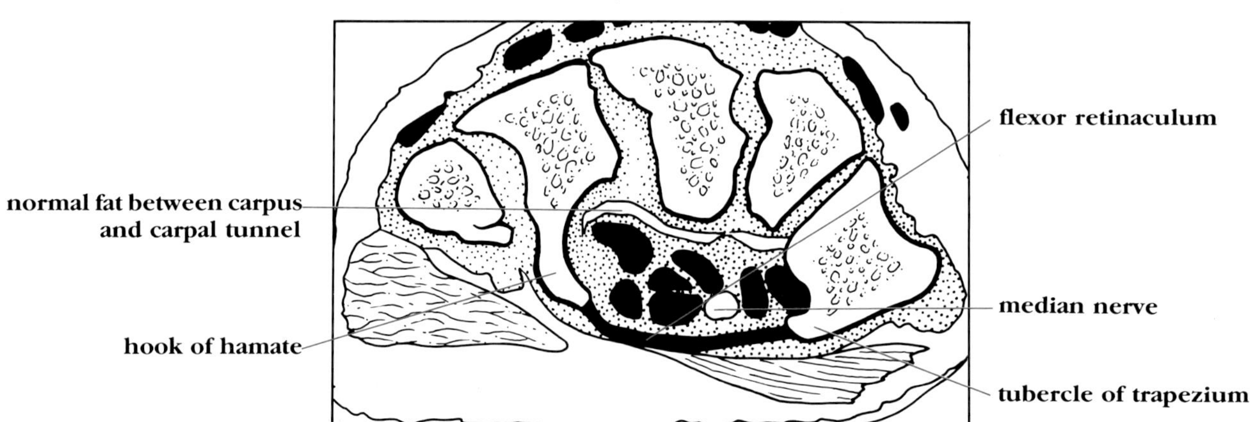

tunnel. The flexor pollicis longus is located in the most radial aspect of the carpal tunnel. The flexor carpi radialis tendon, which is not within the carpal tunnel proper, splits the radial aspect of the flexor retinaculum as it courses more distally in the groove of the trapezium. The midline palmaris longus tendon is seen superficial to the flexor retinaculum. The flexor carpi ulnaris can be noted at the superficial ulnar aspect of the volar wrist.

The lumbricalis muscles attach to the deep flexor tendons of the hand, and are seen as intermediate SI structures which appear contiguous with both the superficial and deep flexor tendons (Fig. 5.11). If the fingers are not extended, the lumbricalis

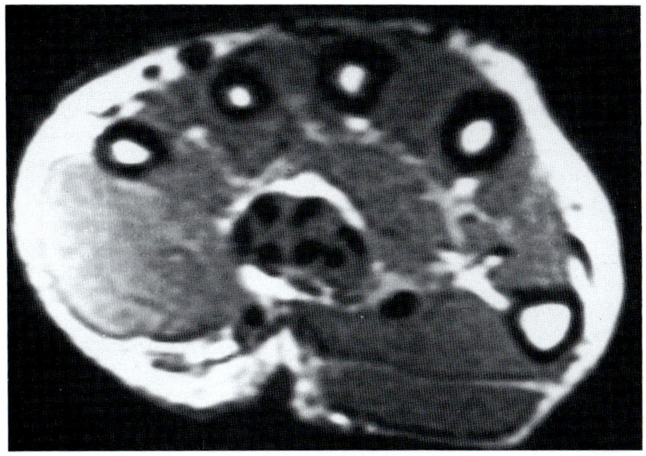

FIGURE 5.11. Normal Hand Anatomy. Axial plane [SE 2000/20]. Proximal aspect of the metacarpal bones in pronation (*plane H* in Fig. 5.3). Note the normal intermediate SI lumbricalis muscles, which appear contiguous with the deep and superficial flexor tendons. This normal appearance is sometimes confused with carpal tunnel pathology.

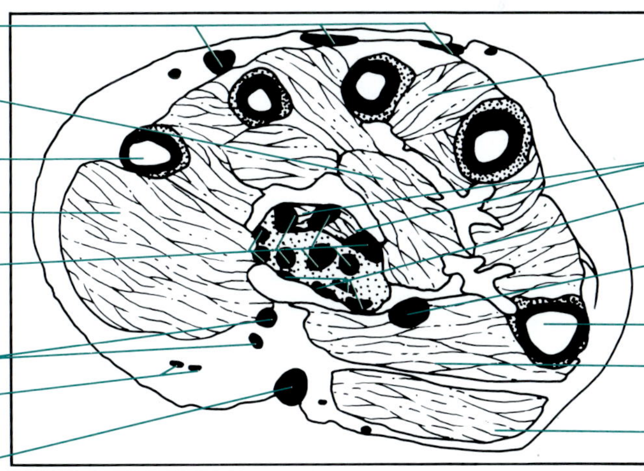

muscles may actually appear within the confines of the carpal tunnel and should not be confused with pathology. The interosseous tendons are shown as vertically oriented low SI bands adjacent to their respective metacarpal and proximal phalanges (Figs. 5.12 and 5.13). Axial images through the digits depict the normal anatomy of the flexor and extensor tendons as they insert in the mid and distal phalanges (Fig. 5.14–5.16).

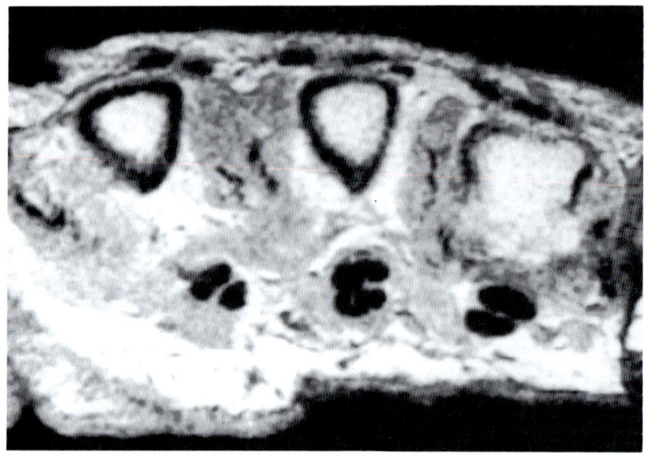

FIGURE 5.12. Normal Hand Anatomy. Axial plane [SE 1500/25]. Midmetacarpal level in pronation (*plane 1* in Fig. 5.3). Note the interossei muscles and tendons.

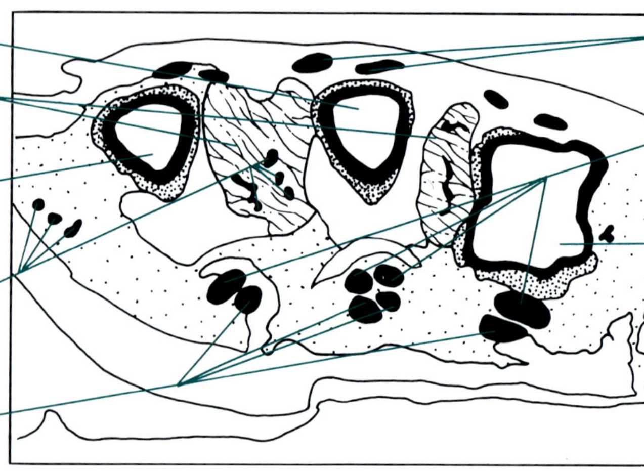

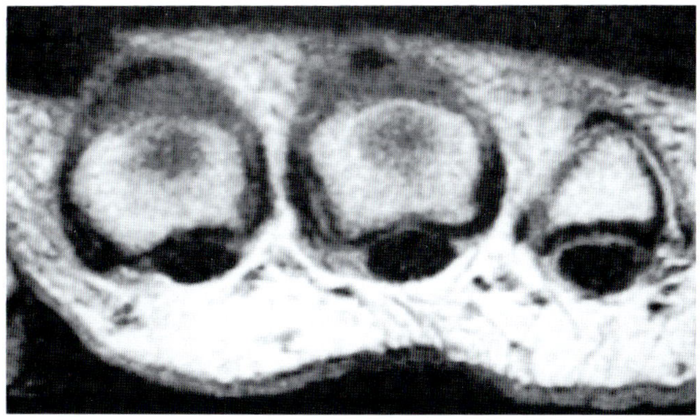

FIGURE 5.13. Normal Hand Anatomy. Axial plane [SE 1500/25]. Distal metacarpal region (*plane J* in Fig. 5.3). Note the distribution of the tendons around the metacarpals.

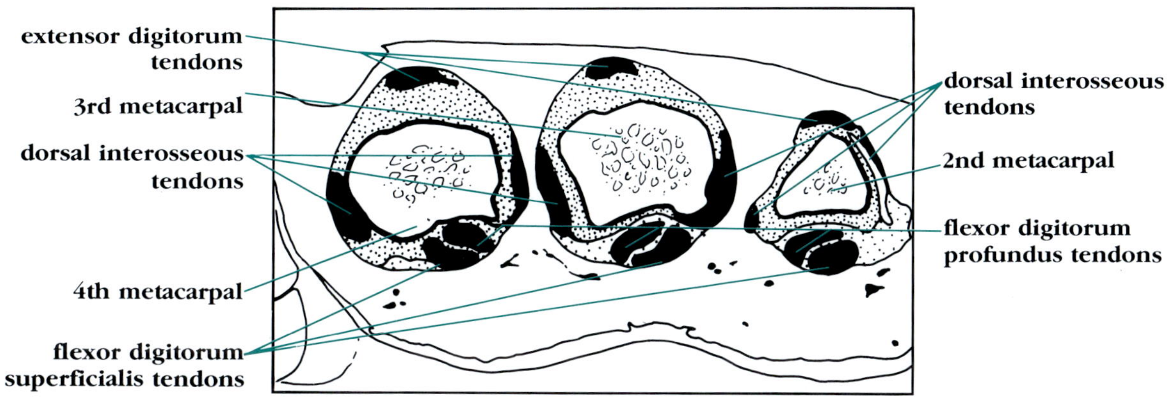

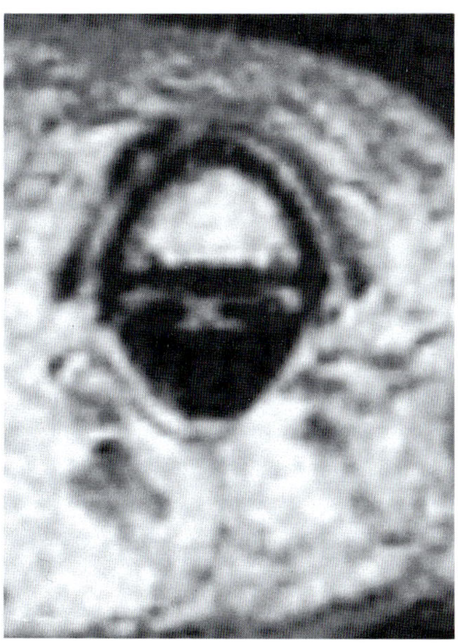

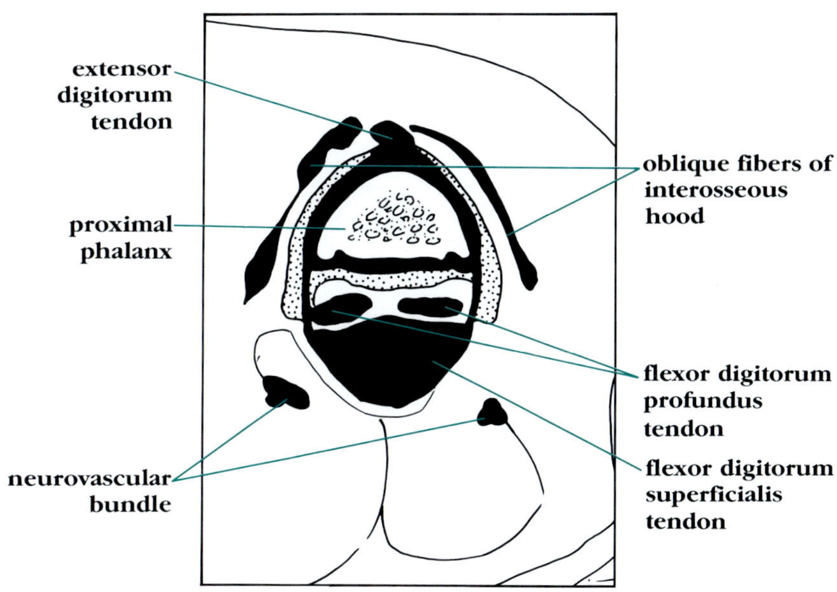

FIGURE 5.14. Normal Finger Anatomy. Axial plane [SE 1500/25]. Proximal phalangeal region in pronation (*plane K* in Fig. 5.3). Note the flexor and extensor mechanism.

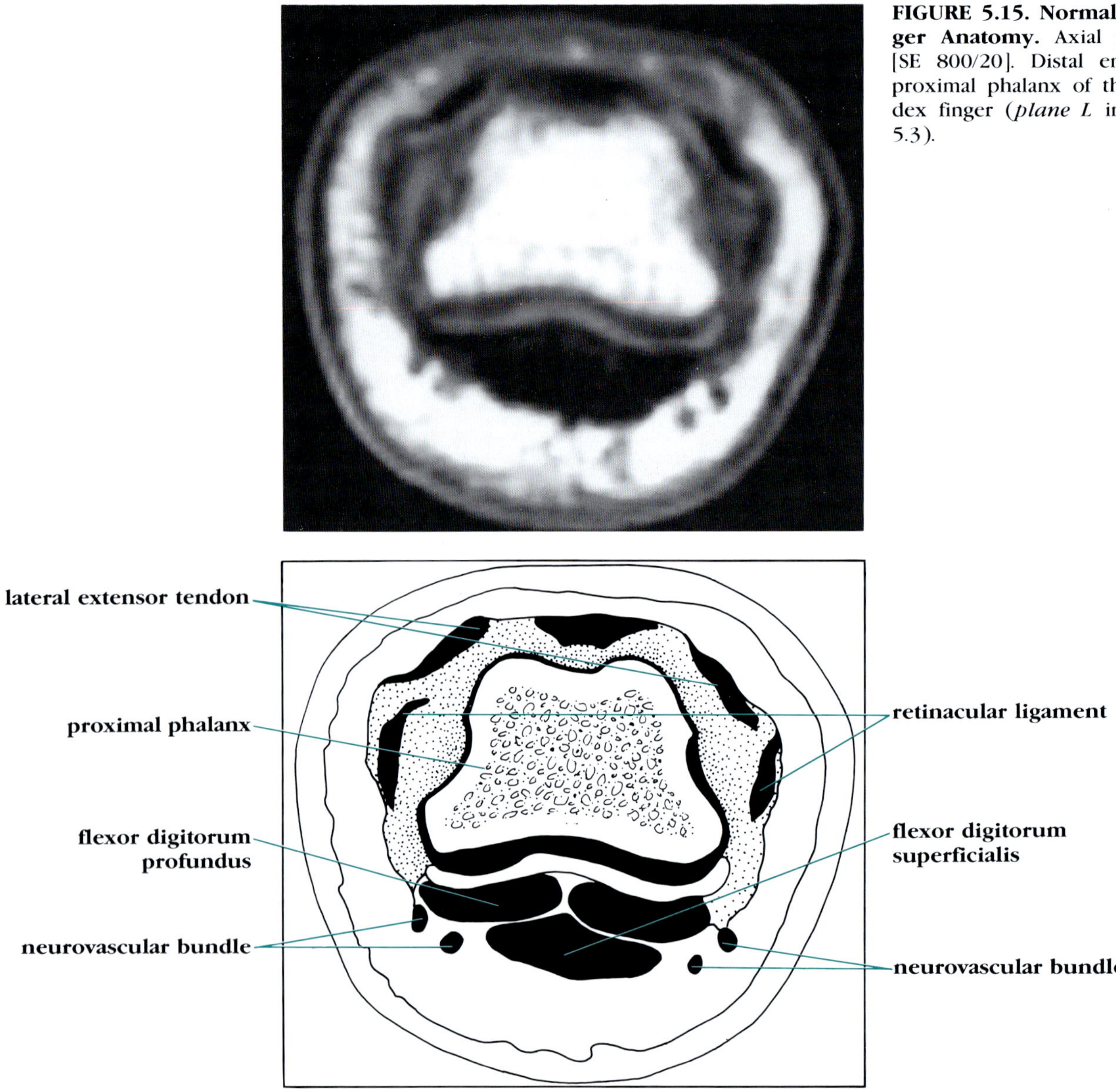

FIGURE 5.15. Normal Finger Anatomy. Axial plane [SE 800/20]. Distal end of proximal phalanx of the index finger (*plane L* in Fig. 5.3).

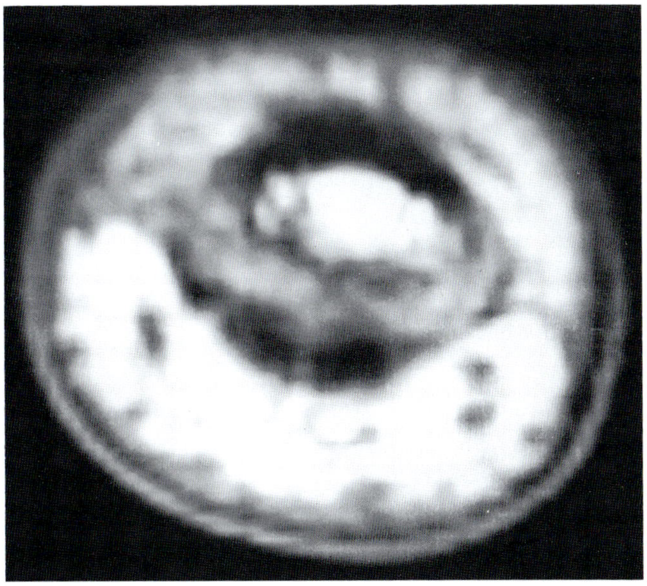

FIGURE 5.16. Normal Finger Anatomy. Axial plane [SE 800/20]. Midportion of the midphalanx of the index finger (*plane M* in Fig. 5.3).

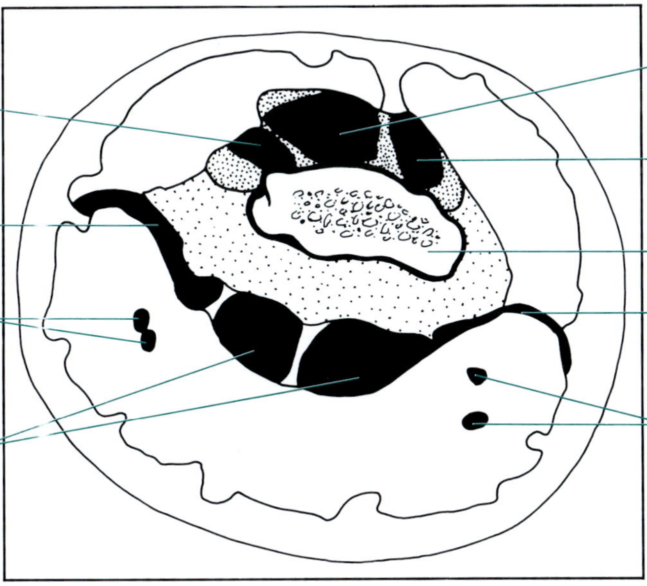

Ligaments

While the detailed anatomy of wrist ligaments has been described in other publications (12,41,49), MRI correlation has only recently been demonstrated by Kellerhouse (23). The schematic drawings in Figs. 5.17 and 5.18 of the volar and dorsal intrinsic and extrinsic ligaments of the wrist provide a review for the normal ligamentous anatomy. In general, the horizontally oriented ligaments are best seen in the axial plane. Ligaments show low SI on all SE pulse sequences. On fast gradient 3DFT, the ligaments may appear of slightly higher SI.

The flexor retinaculum is displayed as a thick, low SI bandlike structure attaching to the hamate and pisiform bone medially, and to the tubercle of the trapezium and scaphoid laterally (Figs. 5.7 and 5.10). Distally, it is thickest at its primary attachments to the hood of the hamate and tubercle of the trapezium, more distally becoming continuous with the palmar aponeurosis (Fig. 5.10).

The triangular fibrocartilage (TFC) is thicker at its anterior and posterior borders, where it bridges the radius at its ulnar aspect and the ulna at its radial aspect (Fig. 5.5). Also known as the anterior and poster or distal radial ulnar ligaments (49), these are seen as thin low SI bands on the axial plane.

The more horizontal radiolunatotriquetral ligament and the dorsal intercarpal ligament may be seen as low SI bands paralleling, respectively, the volar or dorsal cortical outlines of the carpus (Figs. 5.7–5.9). The volar ligaments are interposed between the carpal bones and carpal tunnel. Occasionally, the more vertical radioscaphocapitate ligament, the dorsal radioscapholunate, and the volar radioscapholunate ligament may be visible (Figs. 5.8A and 5.9).

The short interosseous ligaments of the carpal bones, seen occasionally in the axial plane, are best depicted in the coronal plane.

The ligaments of the fingers—including Cleland's ligament, the triangular ligament, and the oblique fibers of the interosseous hood—are apparent in the axial plane as are the tendons and neurovascular bundle (Figs. 5.14–5.16).

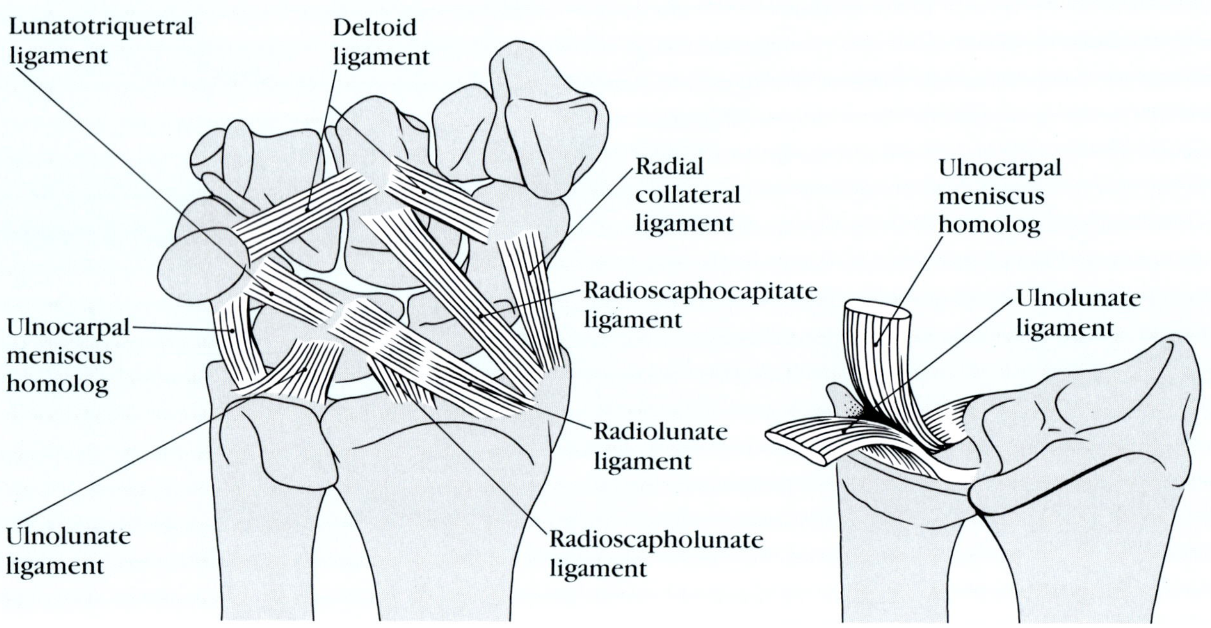

FIGURE 5.17. Normal volar ligaments of the wrist.

Neurovascular Structures

The median nerve is best seen in the axial plane as an intermediate SI round or oval structure easily distinguished from surrounding fat or tendons, such as the flexor pollicis longus radially and the flexor digitorum superficialis ulnarly (Figs. 5.7 and 5.10). Superficially, the median nerve is bounded by the flexor retinaculum, although it may be located deeper, interposed between the flexor pollicis longus and superficial flexor tendons. Its long axis may be perpendicular to the retinaculum rather than parallel, which is its usual orientation (37). The median nerve can be traced distally to its three terminal branches (Fig. 5.11).

The radial artery is generally well viewed proximally, radial to the flexor carpi radialis and volar to the radius. It courses deep, distally to the tendons of the first dorsal compartment adjacent to the trapezium, before terminating in the palmar branches (Figs. 5.4 and 5.7).

▼▼▼ *CORONAL PLANE*

Tendons

In the coronal plane, the flexor and extensor tendons can be well seen longitudinally, extending as low SI tubular structures with their relationship to adjacent structures, particularly the bones, clearly

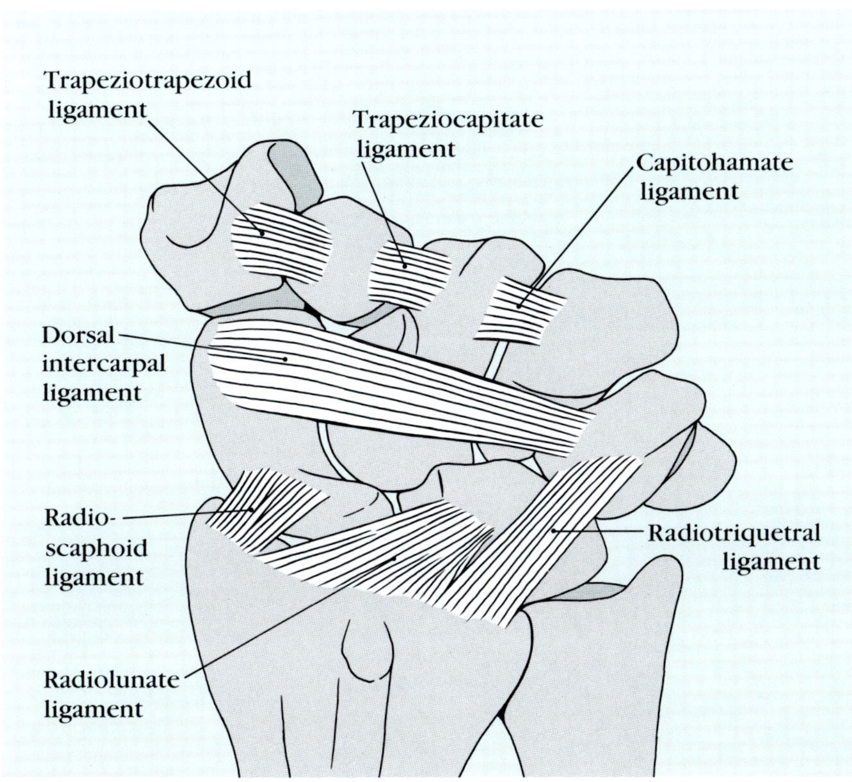

FIGURE 5.18. Normal dorsal ligaments of the wrist.

depicted. This allows a panoramic view of the radial carpal, carpocarpal and carpometacarpal articulations, as well as overlying articular cartilage (Figs. 5.19–5.21). The deep and superficial flexor tendons are especially evident within the confines of the carpal tunnel (Fig. 5.20).

Ligaments

The interosseous ligaments of the proximal carpal row extend from the palmar to dorsal bone surfaces and are best seen in the coronal plane. These low SI structures measure approximately 1–2 mm, ex-

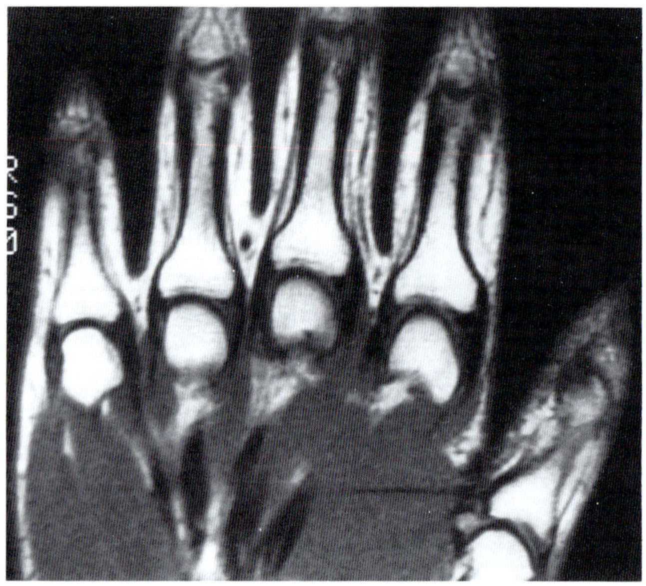

FIGURE 5.19. Normal Hand Anatomy. Coronal plane [SE 800/20] at the level of the metacarpal phalangeal articulation. Note the intermediate SI articular cartilage overlying the low SI subarticular cortex of the distal second metacarpal bone.

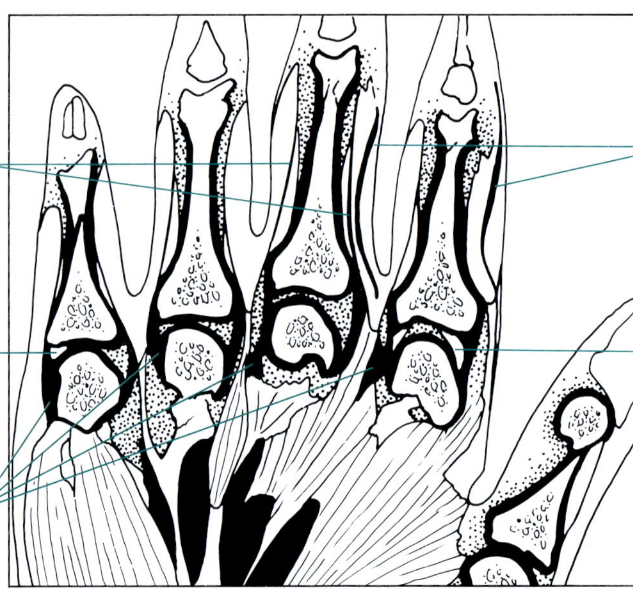

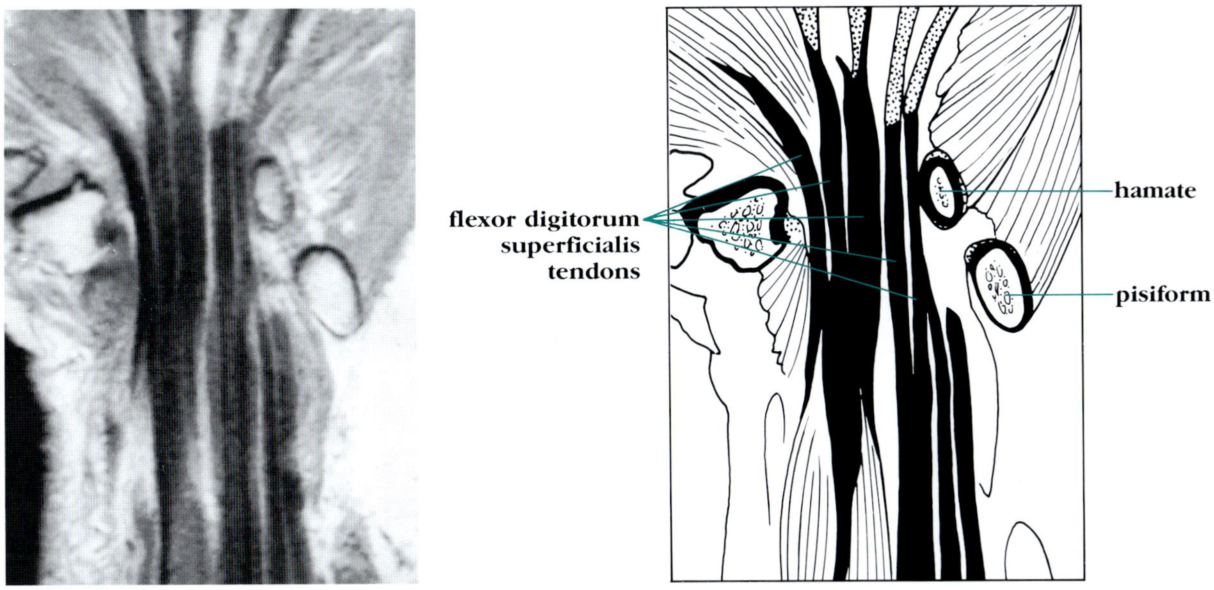

FIGURE 5.20. Normal Hand Anatomy. Coronal plane [SE 800/20] at the level of the wrist. Note the tubular low SI flexor tendons at the level of the carpal tunnel.

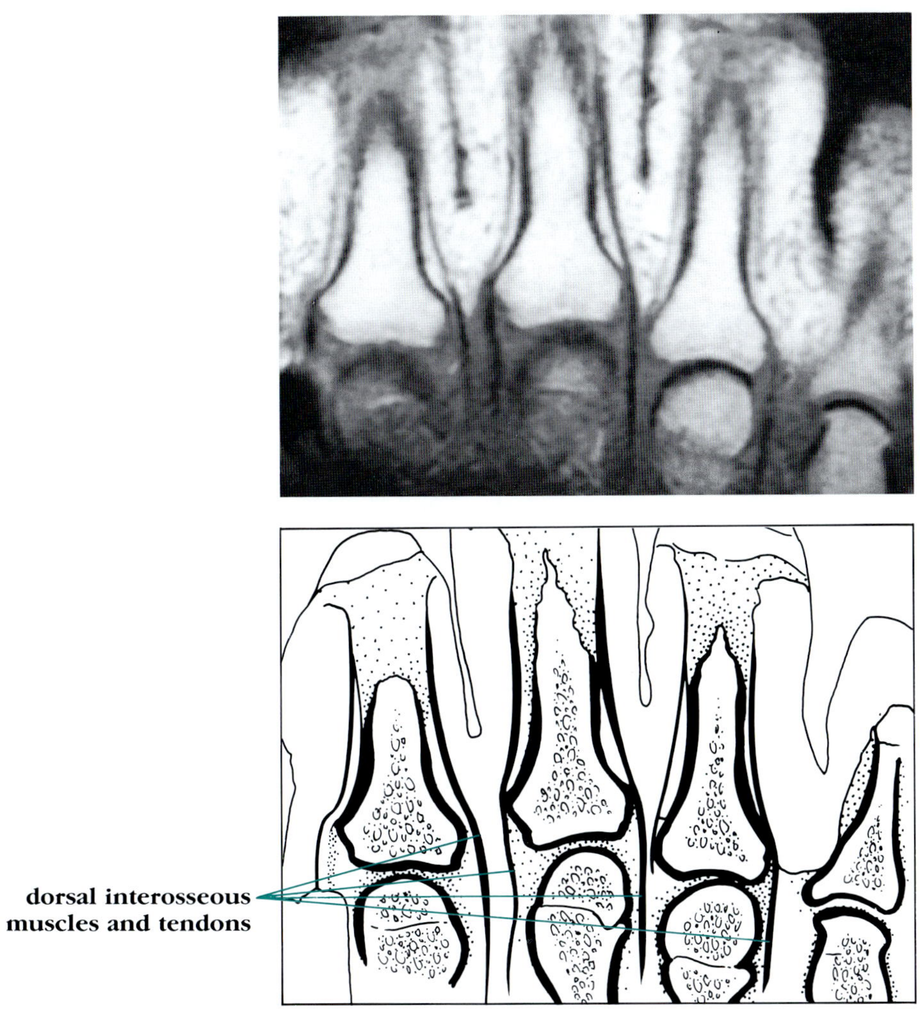

FIGURE 5.21. Normal Hand Anatomy. Coronal plane [SE 600/25] through the metacarpal phalangeal articulation. Note the dorsal interosseous tendons.

tending from the lunate to the triquetrum medially and to the scaphoid laterally (Figs. 5.22–5.24). The scapholunate ligament is seen routinely and has reportedly been visualized in 70 percent of normal T1W1 studies (23), and 90 percent of normal gradient echo images.

The distal carpal interosseous ligaments do not usually extend from volar to dorsal capsules, thereby allowing communication between the midcarpal and carpometacarpal joints. The capitatohamate, the strongest ligament, is seen routinely at its anterior and distal attachments (Fig. 5.24).

The radial collateral ligament extends radially from the radial styloid process to the waist of the scaphoid and to the trapezium, finally blending distally with the transverse carpal ligament and the

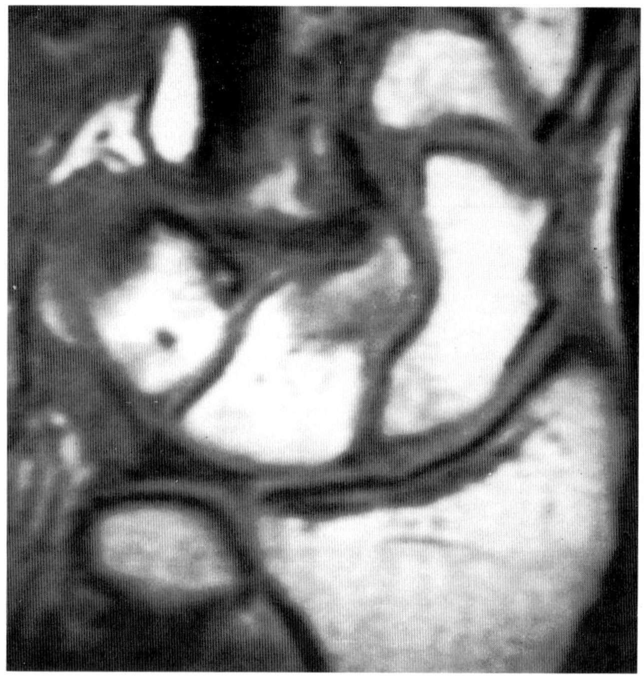

FIGURE 5.22. Normal Wrist Anatomy. Coronal plane [SE 800/20] through the volar radiocarpal articulation. The triangular-shaped low SI TFC appears discontinuous with the ulnar aspect of the radius due to overlying articular cartilage. Attachment to the base of the ulnar is apparent. The interosseous ligaments connecting the lunate to the scaphoid and triquetrum proximally are seen as short SI structures.

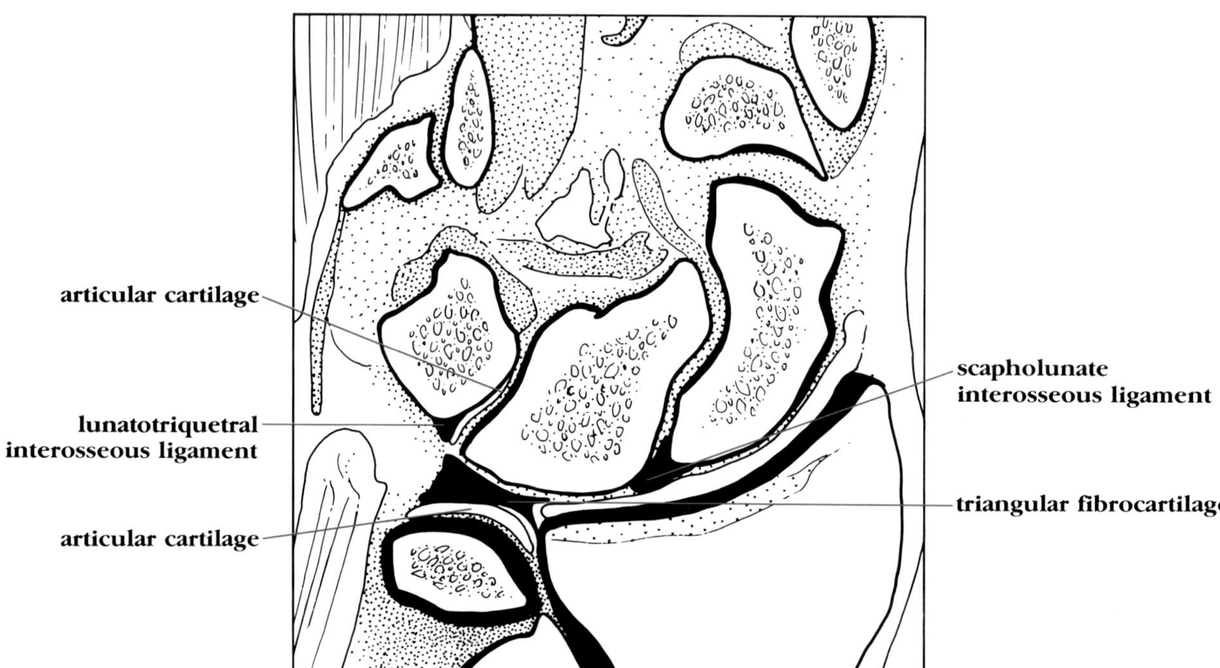

dorsal capsular ligament (Fig. 5.25). The ulnar collateral ligament has recently been questioned as a distinct entity (49). This linear low SI structure attaches to the base and body of the styloid process of the ulna as well as to the anterior and posterior borders of the TFC. It extends distally to the pisiform and transverse carpal ligament, triquetrum, and finally the ulnar border of the hamate and base of the fifth metacarpal (Fig. 5.25).

The functionally importanat TFCC, or ulnar carpal complex, is apparent within the ulnar aspect of the proximal wrist joint (Figs. 5.22, 5.23, and 5.25). The TFCC is composed of the TFC, the ulnocarpal ligaments, and the extensor carpi ulnaris tendon sheath. The low SI TFC and ulnar carpal meniscus (meniscal homolog) originate at the dorsal ulnar aspect of the radius, where they may appear discontinuous due to the articular cartilage covering

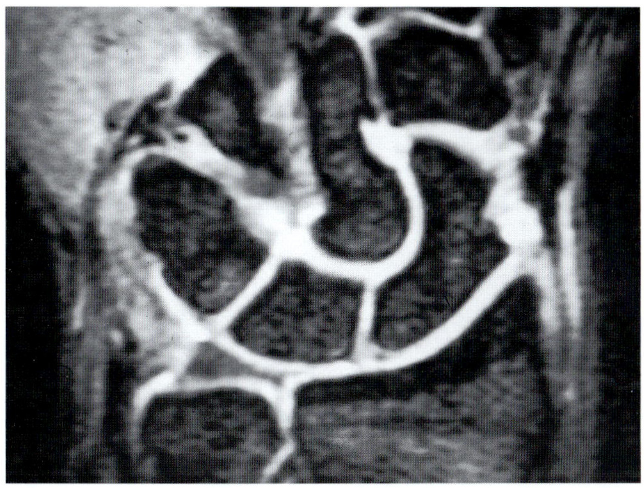

FIGURE 5.23. Normal Wrist Anatomy. Coronal plane [GRASS 1000/12, 20-degree flip angle] at similar level to Fig. 5.22. Due to the positive contrast background, the interosseous ligaments proximally as well as the triangular fibrocartilage are more clearly delineated.

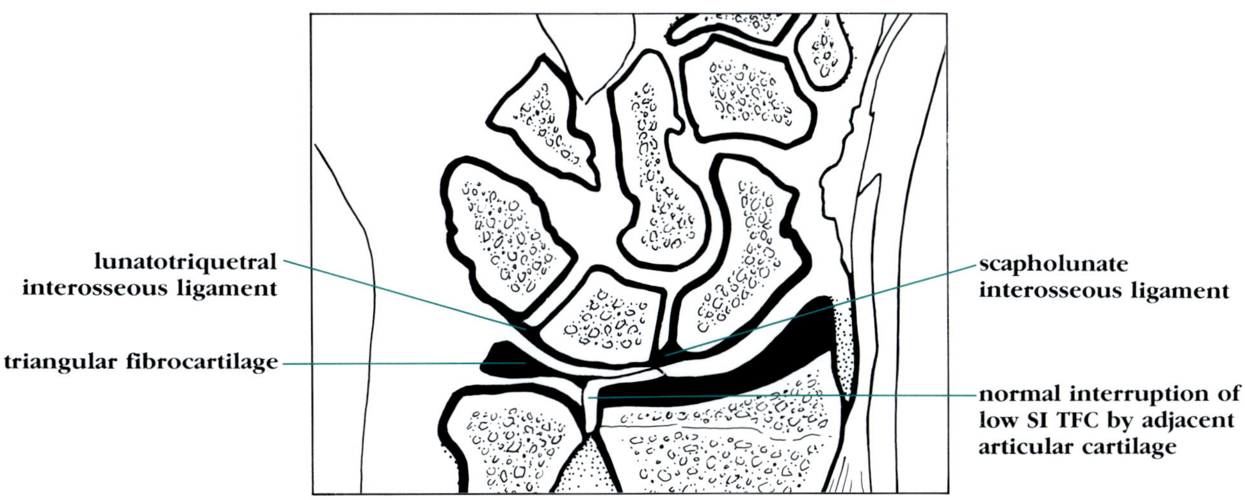

FIGURE 5.24. Normal Wrist Anatomy. Coronal plane [SE 600/20] through the capitate-hamate articulation. The obliquely oriented, low SI interosseous ligament can be seen bridging the capitate and hamate.

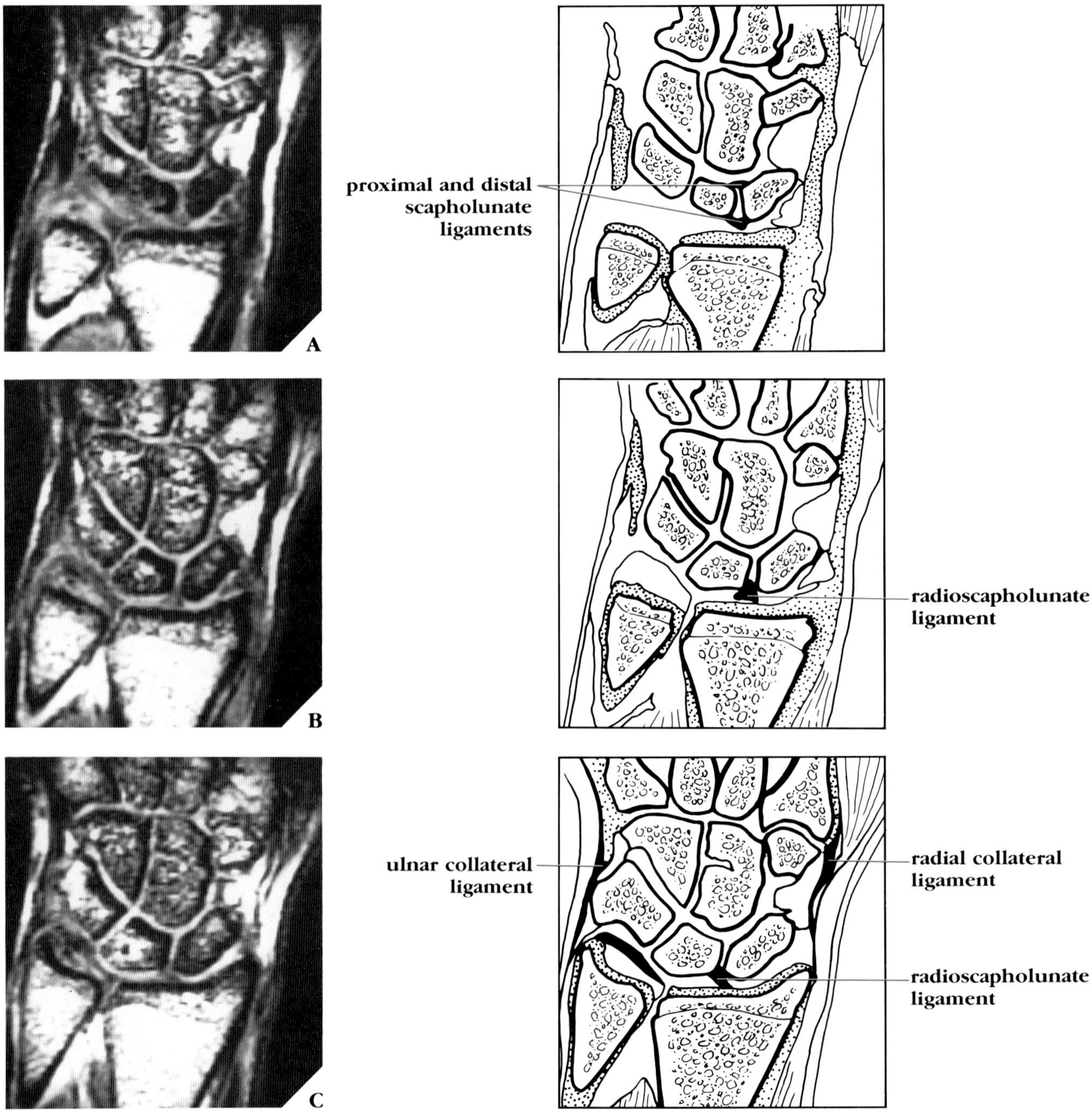

FIGURE 5.25. Normal Wrist Anatomy. Coronal plane [3DFT Grass 50/15, flip angle 40 degrees, 0.8-mm continguous slices]. The imaging sequence extends from the volar (**A**) to the dorsal aspect of the wrist (**F**). The ligaments are seen as slightly

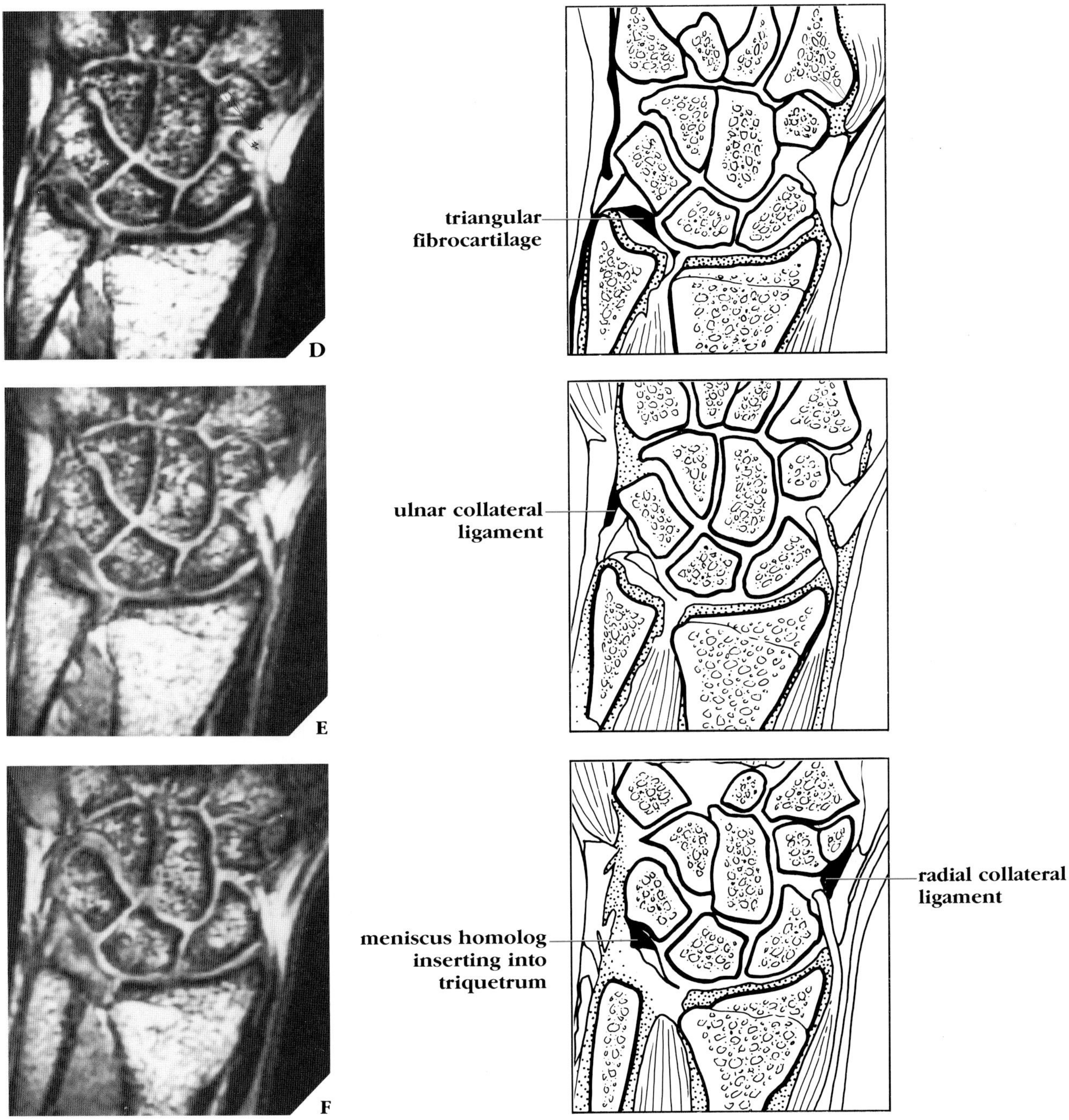

higher SI structures than on routine SE imaging. The radioscapholunate ligament is well seen on multiple sections.

the distal radius. The TFC then extends ulnarly to insert at the base of the styloid process. The meniscus extends ulnarly and volarly to insert on the triquetrum (Fig. 5.25C). The meniscus may or may not be seen as a separate structure anatomically; the TFCC may appear on MRI as an amorphous mass.

Adjacent to the meniscus is the prestyloid recess, a synovial-filled space surrounding the ulnar styloid and typically devoid of ligamentous attachments (22, 42). The TFC is thin at its center and is sometimes perforated, allowing communication with the prestyloid recess. The short vertical radioscapholunate ligament may also be seen in the coronal plane (Fig. 5.25). The volar and dorsal radiocarpal ligaments are seen only infrequently in this plane (Fig. 5.26).

Neurovascular Structures

As with the tendons, the longitudinal course of the radial artery, median nerve, and ulnar artery and nerve can be better appreciated in the coronal plane.

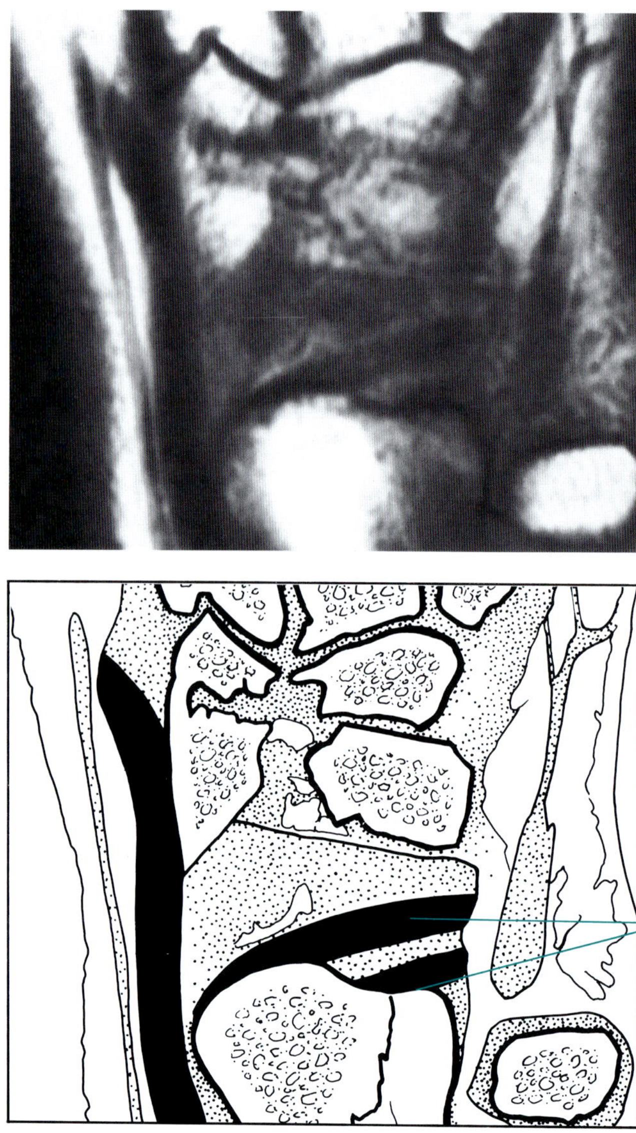

FIGURE 5.26. Normal Wrist Anatomy. Coronal plane [SE 600/20] at the dorsal radiocarpal articulation. The obliquely oriented low SI dorsoradial carpal ligaments are seen extending from the lateral radius ulnarly and distally.

▼▼▼ *SAGITTAL PLANE*

Tendons

All extensor tendons with their respective insertions are clearly depicted in the sagittal plane.

The flexor digitorum profundus tendon is well demonstrated in the midline sagittal plane of the digits as it inserts on the distal phalanx (Fig. 5.27). The flexor digitorum superficialis tendon is seen in the paramidline plane as it inserts on the middle phalanx (Fig. 5.27).

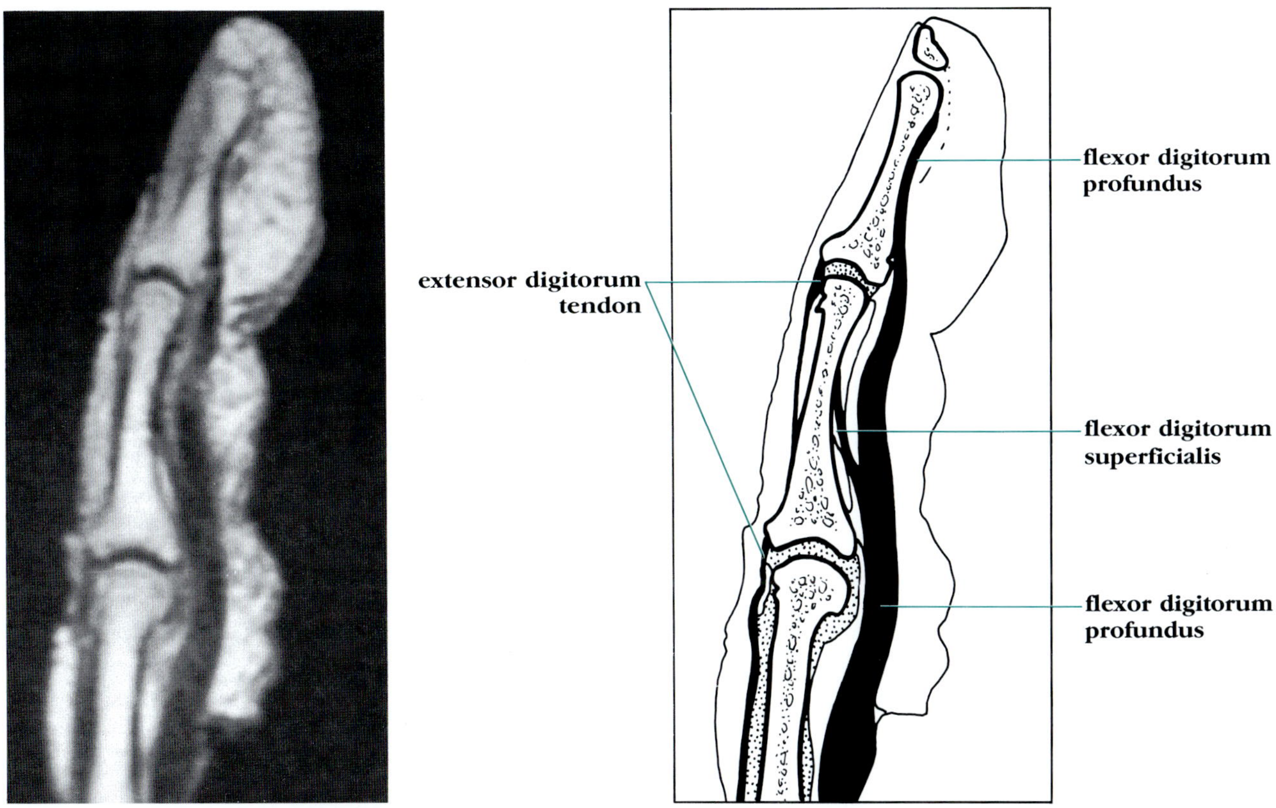

FIGURE 5.27. Normal Finger Anatomy. Sagittal plane [SE 800/20] through the index finger of cadaver specimen. Note the insertion of the flexor digitorum tendons.

Ligaments

The sagittal plane offers the best depiction of the volar carpal ligaments, which are principally responsible for maintaining alignment of the carpal bones (Figs. 5.28 and 5.29). The most lateral of the radial carpal ligaments is the radiocapitate (radioscaphocapitate) ligament. Seen as a low SI band, it originates at the anterior aspect of the radial styloid and courses across the volar concave aspect of the scaphoid, attaching to the volar midcapitate.

Medial to this, and more horizontal, is the radiolunate ligament (radiolunatotriquetral), a thicker bundle of fibers seen more inferiorly as it courses from the base of the radial styloid, across the lunate, to its volar attachment on the triquetral bone. The most medial radioscapholunate, or ligament of Kuenz and Testut, originates at the articular crest of the radius, separating the scaphoid and lunate

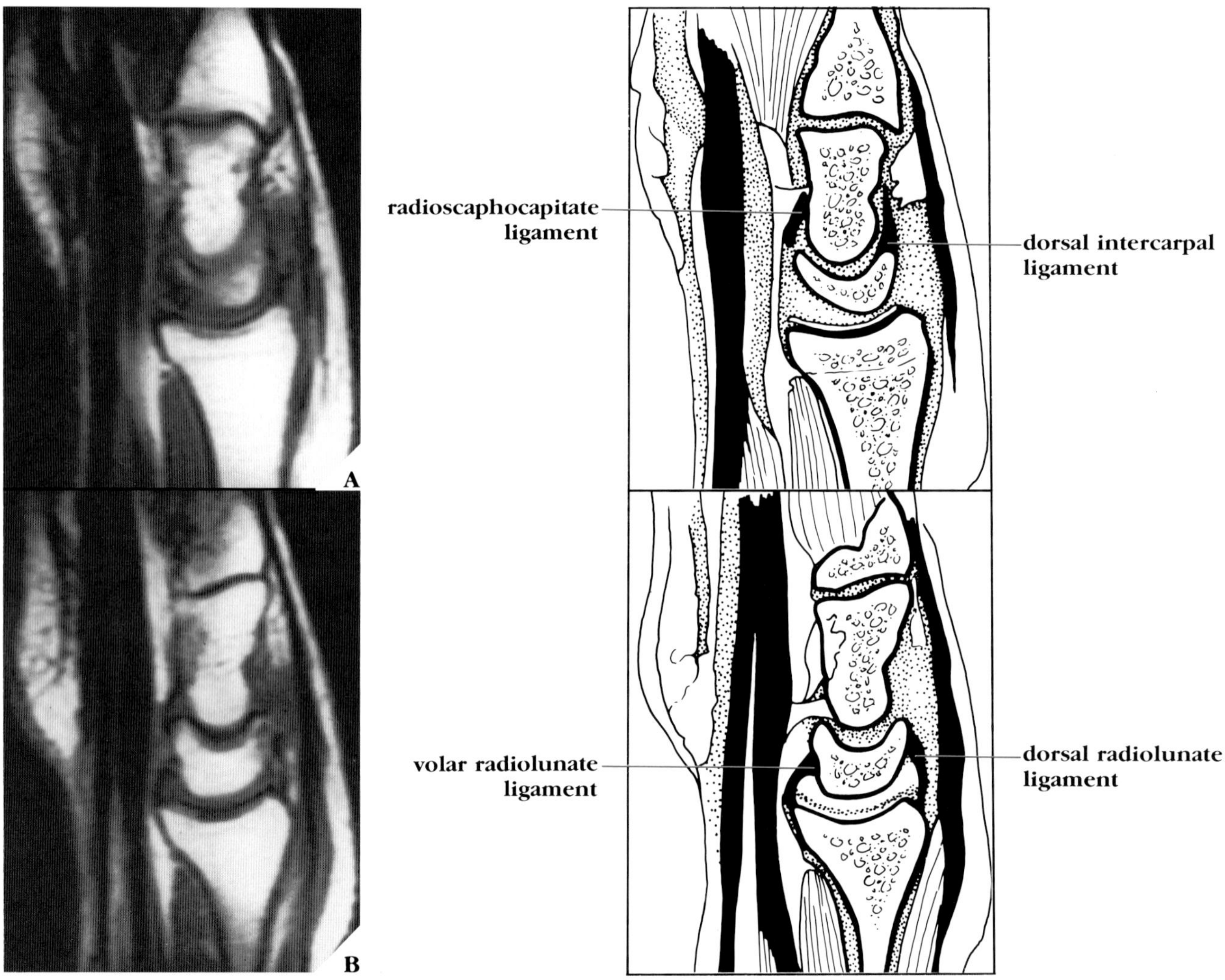

FIGURE 5.28. Normal Wrist Anatomy. Sigittal plane [SE 600/20] through the wrist from the midaspect (**A** and **B**) to the ulnar aspect (**C** and **D**). The volar and dorsal radiolunate components of the radioscapholunate ligaments tether the volar and dorsal poles of the lunate, respectively. The radiolunatotriquetral ligament is seen volar to the capitate lunate articu-

facets. It takes a short vertical course to its insertion at the articular surfaces of the lunate and scaphoid, also covering the interspace (51).

The more superficial "V" or deltoid ligaments are not normally visualized as distinct structures (Fig. 5.28). The TFCC is evident as a biconcave low SI disc interposed between the ulna and triquetrum, with the ulnocarpal ligament seen at its anterior border (Figs. 5.28 and 5.29).

The dorsal carpal ligaments include the radioscaphoid, radiolunate, radiotriquetral, and dorsal intercarpal ligaments. The dorsal intercarpal ligaments are seen dorsal to the midcarpal bones as a low SI vertical band (Figs. 5.28 and 5.29). The dorsal radiolunate is routinely visualized, whereas the radioscaphoid and radiotriquetral are not (Figs. 5.28 and 5.29).

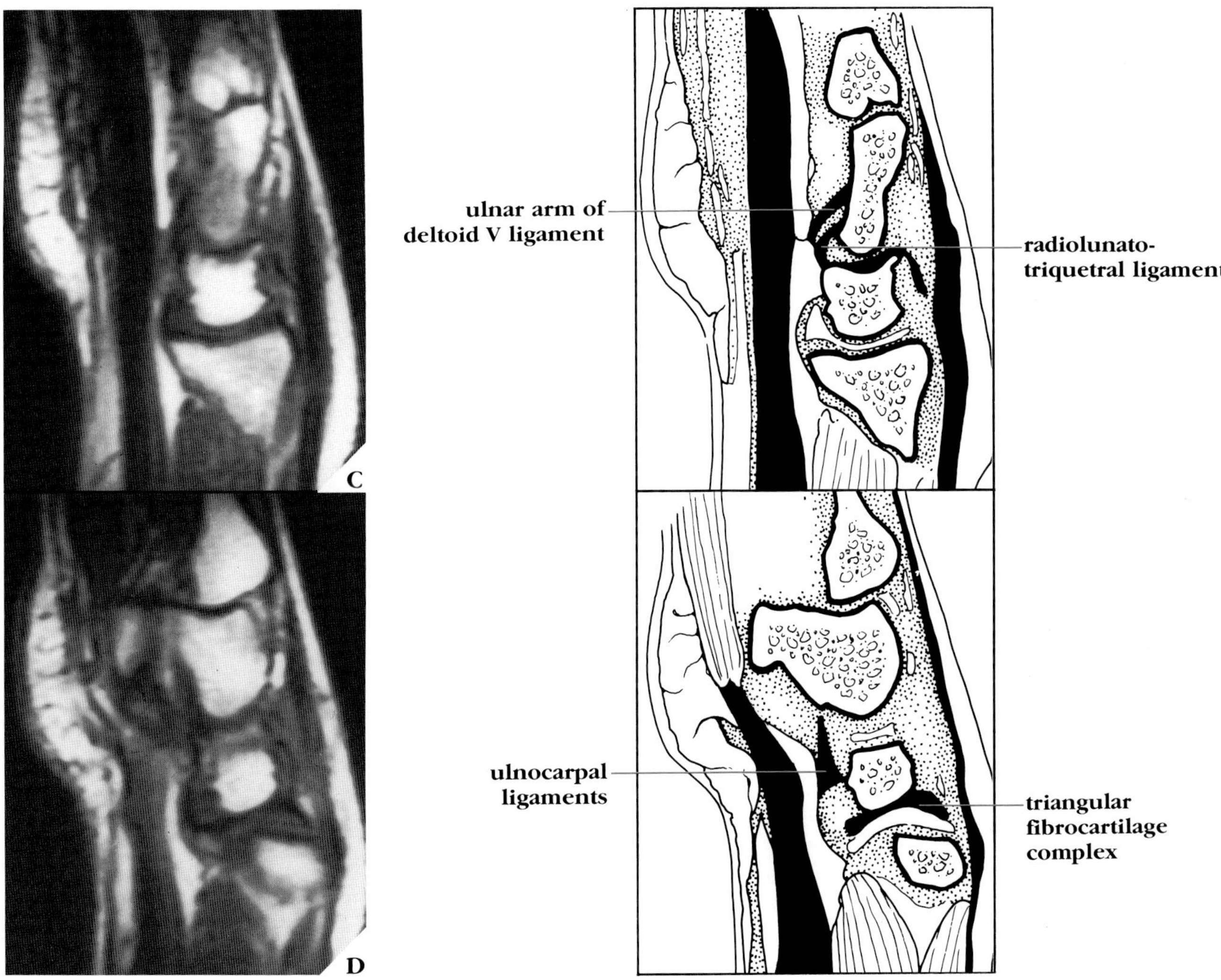

lation. The radioscaphocapitate is seen inserting at the volar and proximal one third of the capitate at its midaspect. The ulnar arm of the deltoid V-shaped ligament is occasionally seen.

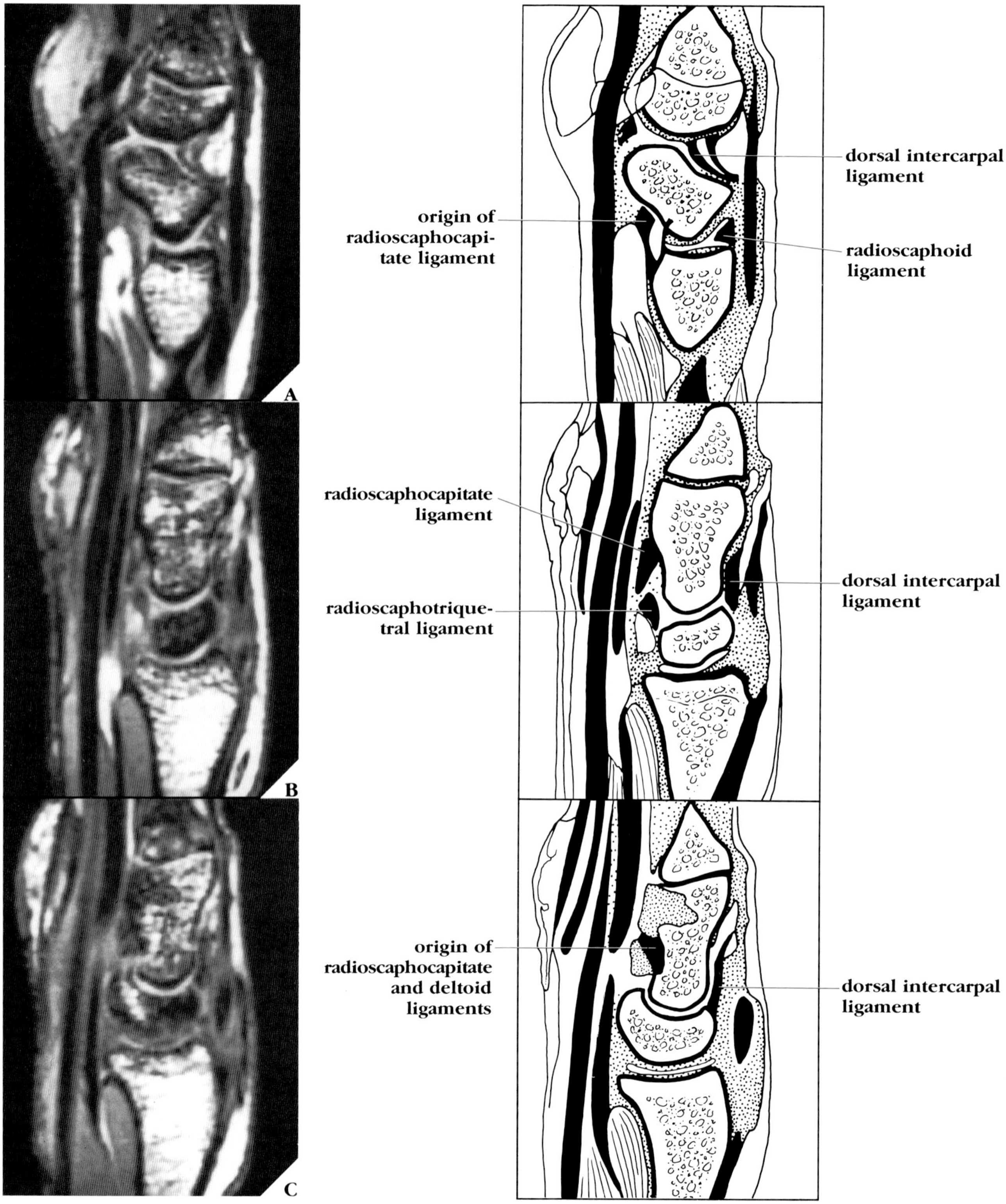

FIGURE 5.29. Normal Wrist Anatomy. Sagittal plane [3DFT GRASS 50/15, 40-degree flip angle, 0.8-mm contiguous slices] from the radial to ulnar aspect of the wrist. This technique demonstrates the ligamentous anatomy more clearly than the

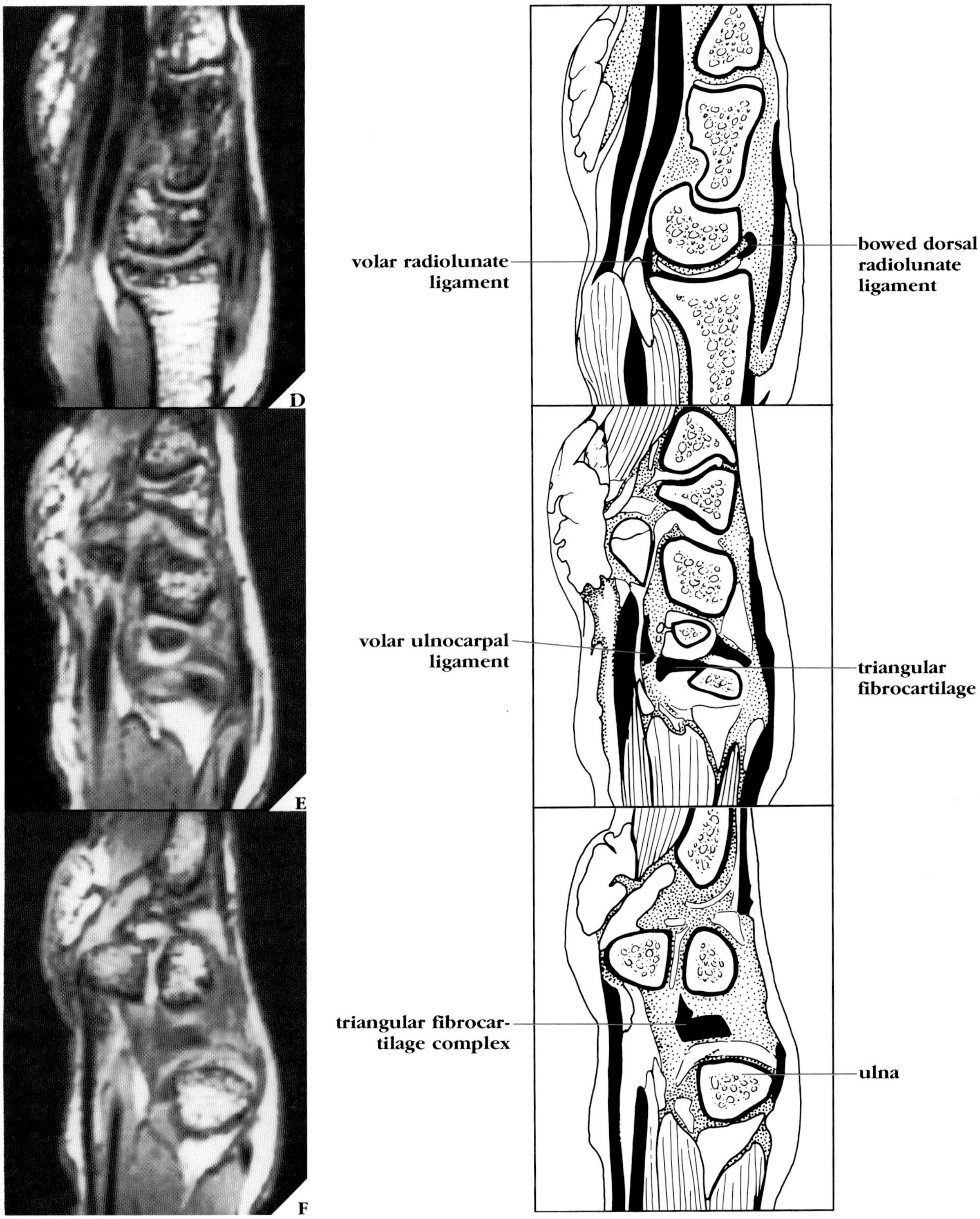

SE sequences seen in Fig. 5.28. Note the intricate detail of the dorsal and volar ligaments, including the individual radioscaphoid ligament tethering the proximal pole of the scaphoid bone (**A** and **B**).

LIGAMENTOUS INJURY

MRI is the only imaging technique that allows direct visualization of both normal and pathologic anatomy of wrist ligaments. This insight into the functional anatomy of the ligaments is crucial to understanding wrist biomechanics as well as the clinically important carpal instability and painful wrist syndromes.

Plain film, tomography, and fluoroscopy are limited to visualization of the osseous structures from which etiologies of soft tissue derangement must be inferred. The contrast and spatial resolution of computerized tomography is inadequate for visualization of the ligaments and for differentiating them from surrounding scar, hematoma, or edema (43). Double and single contrast arthrography, with or without substraction technique, indirectly evaluates ligamentous integrity which may be irrelevant to the clinical problem. Moreover, these are invasive techniques which may be time consuming, particularly with multiple compartment injections, and do not permit visualization of surrounding soft tissues.

MRI has inherently good soft tissue contrast resolution as well as other benefits. When employing the surface coil technique and a small FOV, it is possible to achieve the contrast resolution needed for high resolution detailing of the ligaments. As mentioned, fast gradient 3DFT offers thin contiguous section capability, which is excellent for ligament depiction.

▼▼▼ *KINEMATICS OF THE WRIST*

A detailed discussion of the evolving, complex subject of carpal kinematics and pathomechanics is beyond the scope of this chapter; however, a limited review of several basic theories is presented below.

The kinematics of the wrist represent a dynamic interplay of the geometry of the carpal bones, connecting ligaments, and surrounding tendons in which interruption of one compartment of this system significantly affects the functioning of those remaining undisturbed (18). In 1919, Navarro initiated the columnar theory of the carpus which, in 1976, was modified by Taleisnik to explain recognized patterns of carpal instability (39,47). Gilford and colleagues conceptualized the wrist as a chain, with the distal radius, midcarpal row, and distal carpal row as its links (19). Linscheid and colleagues described the "slider-crank" theory to explain the scaphoid role in preventing collapse (30). Dorsal intercalated segmental instability (DISI) and volar intercalated segmental instability (VISI) were described as two distinct patterns of intercalary instability. Named for the angle formed by the capitate and lunate in the lateral projection, they are important in determining perilunate instability (13,14). The perilunate pattern of injury to the volar ligaments of the wrist (Stage I–IV) was enumerated by Mayfield and co-workers (33,34). Newly recognized patterns of carpal instability, including that of the midcarpal and proximal carpal, have emerged. Lichtman and Martin described the ring concept of the carpal bones and bridging ligaments to explain the wide spectrum of carpal instabilities (23,24).

To date, MRI literature has dealt mainly with the interosseous ligaments, particularly the scapholunate ligament, the ulnar and radial collateral ligaments, and the TFC. Kellerhouse recently expanded the data when he described the ability of MRI to depict the volar radial carpal ligaments as well as identify their role in carpal instability (23).

Scapholunate Ligament Tears

Lack of visualization or interruption of the scapholunate ligament may indicate a tear of the ligament, because it is routinely seen in the coronal plane, and as often as 90 percent of the time on gradient

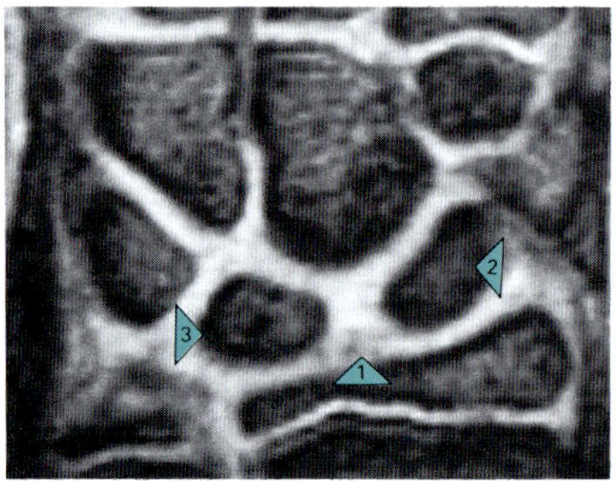

FIGURE 5.30. Posttraumatic Scapholunate Ligament Interruption. Coronal plane [GRASS 200/12] at the level of the scapholunate articulation. The interspace between the lunate and scaphoid is increased and the continuity of the scapholunate ligament is interrupted proximally, indicating posttraumatic disruption. Arrow **1** Interrupted scapholunate ligament. Arrow **2** Scaphoid. Arrow **3** Lunate.

echo images (23) (Fig. 5.30). The significance of this finding, which rests on clinical grounds, may be qualified by normal degeneration due to aging (11).

It has also been demonstrated that interruption of the scapholunate ligament alone may not be sufficient to produce scapholunate instability, a condition classified as radial or lateral instability by Taleisnik and as one of the perilunate instabilities by Lichtman. This most common of the carpal instabilities, usually secondary to a fall on an outstretched hand, materializes as acute radial deviation when the hand is extended and severe compression of the scaphoid space. The volar radioscapholunate ligament and the dorsal radiolunate ligaments appear to play an important role in maintaining the integrity of this joint. The radioscapholunate ligaments must be disrupted before instability with complete subluxation of the scaphoid can occur (11,48,53).

Lunatotriquetral Ligament Ruptures

The lunatotriquetral interosseous ligament is not always visible; thus, the lack of visualization of this ligament does not necessarily indicate disruption. Its visualization can, however, be improved with 3DFT or gradient echo imaging.

Volar Interosseous Ligament Ruptures

The above discussion underscores the importance of identifying the normal and pathologic ligamentous anatomy of the radiocarpal, intercarpal, and interosseous ligaments. The arrangement and integrity of the volar ligaments, the main functional ligaments of the wrist, dictate the complex intercarpal motion (32). The volar radioscapholunate is invariably present, stabilizing the volar pole of the lunate and proximal portion of the scaphoid (Figs. 5.25, 5.28, 5.29). Disruption of this, or of the radioscaphocapitate ligament, can produce a DISI deformity with or without rotatory subluxation of the scaphoid (Fig. 5.31). The radioscaphocapitate is smaller and weaker than the scaphoid, and serves as a rotational sling for the waist of the latter (Figs. 5.8, 5.9, 5.28, 5.29, and 5.31). Of possible significance in injury may be the gap between the radioscaphocapitate, the more inferior radioscaphotriquetral ligament, and the space of Poirer over the volar capitatolunate joint (33,34).

Current MRI experience with disruption of the wrist ligaments is limited. However, as with other joints, complete or incomplete tears may be seen on MRI by lack of visualization, retraction with or without thickening, and surrounding high SI change on T2W1 due to edema, synovitis, or joint fluid (Figs. 5.30–5.33). These findings may present with normal plain films and may, therefore, help direct the clinician to a more specific diagnosis. Laxity of these ligaments can also occur with incomplete tears or following trauma. A wavy MRI contour of an unstressed ligament may be seen normally, since the ligaments are passive inhibitors of carpal motion (Figs. 5.27 and 5.32).

Triangular Fibrocartilage Complex Tears

Similar MRI findings also occur with pathology involving the TFCC at the volar aspect of the wrist, where ligamentous injury could be the cause of VISI deformity, midcarpal instability, or pain at the ulnar aspect of the wrist (Figs. 5.32 and 5.33). Abnormal stress and injury to the TFCC is seen in patients with negative or positive ulnar variance due to changes in axial loading. Typically, the majority of axial loading is through the capitate to lunate and subsequently to the radius. Less than 20 percent of the axial loading forces are normally transmitted to the TFCC at the ulnar aspect of the wrist (42). These relationships change with ulnar lengthening or shortening, causing ulnar lunate impingement and abnormal stress on the TFC with possible disruption.

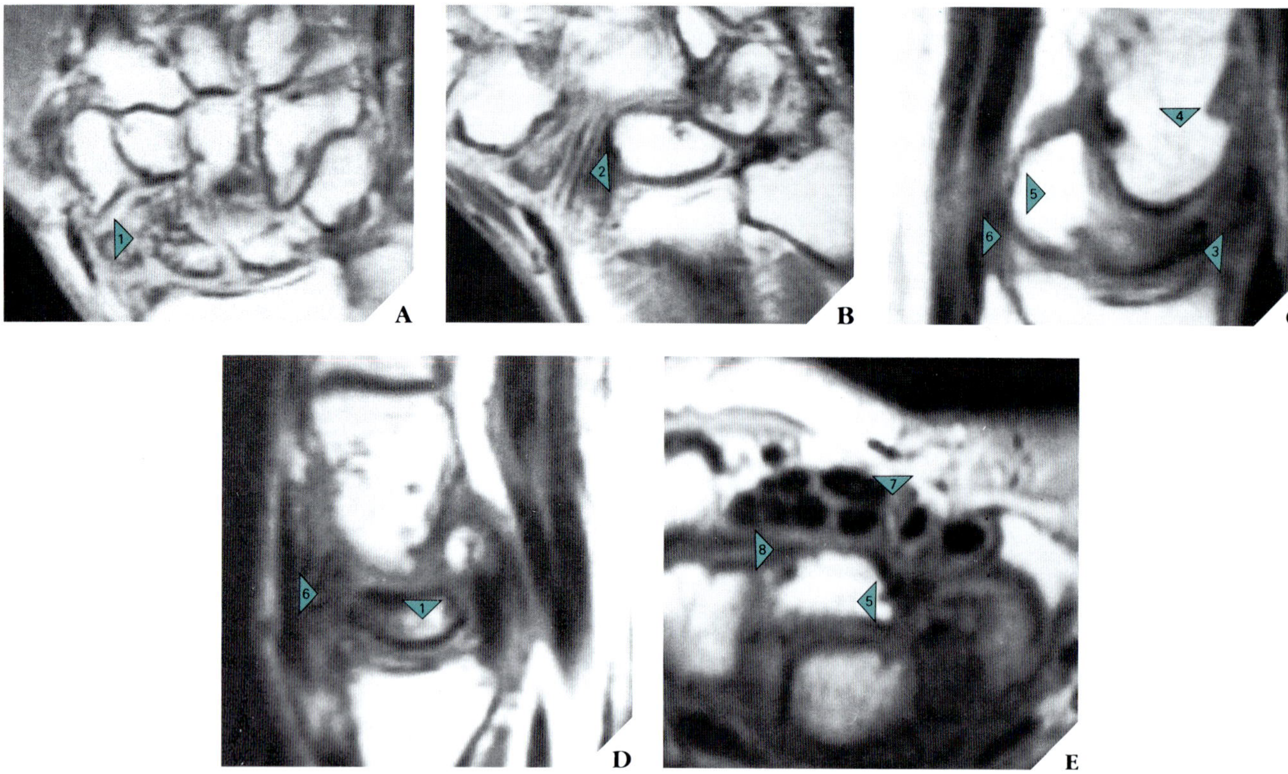

FIGURE 5.31. Nonunited Scaphoid Fracture with AVN and DISI Deformity of the Wrist. (A) Coronal image through the scaphoid and **(B)** coronal image just volar to **A** [SE 1500/20]. **(C)** Sagittal image through capitate lunate articulation and **(D)** sagittal image just radial to **C. (E)** Axial image at the level of the lunate [SE 500/20]. **(A)** Note the midwaist fracture of the navicular with abnormally low SI within the marrow of the scaphoid both distally and proximally, indicating AVN. The intact linear low SI radioscaphocapitate ligament is accentuated as it traverses the waist of the scaphoid **(B)** to attach on the volar capitate. The lunate shows dorsal rotation with proximal migration of the capitate. The volar and dorsal radioscapholunate ligaments appear incomplete with stretching and thickening of the dorsal ligaments (**C** and **D**). The axial image **(E)** shows volar subluxation of the lunate with stretching of the radiolunatotriquetral ligament. Arrows **1** Abnormal scaphoid. Arrow **2** Normal radioscaphocapitate ligament. Arrow **3** Stretched and thickened dorsal radioscapholunate ligament. Arrow **4** Capitate. Arrows **5** DISI of the lunate. Arrow **6** Incomplete volar radioscapholunate ligament. Arrow **7** Vertically oriented median nerve. Arrow **8** Stretched radiolunatotriquetral ligament.

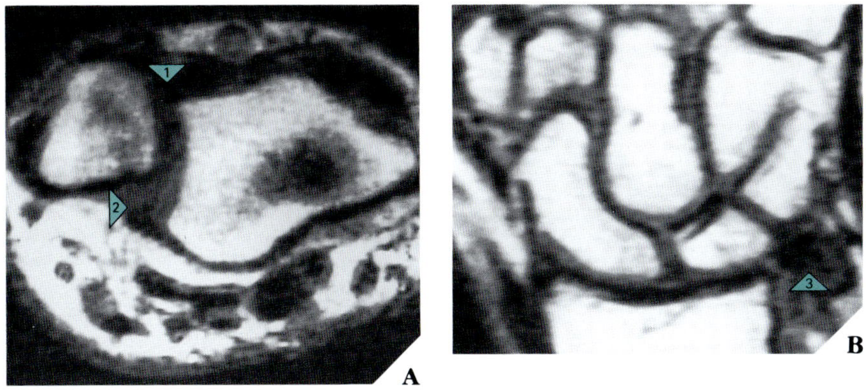

FIGURE 5.32. Posttraumatic Dorsal Subluxation of the Ulna with Interruption of the Volar Distal Radiocarpal Ligament and Triangular Fibrocartilage. (A) Axial plane [SE 800/25] through the distal radioulnar joint, and **(B)** coronal plane [SE 600/25] through the TFC. Note the continuity of the anterior distal radial ulnar ligament and the lack of visualization of the corresponding volar ligament. The distal ulna is subluxed dorsally relative to the sigmoid notch of the radius. The coronal section demonstrates the lack of continuity and the disruption of the triangular fibrocartilage. Arrow **1** Intact dorsal distal radioulnar ligament. Arrow **2** Disrupted volar distal radioulnar ligament. Arrow **3** Fragmented TFC.

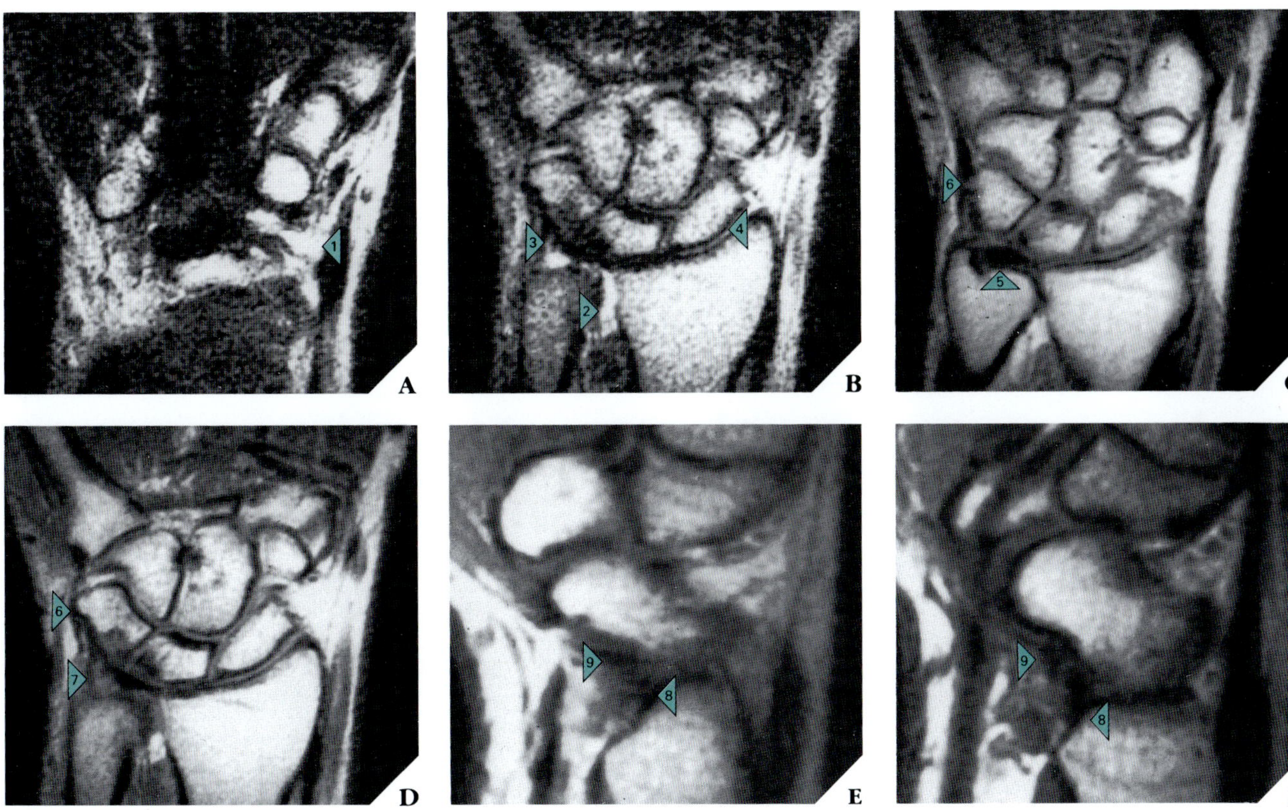

FIGURE 5.33. Posttraumatic Ligamentous Injury. (A,B) Coronal plane [SE 1500/70] through the volar aspect of the wrist, **(C,D)** coronal plane [SE 1500/20] through the dorsal aspect of the wrist, and **(E,F)** sagittal plane [SE 2000/20] through the radial aspect of the wrist. Note the radioulnar joint effusion seen in the coronal T2WI with a linear high SI extension into the TFC **(B).** Note, however, that the TFC appears intact in **C,** indicating that the tear of the TFC is incomplete. A small radiocarpal joint effusion is seen radially; the high SI within the soft tissues of the radial wrist also involves the first extensor compartment. The volar radial carpal ligament attachments appear intact **(D,E)**, as does the radial collateral ligament. Arrow **1** Edema/inflammation volar to radius. Arrow **2** Distal radioulnar joint effusion. Arrow **3** Incomplete tear of TFC. Arrow **4** Radiocarpal joint effusion. Arrow **5** Intact TFC more volarly. Arrows **6** Intact ulnar collateral ligament. Arrow **7** Intact meniscus homolog. Arrows **8** Origin of the radiolunatotriquetral ligament. Arrows **9** Radioscaphocapitate ligament at waist of the scaphoid.

As demonstrated, the TFCC is well defined on MRI as a low SI wedge-shaped structure at the ulnar aspect of the wrist. High SI change within the TFC on T2WI, or discontinuity of this structure, indicates partial or complete tear, respectively, and may correlate with pain at the ulnar side of the wrist (Figs. 5.32 and 5.33). Unfortunately, degenerative changes of the TFCC occur in patients over 50 years of age, producing abnormally shaped TFCC with high SI. Caution should be exercised before diagnosing TFCC tears in this population.

Pathology of the ligaments of the digits can also be assessed well with surface coil MRI (Fig. 5.34).

TENDINOUS LESIONS

Assessment of pathology of the tendons of the wrist, as with other soft tissues, has been limited to ultrasound (17) or computerized tomography. Shortcomings of ultrasound include limited anatomic detail, the ability to image only superficially located tendons, and the inability to image bone. MRI allows evaluation of both the osseous and soft tissue structures, particularly the tendons (7), and excellently depicts the pathologic tendinous anatomy of the wrist and hand. The limitations of computerized tomography have previously been mentioned.

Tendon Rupture

Only a few examples of tendon injuries of the hand as depicted by MRI have been described (7). Complete tendon ruptures are seen on MRI as a gap in the continuity of the tendon (Fig. 5.35). Incomplete ruptures or chronic tears are depicted by MRI as irregular thinning or thickening of the tendon, sometimes with linear areas of high SI on long TR/TE images, within the substance of the tendon (6).

While in most cases tendon ruptures can be diagnosed clinically, the size and location of the tear may not be as evident upon physical examination. MRI can accurately demonstrate the anatomy and help in planning appropriate surgical repair or reconstruction techniques, such as tendon grafting, tendon transfer, and tenolysis.

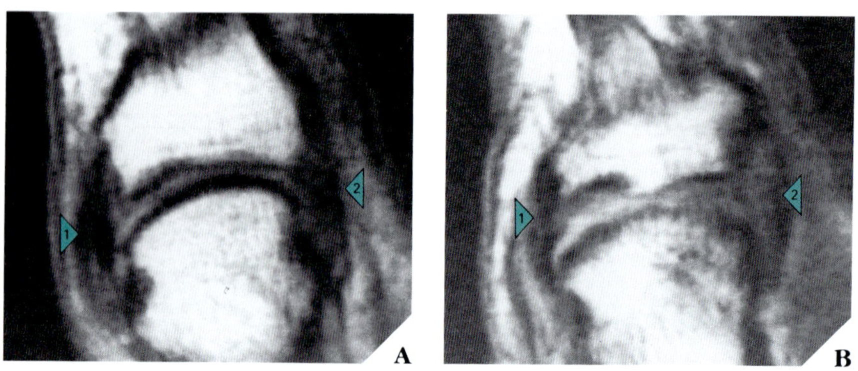

FIGURE 5.34. Normal Medial and Lateral Collateral Ligaments Following Trauma to the Thumb. Sagittal plane [SE 500/20] through the first metacarpal phalangeal joint. The clinical question arose as to whether the medial collateral ligament was ruptured following trauma to the thumb. MRI demonstrates intact medial and lateral collateral ligaments. Arrows **1** Medial collateral ligament. Arrows **2** Lateral collateral ligament.

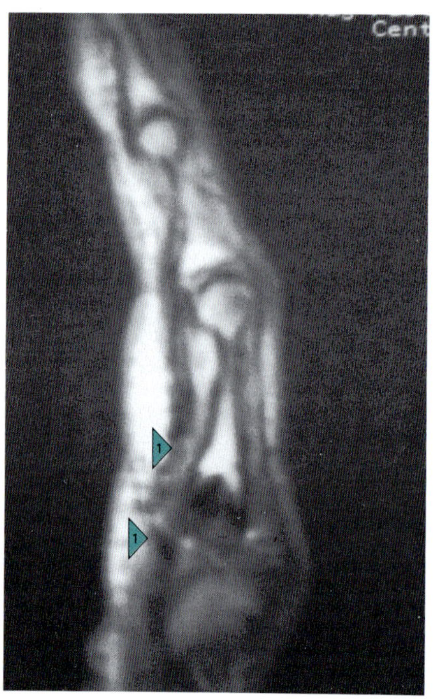

FIGURE 5.35. Rupture of the Flexor Tendon of the Index Finger Following Surgical Repair. Sagittal plane [SE 800/20]. Arrows **1** Segmented ends of reruptured flexor tendon.

Postsurgical Scar

Adhesions and scar tissue formation around the repaired tendon are frequent postoperative complications that lead to functional impairment. Differentiation between re-rupture and scar formation may be difficult on clinical grounds. In these cases, MRI shows irregularity of the tendon surface and low SI scar tissue around the tendon (Fig. 5.36).

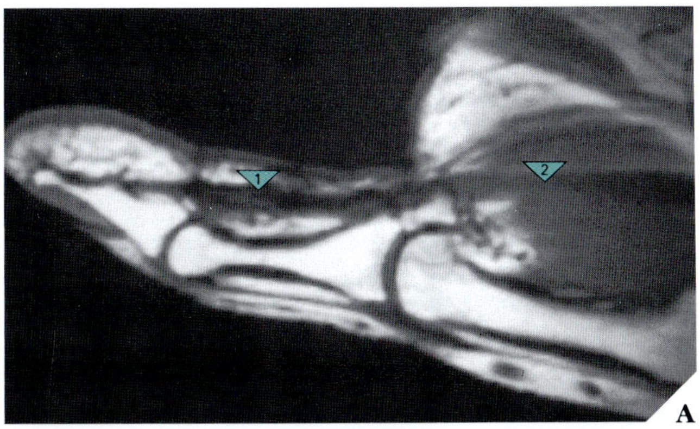

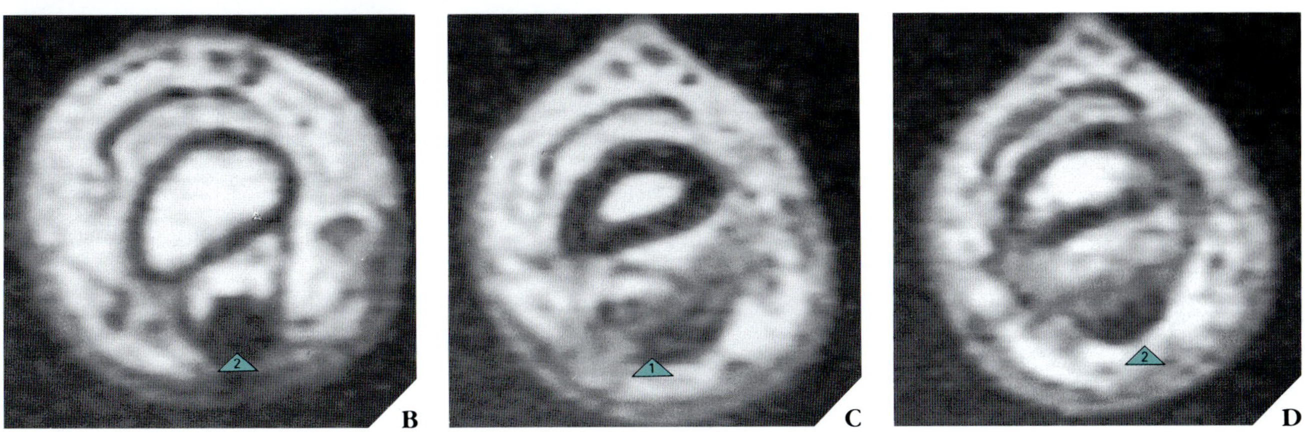

FIGURE 5.36. Postsurgical Scar Following Tendon Repair. (A) Sagittal plane and **(B–D)** axial planes [SE 800/20] through the thumb. There is contour irregularity of the flexor pollicis longus tendon. Axial section through the first metacarpal shaft **(B)** demonstrates normal flexor tendon. Axial section through the scar shows irregularity of the tendon **(C).** The more distal axial section **(D)** demonstrates normal distal flexor pollicis longus tendon. Arrows **1** Contour irregularity of flexor pollicis longus tendon. Arrows **2** Normal tendon.

Tenosynovitis

Most of the available literature dealing with inflammatory conditions of the tendons of the hands and wrist is based on ultrasound examinations. Rheumatoid arthritis, infection, systemic lupus erythematosus, diabetes mellitus, gout, and syphilis can all cause inflammatory changes in the tendons. Only a few cases of tenosynovitis, as seen by MRI, have been described (7,10). Acute tenosynovitis is seen on MRI as high SI fluid on long TR/TE images distending the synovial sheaths, either with normal appearing tendons or with high SI within the tendon substance (Figs. 5.37, 5.43, and 5.48). Chronic tenosynovitis is depicted by MRI as tendon and synovial sheath thickening with fluid in the synovial sheath (Fig. 5.38).

Entrapment syndromes with or without associated tenosynovitis are also accurately assessed with MRI. The diagnosis many times may be apparent clinically; however, MRI can be beneficial in defining the exact location of the entrapment for potential surgical treatment. Entrapment can involve virtually any tendon. When the first extensor tendon compartment is involved, it is called DeQuervain's syndrome.

COMPRESSION NEUROPATHIES

Carpal Tunnel Syndrome

Carpal tunnel syndrome is the most common compression neuropathy involving the upper extremity. There is a 2:1 female to male predominance with 50 percent of the patients being between 40 and 60 years of age (16). The usual symptoms include weakness or clumsiness in the hand with hyperesthesias and paresthesias, aggravated by grasping, in the distribution of the median nerve. Referred pain to the more proximal arm may also be seen. There are many associated conditions that involve the structures within or surrounding the carpal tunnel. Rheumatoid arthritis, thyroid imbalance (particularly myxedema), acromegaly, multiple myeloma, amyloidosis, diabetes mellitus, local trauma to the wrist, alcohol abuse, hemophilia, local tumors (particularly lipomata and ganglions), hormonal changes of menopause, pregnancy, and gout are all conditions that have been associated with CTS. Carpal tunnel syndrome can also be attributed to a variety of anatomic abnormalities, including aberrant vessels, such as a persistent medial artery which may be thrombosed, and aberrant muscles. The diagnosis is usually made with the wrist flexion test, percussion of the medial nerve at the level of the wrist, and objective sensory findings. Diagnosis can be confirmed by electromyographic studies.

Although the clinical diagnosis of carpal tunnel is not difficult, many times the etiology is not demonstrated with standard imaging techniques. Plain

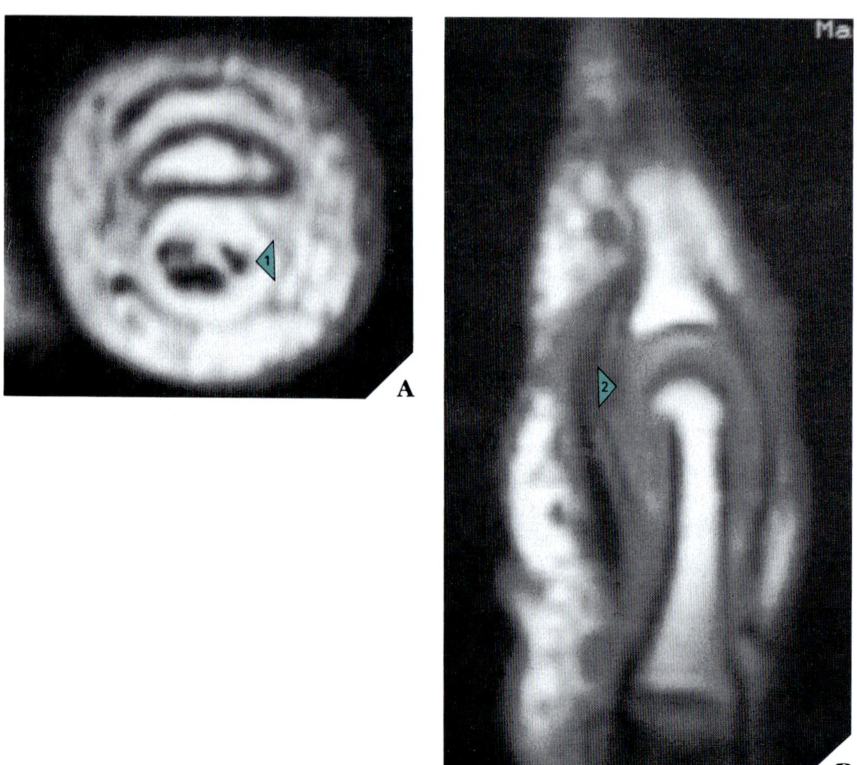

FIGURE 5.37. Acute Noninfectious Tenosynovitis of the Flexor Tendons of the Index Finger with Effusion of the Proximal Interphalangeal Joint. **(A)** Axial plane [SE 1500/60] through the base of the index finger and **(B)** sagittal plane (SE 600/20) through mid-proximal interphalangeal joint. The patient is 6 months old with index finger swelling and pain on flexion. The axial image demonstrates distention of the synovial sheath by fluid. Sagittal image shows distention of the proximal interphalangeal joint with fluid. Aspiration revealed a noninfectious etiology. Arrow **1** Flexor tenosynovitis. Arrow **2** Proximal interphalangeal joint effusion.

films and tomography, useful in evaluating the bony structures, often appear normal. High resolution computerized tomography has emerged, but the lack of soft tissue contrast resolution is a severe limitation (21,56). Ultrasound, helpful in demonstrating tenosynovitis, gives poor definition of the structures within the carpal tunnel as well as a limited FOV.

MRI has overcome these limitations. Its clarity in visualization of the normal and pathologic anatomy of the carpal tunnel has revealed the following in patients presenting with the carpal tunnel syndrome (23,37,54). The most common cause of carpal tunnel syndrome is tenosynovitis of the wrist flexor tendons. This may be seen as a separation of the tendons by thickened synovium and as in-

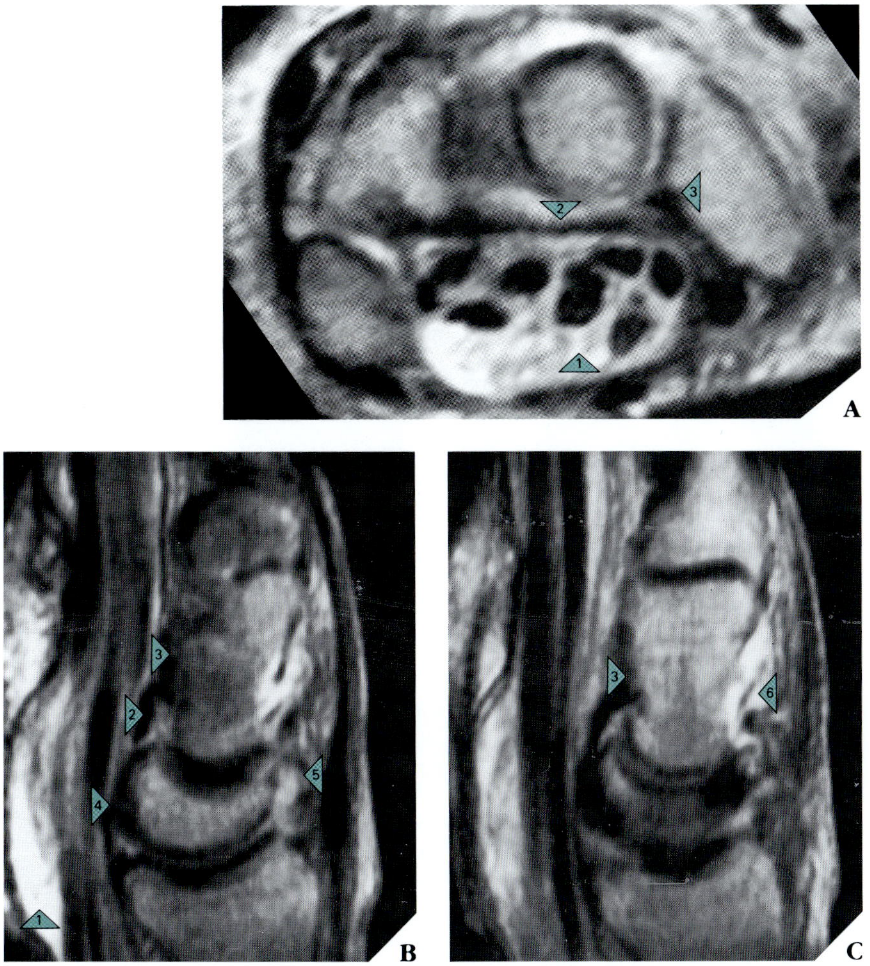

FIGURE 5.38. Chronic Noninfectious Flexor Tenosynovitis. (A) Axial plane [SE 2000/80] through the proximal carpal row and (**B** and **C**) sagittal planes [SE 2000/80] through the midwrist. The synovial sheaths surrounding the flexor tendons of the proximal carpal tunnel are distended with high SI on T2WI, indicating flexor tenosynovitis, which was noninfectious by aspiration. Note that the volar radiocarpal ligaments are abnormally thickened, possibly relating to the presence of chronic flexor tenosynovitis. The intact dorsal radioscapholunate ligament is slightly bowed in response to the slight dorsiflexion of the lunate. The radioscaphocapitate ligament is visualized clearly on the axial plane as it traverses the scaphoid waist; its attachment on the volar midcapitate is seen on the sagittal image. Arrows **1** Flexor tenosynovitis. Arrows **2** Thickened radiolunatotriquetral ligament. Arrows **3** Radioscaphocapitate ligament at the waist of the scaphoid. Arrow **4** Volar radioscapholunate ligament. Arrow **5** Dorsal radioscapholunate ligament. Arrow **6** Dorsal intercarpal ligament.

creased SI on T2WI due to higher water content (Fig. 5.39). The median nerve may be enlarged, either at the level of the flexor retinaculum, or proximal or distal to the flexor retinaculum. Reports have also noted increased SI within the medial nerve on the T2WI; volar bowing of the transverse carpal ligament; distortion of the medial nerve, such as flattening or angulation; and obliteration of the fat interposed between the flexor digitorum tendons and carpal bones (Figs. 5.39 and 5.40). Comparison with the opposite wrist may help assess median nerve size, although bilaterality of carpal tunnel syndrome is not uncommon.

Local masses, inflammatory change, and post-traumatic changes are well demonstrated on MRI. The exact role of MRI in carpal tunnel syndrome is not clearly identified at this point. However, it appears to be suitable not only for diagnosis but for presurgical planning and postoperative evaluation as well (Fig. 5.40). Important to surgeons is identification of the thenar branch of the median nerve, which reportedly shows four variations (15,26). Two of these may potentially be detected by MRI: an aberrant course, and a high division of the thenar branch, which can be separated by aberrant muscle or persistent median artery.

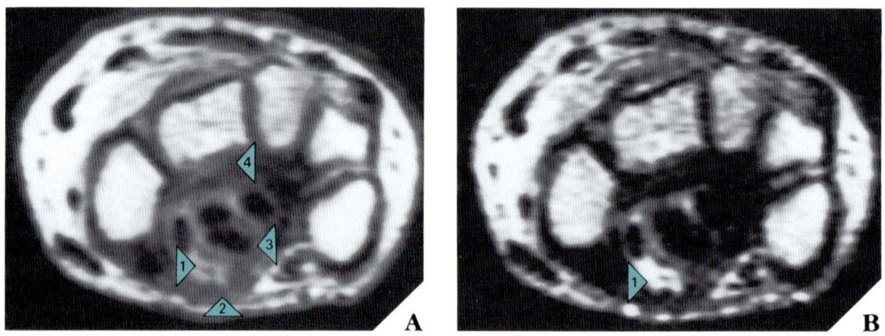

FIGURE 5.39. Carpal Tunnel Syndrome. Axial plane **(A)** [SE 2000/20] and **(B)** [SE 2000/80]. The patient was a 21-year-old female with carpal tunnel syndrome. X-ray films were normal. MRI demonstrates increased SI of the median nerve on T2WI, and thickened synovium surrounding the flexor tendons with increased volume of the carpal tunnel. Note the volar bowing of the flexor retinaculum, and the absent fat plane between the deep flexor tendons and the more dorsal radiolunatotriquetral ligament. Arrows **1** Median nerve. Arrow **2** Bowed flexor retinaculum. Arrow **3** Thickened synovium. Arrow **4** Absent fat plane.

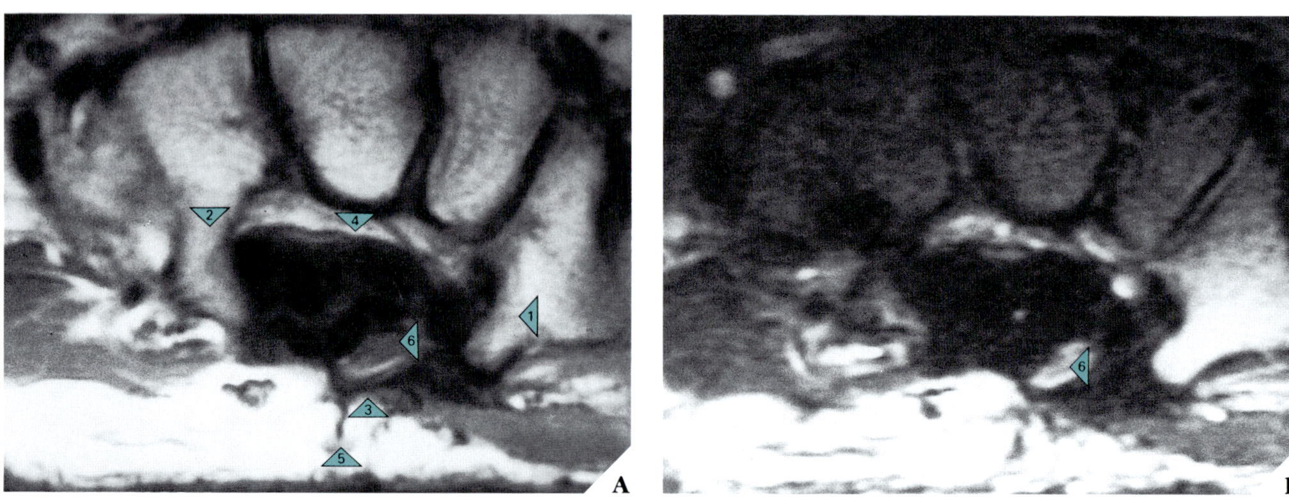

FIGURE 5.40. Carpal Tunnel Syndrome. Status: Post-carpal tunnel release. Axial plane **(A)** [SE 600/25] and **(B)** [SE 2000/80] through the distal carpal row at the level of the hook of the hamate. There is residual increased SI of the median nerve on T2WI **(B)** following carpal tunnel release. The volar midline scar is visible. The flexor retinaculum appears discontinuous; its wavy contour evolved following surgery. The contents of the carpal tunnel have expanded and migrated volarly, and normal fat is interposed between the carpal tunnel and carpal bones. Arrow **1** Tubercle of the trapezium. Arrow **2** Hook of the hamate. Arrow **3** Interrupted flexor retinaculum. Arrow **4** Dorsal fat with volarly displaced carpal tunnel. Arrow **5** Volar scar. Arrows **6** Median nerve.

Guyon's Canal

Another compression neuropathy of the wrist is ulnar nerve entrapment within Guyon's canal. The ulnar nerve may be entrapped within Guyon's canal before dividing into superficial and deep branches, or may become entrapped distal to this bifurcation. The common etiologies of this condition are occupational, an occult tumorous condition such as a ganglion or lipoma, or fracture of the hamate or base of the fourth and fifth metacarpal bones (16). As with the carpal tunnel syndrome, ulnar nerve entrapment is well assessed with MRI.

ARTHRITIS AND SYNOVIAL LESIONS

Rheumatoid Arthritis

Most of the known arthritides may at one point involve the hand and wrist. MRI findings in rheumatoid arthritis can be used as a prototype for all of the inflammatory arthropathies. Rheumatoid arthritis affects 0.5–1 percent of the population in the United States, and may involve 4.5 percent of persons over the age of 55 (35). The wrist is the initial site of rheumatoid arthritis synovitis in only 2.7 percent of patients, originating there six times less frequently than in the hand (46). However, in 95 percent of the cases of rheumatoid arthritis, the wrist eventually becomes involved (19).

The role of radiography is threefold: first, to diagnose, or confirm diagnosis, of a particular inflammatory arthropathy and define its extent; second, to monitor the response to therapy and document the lack of progression of the disease process; and third, to define the extent of bony and soft tissue involvement for possible operative intervention and preoperative planning.

The early histologic manifestations of rheumatoid arthritis principally involve soft tissues, including joint effusion, periarticular soft tissue edema, and hypertrophied synovium with early pannus formation. The disease must be advanced before small changes in bone and cartilage become apparent on plain films (25).

As the disease progresses, cartilage thinning with diminished joint space height, bone erosions, and subchondral cyst formation become more apparent. These changes are well demonstrated on plain films. However, determining the effectiveness of therapy may be difficult, because, once the disease has progressed, subtle body changes may not be apparent (5,54). Additionally, plain film does not permit direct visualization of soft tissue pathology. Computerized tomography has had no role in diagnosing rheumatoid arthritis or evaluating its response to therapy.

MRI improves radiographic evaluation of rheumatoid arthritis in all three aspects: diagnosis, soft tissue pathology, and therapy effectiveness. As discussed previously, the soft tissue resolution of MRI is superior to previous radiographic imaging modalities and, with the improved spatial resolution of surface coil technique, the early soft tissue changes of rheumatoid arthritis can be seen long before they would be apparent with plain film (5).

Early joint effusions are characterized by their long T1 and T2 relaxation parameters, with the SI on T2WI readily seen (8) (Fig. 5.41). Since synovial proliferation (pannus) is also characterized by long

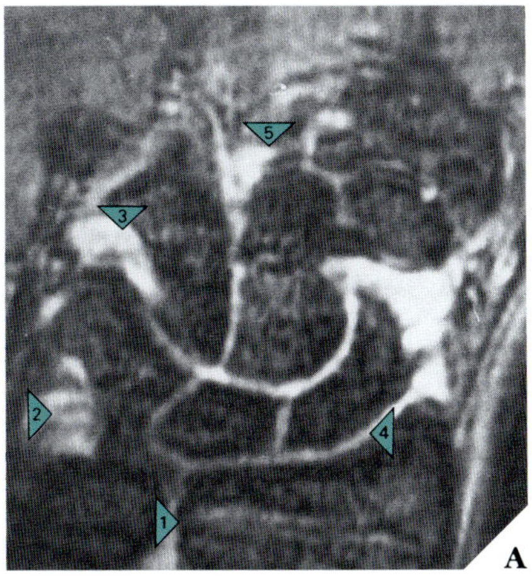

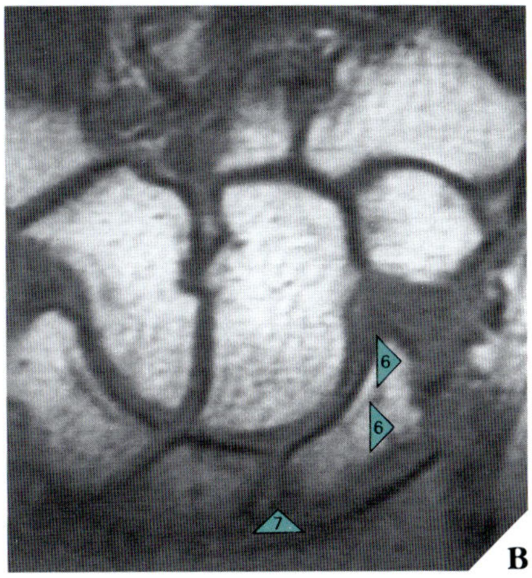

FIGURE 5.41. A,B. Rheumatoid Arthritis of the Wrist. (A) Coronal plane [GRASS 1000/20, 20-degree flip angle] and **(B)** [SE 1000/20]. Plain film examination was normal in this patient with known rheumatoid arthritis of the wrist. MRI demonstrates marginal scaphoid erosions with adjacent fluid and/or pannus. Joint fluid is also seen in the radioulnar, prestyloid recess, midcarpal, radiocarpal, and carpal metacarpal joints. Intact scapholunate ligaments are seen. Arrow **1** Distal radioulnar joint effusion. Arrow **2** Prestyloid recess effusion. Arrow **3** Midcarpal joint effusion. Arrow **4** Radiocarpal joint effusion. Arrow **5** Carpometacarpal joint effusion. Arrows **6** Scaphoid erosion with adjacent pannus. Arrow **7** Intact scapholunate ligament.

T2 relaxation with low SI on T1WI and high SI on T2W1 (Figs. 5.41–5.43), it may occasionally be difficult to distinguish from synovial fluid. It has been found, however, that longstanding pannus may demonstrate low signal on both T1WI and T2WI; this signal reflects hemosiderin deposition, which may be due to the leakage of blood in the hypertrophied synovium (5,38,54).

As with other joints, the articular cartilage and fibrocartilage are well demonstrated by MRI. The articular cartilage is of intermediate SI, whereas the fibrocartilage is of low SI on all imaging sequences. Thinning or destruction of the cartilage can also be demonstrated (38) (Fig. 5.42). On T2WI, bony erosions and adjacent high SI pannus become evident, offering a particular advantage in monitoring the response to therapy, for such subtle abnormalities may not be detected on plain films (Figs. 5.41–5.44).

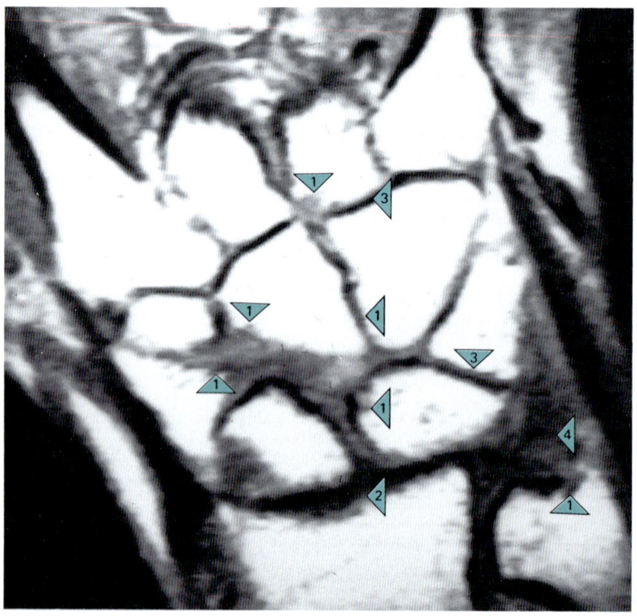

FIGURE 5.42. Rheumatoid Arthritis of the Wrist. Coronal section [SE 800/20]. The patient is a 40-year-old female with a 2-year history of rheumatoid arthritis. Plain film studies demonstrated osteopenia and a small bone erosion over the ulnar styloid process and base of the fourth metacarpal bone. MRI demonstrates multiple bone erosions, narrowing of the radiocarpal joint space, articular cartilage thinning, and destruction of the TFC. Arrows **1** Multiple marginal erosions. Arrow **2** Radiocarpal joint space narrowing. Arrows **3** Articular cartilage thinning. Arrow **4** Destruction of the TFC.

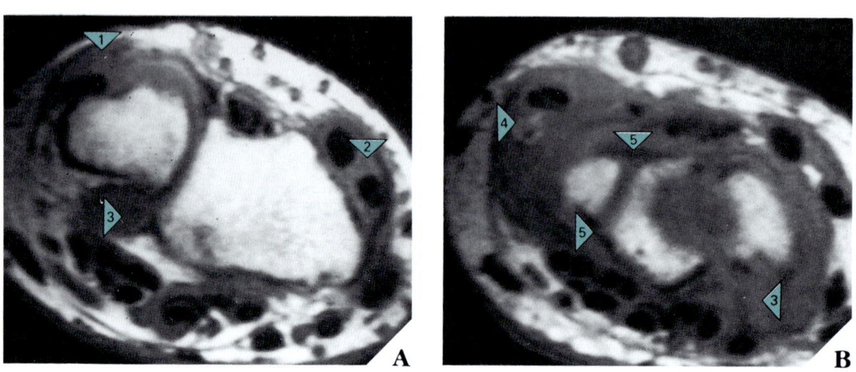

FIGURE 5.43. Rheumatoid Arthritis of the Wrist. Axial plane [SE 800/20] through **(A)** the distal radioulnar joint and **(B)** the radiocarpal joint. The intermediate SI on this T1WI reveals the distended synovial sheaths that involve the extensor carpi ulnaris ulnarly and the extensor carpi radialis longus and brevis tendons radially. Clinically, this manifests as the "hourglass" deformity at the dorsum of the wrist. Fluid is seen in the radioulnar and radiocarpal joints, and intact distal radioulnar ligaments are visible. Multiple erosions and adjacent pannus about the ulnar styloid are also appreciated. Arrow **1** Tenosynovitis of the extensor carpi ulnaris tendon. Arrow **2** Extensor carpi radialis longus and brevis tenosynvitis. Arrow **3** Joint effusion. Arrow **4** Erosions of the ulnar styloid. Arrows **5** Intact distal radioulnar ligaments.

Involvement of the tendons and ligaments of the hand and wrist by the inflammatory process of rheumatoid arthritis is well known. Prior to the advent of MRI, these derangements could only be inferred, for the imaging techniques in use did not permit direct visualization of the soft tissues. Tenosynovitis is a common manifestation of rheumatoid arthritis. It is clinically detected earlier in the extensor compartments, particularly in the ulnar three dorsal compartments where it manifests as the typical "hourglass" deformity (Fig. 5.43). Flexor tenosynovitis may not be as easy to identify as extensor tenosynovitis. However, it tends to be less well tolerated by the patient than inflammation of the extensor tendons, due to the contained space of the carpal tunnel (49). The manifestations of tenosynovitis, particularly distension of the synovial sheath with surrounding high SI fluid on T2WI, are easily discernible with MRI (Fig. 5.44).

Tendon ruptures or impending ruptures occur frequently with rheumatoid arthritis. Extensor tendor rupture of the digiti quinti minimi, although common, may be quite difficult to detect clinically (49). Studies have documented the ability of MRI to detect tendon ruptures (7). Mallet finger, swan neck deformity, and boutonniere deformity are all manifestations of tendinous abnormalities. MRI is useful for defining the extent of tendon abnormality and it can be employed in preoperative planning for procedures such as tendon transfers, relocations, and repairs.

In cases of longstanding rheumatoid arthritis, there is often deformity of the hand and wrist. Taleisnik describes carpal supination due to loss of the ulnar carpal complex. Loss of radial carpal height, with secondary palmar flexion of the scaphoid, occurs after palmar ligament support is destroyed. Grooving and notching at the waist of the scaphoid, where the radioscaphocapitate ligament crosses, occurs secondary to synovitis (Fig. 5.44). This may result in VISI deformity, with possible palmar dislocation of the carpus, and ulnar translocation following loss of the dorsal and palmar ligament support (49). An MRI assessment of ligamentous

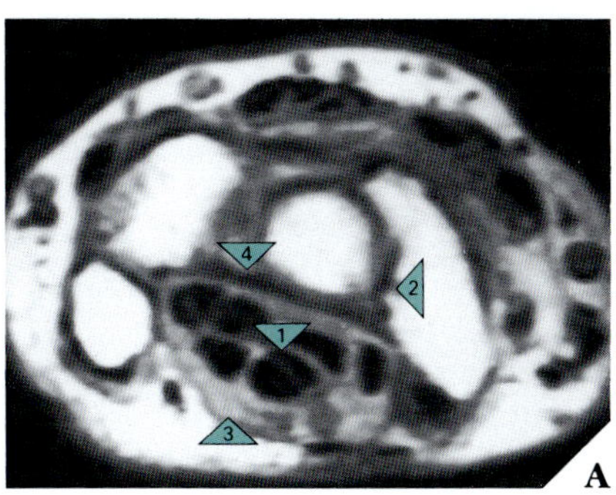

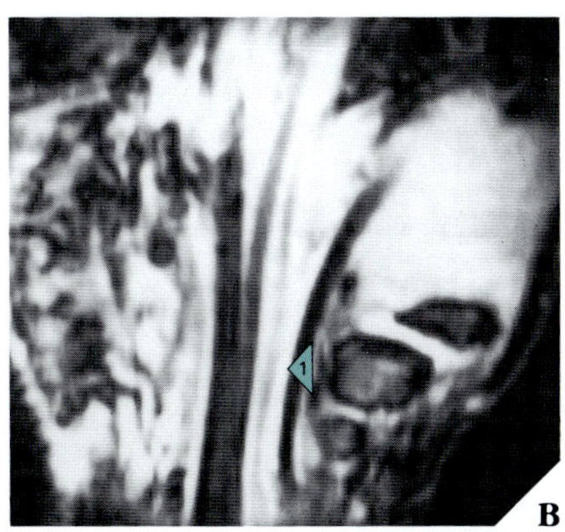

FIGURE 5.44. Rheumatoid Arthritis with Flexor Tenosynovitis and Scaphoid Erosions Adjacent to the Radioscaphocapitate Ligament. (A) Axial plane [SE 800/20] through the proximal carpal row and **(B)** coronal plane [GRASS 1000/12, 20-degree flip angle] through the carpal tunnel. Note the thickened synovium adjacent to the flexor tendons with corresponding high SI on the GRASS image as well as slight palmar bowing of the flexor retinaculum. Several small erosions can be identified at the waist of the scaphoid adjacent to the radioscaphocapitate ligament. Arrows **1** Flexor tendon with adjacent tenosynovitis. Arrow **2** Marginal carpal erosion at the waist of the scaphoid adjacent to the scaphocapitate ligament. Arrow **3** Bowed flexor retinaculum. Arrow **4** Radiolunatotriquetral ligament.

integrity may be useful for preoperative planning, including capsular or ligamentous repair and reconstruction (Fig. 5.45).

Miscellaneous Arthropathies

The MRI capabilities described above are also applied to other inflammatory arthropathies of the hand and wrist, including, for example, gout, psoriasis, and hemophiliac arthropathy. MRI can also demonstrate the changes in degenerative joint disease as well as calcium pyrophosphate deposition disease. Although the bony changes in these disease entities are depicted on plain films, MRI may be beneficial in demonstrating the degree of articular cartilage destruction, associated bursal inflammation, or mechanical impingement of the tendinous structures.

Ganglion Cyst

Ganglion and synovial cysts of the hand and wrist are included as synovial lesions. Infections and tumorous abnormalities of the hand and wrist are discussed in Chapters 10 and 11.

Ganglions compromise 50 to 70 percent of all soft tissue tumors of the hand. There is a 3:1 female predilection and 70 percent present between the second and fourth decades. They are usually single and in very specific locations, but have been re-

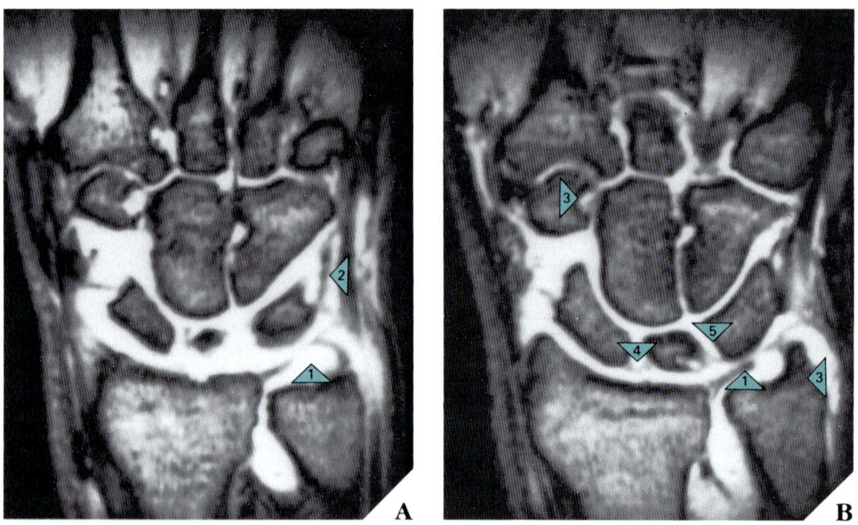

FIGURE 5.45. Rheumatoid Arthritis. Coronal plane [GRASS 1000/12, 20-degree flip angle] through the middorsal wrist. Note the excellent depiction of the interosseous ligaments, TFCC, and radial collateral ligament. This clarity is due to the positive contrast background accentuated by the GRASS technique and by the presence of joint fluid. Bone erosions and multiple compartment effusions are present. Arrows **1** TFCC. Arrow **2** Radial collateral ligament. Arrows **3** Bone erosions. Arrow **4** Scapholunate ligament. Arrow **5** Lunatotriquetral ligament.

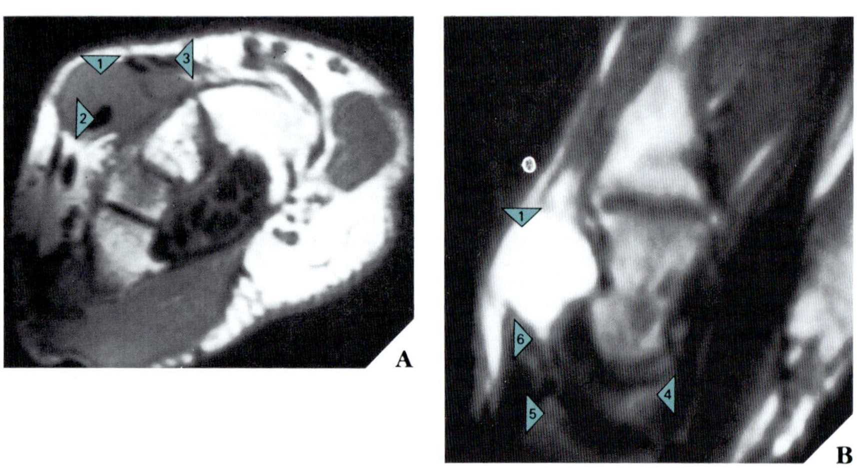

FIGURE 5.46. Ganglion Cyst. (A) Axial plane [SE 800/20] through the distal carpal row and **(B)** sagittal plane [SE 2000/80] through the midwrist. In the axial image **(A)**, note the intermediate SI ganglion at the dorsal aspect of the distal carpal bones. The mass elevates the extensor communis tendons and extensor carpi radialis brevis tendon; it extends between these tendon groups and partially surrounds the extensor carpi radialis brevis. The sagittal T2WI **(B)** demonstrates high SI, indicating the cystic nature of the mass. The typical tail of the cyst is seen inferiorly, originating at the dorsal aspect of the articulation of the lunate and radius. Arrows **1** Ganglion cyst. Arrow **2** Extensor carpi radialis brevis tendon. Arrow **3** Extensor communis tendon. Arrow **4** Lunate. Arrow **5** Radius. Arrow **6** Tail of ganglion cyst.

ported in almost every joint of the hand and wrist, usually attached to an adjacent underlying joint capsule, tendon, or tendon sheath. The presenting symptoms are generally a palpable mass, pain or weakness, and, in 10 percent of patients, a history of previous trauma. Ganglions have no malignant potential (2). Plain film radiographs are typically normal. Conservative management is poor, but they may regress or resolve spontaneously. Recurrence is rare with proper excision; however, over 50 percent will recur if incompletely excised (3).

The role of radiology in complete surgical removal of the ganglion is twofold: first, it aids in determining the correct diagnosis, and, second, it reveals the precise location and extent of the mass within the complex anatomy of the hand and wrist.

The full extent of ganglions is well demonstrated by MRI, owing to the SI changes reflecting their histology (54). They are usually single or multiloculated cystic structures that contain mucin, and they are often found not to have epithelial or synovial lining (2). Mucinous content gives the ganglion its long T1 and T2 values, which are manifested on MRI by low SI on T1WI and high SI T2WI. Ordinarily, the ganglion is clearly demarcated from adjacent nerves and tendons, and can be easily traced to its point of origin, sometimes with a typical tail (Fig. 5.46).

AVASCULAR NECROSIS AND TRAUMA

Wrist, hand, and finger injuries account for 38.7 percent of all occupational accidents treated in the emergency room. Fractures of the scaphoid are second only to fractures of the distal radius, and, overall, are the most common fractures seen in athletes and military personnel (27,49).

Fractures

Simple fractures are generally revealed by plain film radiography with clinical correlation. Computerized tomography is also suited to evaluate fractures, unless the fracture line is in the plane of section. Occult fractures are easily demonstrated on MRI as SI lines through the bony cortex or marrow of the respective bone (Fig. 5.47). Bone scintigraphy can demonstrate a positive scan, but it lacks specificity and spatial resolution.

Scaphoid nonunion is seen most frequently in young men in their second or third decades of life. It occurs commonly as a result in a delayed diagnosis, and in cases requiring immobilization in which activity was begun too soon. The sequelae of a scaphoid nonunion are significant, with advanced degenerative arthritis developing in virtually all cases within three decades (31,47). Consequently, early recognition and treatment is imperative, and MRI can be helpful when plain film or other radiographic imaging modalities are inconclusive.

Avascular Necrosis

Avascular necrosis (AVN) is typically manifest in the scaphoid and lunate bones of the wrist. It is a common sequela of a fracture through the mid or proximal pole of the scaphoid, which are the most frequent types of scaphoid fractures. In many individuals, it is due to a more distal entrance of the blood supply in the scaphoid bone (40). In promptly treated fractures, AVN is potentially a reversible condition (45).

Plain film findings of AVN lag behind the onset of the disease. MRI has been reported to be as sensitive as bone scintigraphy, with the benefit of increased specificity (44,54). The increased sensitivity and specificity are due to the high SI yellow

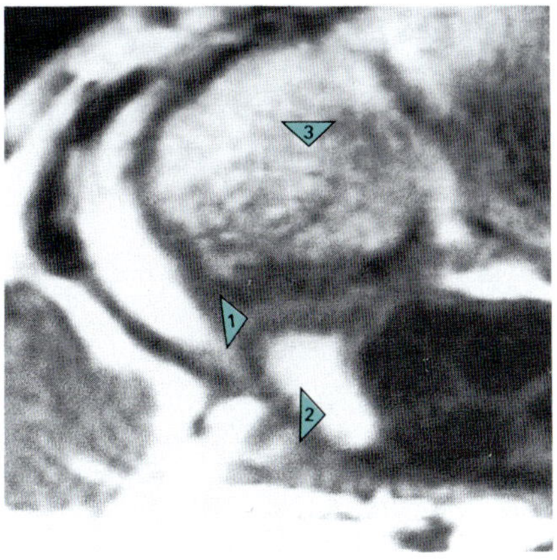

FIGURE 5.47. Hook of the Hamate Fracture. Axial plane [SE 2500/25] through the hook of the hamate. The fracture line through the base of the hook of the hamate is well visualized. The high SI from the bone marrow of the hook of the hamate, accentuated by proximity to the surface coil, indicates viable bone. Arrows **1** Fracture through the base of the hook of the hamate. Arrow **2** Hook of the hamate. Arrow **3** Hamate.

marrow within the carpal bones of T1WI. With AVN, this high SI is lost and abnormal SI is seen in the avascular areas (Figs. 5.31 and 5.48).

Kienböck's Disease

The relatively rare Kienböck's disease (lunatomalacia) represents avascular necrosis of the lunate and usually occurs in 20- to 40-year-old patients with a 2:1 predominance of males to females. Typical presenting symptoms are pain and stiffness in the wrist. In its early stages, the disease is radiographically and clinically indistinguishable from other causes of wrist pain (55). The etiology is uncertain. Hulten noted negative ulnar variance in 78 percent of his patients, corresponding to 23 percent of the normal population. This possibly predisposes the lunate to shear forces with subsequent development of AVN (20,29).

Based on plain films, this entity can be classified as Stages I–IV (14). In early Kienböck's disease X-ray films are often normal. This is important, in that the condition, if identified early, could be reversed simply by immobilization of the involved wrist (1). As the disease progresses through Stages III and IV, there is increased density, with and without fractures, and subsequent collapse of the lunate with fragmentation and resorptive changes. When this occurs, the capitate migrates proximally with wrist shortening, predisposing to rotatory subluxation of the scaphoid with scapholunate dislocation. This may cause carpal tunnel syndrome (23).

MRI is useful not only in detecting the potentially reversible Stage I Kienböck's disease, but also in determining the degree of involvement, since there are numerous treatment options depending on the extent of the process (Fig. 5.49).

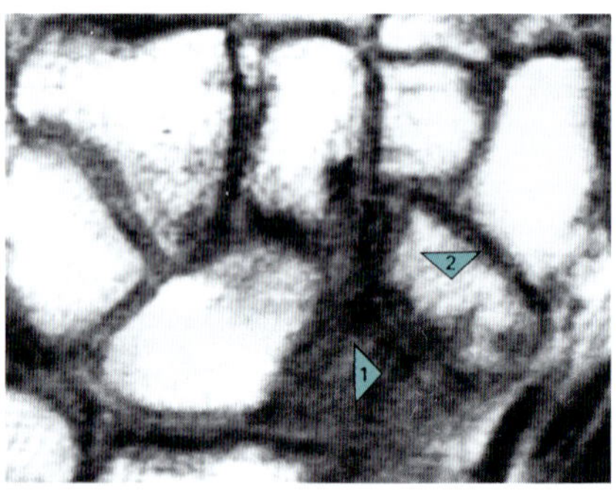

FIGURE 5.48. Avascular Necrosis of Scaphoid. Coronal plane [SE 2000/25] through the scaphoid. The distal viable pole of the scaphoid shows normal high SI in the bone marrow as compared with the low SI from the proximal half of the scaphoid, indicating avascular necrosis. Arrow **1** AVN. Arrow **2** Normal distal pole of scaphoid.

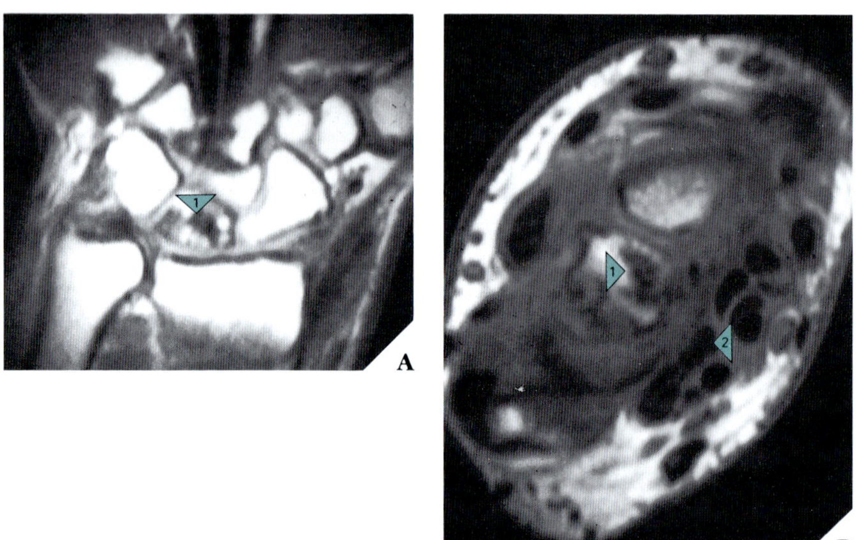

A

B

FIGURE 5.49. Kienböck's Disease. (A) Coronal plane through the carpal bones [SE 800/20] and **(B)** axial plane [SE 800/20] through the lunate bone. This coronal T1WI demonstrates low SI changes within the bone marrow of the lunate as compared with the normal high SI bone marrow of adjacent carpal bones. The height of the lunate is preserved; however, there is volar subluxation and dorsiflexion of the lunate with subsequent mild wrist shortening and compression of the deep flexor tendons as seen on the axial image **(B).** There is distension of the synovial sheaths of the flexor tendons. Arrows **1** Lunate AVN. Arrow **2** Flexor tendons compression and tenosynovitis.

REFERENCES

1. Alexander HA, Lichtman DM: Kienböck's disease. In: Lichtman DM, ed. The wrist and its disorders. Philadelphia, WB Saunders, 1988.
2. Angelides A: Ganglions of the hand and wrist. In: Green DP, ed. Operative Hand Surgery, 2nd ed. New York, Churchill Livingstone, 1988.
3. Angelides AC, Wallace P: The dorsal ganglion of the wrist: its pathogenesis, gross and microscopic anatomy, and surgical treatment. Hand Surg, 1976;1:228–235.
4. Baker, LL, Hajek PC, Bjorkengren MD, et al: High-resolution magnetic resonance imaging of the wrist: normal anatomy. Skeletal Radiol 1987;6:128–132.
5. Beltran J, Caudill J, Herman LS, et al: Rheumatoid arthritis: MR imaging manifestations. Radiology 1987;165:153–157.
6. Beltran J, Mosure J: High field surface coil MRI of the tendons. Crit Rev Radiol (in press), 1989.
7. Beltran J, Noto AM, Herman LS, et al: Tendons: high field strength, surface coil MR imaging. Radiology 1967;162:735–740.
8. Beltran J, Noto AM, Herman LS, et al: Joint effusions: MR imaging. Radiology 1986;158: 133–137.
9. Beltran J, Noto AM, Mosure JC, et al: The knee: surface coil MR imaging at 1.5 T. Radiology 1986;159:747–751.
10. Beltran J, Noto AM, Mosure JC, et al: Ankle: surface coil MR imaging at 1.5 T. Radiology 1986;161:203–209.
11. Blott G: Scapholunate instability. In: Lichtman DM, ed. The Wrist and Its Disorders. Philadelphia, WB Saunders, 1988.
12. Bogumill GP: Anatomy of the wrist. In: Lichtman DM, ed. The Wrist and Its Disorders, Philadelphia, WB Saunders, 1988.
13. Brown DE, Lichtman DM: Midcarpal instability. In: Taleisnik J, ed. Management of Wrist Problems. Hand Clin 1987; III:1–184.
14. Decoulx P, March M, Minet P, et al: La maladie de Kienböck chez le minieur, etude clinique et pathogenique (avec de 1330 radios du poignet). Lille Chir 1957;2:65.
15. Eiken O, Carstram N, Eddeland A: Anomalous distal branching of the medial nerve. Scand J Plast Reconstr Surg 1971;5:149–152.
16. Eversman WW: Entrapment and compression neuropathies. In: Green DP, ed. Operative Hand Surgery, 2nd ed. New York, Churchill Livingstone, 1988.
17. Fornage BD, Rifkin MD: Ultrasound examination of the hand and foot. Radiol Clin North Am 1988;26:109.
18. Gilford W, Boltan R, Lambrinudi C: The mechanism of the wrist joint. Guy's Hospital Report 1943;92:52–59.
19. Hooper J: The surgery of the wrist in rheumatoid arthritis. Aust N Z J Surg 1972;42:135.
20. Hulten O: Uber anatomische variationen der Handgelenkknochen. Acta Radiol Scand 1928; 9:155.
21. John V, Nan HE, Nahsen HK, et al: CT of carpal tunnel syndrome. AJNR 1983;4:770–772.
22. Kauer JMG: The articular disc of the hand. Acta Anat 1975;93:590.
23. Kellerhouse, LE: Magnetic resonance imaging of the wrist. Presented at the first meeting on Issues in Musculoskeletal Imaging. Newport Beach, California, March 1989.
24. Koenig H, Dietrich L, Meissner R: The wrist: a preliminary report on high resolution MR imaging. Radiology 1986;160:463–476.
25. Lannuzzi L, Dawson N, Zeirn, et al: Does drug therapy slow radiographic deterioration in rheumatoid arthritis? N Engl J Med 1983; 309:1023–1028.
26. Lanzu U: Anatomical variations of the median nerve in the carpal tunnel: J Hand Surg 1977;2:44–53.
27. Leads from the MMWR: Occupational finger injuries: US JAMA 1983;250(22):3029.
28. Lichtman DM, Martin RA: Introduction to the carpal instabilities. In: Lichtman DM, ed. The Wrist and Its Disorders. Philadelphia, WB Saunders, 1988.
29. Linscheid R: Ulnar lengthening and shortening. In: Taleisnik J, ed. Management of Wrist Problems. Hand Clinics Feb 1987;III:69–79.
30. Linscheid R, Dobyns J, Beabout J, et al: Traumatic instability of the wrist. J Bone Joint Surg 1972;54A:1612.
31. Mack GR, Lichtman DM: Scaphoid nonunion. In: Lichtman DM, ed. The Wrist and Its Disorders. Philadelphia, WB Saunders, 1988.
32. Mayfield JK: Pathogenesis of wrist ligament instability. In: Lichtman DM, ed. The Wrist and Its Disorders. Philadelphia, WB Saunders, 1988.
33. Mayfield JK, Johnson RP, Kilcoyne RF: The ligaments of the human wrist and their functional significance. Anat Rec 1976;186:412–428.

34. Mayfield JK, Johnson RP, Kilcoyne RF: Carpal dislocations: pathomechanics and progressive perilunar instability. J Hand Surg 1980; 5:226–241.
35. McDuffie FC: Morbidity impact of rheumatoid arthritis on society: Am J Med 1985;78:1–5.
36. Mesgarzedeh M, Schenck CD, Bonakdarpour A: MR imaging of the wrist and carpal tunnel: correlation with normal anatomy. Radiology 1985;160:143–146 (abstr).
37. Middleton WD, Kneeland JB, Kellman GM, et al: MR imaging of the carpal tunnel: normal anatomy and preliminary findings in the carpal tunnel syndrome. AJR 1987;148:307–316.
38. Muirden KA, Sanator GB: Iron in the synovial membrane in rheumatoid arthritis and other joint diseases. Ann Rheum Dis 1968;27:38–48.
39. Navarro A: 91 Movimiento de rotacion en el carpo y su relacion con la fisiopatologia de sus lesiones tramaticas. Bol Trabajos Soc Argentina, Ortoped Traumatol 1969;34:337.
40. Obletz BE, Halbstein BM: Non-union of fractures of the carpal navicular. J Bone Joint Surg 1938;20:424–428.
41. Palmer AK: The distal radioulnar joint; anatomy biomechanics, triangular fibrocartilage complex abnormalities. In: Taleisnik J, ed. Management of Wrist Problems. Hand Clinics 1987; III:1–184.
42. Palmer AK, Werner FW: The triangular fibrocartilage complex of the wrist: anatomy and function. J Hand Surg 1981;6:153.
43. Reinig JW, Dorwant RH, Roden WC: MR imaging of a ruptured achilles tendon: case report. J Comput Assist Tomogr 1985;9:1131.
44. Reinus WR, et al: Carpal avascular necrosis: MR imaging. Radiology 1986;160:689.
45. Russe O: Fracture of the carpal navicular, diagnosis, nonoperative treatment, and operative treatment. J Bone Joint Surg 1960;42A: 759–768.
46. Short CL, Bauer E, Reynolds WE: Rheumatoid arthritis. A definition of the disease and a clinical description based on a numerical study of 293 patients and controls. Harvard University Press, Cambridge, MA, 1957.
47. Taleisnik J: Post-traumatic carpal instability. Clin Orthop 1980;149:73.
48. Taleisnik J: Scapholunate dissociation. In: Strickland JW, Stechen JB, eds. Difficult Problems in Hand Surgery. St. Louis, Mosby, 1982.
49. Taleisnik J: The Wrist. New York, Churchill Livingstone, 1985.
50. Taleisnik J: Pain on the ulnar side of the wrist. In: Taleisnik J, ed. Management of Wrist Problems. Hand Clinics, 1987;III:51–68.
51. Testut L, Latarjet A: Tratado de anatomia Humana, Vol I, 9th ed (revised). Buenos Aires, Salvat Editores, 1951.
52. Tyrrell RL, Gluckert K, Patharia M, et al: Three dimensional MR imaging of the knee: comparison with arthroscopy. Radiology 1988; 166:865–872.
53. Watson KH, Black DM: Instabilities of the wrist: In: Taleisnik J, ed. Management of Wrist Problems, Hand Clinics 1987;III:103–111.
54. Weiss KL, Beltran J, Lubbers LM: Highfield MR surface coil imaging of the hand and wrist. Part II: Pathologic correlations and clinical relevance. Radiology 1986;160:147–152.
55. Weiss KL, Beltran J, Shanan OM, et al: Highfield MR surface coil imaging of the hand and wrist. Part I: Normal anatomy. Radiology 1986; 160:143–146.
56. Zucker-Pinchoff B, Herman G, Srinivasan R: Computed tomography of the carpal tunnel: a radioanatomical study. J Comput Assist Tomogr 1982;5:525–528.

FIGURE CREDITS

Figure 5.17. Adapted from Taleisnik J: The Wrist. New York Churchill Livingstone, 1985.

Figure 5.18. Adapted from Obletz BE, Halbstein BM: Non-union of fractures of the carpal navicular. J Bone Joint Surg 1983;20:424–428.

Figure 5.19. From Weiss KL, Beltran J, Lubbers LM: Highfield MR surface coil imaging of the hand and wrist. Part I: Normal anatomy. Radiology 1986;160:143–146.

Figures 5.27, 5.35–5.37. From Beltran J, Noto AM, Herman LS, et al: Tendons: high field strength, surface coil MR imaging. Radiology 1967;162:735–740.

Figure 5.42. From Beltran J, Caudill J, Herman LS, et al: Rheumatoid arthritis: MR imaging manifestations. Radiology 1987;165:153–157.

Figure 5.48. From Weiss KL, Beltran J, Lubbers LM: Highfield MR surface coil imaging of the hand and wrist. Part II: Pathologic correlations and clinical relevance. Radiology 1986;160:147–152.

CHAPTER SIX

Javier Beltran

The Pelvis and Hip

TECHNICAL CONSIDERATIONS

IMAGING PLANES AND PULSE SEQUENCES

There are different approaches to pelvic and hip imaging, depending on the type of abnormality to be studied. Axial sections using short TR and short TE pulse sequences (T1WI), followed by T2- and PD-weighted coronal sections through both hips, with long TR and short and long TE, are adequate for routine imaging of the hip joints in most cases. This technique can be tailored to specific clinical situations (11). To image pelvic and sacral lesions, in addition to sagittal sections, the coronal sections can be expanded to cover the entire pelvis in the anterior-posterior dimension. Patients with suspected stress fractures of the femoral neck can be rapidly evaluated using only T1-weighted coronal sections (12). To assess the extent of cartilage or femoral head damage in patients with avascular necrosis or inflammatory joint conditions, sagittal sections may be helpful using PDWI and T2WI (35). Short inversion time pulse sequences (STIR) are useful in assessing bone marrow lesions.

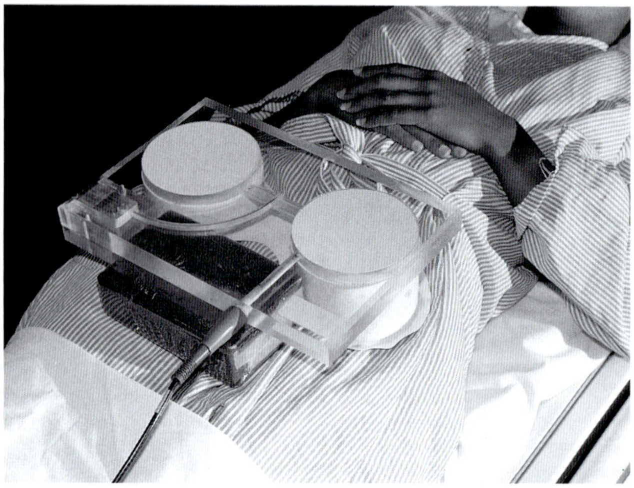

FIGURE 6.1. **Dual Surface Coils.** Butterfly surface coils in place for bilateral hip imaging.

FIELD OF VIEW

A large FOV of 32 to 40 cm can be used for routine imaging of both hips including the pelvis and proximal femora. Smaller FOV (16 cm) with surface coil imaging can be used in selected cases to improve spatial resolution.

ACQUISITION MATRIX AND SLICE THICKNESS

A matrix of 128 × 256 provides good signal-to-noise and contrast-to-noise ratios within an acceptable acquisition time. For most cases, 5-mm slice thickness with 1-mm interslice gaps is adequate.

▼ ▼ ▼ *SURFACE COILS*

Most pelvic and hip studies can be obtained using the body coil. Single or dual (butterfly) surface coils can be used in specific clinical circumstances (Fig. 6.1) (36).

NORMAL ANATOMY

The complex normal anatomy of the pelvis and hips is well depicted by MRI. Muscles show intermediate-to-low SI in all pulse sequences. Tendons, ligaments, acetabular labrum, and cortical bone show practically no signal on any pulse sequence. Articular cartilage is seen as intermediate-to-high SI compared with the muscle on T1WI, with a slight decrease in SI on T2WI. Vascular structures can be seen as low SI (flow void phenomenon) or high SI, depending on the velocity and direction of the flow. Pulsating vessels produce a characteristic artifact in the direction of the phase-encoding gradient (31).

Fat planes and fatty bone marrow can be easily identified because of the high SI on T1WI and low SI on T2WI. Red marrow demonstrates intermediate-to-low SI, depending on its cellularity and protein content. The relative distribution of the fatty (yellow) marrow and hematopoietic (red) marrow depends on the age of the individual (Fig. 6.2) (41). At birth, red marrow occupies almost the entire bone marrow space in the skeleton, with the exception of the nonossified epiphyses and apophyses. Fatty infiltration of the red marrow begins shortly after birth in the small bones of the hands and feet, progressing proximally. In the long bones, fatty infiltration begins in the epiphyses, apophyses, and middiaphyseal regions, progressing proximally and distally. By the age of 15–18 years, most of the bone marrow space of the diaphyses of the long bones is occupied by yellow marrow. After the age of 25–30 years, only a small amount of red marrow is present in the proximal metaphyses of the femora and humeri (see Fig. 6.9). Parts of the skull, ster-

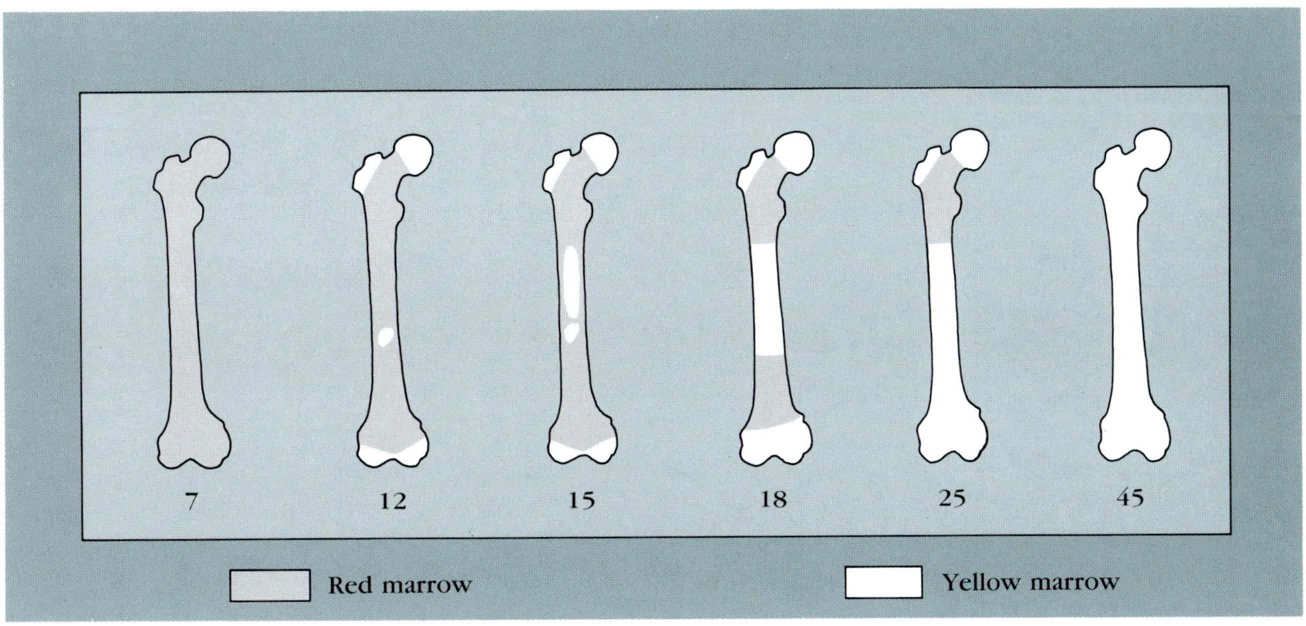

FIGURE 6.2. Red–Yellow Marrow Changes with Age. Schematic drawing of the progressive replacement of red bone marrow by fatty marrow in the long bones. Ages (in years) appear beneath each image.

num, spine, ribs, and pelvis show the presence of hematopoietic bone marrow well into adult life, with some degree of fatty infiltration (20,41).

Depending on the amount of fat, the residual islands of red marrow in the spine, pelvis, and proximal femora may demonstrate an inhomogeneous pattern on MRI, not to be confused with tumoral infiltration (Fig. 6.3).

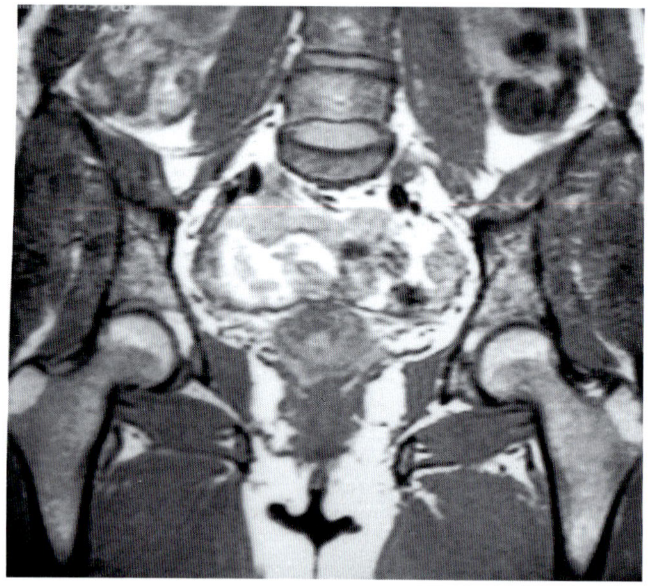

FIGURE 6.3. Normal Pelvis. Coronal section [SE 800/20] through the hip joints in a 32-year-old female. There is low SI hematopoietic marrow in the femoral necks, pelvic bones, and lower lumbar spine. The irregular pattern of the pelvis is due to inhomogeneous fatty replacement. There is high SI fatty marrow in the femoral epiphyses and both greater trochanters. A small amount of fat is present in both acetabular fossae.

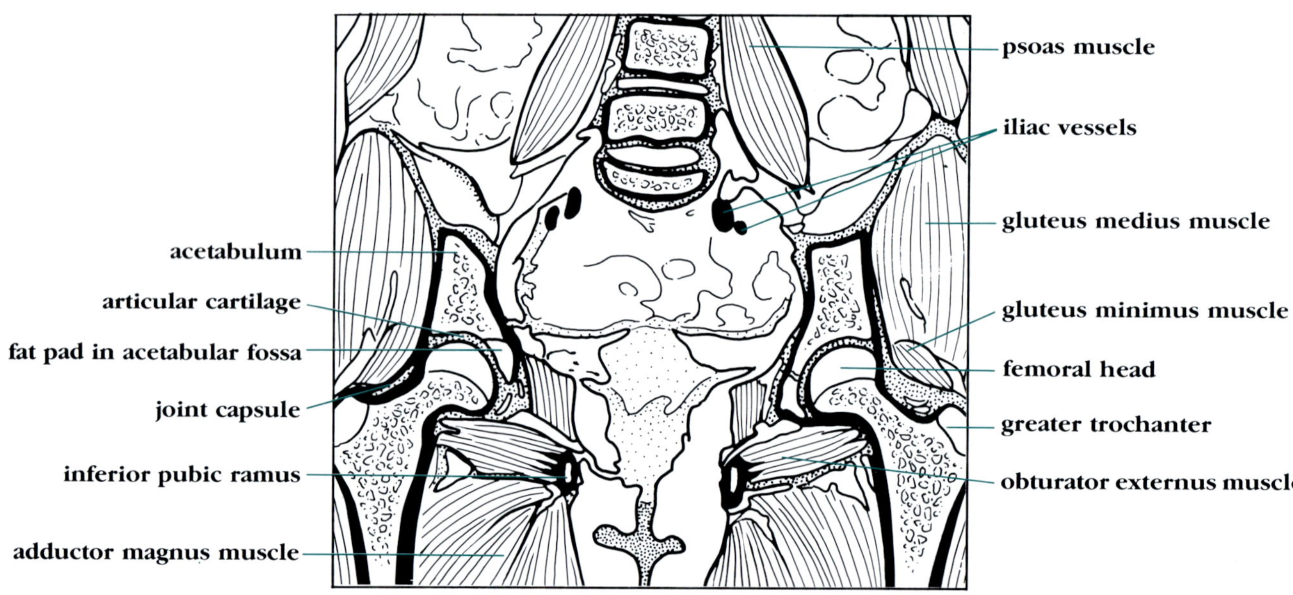

▼▼▼ *AXIAL PLANE*

Axial sections are useful for evaluating the sacroiliac joints (Fig. 6.4), pelvic and gluteal muscles (Figs. 6.5, 6.6), and proximal femora, including hip joints (Fig. 6.7).

▼▼▼ *CORONAL PLANE*

Posterior coronal sections demonstrate the sacroiliac joints and the hamstring tendons inserting in the ischial tuberosity (Fig. 6.8). The hamstring muscles contribute to the flexion of the leg and

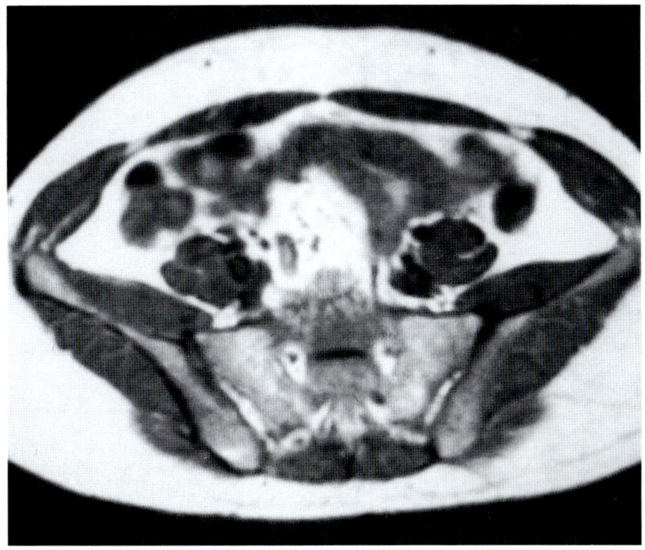

FIGURE 6.4. Normal Pelvis. Axial section [SE 800/20]. The sacroiliac joints are well depicted in this section.

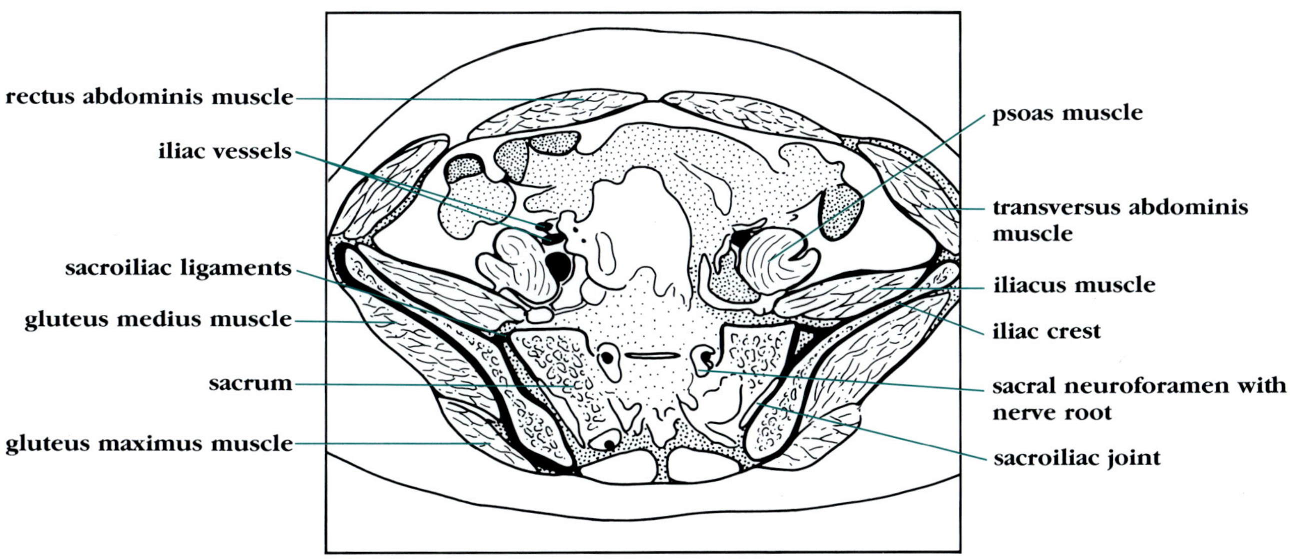

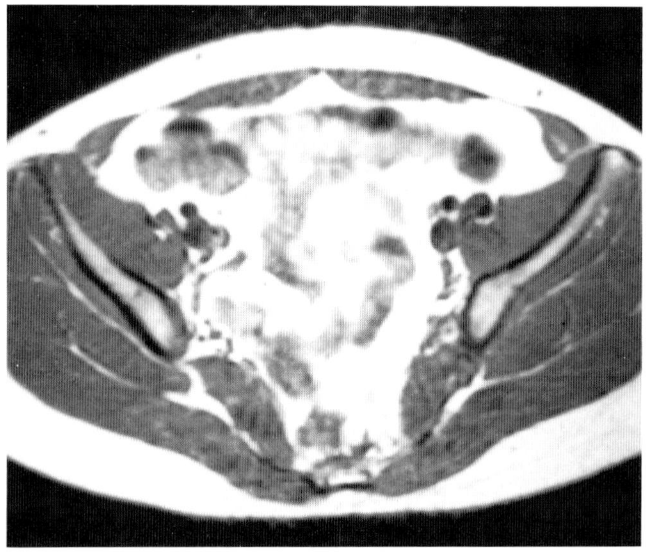

FIGURE 6.5. Normal Pelvis. Axial section [SE 800/20] through iliac wings. The glutei, piriformis, and iliopsoas muscles are displayed in this section.

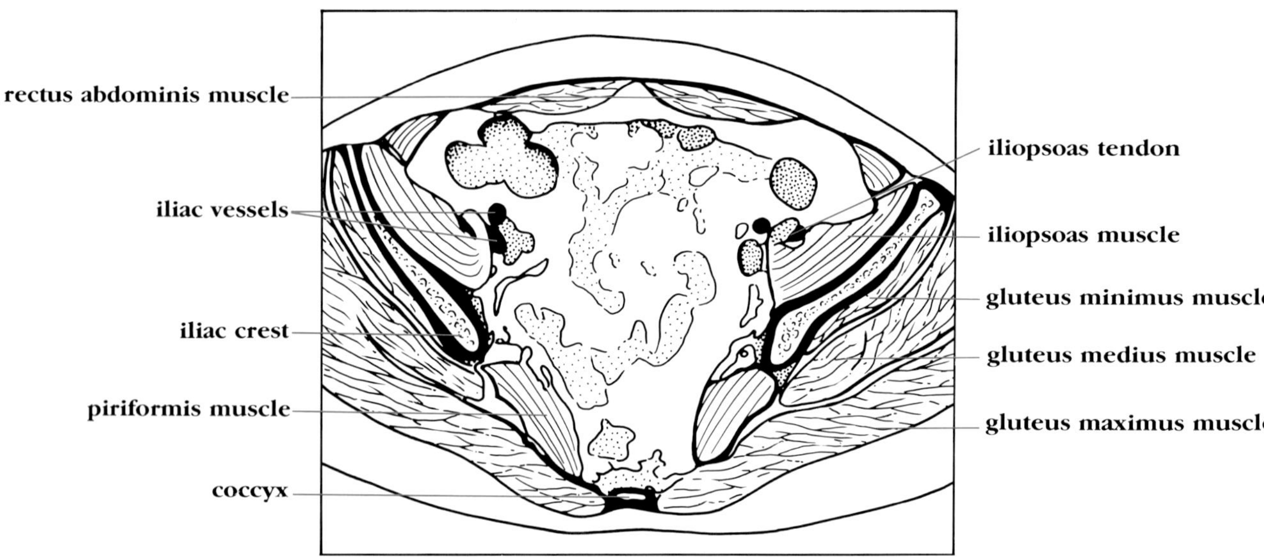

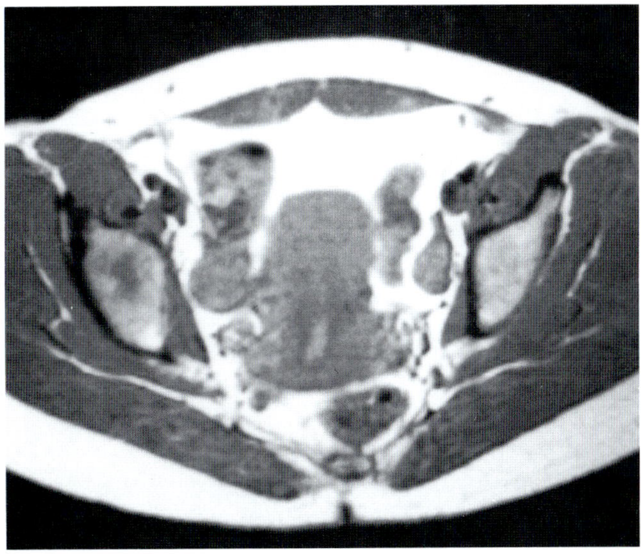

FIGURE 6.6. Normal Pelvis. Axial section [SE 800/20] through supraacetabular area.

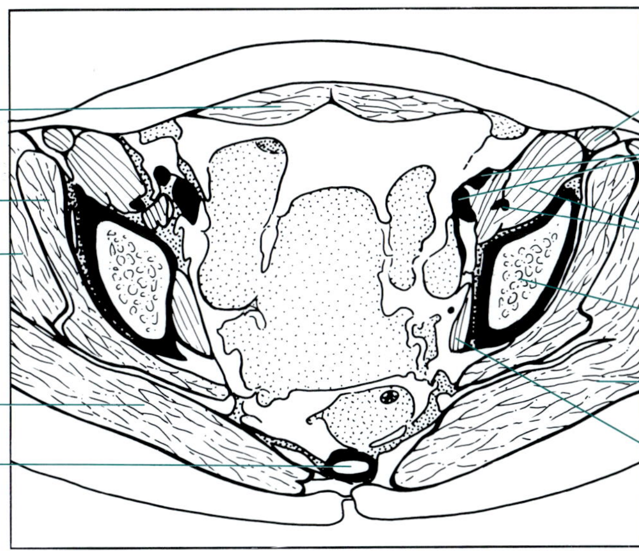

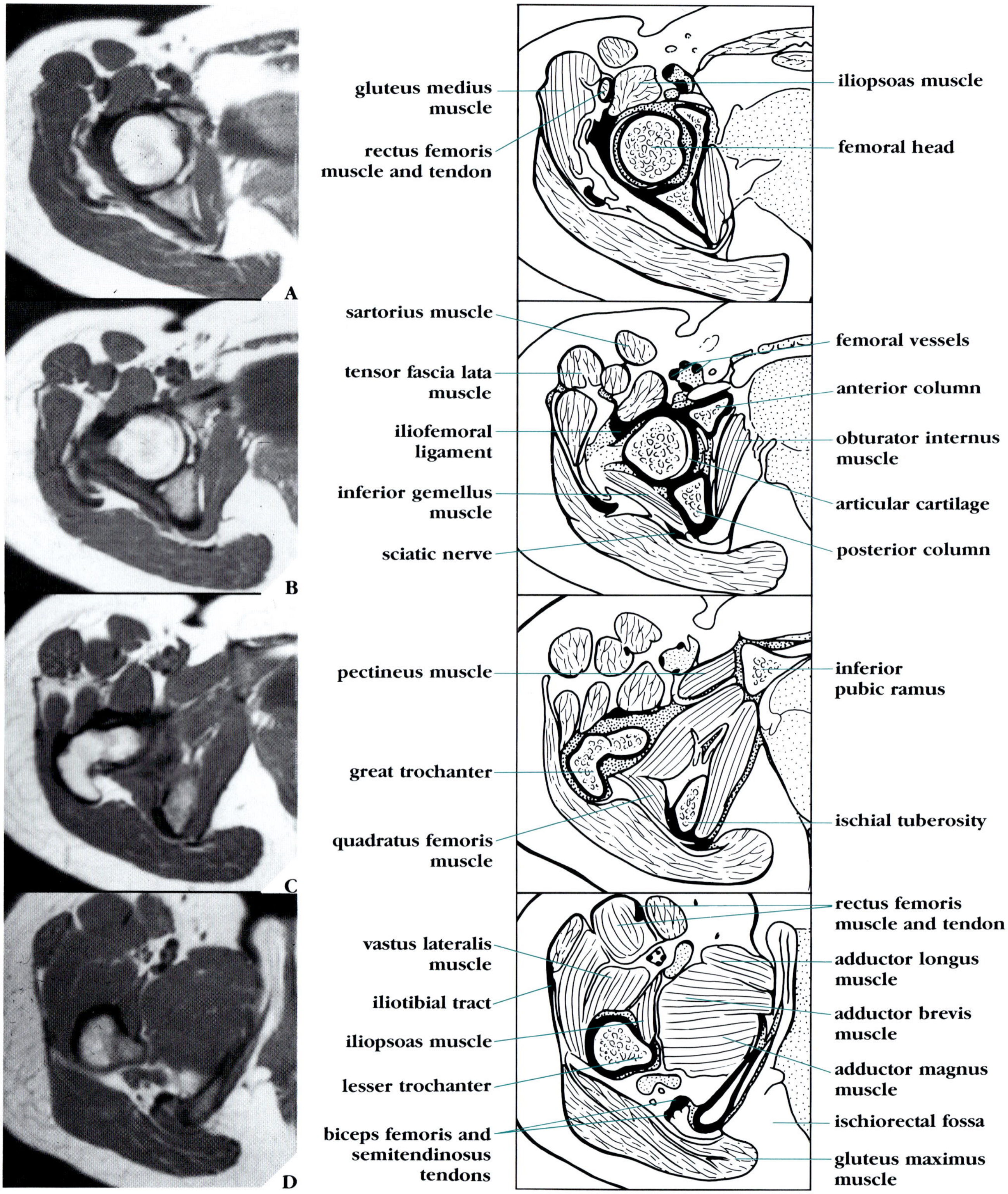

FIGURE 6.7. Normal Hip. A–D Four contiguous axial sections [SE 800/20] through the femoral head and neck.

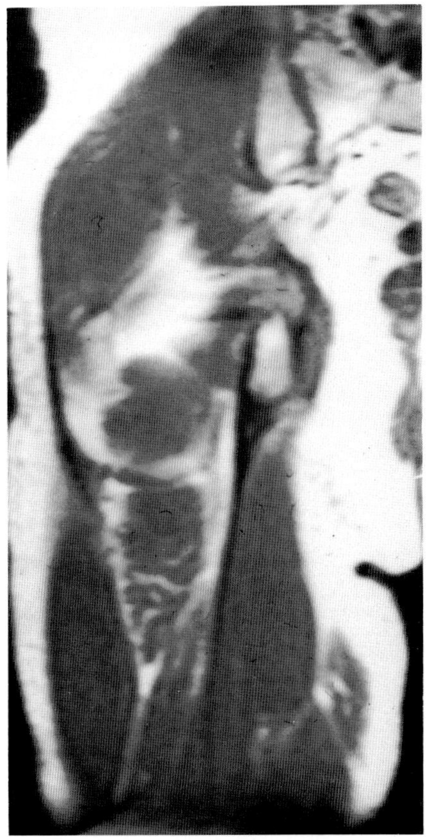

FIGURE 6.8. **Normal Pelvis.** Coronal section [SE 800/20] through the sacroiliac joint and ischial tuberosity. The hamstring tendons are well shown in this section.

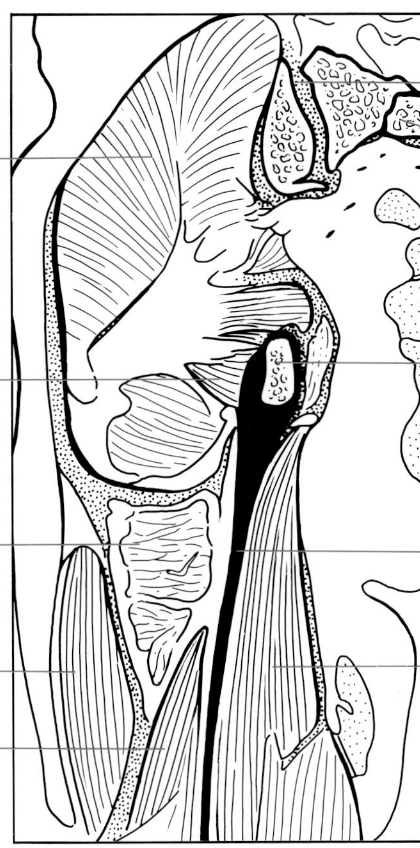

extension of the thigh, with minimal rotary action. The muscles composing the hamstring group include the semimembranosus, semitendinosus, biceps femoris, and the ischiocondylar portion of the adductor magnus (28).

The hip joints and proximal femora are well seen in more anterior coronal sections (Figs. 6.9–6.12). The distribution of red and yellow marrow in the proximal femora can be assessed. In young individuals, the primary compression trabeculae of the

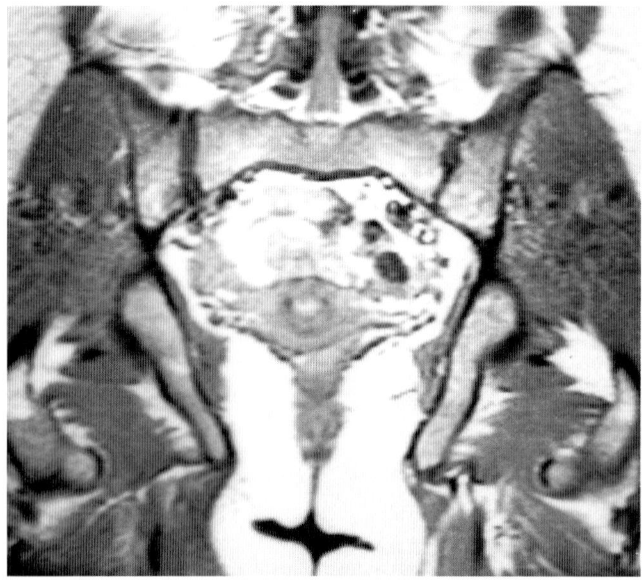

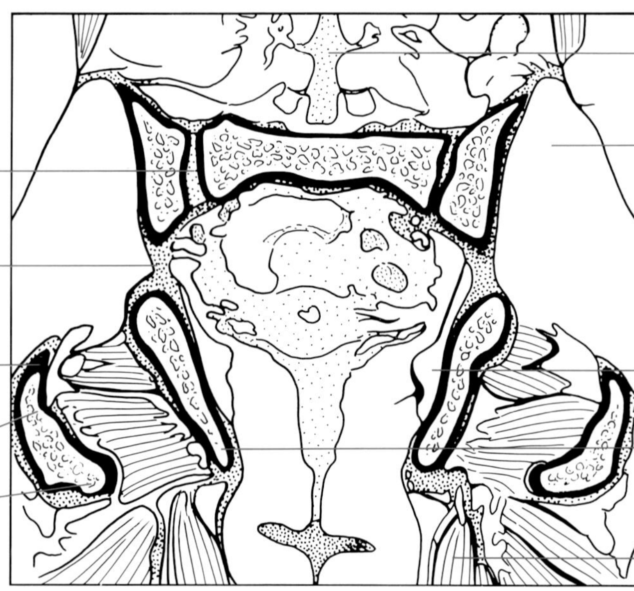

FIGURE 6.9. Normal Pelvis. Coronal section [SE 800/20] through the sacroiliac joints.

femoral head and neck may produce a low SI vertical band across the femoral head, not to be confused with avascular necrosis or tumor (Fig. 6.12). A small amount of fat is present in the acetabular fossa, producing an area of high SI on T1WI and PDWI (Fig. 6.3). The femoral vessels are identified in more anterior sections (Fig. 6.11).

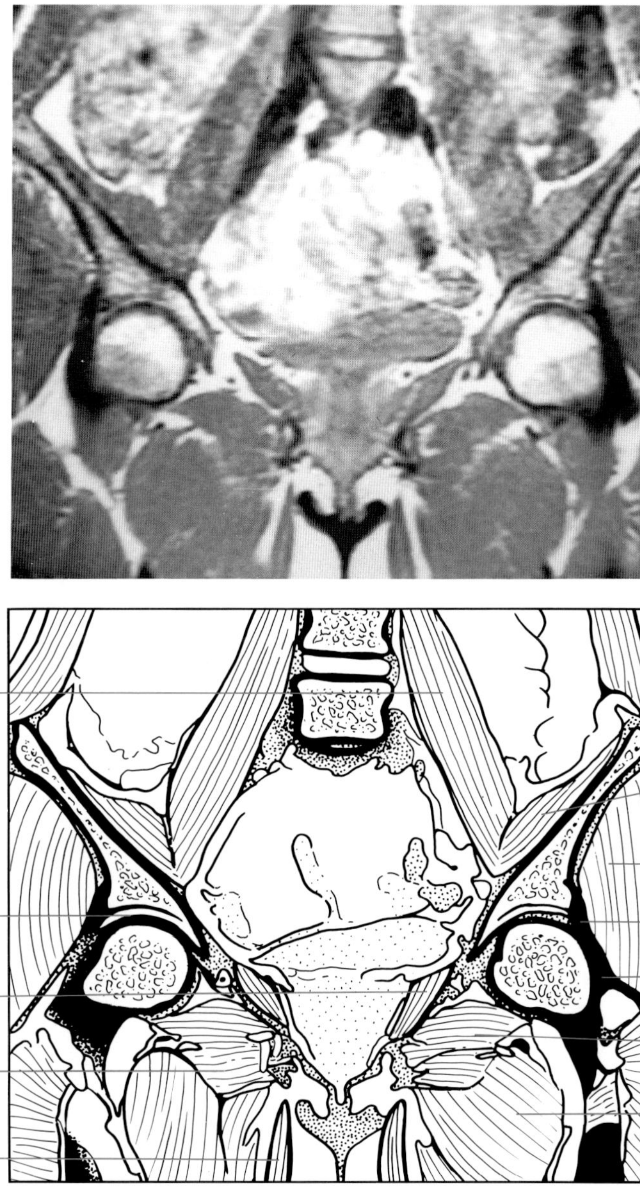

FIGURE 6.10. Normal Hips. Coronal section [SE 800/20] through both hip joints. The articular cartilage thickness can be assessed in this section.

FIGURE 6.11. Normal Pelvis. Coronal section [SE 800/20] through both inguinal areas demonstrating the femoral vessels.

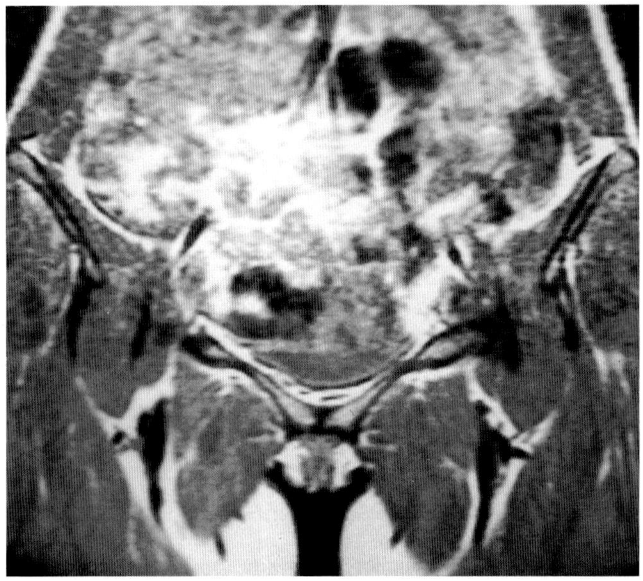

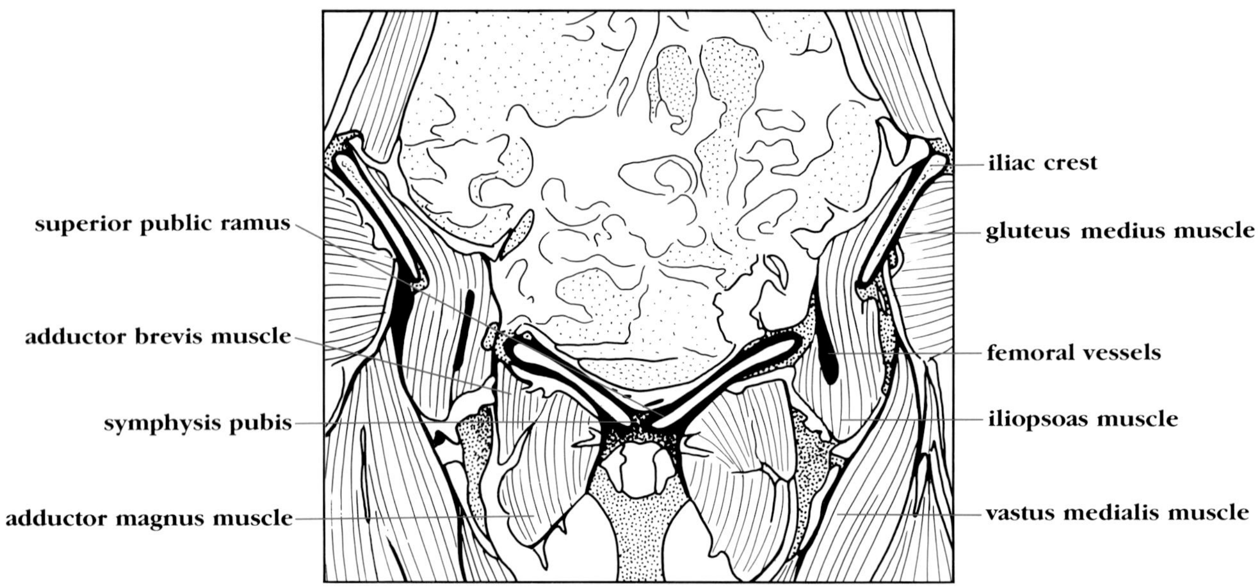

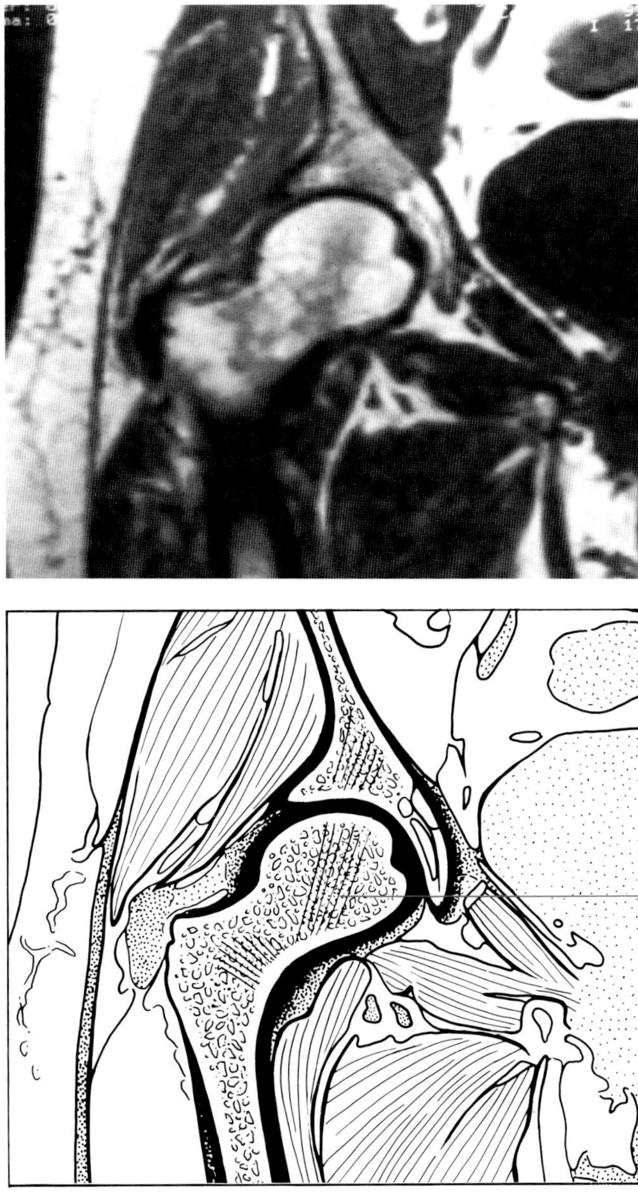

FIGURE 6.12. Normal Hip Joint. Coronal section [SE 800/20] through the right hip joint. There is a low SI, vertically oriented band across the femoral head and neck representing the primary compression trabeculae.

SAGITTAL PLANE

The sagittal plane is ideal for evaluating the articular cartilage (Figs. 6.13, 6.14). The acetabular labrum is seen as a low SI triangle in the anterior margin of the acetabulum (Fig. 6.14), separated from the iliopsoas tendon by a thin fat plane. The transverse acetabular ligament is seen as a low SI band between the anterior and posterior margins of the inferior aspect of the acetabular fossa (Fig. 6.14). The fovea of the femoral head for the insertion of the

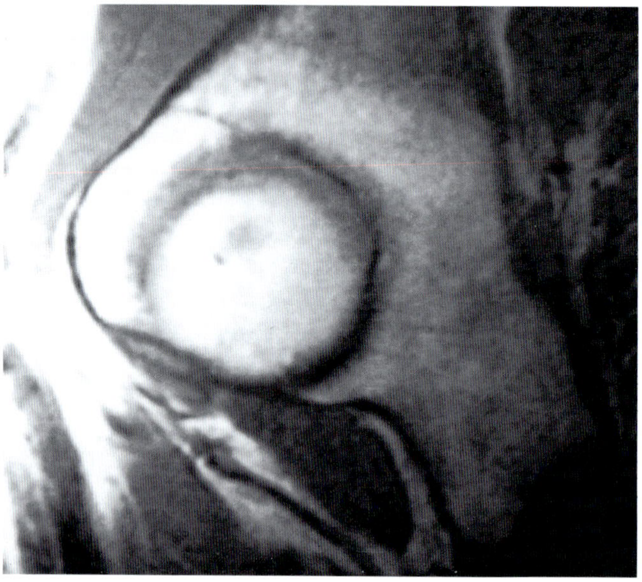

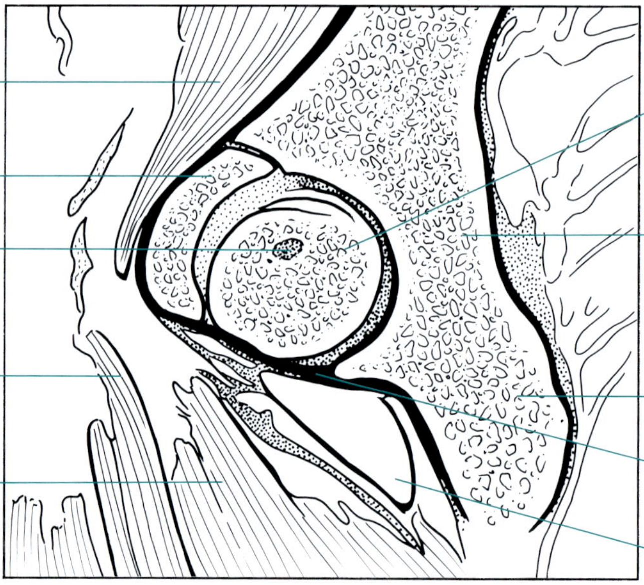

FIGURE 6.13. Normal Hip Joint. Surface coil sagittal section [SE 2000/20] through the medial aspect of the femoral head. The low SI dot in the femoral head represents the fovea for the insertion of the teres ligament.

teres ligament is seen in more medial sections as a low SI dot. This should not be confused with a femoral head lesion (Fig. 6.13). Due to its oblique orientation, the teres ligament is seldom seen.

The tendon insertions of the hamstring muscles can be seen in the posterior aspect of the thigh, in the region of the ischial tuberosity (Fig. 6.15).

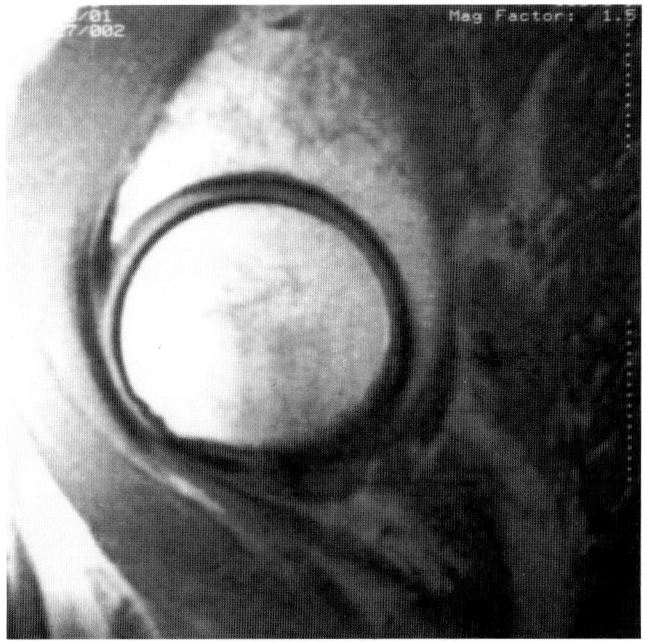

FIGURE 6.14. Normal Hip Joint. Surface coil sagittal section [SE 2000/20] through the acetabulum. The acetabular labrum and articular cartilage are well demonstrated.

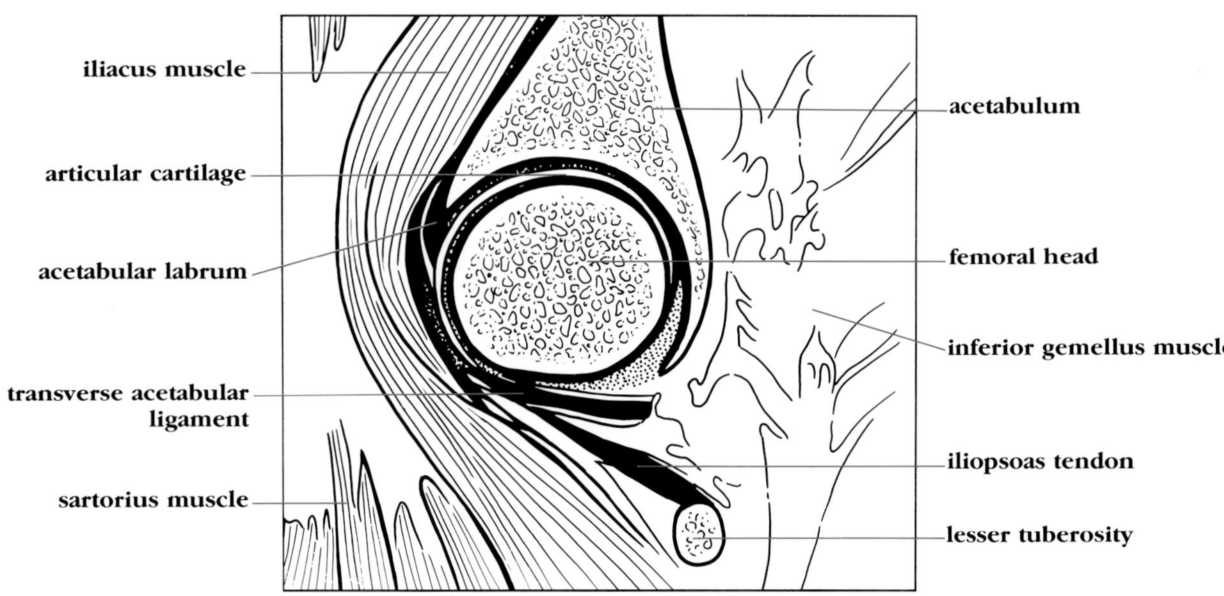

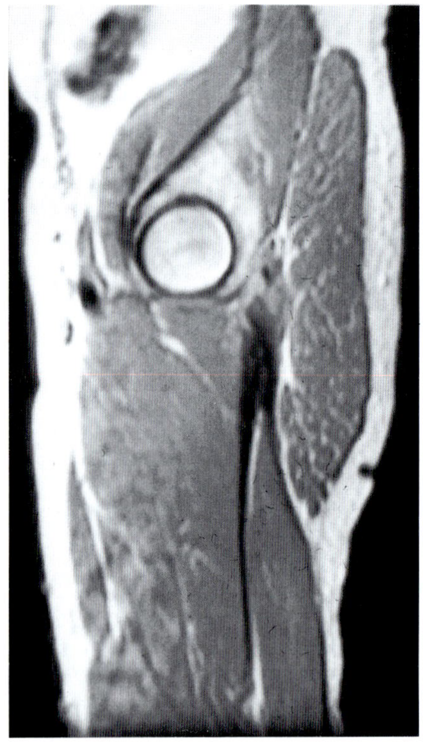

FIGURE 6.15. Normal Pelvis. Sagittal section [SE 2000/20] through the hip joint demonstrating the hamstring tendons.

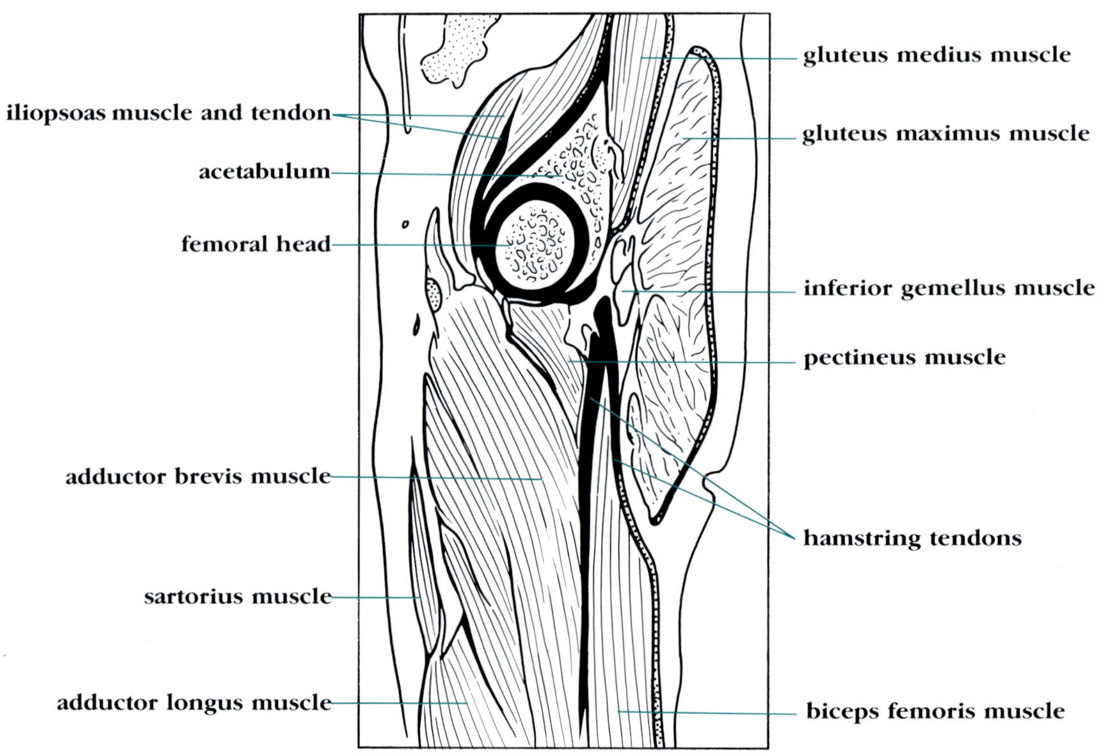

AVASCULAR NECROSIS AND OTHER BONE MARROW LESIONS

BONE MARROW ISCHEMIA (AVASCULAR NECROSIS)

It is generally agreed that ischemic or avascular necrosis (AVN) of the femoral head is secondary to vascular obstruction leading to bone necrosis, and subsequent development of collapse. With the exception of trauma, the mechanism by which the vascular occlusion takes place is poorly understood. Several conditions have been associated with AVN of the femoral head, including steroid use, pancreatitis, alcoholism, rheumatoid arthritis, sickle cell anemia, Gaucher's disease, caisson disease, gout, and systemic lupus erythematosus, among others.

There is still controversy regarding the appropriate management of AVN. Treatment options include core decompression, muscle pedicle graft, rotational osteotomy, joint fusion, and total joint replacement. If indicated, core decompression should be performed during the early stages of the disease, before subchondral fracture takes place; therefore, early diagnosis of AVN is important.

Noninvasive diagnostic techniques include plain radiographs, bone scintigraphy, single photon emission computed tomography (SPECT), CT, and MRI (4,15,18).

Since its clinical introduction, MRI has been shown to be a highly sensitive and specific test for early detection of AVN (17,26,27) with greater accuracy than other existing modalities. MRI patterns of AVN reflect the size of the lesions and roughly the stage of the disease (2,21,26). Small infarcts are seen as low SI bands across the weight-bearing portion of the femoral head (Fig. 6.16). Often these small areas of AVN produce no symptoms.

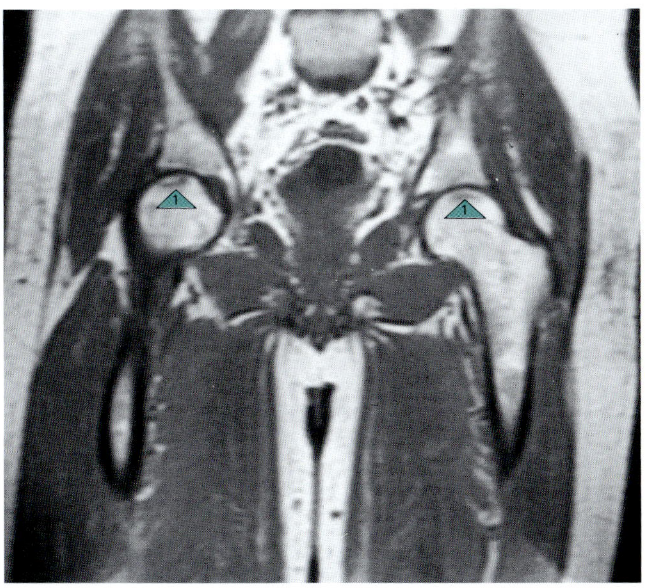

FIGURE 6.16. Avascular Necrosis. Coronal section of hips [SE 2000/20]. There are bilateral small crescent-shaped infarcts in the weight-bearing segments of both femoral heads. Radiographs were normal and the patient was asymptomatic. Arrows **1** Avascular necrosis.

Larger lesions are seen on MRI as small or large rings or bands in the femoral head (Figs. 6.17, 6.18). Pathologically, the demarcation between the infarct and the normal bone marrow is called the reactive interphase. This area is composed of an outer layer of new bone formation with osteoblastic activity, and an inner layer of hypervascularity and granulation tissue. Although the concept is still controversial (14,25), this change can be seen on MRI. T1WI demonstrates a thick band of decreased SI. On T2WI, the low SI band becomes thinner and a second band, or line of increased SI, can be seen. The combination of these two bands has been termed the "double-line" sign and it is believed to be highly specific for AVN (Fig. 6.18). Whether the double line reflects the pathologic changes occurring at the reactive interphase or a chemical shift artifact is debatable.

As the disease progresses, with continuous weight-bearing, a small area of subchondral fracture develops. This is clearly depicted on plain films and on MRI as a high SI line in the subchondral cortex (Fig. 6.19). Subsequently, in the late stages of the disease, femoral head collapse and secondary degenerative changes develop (Figs. 6.20, 6.21).

The SI changes or patterns seen within the infarcted segment roughly reflect the pathophysiologic events taking place in the region of cellular death. During the early stages (class A pattern), high SI on T1WI and low SI on T2WI are seen within the area of necrosis of the femoral head. These SI

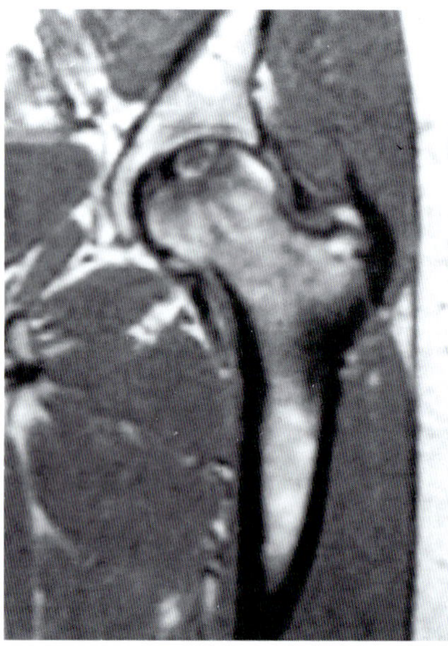

FIGURE 6.17. Avascular Necrosis. Coronal section [SE 2000/20]. Small ring pattern of avascular necrosis.

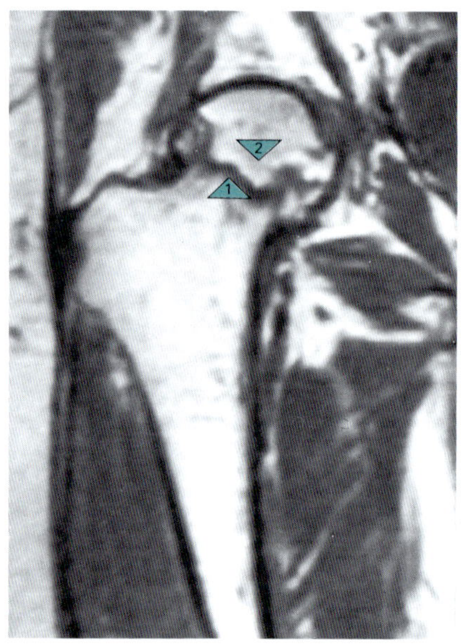

FIGURE 6.18. Avascular Necrosis. Coronal section [SE 2000/20]. Large ring pattern. "Double-line" sign. Arrow **1** Low SI bone forming interphase. Arrow **2** High SI hypervascular interphase.

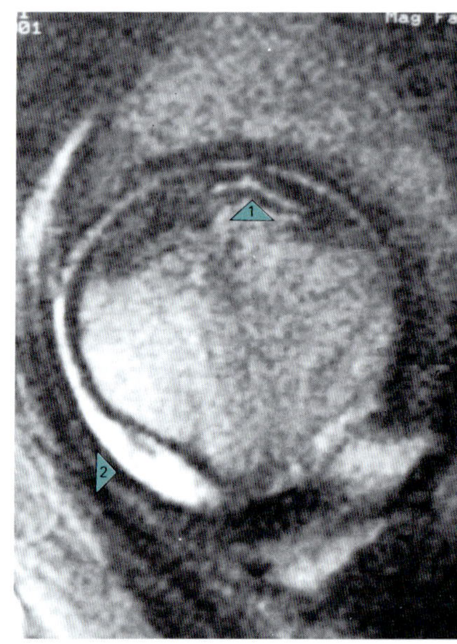

FIGURE 6.19. Avascular Necrosis. Sagittal surface coil section [SE 2000/80]. There is a small subchondral fracture shown as a high SI line in the femoral head. Plain film findings showed a subchondral "crescent." There is an associated joint effusion anteriorly. The contour of the femoral head is preserved, without evidence of collapse. The articular cartilage thickness is normal. Arrow **1** Subchondral fracture. Arrow **2** Joint effusion.

changes are similar to those of the normal fatty marrow (Fig. 6.18). In a more advanced stage, when subacute hemorrhage develops, high SI can be seen on both T1WI and T2WI (class B pattern). As more fluid accumulates within the necrotic segment of bone, a fluid-like pattern develops, with low SI on T1WI and high SI on T2WI (class C pattern) (Fig. 6.22). During the late stages of the disease, fibrosis and sclerosis predominate, demonstrating low SI on T1WI and T2WI (class D pattern) (26).

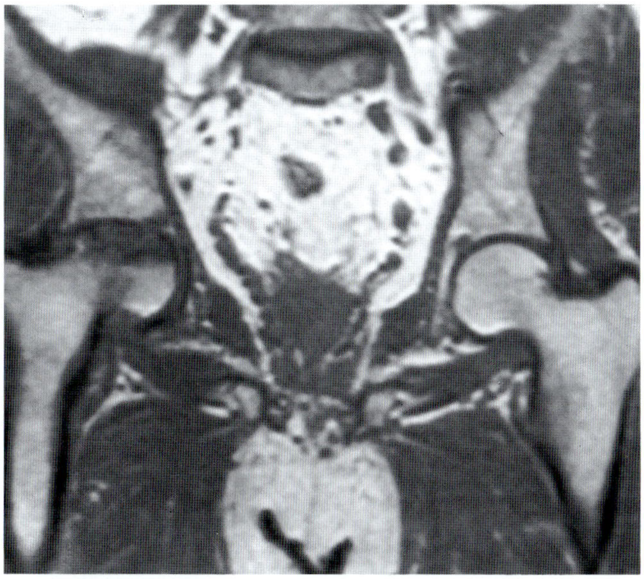

FIGURE 6.20. Avascular Necrosis. Coronal section [SE 2000/20]. There is avascular necrosis of the right femoral head, with flattening of the superior surface. The acetabulum is normal. Radiographs were also abnormal. MRI was obtained to exclude contralateral avascular necrosis.

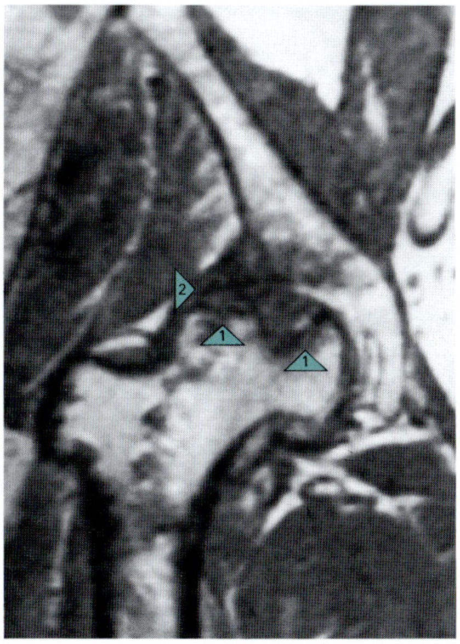

FIGURE 6.21. Avascular Necrosis. Coronal section [SE 2000/20]. Advanced avascular necrosis with collapse of the femoral head and degenerative changes of the acetabulum with joint space narrowing. Arrows **1** Femoral head AVN. Arrow **2** Acetabulum.

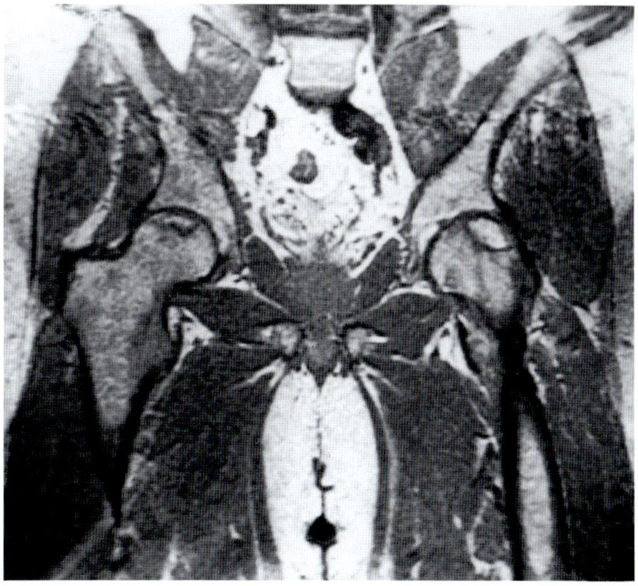

FIGURE 6.22. Avascular Necrosis. Coronal section [SE 2000/80]. High SI areas of avascular necrosis involving both femoral heads on this T2WI.

This pattern classification helps to understand the pathologic events occurring in the necrotic segment of bone, but it is only a simplified representation of these changes, since differentiation between viable and nonviable tissue cannot be made on the basis of the SI changes seen on MRI (21).

Despite its high sensitivity, a negative MRI study does not fully exclude AVN. Several examples of false negative results have been reported (4). Bone scintigraphy can be used to complement a negative MRI study in high risk patients. Figure 6.23 is a suggested algorithm for patients suspected of having AVN of the femoral heads.

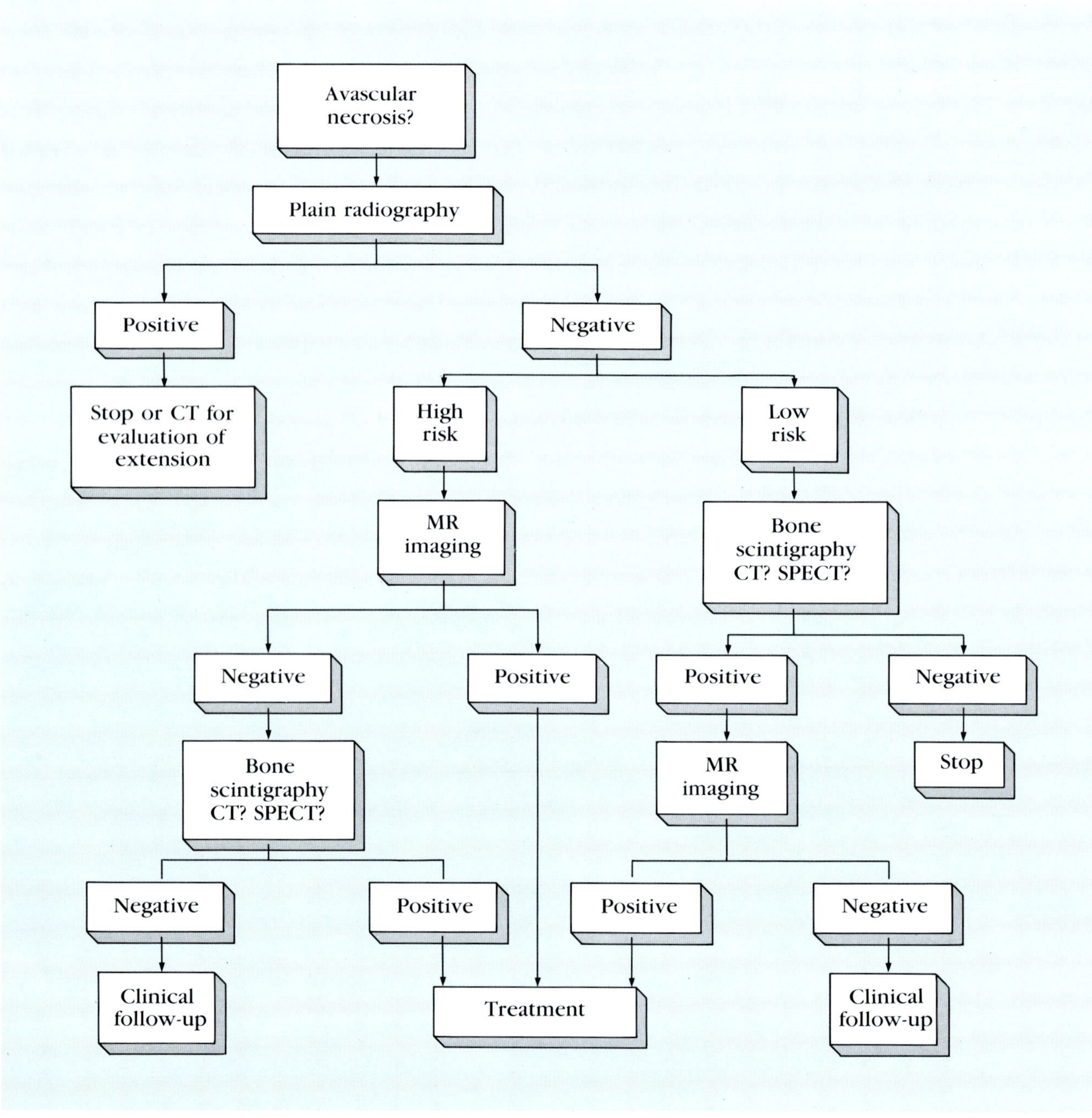

FIGURE 6.23. Suggested Algorithm for Patients Suspected of Having Avascular Necrosis of the Femoral Head. The exact role of computed tomography and single photon emission computed tomography in patients with avascular necrosis has not yet been defined. Bone scintigraphy complements a negative MRI in high risk patients.

Legg–Perthes Disease

Legg–Perthes disease can be included within the group of entities related to AVN. Although the causal mechanisms are not fully understood, the basic process in Legg–Perthes disease is osteonecrosis. There is still some controversy regarding the treatment of this disease, but most authors agree that splint therapy is indicated if no serious flattening of the femoral head is present, and containment of the femoral head by the acetabulum should be achieved with treatment. Femoral head containment and joint congruency can be assessed with arthrography. However, recent publications indicate that MRI can provide the same information noninvasively (9,35).

Since the articular cartilage covering the epiphyseal ossification center of the femoral head is not seen radiographically, measurements of femoral head containment based on plain films may be misleading. Coronal MRI sections (Fig. 6.24A) can demonstrate the degree of femoral head coverage by the acetabulum better than plain films (9). Sagittal sections are ideal for assessment of congruity or lack of collapse of the articular cartilage (Fig. 6.24C).

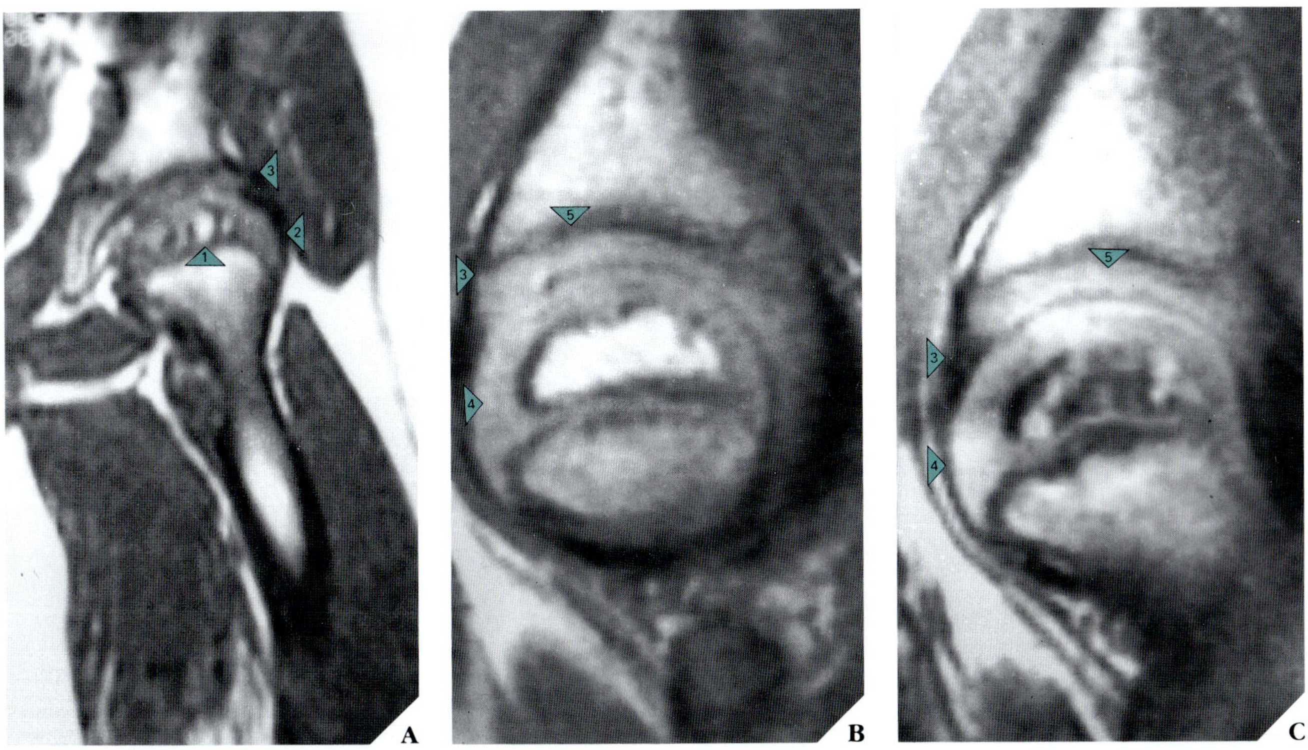

FIGURE 6.24. Legg–Perthes Disease. A Coronal section [SE 2000/20]. There is irregularity and fragmentation of the left femoral ossification center in this 4-year-old patient. The left femoral epiphysis is covered by the acetabulum and labrum for an extension of about 70 percent. **B** Surface coil sagittal section [SE 2000/20] through the normal right hip. Note the normal shape and SI of the ossification center and the normal, congruent articular cartilage of the femoral head and acetabulum. **C** Surface coil sagittal section [SE 2000/20] through the left femoral head. Fragmentation of the ossification center is redemonstrated. There is complete congruency of both articular surfaces. Arrow **1** Fragmented ossification center. Arrow **2** Nonossified cartilage of the left femoral epiphysis. Arrows **3** Acetabular labrum. Arrows **4** Articular cartilage of the femoral head. Arrows **5** Articular cartilage of the acetabulum.

Other findings occasionally seen in patients with Legg–Perthes disease include acute chondrolysis, cartilage hypertrophy, and synovitis with pannus formation (9,10,13). All of these abnormalities can be demonstrated with MRI.

BONE MARROW RECONVERSION

Reconversion of fatty marrow to red marrow occurs in individuals with increased demand for hematopoiesis, such as sickle cell anemia, and diffuse marrow replacement disorders including metastatic disease, myeloma, and myelofibrosis (41). The reconversion process begins in the spine and flat bones of the pelvis, followed by the extremities, from proximal to distal, from metaphysis to diaphysis. The epiphyseal ends of the long bones are the latest to reconvert to red marrow.

Red marrow is normally hypointense in relation to the fatty marrow, probably due to its increased cellularity and protein content. MRI in patients with red marrow reconversion shows diffuse hypointense marrow on T1WI and T2WI similar to the SI of the muscle (33,41) (Fig. 6.25). The degree of hypointensity of the bone marrow roughly reflects the chronicity and severity of the reconversion process. In patients with long-standing sickle cell anemia, hemosiderin deposition in the bone marrow contributes to further SI decrease (33).

Sickle cell anemia is a well-known predisposing condition for AVN. Avascular necrosis can involve practically any area of the skeleton but it is more frequent in the femoral heads (Fig. 6.25). Acute infarctions may demonstrate hyperintensity on T2WI, probably due to the edema and hemorrhage. Older infarcts may be seen as areas of lower SI on T2WI because of fibrotic infiltration (32).

BONE MARROW REPLACEMENT

Leukemia, lymphoma, myeloma, metastasis, myelofibrosis, osteomyelitis, and Gaucher's disease are examples of diseases producing bone marrow replacement by neoplastic or inflammatory cellular infiltrates. In most cases the MRI changes are nonspecific, showing decreased SI on T1WI and high SI on T2WI. (The reader is referred to Chapters 9–11 for a more detailed description of the MRI findings in bone marrow replacement disorders.)

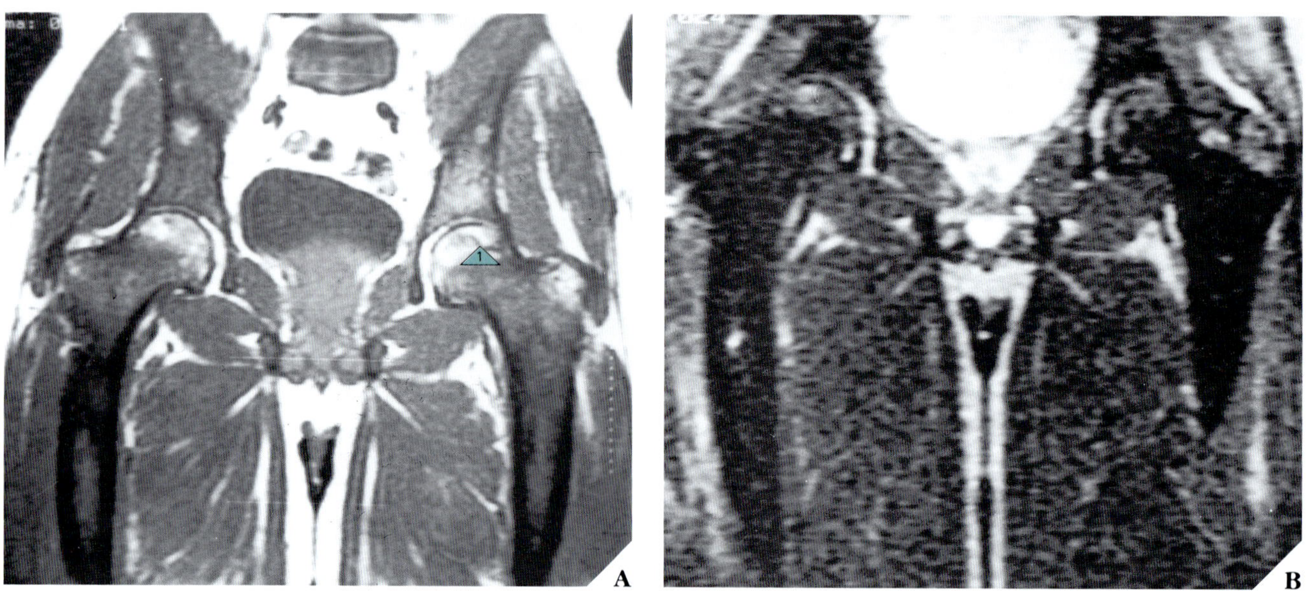

FIGURE 6.25. Sickle Cell Anemia. A Coronal section [SE 800/20]. There is low SI in the bone marrow spaces of the pelvis and proximal femora, with the exception of the femoral epiphyses and greater trochanters, in this T1WI. The SI of the bone marrow is similar to that of the muscle and indicates fatty marrow reconversion to red marrow. The right femoral epiphysis shows inhomogeneous SI probably due to the presence of mixed yellow and red marrow. There is an area of avascular necrosis in the left femoral head. **B** Coronal section [SE 2000/80]. This T2WI in the same patient obtained 2 years later demonstrates further SI decrease of the bone marrow spaces, probably representing hemosiderosis. The bright spot in the right femoral diaphysis probably represents a small infarct. Arrow **1** Avascular necrosis.

▼▼▼ BONE MARROW DEPLETION

Myeloid depletion results in decreased cellularity of the hematopoietic bone marrow and fatty infiltration. This phenomenon occurs in aplastic anemia, postradiation therapy, and postchemotherapy. The normal intermediate-to-low SI of the hematopoietic red marrow present in the spine and pelvis in normal adults becomes diffusely replaced by fatty marrow in aplastic anemia and postchemotherapy, or locally replaced postradiation therapy (see Chapter 9). Islands of low SI hematopoietic marrow can be seen in patients with aplastic anemia following treatment (41).

▼▼▼ BONE MARROW EDEMA

Bone marrow edema can occur as a result of several pathophysiologic mechanisms including trauma, tumor, stress, reflex sympathetic dystrophy, transient osteoporosis, ischemia, and infection. Increased interstitial water and hypervascularity are the common denominators in these situations. MRI findings are nonspecific, showing low SI on T1WI and increased SI on T2WI (41). Distribution of the abnormal SI and, indeed, clinical symptoms are helpful in the differential diagnosis.

Transient Osteoporosis of the Hip

The term transient osteoporosis refers to a clinical entity originally described in pregnant females but seen more often in middle-aged males. It is characterized by the insidious onset of hip pain lasting for several months, and it is self-limited, although some patients may eventually develop avascular necrosis. Conservative treatment is recommended once the correct diagnosis is made.

Patients with transient osteoporosis of the hip are often evaluated with multiple tests, including plain films, bone scintigraphy, CT, MRI, and biopsy. Radiographs are often normal or reveal mild osteoporosis. Bone scintigraphy shows increased uptake in the region of the hip, and CT may demonstrate osteoporosis and joint effusion. Low SI on T1WI and high SI on T2WI involving the femoral head and neck, with ill-defined margins, are the characteristic findings on MRI (8,42) (Fig. 6.26).

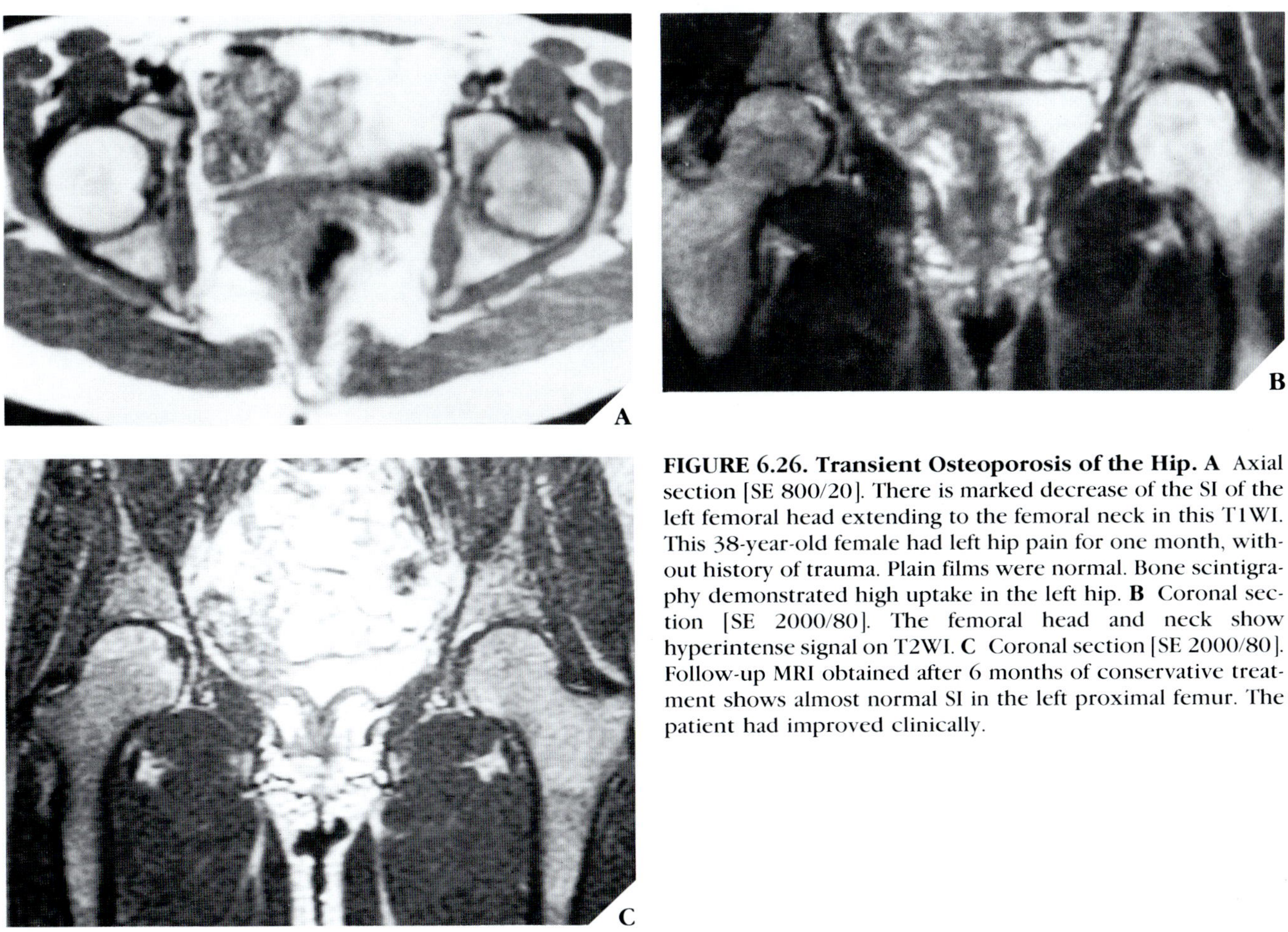

FIGURE 6.26. Transient Osteoporosis of the Hip. A Axial section [SE 800/20]. There is marked decrease of the SI of the left femoral head extending to the femoral neck in this T1WI. This 38-year-old female had left hip pain for one month, without history of trauma. Plain films were normal. Bone scintigraphy demonstrated high uptake in the left hip. **B** Coronal section [SE 2000/80]. The femoral head and neck show hyperintense signal on T2WI. **C** Coronal section [SE 2000/80]. Follow-up MRI obtained after 6 months of conservative treatment shows almost normal SI in the left proximal femur. The patient had improved clinically.

The differential diagnois of transient osteoporosis of the hip on MRI has to be made with tumor, osteomyelitis, septic arthritis, stress fracture, and AVN. Tumor and osteomyelitis may have a very similar MRI pattern. Joint effusion can be an associated finding in all these conditions. However, in the proper clinical setting there are some MRI features that can help in differentiating some of these entities from transient osteoporosis. Transient osteoporosis tends to involve only the metaphysis and epiphysis of the femur. Stress fractures tend to be localized in the neck and are often seen as low SI lines or bands in all pulse sequences. A low SI band surrounding a high SI line on T2WI has been considered pathognomonic of AVN. This double-line sign is not seen in transient osteoporosis. Finally, joint aspiration may be necessary to exclude septic arthritis because it may be impossible to differentiate both entities on the basis of the MRI findings.

Reflex sympathetic dystrophy syndrome (Sudeck's atrophy) and regional migratory osteoporosis are two entities closely related to transient osteoporosis of the hip which may have the same radiographic, scintigraphic, and MRI manifestations. In fact, some authors consider transient osteoporosis to be part of the reflex sympathetic dystrophy syndrome (8,23).

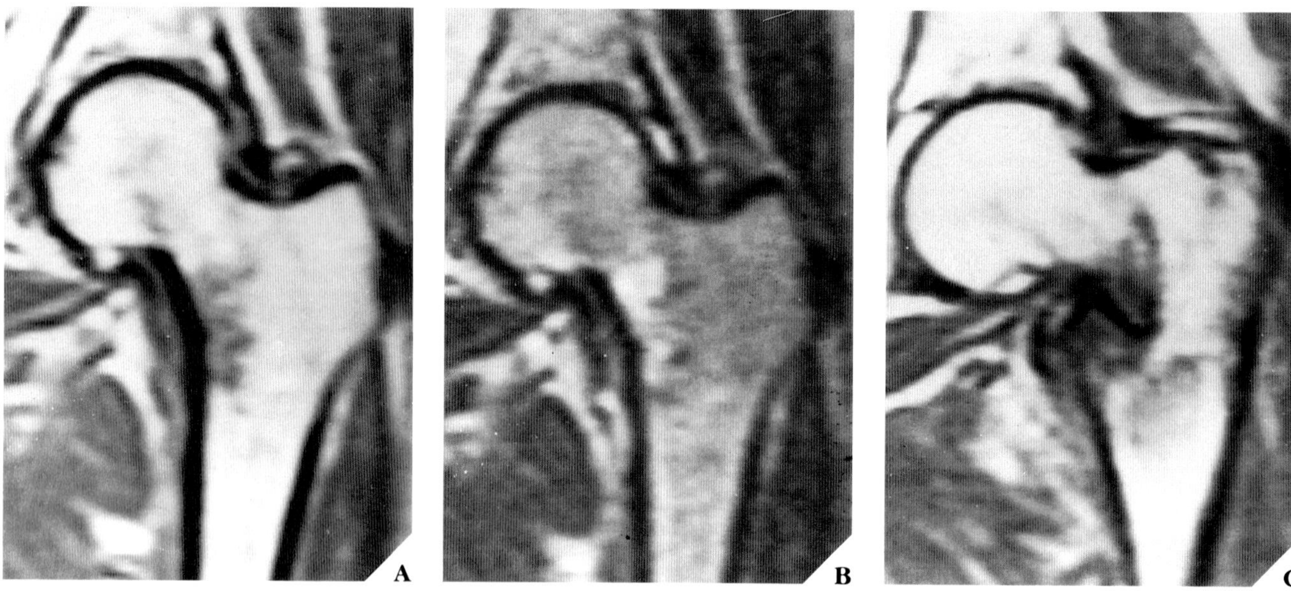

FIGURE 6.27. Insufficiency Fracture of the Femoral Neck. A Coronal section [SE 800/20]. There is an irregular area of low SI in the medial portion of the femoral neck in this elderly female, following minimal trauma. Plain films were normal. **B** Coronal section [SE 2000/80]. The same area of the femoral neck shows high SI. The patient was sent home. **C** Coronal section [SE 800/20] obtained several days later demonstrates collapse of the femoral neck.

Fractures

Stress and insufficiency fractures are frequent in the hip and pelvis. The incidence of hip fractures in Caucasian women after the age of 50 increases dramatically. History of trauma is often missing, and the patient presents with only pain and limitation of motion. Radiographic examination is positive in the vast majority of cases, but in some patients the initial radiographs may be normal. Bone scintigraphy is highly sensitive for early detection of stress and occult fractures in young individuals, but in the elderly there may be a delay between the time of the fracture and the appearance of increased activity on the bone scan. In addition, synovitis, arthritis, and osteoarthritis can all produce increased activity on bone scintigraphy. Conventional tomography and CT may also provide false negative results in some cases (12).

Initial experience with MRI in the evaluation of occult and stress fractures is encouraging. Replacement of the normal hyperintense signal of the fatty marrow on T1WI by low SI edema and hemorrhage related to the fracture can be easily seen during the first 24 hours of onset of symptoms. Increased SI on T2WI also reflects the area of bone contusion or edema (12,22,37,43) (Figs. 6.27 and 6.28).

Some MRI features can help differentiate between occult intraosseous fractures and stress fractures (42). Occult intraosseous fractures tend to be lo-

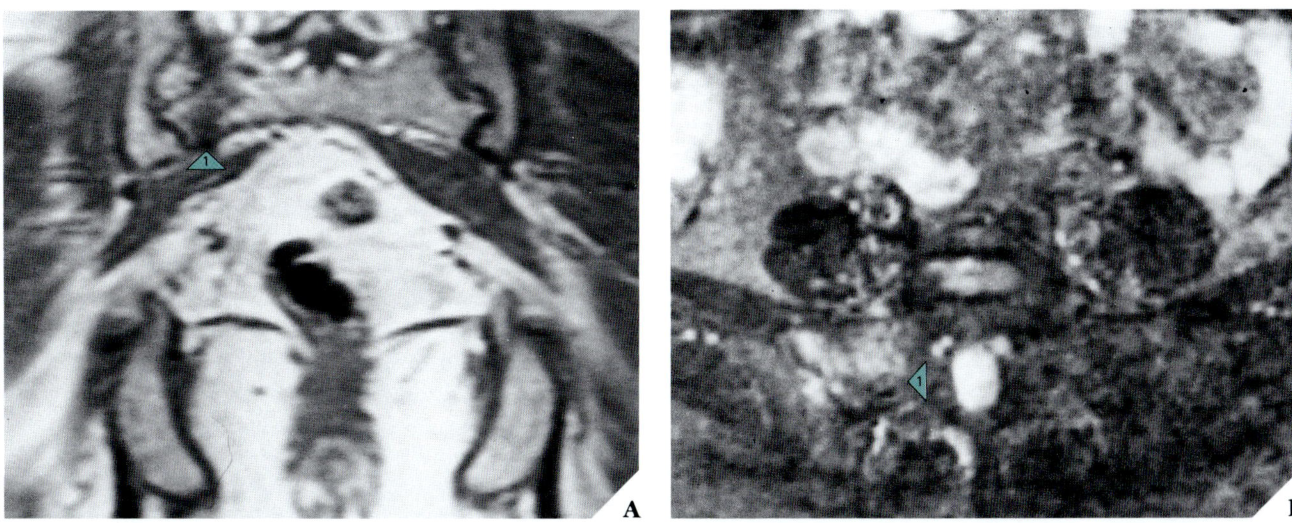

FIGURE 6.28. Insufficiency Fracture of the Sacrum. **A** Coronal section [SE 800/20]. There is a low SI band across the right side of the sacrum, parallel to the sacroiliac joint, consistent with insufficiency fracture in this elderly, osteoporotic female. Plain films were normal. Arrow **1** Insufficiency fracture. **B** Axial section [SE 2000/80]. The area of the insufficiency fracture demonstrates high SI (arrows **1**).

cated in the epiphysis whereas stress fractures are more often seen in the diaphysis or metaphysis. Stress fractures show prominent bands of low SI extending to the cortex, whereas occult intraosseous fractures demonstrate larger areas of abnormal signal within the medullary space. Periosteal elevation with fluid has been described only in stress fractures.

ARTICULAR CARTILAGE AND SYNOVIAL LESIONS

Joint Effusion

Joint effusions of the hip are associated with a number of conditions including trauma, infection, tumor, arthritis, avascular necrosis, and degenerative diseases. MRI demonstrates fluid with low SI on T1WI and high SI on T2WI in the distribution of the joint space. Since fluid tends to accumulate medially and anteriorly, anterior sections are better for diagnosing joint effusions (5) (Fig. 6.29).

Bursitis

Bursitis is a relatively frequent condition in the hip area. The MRI appearance of bursitis is that of fluid collection with low SI on T1WI and high SI on T2WI, in the anatomical location of the bursa. The differential diagnosis with soft tissue abscess may be difficult. Infected bursa or soft tissue abscess should be considered when clinical and/or laboratory signs of infection are present. The SI changes seen on MRI are nonspecific. Infected and noninfected fluid collections have a similar MRI pattern.

There are four bursae present around the hip joint: the iliopsoas, the iliopectineal, the ischiotrochanteric, and the trochanteric bursae (29,38).

The iliopsoas bursa is the largest bursa of the hip joint, and is present in over 90 per cent of adults. This bursa is located between the iliopsoas muscle anteriorly and laterally, and the pectineus muscle and femoral vessels medially and anteriorly. Posteriorly, the bursa is closely related to a thin area of the joint capsule located between the iliofemoral and pubofemoral ligaments.

The iliopsoas bursa can become distended by fluid in different inflammatory conditions, including rheumatoid arthritis, osteoarthritis, synovial chondromatosis, and pigmented villonodular synovitis. Occasionally, the synovial fluid may become infected. In most cases of bursitis there is communication with the hip joint through the normal anterior weak area.

Iliopsoas bursitis may present clinically with a palpable inguinal soft tissue mass. If associated inflammatory changes are present, the symptoms

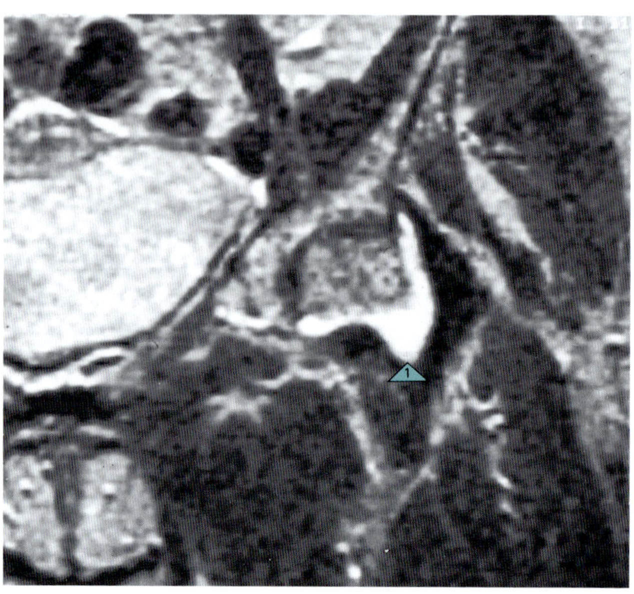

FIGURE 6.29. Hip Joint Effusion. Coronal section [SE 2000/80] through the left hip. There is a high SI effusion better demonstrated in this anterior section. Arrow **1** Joint effusion.

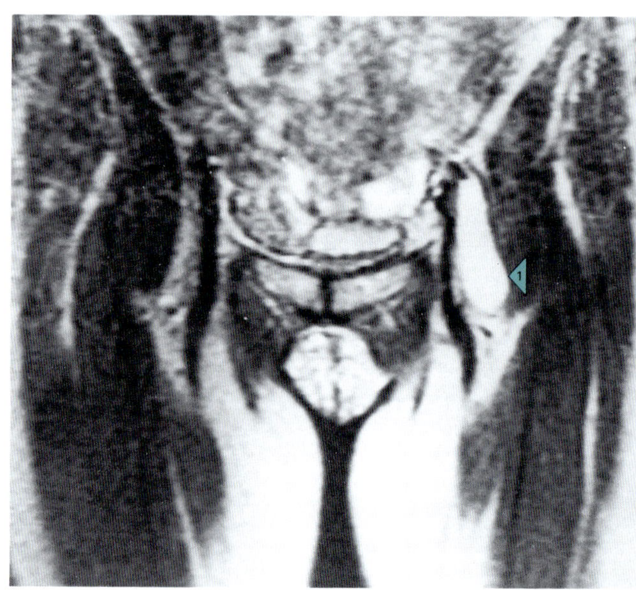

FIGURE 6.30. Iliopsoas Bursitis. Coronal section [SE 2000/80] through the anterior pelvis. There is an elongated fluid collection lateral to the femoral vessels, and medial to the Iliopsoas muscle, extending above the left pubic bone. Arrow **1** Iliopsoas bursitis.

may include pain (sometimes referred to distant regions such as the abdomen, thigh, and knee), paresthesias, neuropathy, urinary frequency due to bladder compression, and edema of the extremity due to compression of the femoral vessels.

Clinically, differential diagnosis includes soft tissue tumors, inguinal hernia, aneurysm, undescended testes, varices, and inguinal adenopathy. Preoperative diagnosis can be obtained by arthrography, if there is communication with the joint, direct puncture, and CT. MRI can provide all the information necessary for preoperative evaluation, including the cystic nature of the mass, its relationship with the surrounding structures, and the lack of communication between the bursa and joint space (Fig. 6.30).

Trochanteric bursitis occurs frequently following total hip replacement. It is more often seen around the knots of the trochanteric wires. Communication with the hip joint space in postsurgical trochanteric bursitis is frequent, and can be demonstrated with arthrography. Trochanteric bursitis is also a relatively frequent cause of hip pain in patients without previous surgery or pathologic conditions of the hip. In these cases, communication with the joint space is rare, hence the absence of joint effusion (29) (Fig. 6.31).

Osteoarthritis

Osteoarthritis of the hip is a common disease of adult males. It is occasionally seen in younger individuals, and can be unilateral. Slipped femoral capital epiphysis during childhood has been suggested as a possible etiology in these cases (34).

The pathologic changes occurring in the hip joint in osteoarthritis include narrowing of the joint space, with superior migration; osteophyte formation; subchondral cysts; and sclerosis. Fragmentation of osteophytes can lead to intraarticular loose bodies (34).

Radiographs accurately depict the articular and bone changes occurring with osteoarthritis, but occasionally an MRI may be requested to exclude other conditions such as AVN. MRI can provide an accurate evaluation of cartilage thickness, subchondral cysts, and intraarticular loose bodies, and

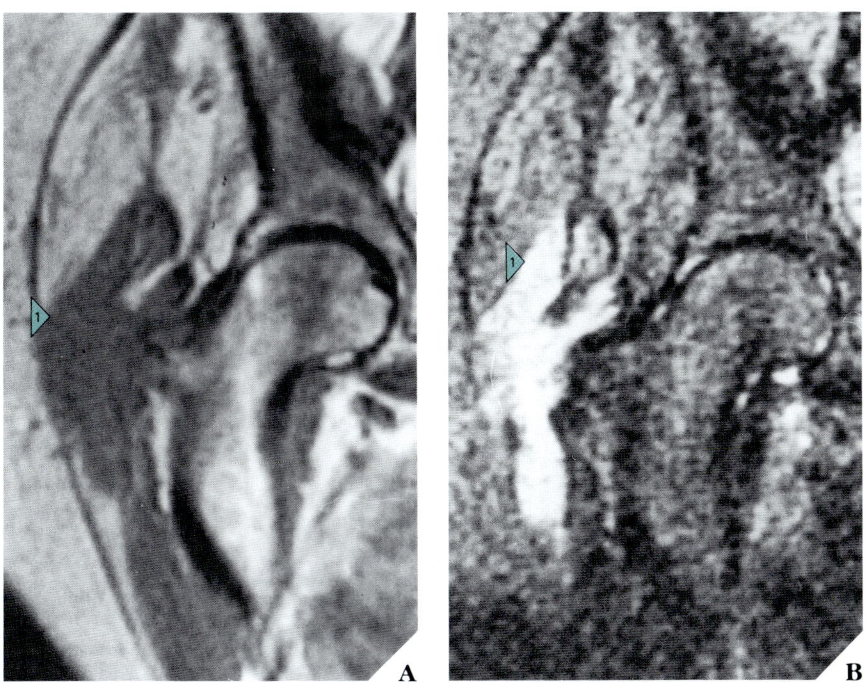

FIGURE 6.31. Trochanteric Bursitis. **A** Coronal section [SE 500/20]. There is a low SI irregular fluid collection adjacent to the greater trochanter. **B** Coronal section [SE 2000/80]. The same area shows high SI on this T2WI. There is no significant joint effusion. Arrows **1** Trochanteric bursitis.

can also help in excluding other conditions (Fig. 6.32). The characteristic findings of AVN of the femoral head have been well described and include areas of varying SI in the femoral head, surrounded by low SI rims, sometimes associated with the double-line sign of high SI bands on T2WI. These findings are different from those of osteoarthritis. Narrowing of the joint space and subchondral cysts are not a feature of AVN, unless secondary degenerative changes are present during the late stages of the disease.

Synovial Osteochondromatosis

Synovial osteochondromatosis is a chronic, progressive, monoarticular disorder most likely due to metaplasia of the synovial membrane with formation of cartilaginous intraarticular nodules. This disorder is more common in males than females and it is found between the third and fifth decades of life. The hip and the knee are the most frequently involved joints, but any joint can be affected. Chronic pain and decreased range of motion are the most common clinical symptoms. Rarely, frac-

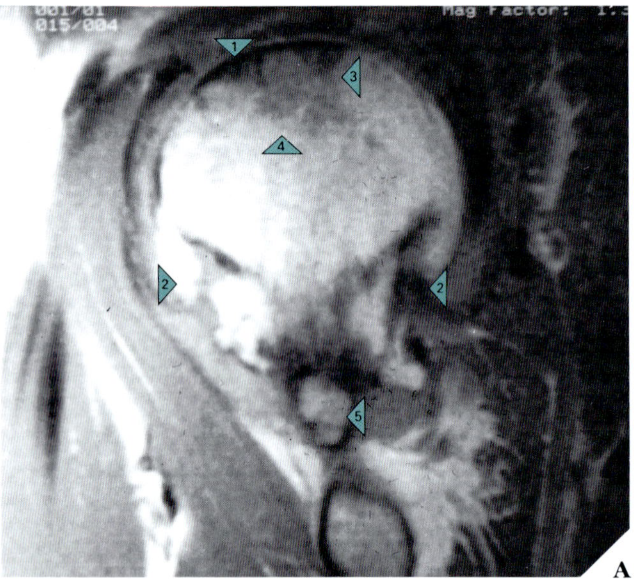

A

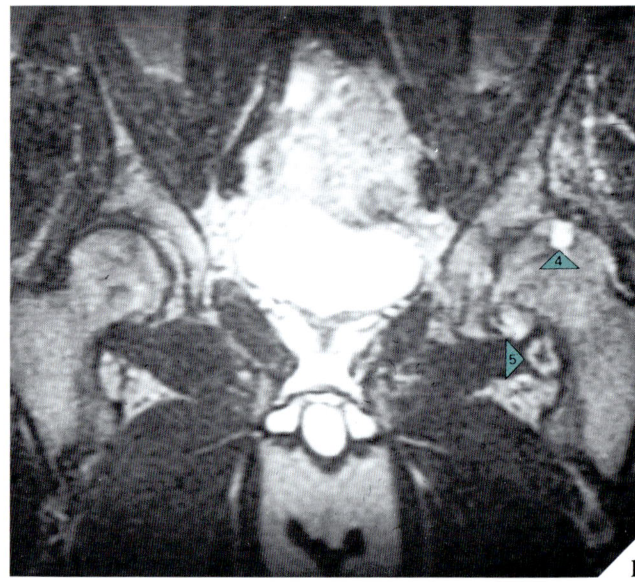

B

FIGURE 6.32. Osteoarthritis. A Surface coil sagittal section [SE 2000/20]. **B** Coronal section [SE 2000/80]. The cystic nature of the subchondral cyst is better demonstrated in this T2WI. Arrow **1** Thinning of the articular cartilage. Arrows **2** Osteophytes. Arrows **3** Subchondral sclerosis. Arrows **4** Subchondral cysts. Arrows **5** Intraarticular loose bodies.

tures and malignant transformation to chondrosarcoma can occur.

Radiologic findings include cartilaginous-type calcifications in the vicinity of a joint, bony erosions, and secondary degenerative osteoarthritis. The intraarticular nodules may be solitary or multiple and may range in size from a few millimeters to several centimeters. Occasionally, these nodules do not calcify and arthrography may be required to make the diagnosis. Joint effusion is an almost constant finding.

MRI is helpful in demonstrating the number, size, and location of the nodules, the presence of joint effusion, and the internal composition of the nodules (39) (Fig. 6.33).

Rheumatoid Arthritis

Rheumatoid arthritis of the hip is relatively rare. With the exception of the soft tissue abnormalities, plain radiographs can accurately demonstrate the bone changes of rheumatoid arthritis; however, MRI

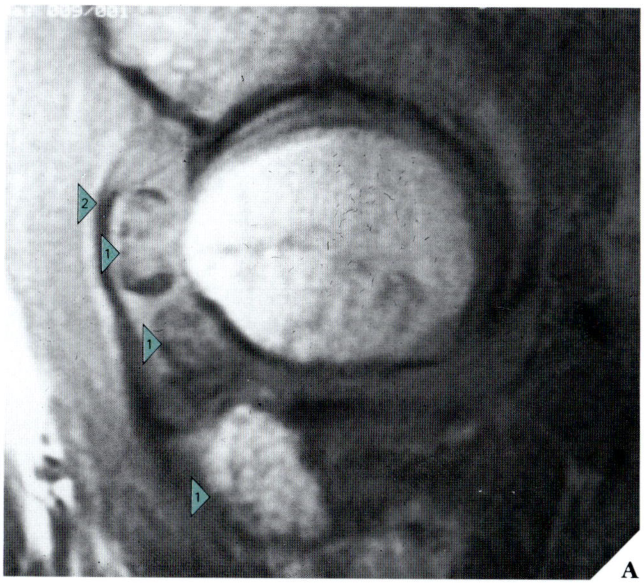

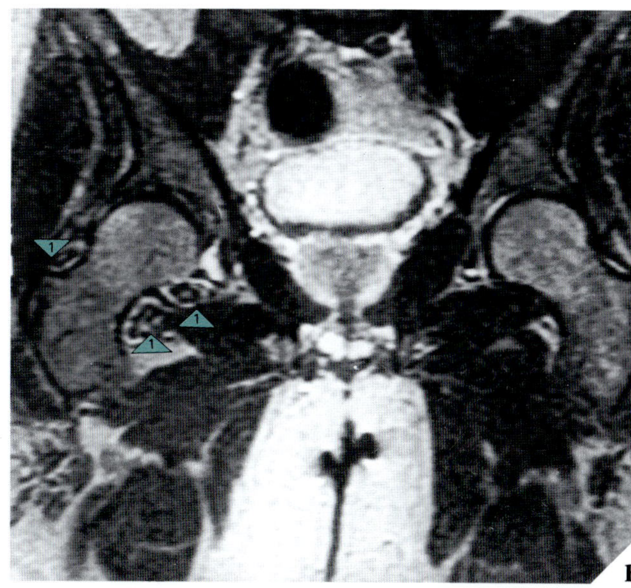

FIGURE 6.33. Synovial Osteochondromatosis. A Surface coil sagittal section [SE 2000/20]. There are two inhomogeneous low SI osteochondromas anterior to the femoral head. A larger third osteochondroma projected inferior to the femoral head shows higher SI, probably reflecting bone marrow formation. The joint capsule is anteriorly displaced. **B** Coronal section [SE 2000/80]. Three osteochondromas are demonstrated in this section. Arrows **1** Synovial osteochondromatosis. Arrow **2** Joint capsule.

can occasionally provide information regarding the extent and severity of cartilage damage and the associated soft tissue lesions (3) (Fig. 6.34).

Osteochondral Fracture

Osteochondritis dissecans or osteochondral fracture is a posttraumatic lesion involving the articular cartilage and subchondral bone. When the lesion involves only the articular cartilage, the term "transchondral fracture" is frequently used. Pathologically, the bony fragment is frequently necrotic, the cartilaginous layer covering the osteochondral fragment is thinned, and a healing response can be seen at the site of detachment.

The osteochondral fragments may remain in the same area or they may become detached. Sometimes revascularization of the fragment occurs, with subsequent healing. If the osteochondral fragment is separated from the bone, it may become a loose intraarticular body, it may be reabsorbed, or it may become imbedded in the synovium at a distant site.

The knee and the ankle are the most frequent sites of osteochondritis dissecans or osteochondral fractures, but this lesion can occur in any traumatized articular surface.

Arthrography has been used to assess the degree of detachment of the osteochondral fragment from the adjacent bone, but recently MRI has become a

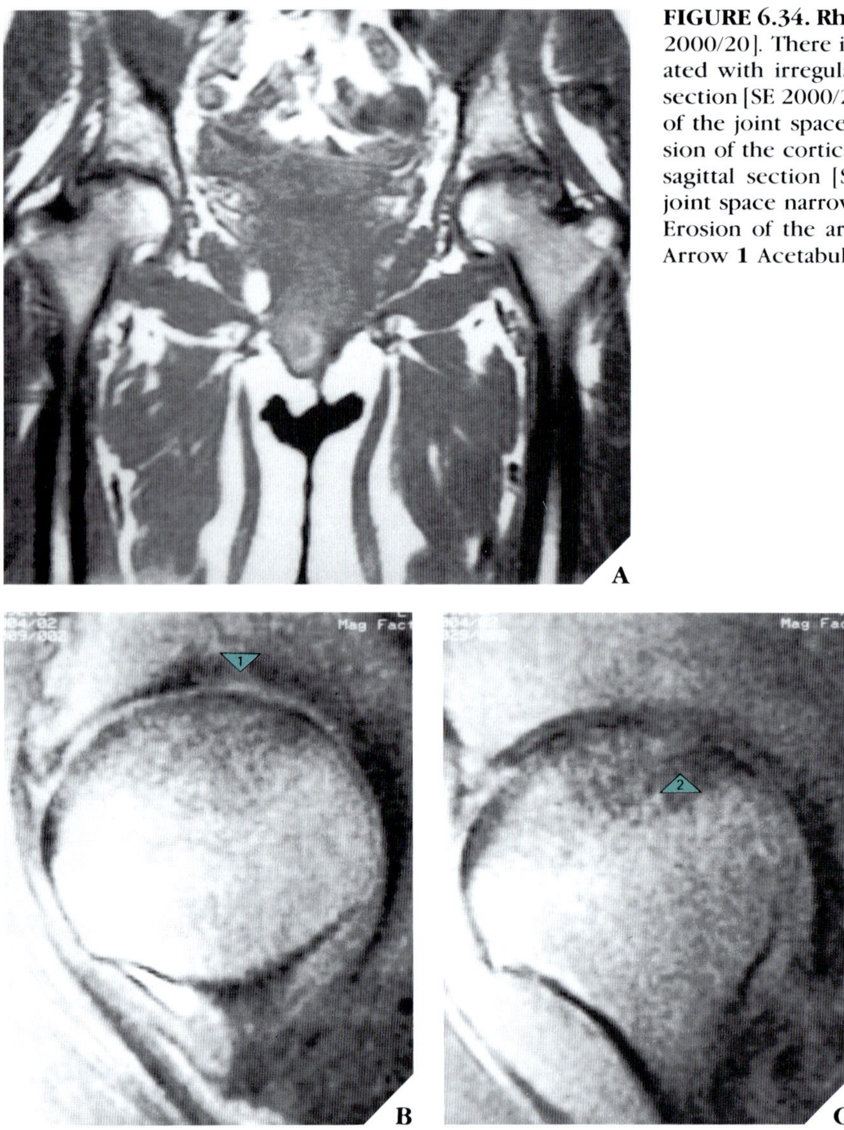

FIGURE 6.34. Rheumatoid Arthritis. A Coronal section [SE 2000/20]. There is narrowing of both hip joint spaces, associated with irregular articular surfaces. **B** Surface coil sagittal section [SE 2000/20] through the right hip. Uniform narrowing of the joint space is demonstrated. There is an irregular erosion of the cortical surface of the acetabulum. **C** Surface coil sagittal section [SE 2000/20] through the left hip. Uniform joint space narrowing is more pronounced than on the right. Erosion of the articular cortex of the femoral head is seen. Arrow **1** Acetabular erosion. Arrow **2** Femoral head erosion.

noninvasive technique for the evaluation of these lesions (24) (Fig. 6.35). Most osteochondral fractures are seen as low SI lesions involving the articular surface of the bone. If the fragment is completely detached, a high SI band surrounding the bony fragment can be seen on PDWI and T2WI, representing either fluid or healing granulation tissue. The viability of the osteochondral fragment is difficult to evaluate with MRI, since SI changes do not exactly reflect the pathophysiologic events taking place.

The main differential diagnosis of osteochondral fractures should be made with AVN. This diagnosis may occasionally be difficult when based solely on the MRI findings. The presence of a double-line sign should favor AVN. However, an osteochondral fragment partially detached can produce similar MRI findings. Localized thinning of the cartilage covering the lesion would favor osteochondral fracture, since cartilage thinning occurs only in the late stages of AVN, and then it involves diffusely the articular cartilage with associated degenerative changes.

TENDON LESIONS

Avulsion fractures in the area of the pelvis or proximal femora are frequently seen in young patients involved in sports (6). Clinically, they present with pain and little external evidence of injury. Treatment is conservative, with bed rest and immobilization of the leg. Plain films demonstrate the avulsed fragment in most cases (40), but MRI may be required to exclude underlying pathology.

Tendon avulsion and rupture can be accurately diagnosed by ultrasound when the lesion is superficially located (16), but deep injuries such as hamstring tendon avulsion or other tendinous lesions around the hip joint may be beyond the reach of ultrasound examination. MRI in these cases can provide useful information including confirmation of the clinical diagnosis and assessment of the extent of the lesion (6).

▼▼▼ *HAMSTRING TENDON AVULSION*

Forceful contraction of the hamstring muscles can produce an avulsion of the ischial apophysis in young patients, before closure of the growth plate. In the adult patient, however, the weak area may not be at the epiphyseal plate but at the musculotendinous unit. Radiographs in these cases may be completely normal.

Most hamstring injuries occur while running or sprinting (1,19). Clinical symptoms of avulsion of the hamstring muscles include sudden onset of buttock pain, difficulty on standing after trauma, palpable defect and tenderness in the region of the ischial tuberosity, and weakness of knee flexion. Treatment of hamstring avulsion includes rest and immobiliza-

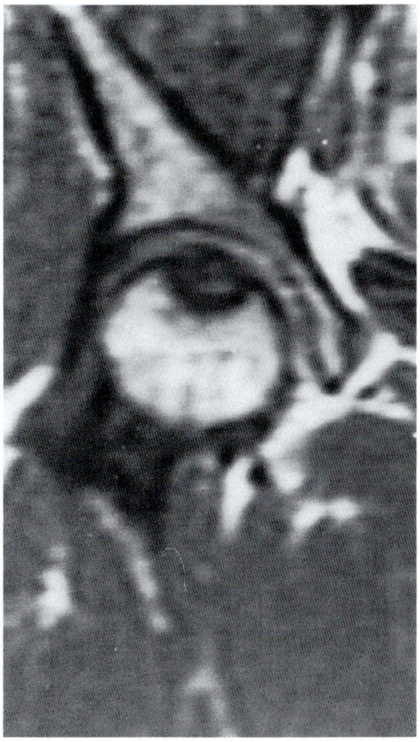

FIGURE 6.35. Osteochondral Fracture. Coronal section [SE 2000/20]. There is an osteochondral fragment partially detached from the femoral head. The articular cartilage at this level is narrowed. Without history of trauma and without localized cartilage narrowing, distinction from avascular necrosis would be very difficult (cf. Fig. 6.17).

tion following injury, with a gradual return to activity. Large ruptures may require surgical intervention. Lack of visualization of the normal tendons of the hamstring muscles in the ischial tuberosity, and inflammatory changes with high SI on T2WI, are the diagnostic findings on MRI (Fig. 6.36).

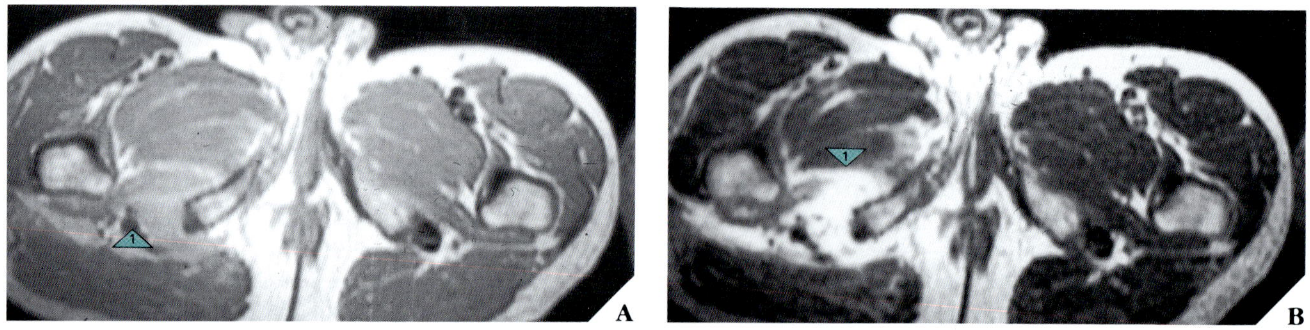

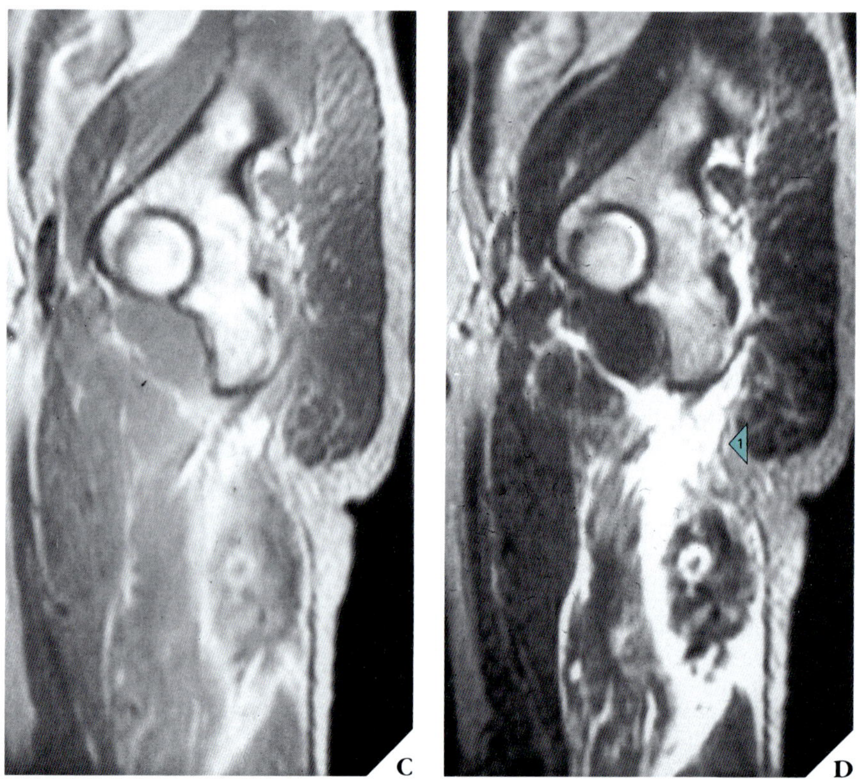

FIGURE 6.36. Hamstring Tendon Avulsion. A Axial section [SE 2000/20] through the ischial tuberosities. There is separation of the hamstring tendons from the right ischial tuberosity. Compare with the normal left side. **B** Axial section [SE 2000/80] at the same level as in **A.** There is high SI in the region of the avulsion due to hematoma. **C** Sagittal section [SE 2000/20] through the right ischial tuberosity. The normal hamstring tendons are not identified (cf. Fig. 6.15). **D** Sagittal section [SE 2000/80] at the same level as in **C.** There is a high SI hematoma at the avulsion site. Arrows **1** Avulsion site.

Iliopsoas Tendon Avulsion

Avulsion of the lesser trochanter at the iliopsoas tendon insertion is rare in adults, and when present is frequently related to a pathologic fracture, most commonly due to metastatic carcinoma (7,30) (Fig. 6.37). Frequently, there is no history of trauma. Occasionally, patients may relate the acute onset of pain following minimal activities such as coughing or walking.

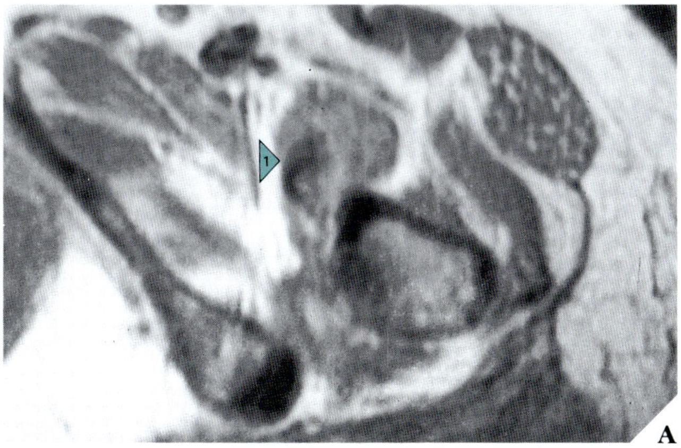

A

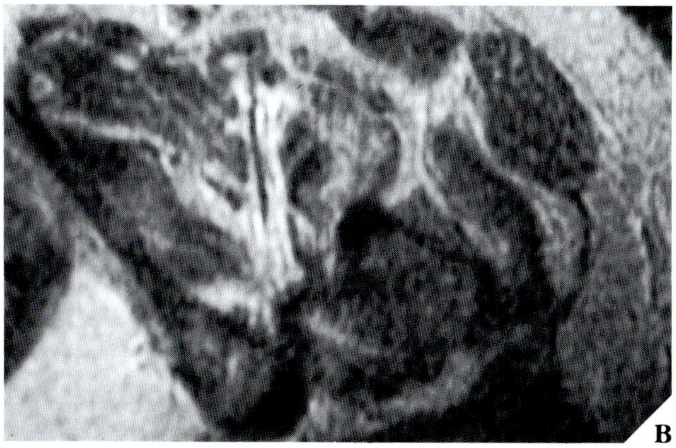

B

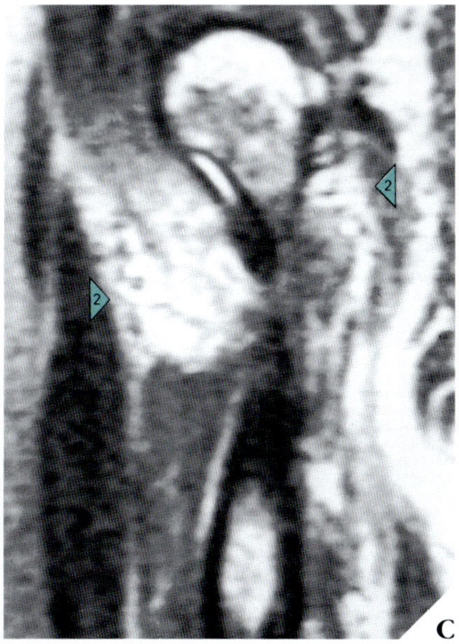

C

FIGURE 6.37. Avulsion of the Lesser Trochanter Due to Metastasis. **A** Axial section [SE 2000/20] through the region of the lesser trochanter. The lesser trochanter is avulsed anteriorly. Bone metastases were present at other levels. **B** Axial section [SE 2000/80] through the same level as in **A.** There is moderate SI increase within the iliopsoas muscle, due to hematoma, in this T2WI. **C** Sagittal section [SE 2000/80]. Increased SI in the region of the iliopsoas muscle is again identified. Arrow **1** Avulsed lesser trochanter. Arrows **2** Iliopsoas muscle.

REFERENCES

1. Agre JC: Hamstring injuries. Proposed aetiological factors, prevention, and treatment. Sports Med 1985;2:21–23.
2. Beltran J, Burk JM, Herman LJ, et al: Avascular necrosis of the femoral head: early MRI detection and radiological correlation. Magn Reson Imag 1987;5:431–442.
3. Beltran J, Caudill JL, Herman LA, et al: Rheumatoid arthritis: MR imaging manifestations. Radiology 1987;165:153–157.
4. Beltran J, Herman LJ, Burk JM, et al: Femoral head avascular necrosis: MR imaging with cinical–pathologic correlation. Radiology 1988;166:215–220.
5. Beltran J, Noto AM, Herman LJ, et al: Joint effusions: MR imaging. Radiology 1986;158:133–137.
6. Beltran J, Noto AM, Herman LJ, Lubbers LM: Tendons: high-field-strength, surface coil MR imaging. Radiology 1987;162:735–740.
7. Bertin KC, Horstman J, Coleman SS: Isolated fracture of the lesser trochanter in adults: an initial manifestation of metastatic malignant disease. J Bone Joint Surg 1984;66:770–773.
8. Bloem JL: Transient osteoporosis of the hip: MR imaging. Radiology 1988;167:753–755.
9. Bluemm RH, Falke THM, de Plantes Z: Early Legg–Perthes disease (ischemic necrosis of the femoral head) demonstrated by magnetic resonance imaging. Skeletal Radiol 1985;14:95–98.
10. Boal DKB, Swenkter EP: Infant hip: assessment with real time US. Radiology 1985;157:667–672.
11. Burk DL, Dalinka MK, Schieber ML, et al: Strategies for musculoskeletal magnetic resonance imaging. Radiol Clin North Am 1988;26:653–672.
12. Deutsch AL, Mink JH, Waxman AD: Occult fractures of the proximal femur: MR imaging. Radiology 1989;170:113–116.
13. Dominguez R, Oh KS, Young LW, Goodman M: Acute chondrolysis complicating Legg–Calve–Perthes diseases. Skeletal Radiol 1987;16:377–382.
14. Ehman RL, Berquist TH, McLeod RA: MR imaging of the musculoskeletal system: a 5-year reappraisal. Radiology 1988;166:313–320.
15. Ficat RP: Idiopathic bone necrosis of the femoral head. Early diagnosis and treatment. J Bone Joint Surg (Am) 1986;68:1313–1319.
16. Fornage BD, Rifkin MD: Ultrasound examination of the tendons. Radiol Clin North Am 1988;26:87–107.
17. Glickstein MF, Burk DL, Schiebler ML, et al: Avascular necrosis versus other diseases of the hip: sensitivity of MRI. Radiology 1988;169:213–216.
18. Hungerforth DS: Bone marrow pressure, venography, and core decompression in ischemic necrosis of the femoral head. In: Hungerforth DS, ed. The Hip: Proceedings of the 7th Open Scientific Meeting of The Hip Society, 1979, vol. 10. St. Louis, Mosby, 1979:218–237.
19. Ishikawa K, Kai K, Mizuta H: Avulsion of the hamstring muscles from the ischial tuberosity. A report of two cases. Clin Orthop 1988;232:153–155.
20. Kricum ME: Red-yellow marrow conversion: its effect on the location of some solitary bone lesions. Skeletal Radiol 1985;14:10–19.
21. Lang BA, Jergesen HE, Moseley ME, et al: Avascular necrosis of the femoral head: high-field-strength MR imaging with histologic correlation. Radiology 1988;169:517–524.
22. Lee JK, Yao L: Stress fractures: MR imaging. Radiology 1988;169:217–220.
23. Lequesne M: Transient osteoporosis of the hip: a nontraumatic variety of Sudek's atrophy. Ann Rheum Dis 1968;27:463–471.
24. Mesgarzadeh M, Sapega AA, Bonakdarpour A, et al: Osteochondritis dissecans: analysis of mechanical stability with radiography, scintigraphy and MR imaging. Radiology 1987;165:775–780.
25. Mitchell DG, Joseph PM, Fallon M, et al: Chemical-shift MR imaging of the femoral head: an in-vitro study of normal hips and hips with avascular necrosis. AJR 1987;148:1159–1164.
26. Mitchell DG, Rao VM, Dalinka MK, et al: Femoral head avascular necrosis: correlation of MR imaging, radiographic staging, radionuclide imaging, and clinical findings. Radiology 1987;162:709–715.
27. Mitchell MD, Kundell HA, Steinberg ME, et al: Avascular necrosis of the hip: comparison of MR, CT and scintigraphy. AJR 1986;147:67–71.
28. Netter FH: The Ciba collection of medical illustrations. Vol. 8: Musculoskeletal system. Summit, NJ, Ciba-Geigy Corporation, 1987.
29. Peters JC, Coleman BG, Turner ML, et al: CT evaluation of enlarged iliopsoas bursa. AJR 1980;135:392–394.

30. Phillips CD, Pope TL, Jones JE, et al: Nontraumatic avulsion of the lesser trochanter: a pathognomonic sign of metastatic disease? Skeletal Radiol 1988;17:106–110.
31. Porter BA, Hastrup W, Richardson ML, et al: Classification and investigation of artifacts in magnetic resonance imaging. Radiographics 1987;7:271–287.
32. Rao VM, Fishman M, Mitchell DG, et al: Painful sickle cell crisis: bone marrow patterns observed with MR imaging. Radiology 1986;161:211–215. 1986;161:211–215.
33. Rao VM, Mitchell DG, Steiner RM, et al: Femoral head avascular necrosis in sickle cell anemia: MR characteristics. Magn Reson Imag 1988;6:661–667.
34. Resnick D, Niwayama G: Degenerative diseases of extraspinal location. In Resnick D, Niwayama G (eds): Diagnosis of Bone and Joint Disorders, ed 2. Philadelphia, W.B. Saunders, 1988,1364–1479.
35. Rush BH, Bramson RT, Ogden JA. Legg–Calve–Perthes disease: detection of cartilaginous and synovial changes with MR imaging. Radiology 1988;167:473–476.
36. Schuman WP, Castagno AA, Baron RL, Richardson ML: MR imaging of avascular necrosis of the femoral head: value of small-field-of-view sagittal surface-coil images.
37. Stafford SA, Rosenthal DI, Gebhardt MC, et al. MRI in stress fracture. AJR 1986;147:553–556.
38. Steinbach LS, Schneider R, Goldman AB: Bursae and abscess cavities communicating with the hip. Diagnosis using arthrography and CT. Radiology 1985;156:303–307.
39. Sundaram M, McGuire MH, Fletcher J, et al: Magnetic resonance imaging of lesions of synovial origin. Skeletal Radiol 1986;15:110–116.
40. Tehranzadeh J: The spectrum of avulsion and avulsion-like injuries of the musculoskeletal system. Radiographics 1987;7:945–974.
41. Vogler JB, Murphy WA: Bone marrow imaging. Radiology 1988;168:679–693.
42. Wilson AJ, Murphy WA, Hardy DC, Totty WG: Transient osteoporosis: transient bone marrow edema? Radiology 1988;167:757–760.
43. Yao L, Lee JK. Occult intraosseous fracture: detection with MR imaging. Radiology 1988;167:749–751.

FIGURE CREDITS

Figure 6.2. Adapted from Kricum ME: Red-yellow marrow conversion: its effect on the location of some solitary bone lesions. Skeletal Radiol 1985;14:10–19.

Figures 6.27, 6.30, and 6.35. Courtesy of J. Mink, M.D., Cedars of Sinai Medical Center, Los Angeles, CA.

CHAPTER SEVEN

Javier Beltran

The Knee

TECHNICAL CONSIDERATIONS

IMAGING PLANES AND PULSE SEQUENCES

T1-weighted images with short TR/TE are obtained in the axial and coronal planes (Fig. 7.1). Proton density (long TR, short TE) and T2WI with long TR/TE are obtained in the sagittal plane. TR and TE are adjusted to allow the maximum number of slices for each plane to cover the entire knee joint. The examination, including positioning and prescanning, can be performed in approximately 45 minutes.

The sagittal plane is the most sensitive plane for detection of meniscal, cruciate ligament, and cartilaginous lesions, as well as for visualization of the femoropatellar joint. However, use of this plane alone is strongly discouraged because significant lesions, such as bucket-handle tears and collateral ligament ruptures, can be missed.

The use of short TR/TE pulse sequences in the sagittal and coronal planes is an acceptable protocol for studying patients who present with knee pain but who have no history of acute injury (34). This technique saves imaging time and provides excellent sensitivity for the detection of meniscal lesions and collateral ligament tears. However, the use of the long TR/TE technique in the sagittal plane permits better evaluation of cruciate ligament tears, articular cartilage lesions, bone marrow lesions, and other tumorous and inflammatory knee disorders.

Fast scanning techniques (GRASS, FLASH, FISP) with three-dimensional Fourier transform (3DFT) (17,20,22) have several advantages over routine SE 2DFT: (a) data of the entire volume of the knee can be acquired in about 10 minutes, (b) image reconstruction can be done in practically any plane of 1-mm contiguous slices, (c) excellent visualization of the articular cartilage of the knee can be obtained, and (d) meniscal lesions can be detected with nearly the same accuracy as that achieved with regular SE 2DFT sequences (47,52,54) (Fig. 7.2).

The disadvantages of fast scanning techniques with 3DFT are as follows: (a) clear separation between joint fluid and hyaline cartilage may be difficult, unless the flip angle is carefully adjusted, (b) the detectability of bone marrow lesions may be decreased, and (c) an independent computer is necessary to evaluate the multiple images and planes obtained with 3DFT. There is potential for fast scanning with 3DFT to be used as a screening technique, for it can sufficiently detect meniscal or ligamentous tears in the majority of patients. The more accurate technique, classic SE 2DFT, can be reserved to study synovial, bone marrow, and mass lesions.

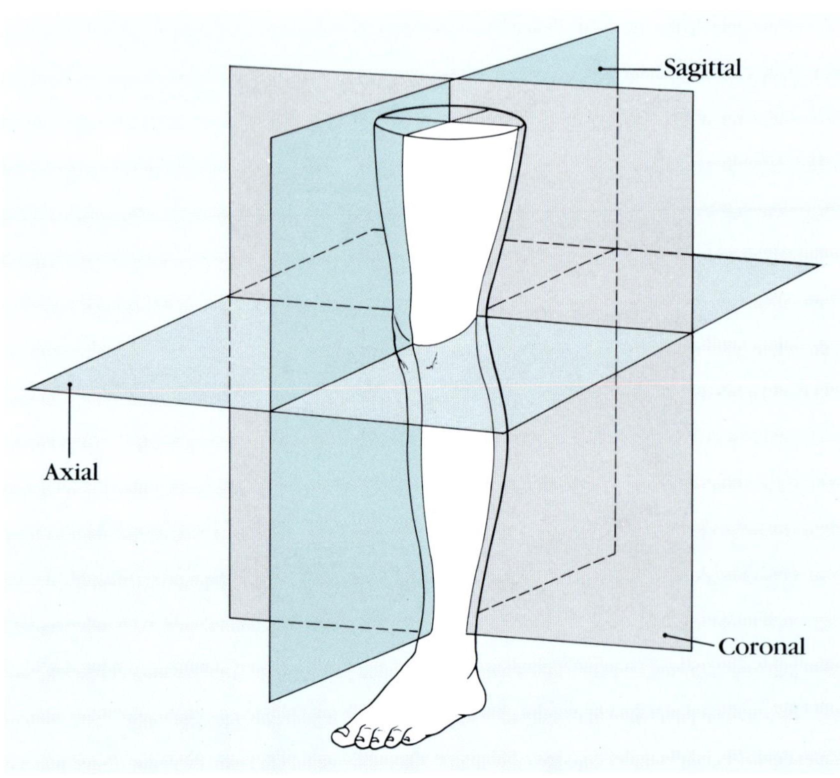

FIGURE 7.1. Routine Planes of Section for Knee Imaging.

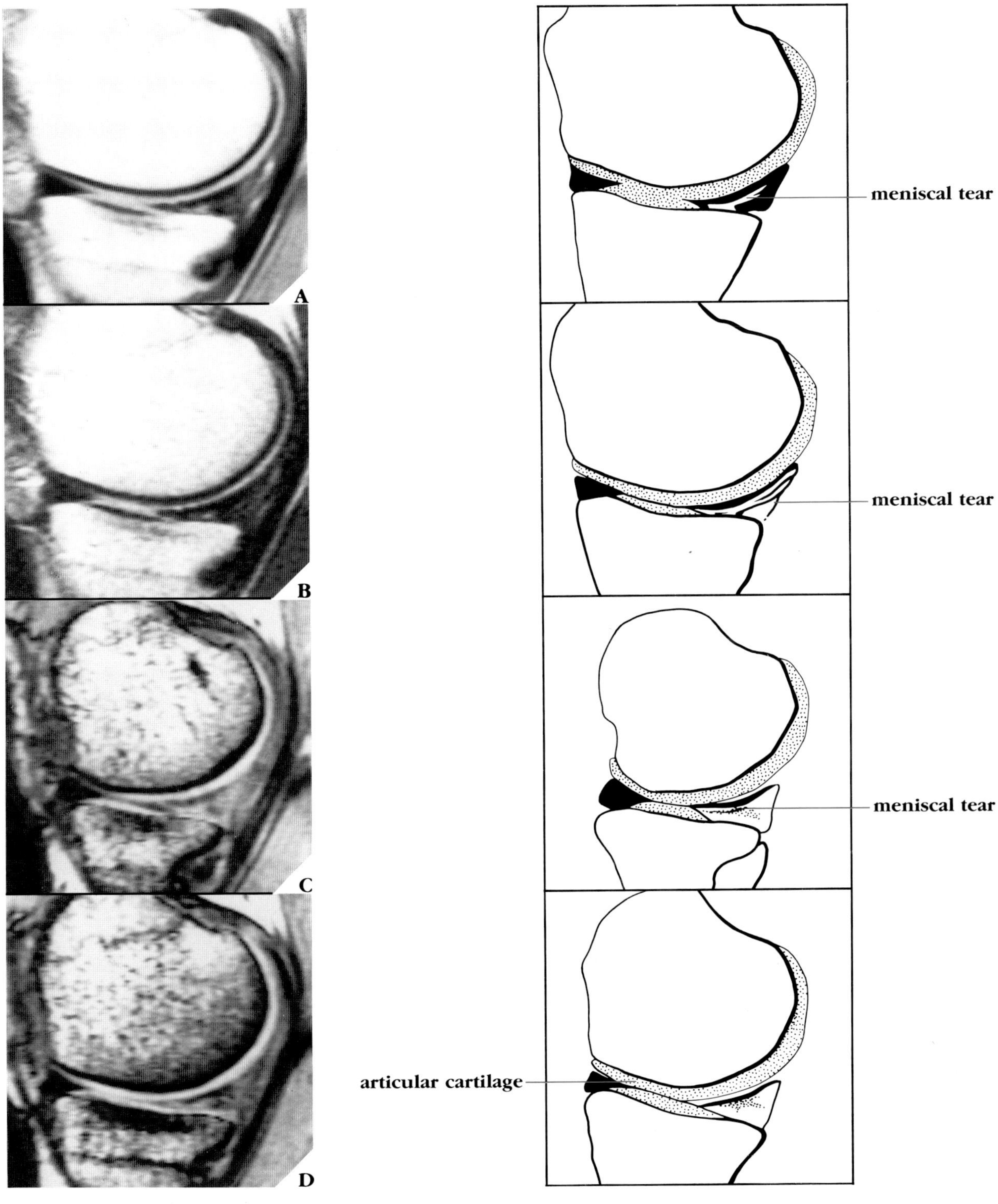

FIGURE 7.2. Tear of the Posterior Horn of the Medial Meniscus. A and **B** 2DFT [SE 2000/20], **C** and **D** 3DFT. The 3DFT images were obtained with GRASS technique, using 128 phase-encoding steps in the anterior to posterior direction, which provided 64 contiguous slices. The flip angle is 30 degrees, TR is 66 msec, and TE is 15 msec with one NEX. The articular cartilage is well-defined on 3DFT as high SI tissue over the medial femoral condyle and tibial plateau.

▼▼▼ *FIELD OF VIEW*

A FOV of 16 cm is sufficient to cover the entire area in most knees, but it can be reduced in pediatric cases. A smaller FOV decreases S/N by decreasing the pixel size and the number of protons per voxel. Small FOV's may also produce more aliasing (wrap-around) artifact (1, 40, 42, 43).

The area covered in the axial plane extends from the suprapatellar region to the tibial tuberosity. The area between the medial and lateral margins of the tibial plateau is covered in the sagittal plane. The patella to the posterior margin of the femoral condyles is covered in the coronal plane.

▼▼▼ *POSITIONING*

The leg is positioned foot first, supine, in the isocenter of the magnetic field with maximum extension and about 10 degrees of external rotation. This allows routine visualization of the anterior cruciate ligament. In the relaxed position, the knee is already 5–10 degrees externally rotated. The use of positioning sponges helps secure and stabilize both the knee and the ankle.

▼▼▼ *ACQUISITION MATRIX*

A fixed acquisition matrix of 256 pixels along the frequency-encoding gradient is obtained in all planes. A matrix of 128 pixels is obtained in the phase-encoding gradient in the sagittal plane (long TR) to keep imaging time short. A high resolution matrix of 256 × 256 pixels can be obtained in the coronal plane using a short TR (11).

▼▼▼ *SLICE THICKNESS*

A slice thickness of 5 mm with 1-mm interslice gaps should be used in all series. This technique will provide adequate S/N without compromising lesion detection. Unless interleaving is used, contiguous slices will produce image degradation by cross talk between the slices.

▼▼▼ *GRADIENTS*

A frequency-encoding gradient, placed in the superior to inferior direction, helps to avoid aliasing artifacts in the coronal and sagittal planes. Presaturation with a 90-degree RF pulse outside of the region of interest decreases flow-related artifacts, originating from the popliteal vessels (16).

Some degree of chemical shift artifact occurs in the direction of the frequency-encoding gradient at the interface between fatty bone marrow and cortical bone. This produces the false appearance of femoral cartilage thinning (34). Knowledge of this artifact or change in the direction of the gradient can help to prevent interpretation errors.

▼▼▼ *SURFACE COILS*

Transmit—receive, wrap-around surface coils are used routinely in all knee examinations (32) (Fig 7.3). This coil provides uniform SI and high spatial resolution across the knee with excellent S/N. The circumferential coil design helps to minimize popliteal artery pulsation artifacts.

Loop–gap resonator coils (Fig. 7.4) can be used instead of the transmit–receive extremity coil, with

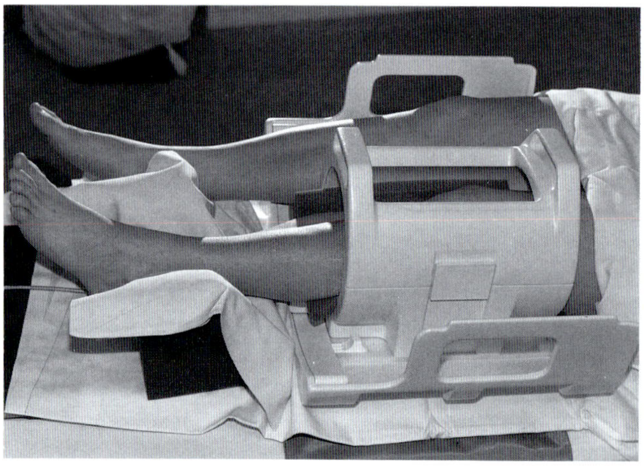

FIGURE 7.3. Transmit–Receive Extremity Coil Positioned for Knee Imaging.

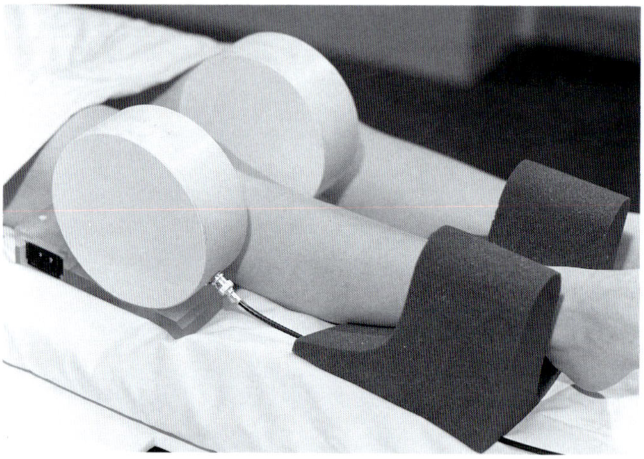

FIGURE 7.4. Receive-Only Loop–Gap Resonator Coils.

similarly uniform distribution of the signal across the knee (25). Loop–gap resonator coils are receive-only coils and must be tuned prior to patient imaging. Flat rectangular or circular surface coils are not recommended because of the drop-off in SI in regions distant from the coil.

NORMAL ANATOMY

SAGITTAL PLANE

The Menisci

The menisci are seen on MRI as wedge-shaped structures of uniformly low SI in practically all pulse sequences. This contrasts with the relatively high SI of the surrounding articular cartilage superiorly and inferiorly and the pericapsular fat peripherally (7,44). On peripheral sections, the menisci are seen as "bow tie" structures (Figs. 7.5–7.7). In about 30 per cent of cases, a horizontal, high SI line can be seen within the substance of the meniscus at the periphery. This line does not represent a meniscal tear. Rather, it is thought to be related to volume averaging of the high SI tissue at the concave meniscal edge on PD (24) (Figs. 7.7 and 7.8).

On the more internal sections, the anterior and posterior horns of the menisci are seen as triangular structures (Figs. 7.5–7.7). A persistent "bow tie" shape is seen in the discoid meniscus (34). At the tibial attachments of the anterior horns, the transverse ligament can be seen separated from the meniscus by a hyperintense line. This is not to be confused with a meniscal tear (Fig. 7.9). The horizontal "V"-shaped structure represents the transverse ligament superiorly and, the attachments of the meniscus inferiorly.

The popliteal tendon is seen as a low SI structure, separated from the posterior horn of the lateral meniscus by a hyperintense line, not to be confused with a peripheral meniscal lesion (24) (Fig. 7.6).

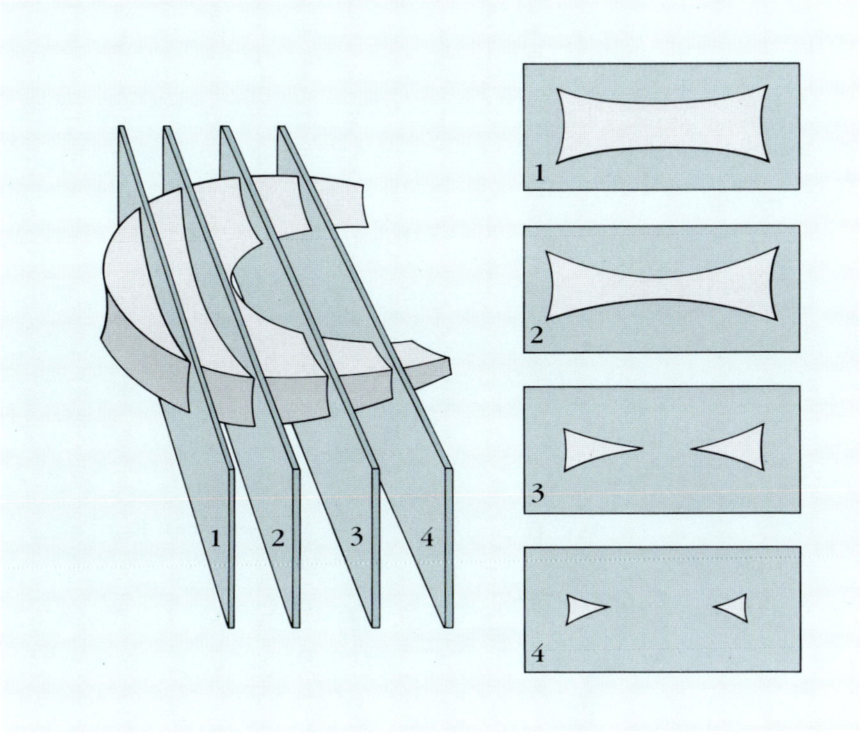

FIGURE 7.5. Normal Meniscal Shape in the Sagittal Plane. The shape of the meniscus in sagittal sections changes depending upon the level of the section. A "bow tie" shape is seen in peripheral sections and a triangular shape is seen in internal sections.

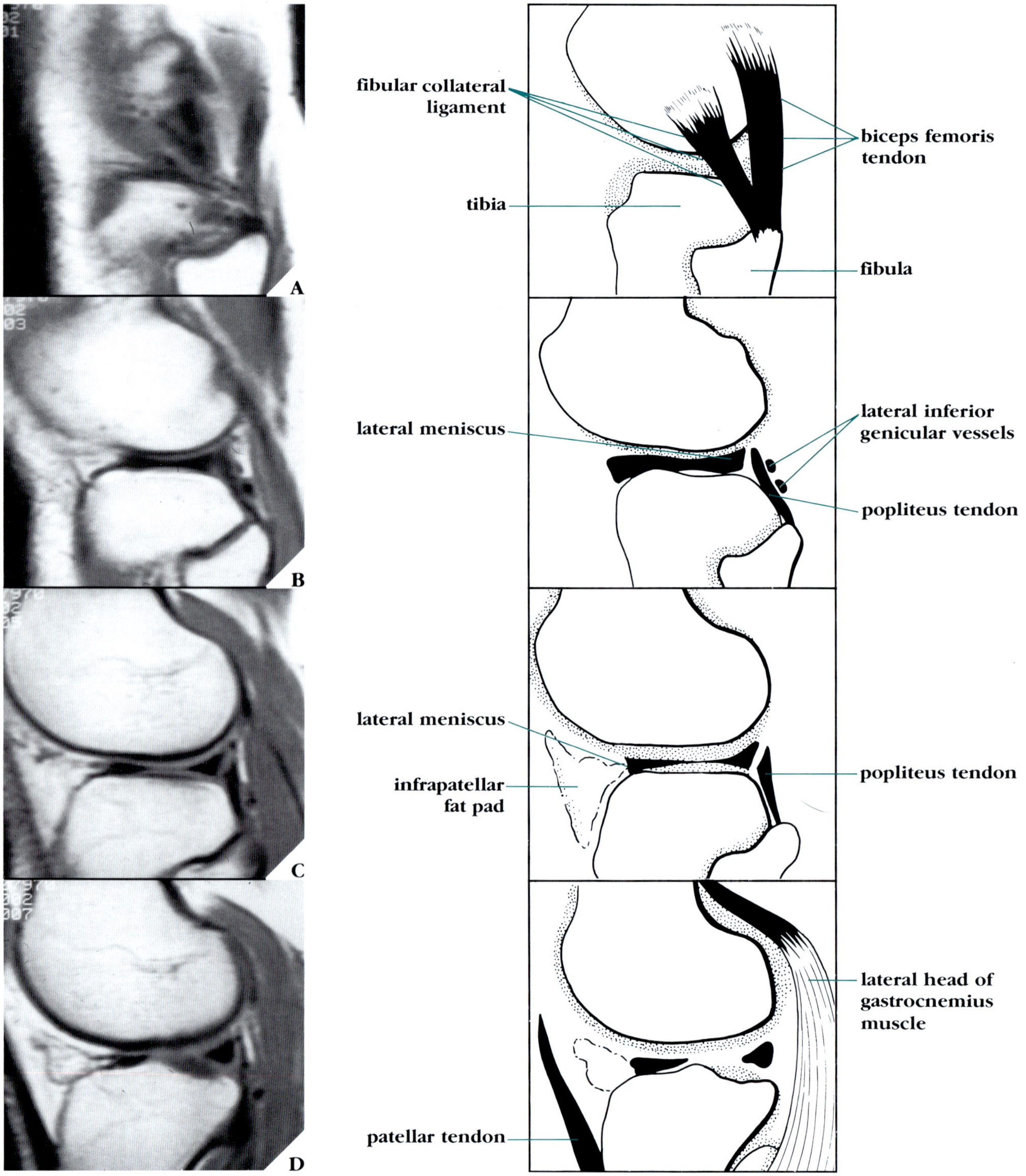

FIGURE 7.6. Normal Lateral Meniscus. Sagittal sections [SE 2000/20]. Note the confluence of the fibular collateral ligament and the biceps femoris tendon in the more peripheral section **(A).** The popliteus tendon is separated from the posterior horn of the lateral meniscus by intermediate SI tissue simulating a vertical tear **(B** and **C).**

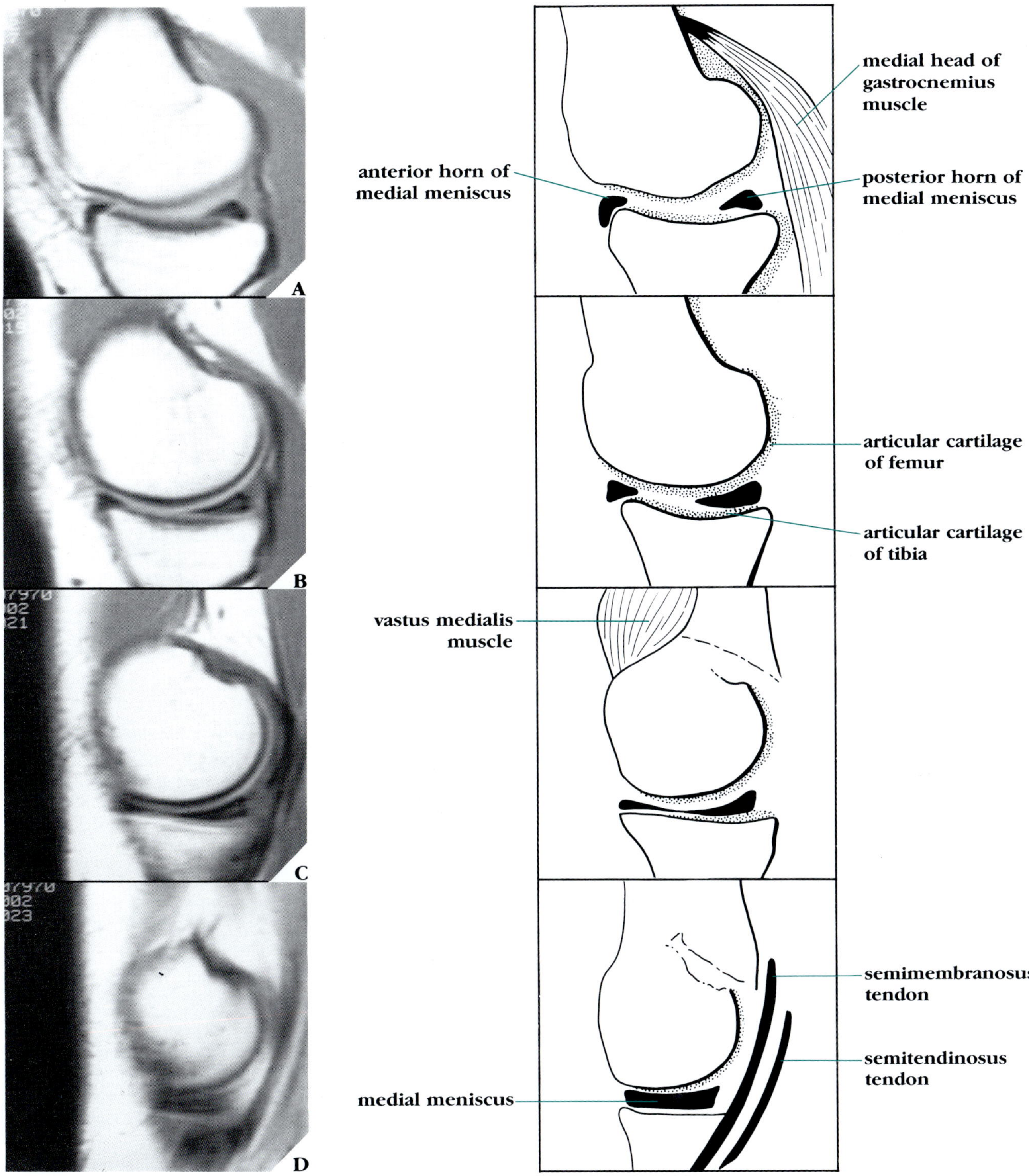

FIGURE 7.7. Normal medial meniscus. Sagittal sections [SE 2000/20]. Note the horizontal hyperintense line through the periphery of the meniscus, produced by tissue averaging with high SI tissue in the concave meniscal edge.

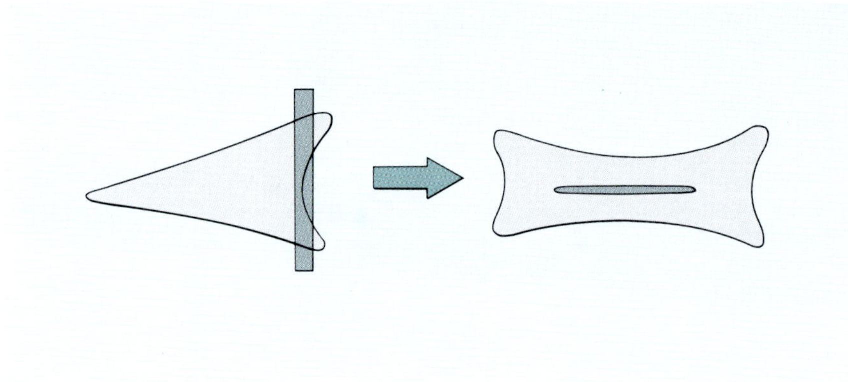

FIGURE 7.8. High SI Tissue Within the Concave Meniscal Edge Simulating a Horizontal Tear. A 5-mm-thick section through the periphery of the meniscus.

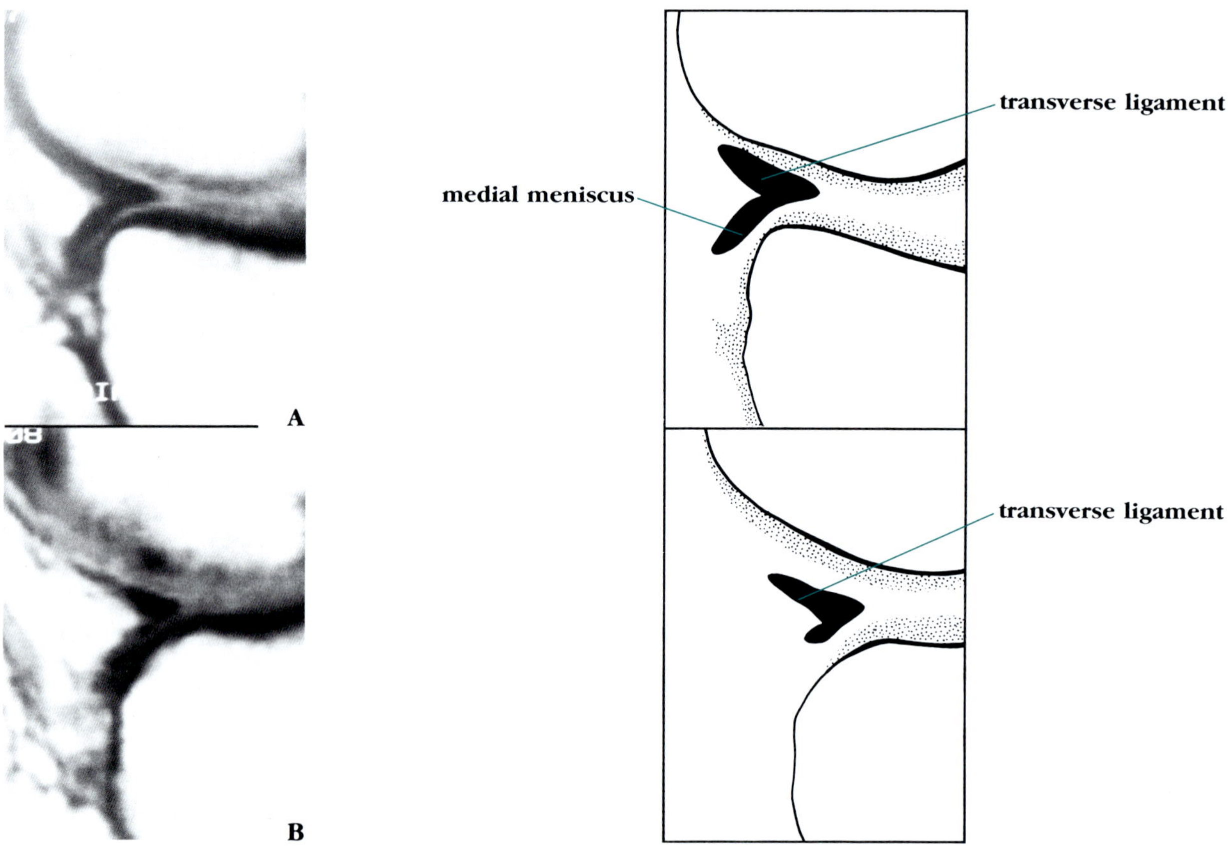

FIGURE 7.9. Normal Tranverse Ligament. A and **B** anterior horn of the medial meniscus; **C** and **D** intercondylar notch; **E** and **F** anterior horn of the lateral meniscus. Consecutive sagittal sections [SE 2000/20]. The transverse ligament attaches to the superior aspect of the anterior meniscal horns. The horizontal "V" at the level of the medial meniscus is formed by the transverse ligament superiorly and the anterior medial meniscal horn inferiorly **(A).** The vertical line at the junction between the transverse ligament and the anterior horn of the lateral meniscus **(F)** simulates a vertical tear (arrow **1**).

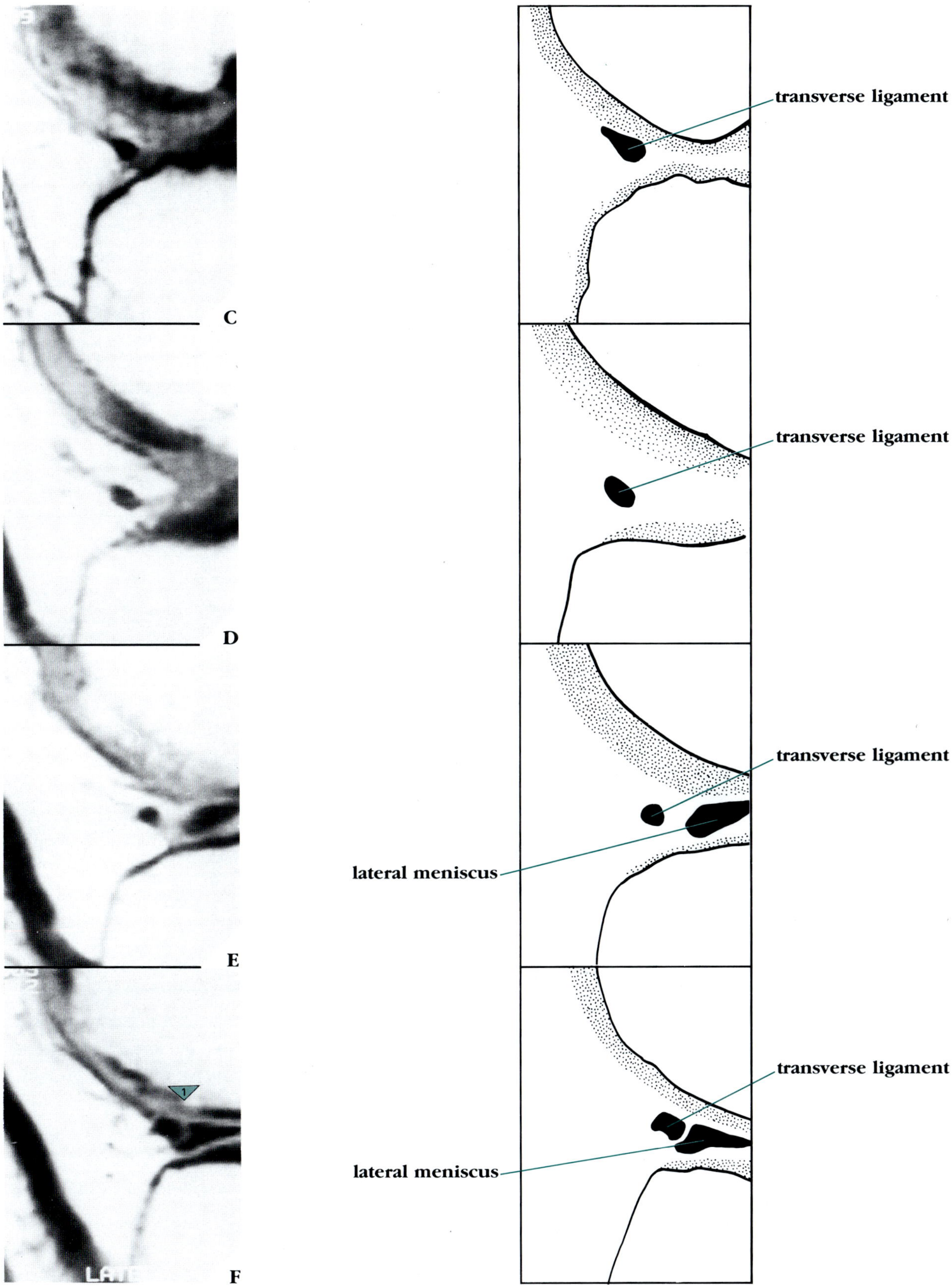
C
transverse ligament
D
transverse ligament
E
transverse ligament
lateral meniscus
F
transverse ligament
lateral meniscus

The Cruciate Ligaments

The cruciate ligaments, like the menisci, are seen as low SI structures in all SE sequences (7,44).

Sagittal sections obtained with 10–15 degrees of external rotation can demonstrate the anterior cruciate ligament (ACL) in over 95 per cent of the cases (34,44). The posterior cruciate ligament (PCL) is seen in practically all normal knees. The ACL is seen as a fan- or band-like structure, extending obliquely from its insertion at the internal aspect of the lateral femoral condyle to the anterior aspect of the tibial spine (Fig. 7.10).

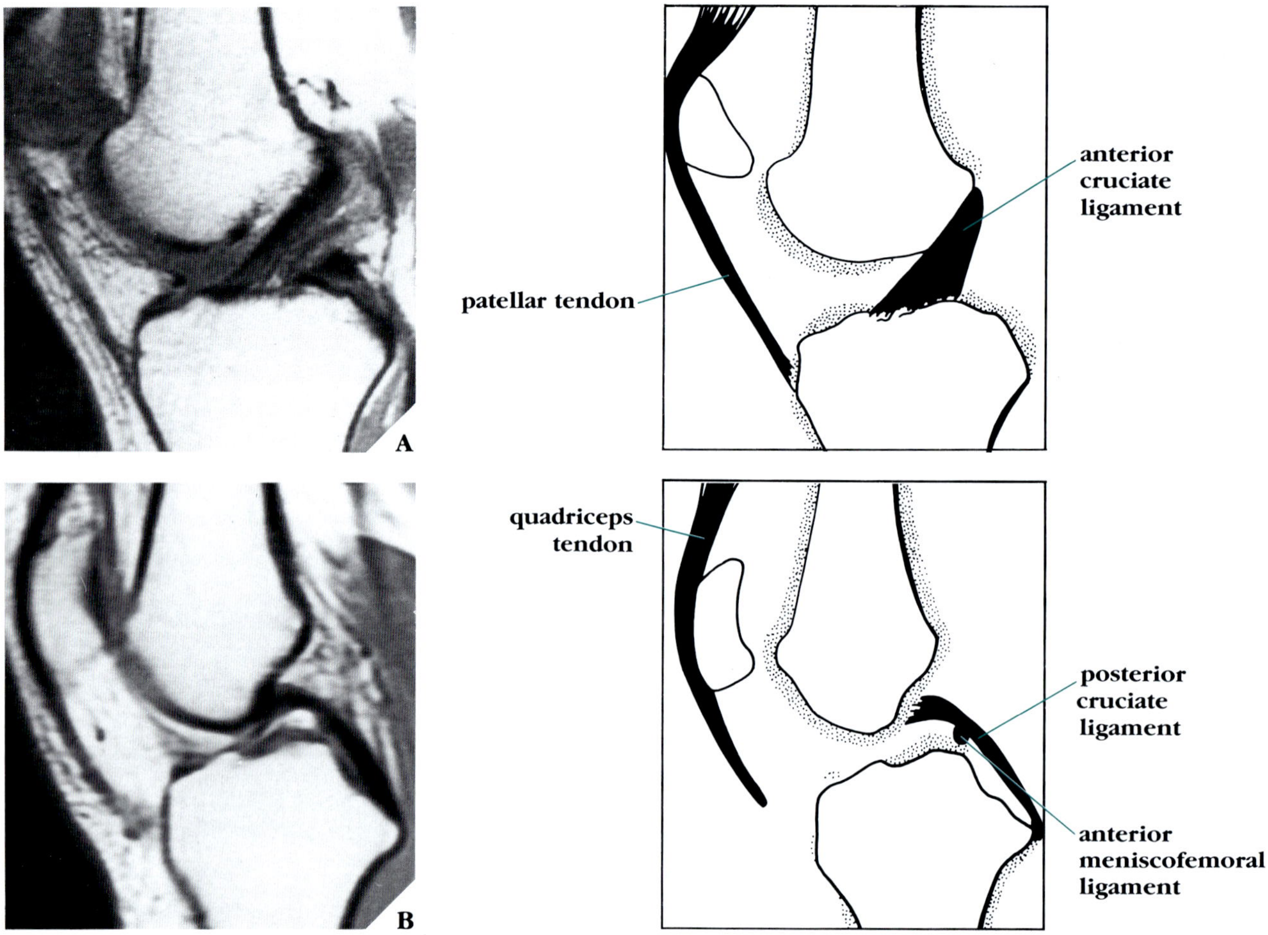

FIGURE 7.10. Normal Cruciate Ligaments. A anterior and **B** posterior cruciate ligament. The anterior margin of the anterior cruciate ligament is straight and well-defined. Sagittal sections [SE 2000/20]. The posterior margin is ill-defined due to the oblique orientation of the ligament. The posterior cruciate ligament is normally seen in its entirety, in one plane, from the femoral to the tibial attachments. Observe the anterior small bulge produced by the anterior meniscofemoral ligament.

The PCL is seen as a smoothly curved, low SI band, extending from the internal aspect of the medial femoral condyle to the posterior aspect of the tibia (Fig. 7.10). The anterior meniscofemoral ligament of Humphrey produces a localized anterior thickening of the PCL. This ligament may separate from the PCL and be easily confused with an intraarticular loose body (Fig. 7.11). Occasionally, the posterior meniscofemoral ligament of Weinberg produces a localized bulge in the posterior aspect of the PCL (Fig. 7.11). The meniscofemoral ligaments can also be identified in the coronal plane.

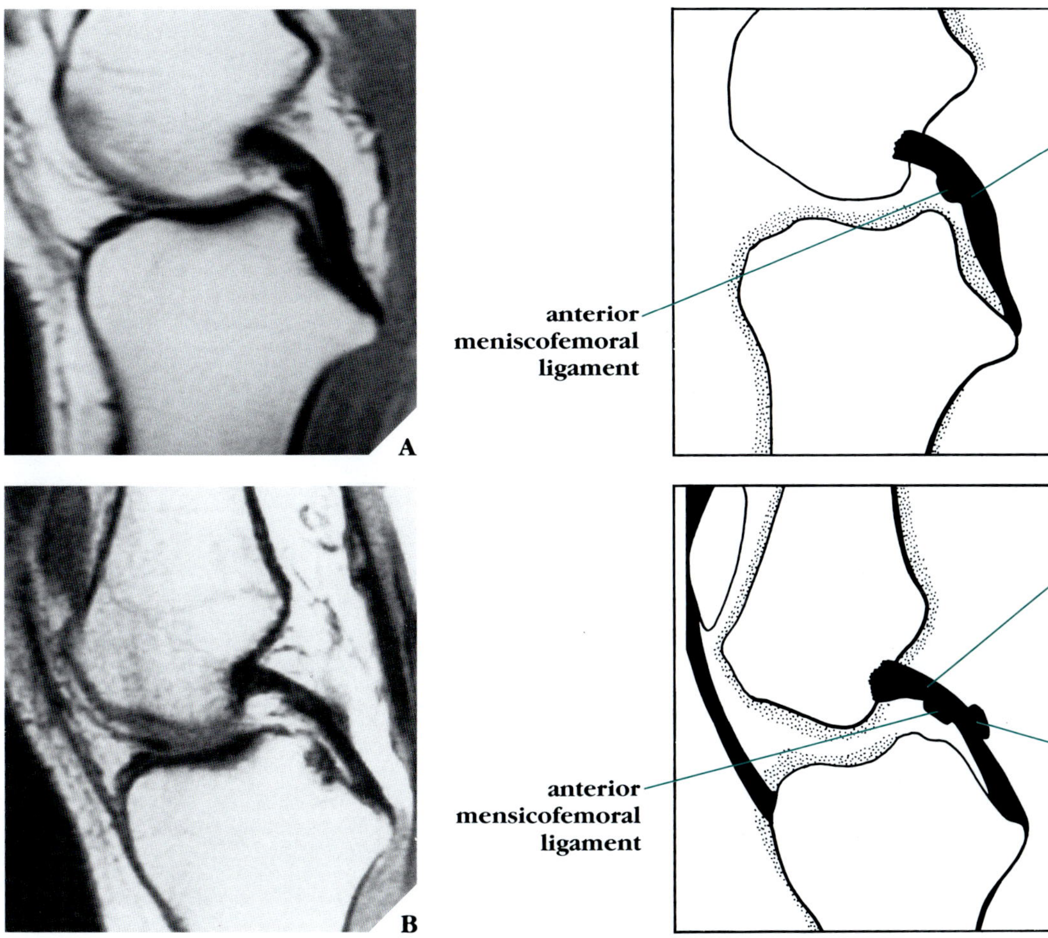

FIGURE 7.11. Variants in Configuration of the Posterior Cruciate Ligament. Sagittal sections [SE 2000/20]. In **A** the anterior meniscofemoral ligament is very prominent, simulating a loose body or meniscal fragment. The posterior meniscofemoral ligament is not present. In **B**, the anterior and posterior meniscofemoral ligaments are present and they are average in size.

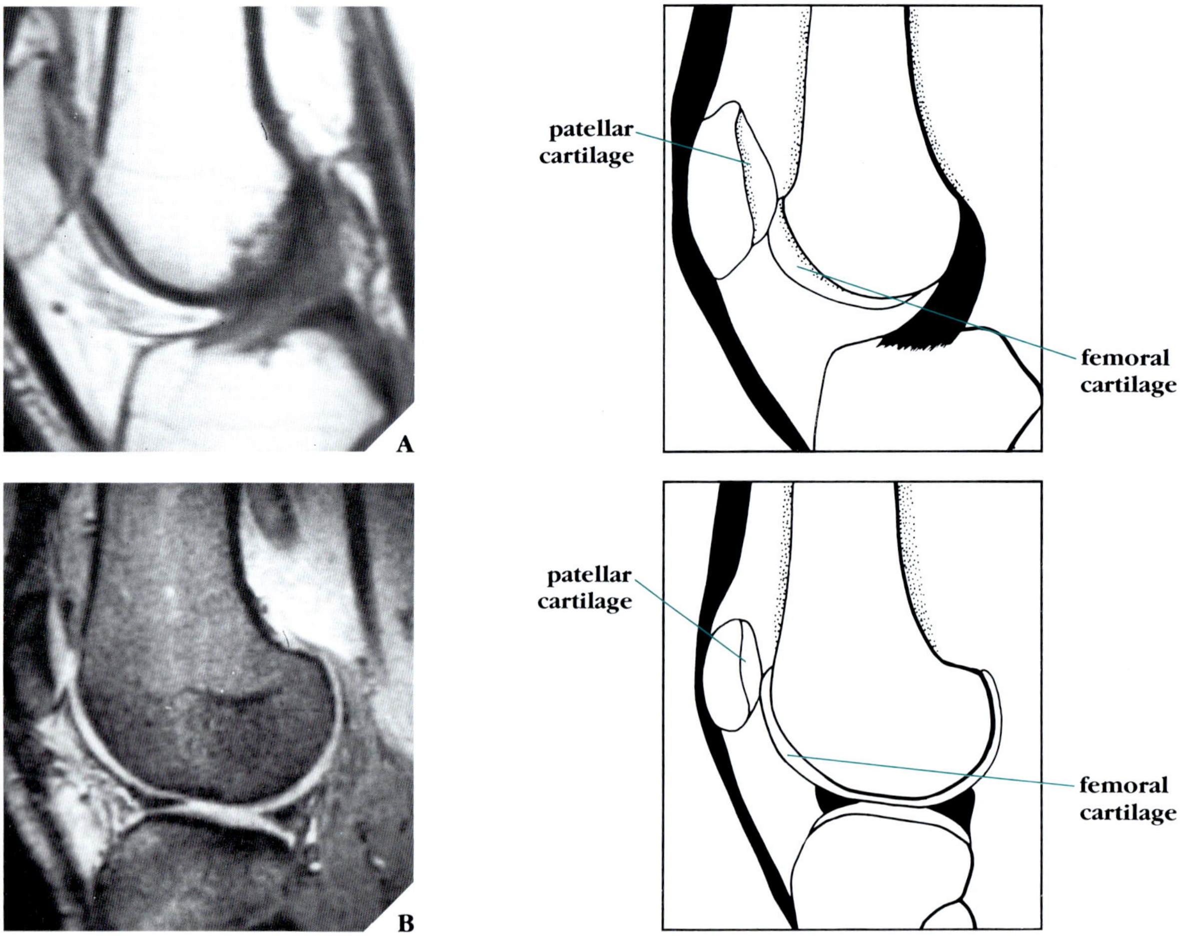

FIGURE 7.12. Normal Femoropatellar Joint Imaged with Two Different Techniques. A SE 2000/20; **B** GRASS Image. **A** Observe the intermediate SI of the articular cartilage in the patella and femoral condyle. The patellar cartilage is normally thickened in the midportion of the patella. The integrity of the quadriceps and patellar tendons is better evaluated with SE technique. **B** The hyperintense articular cartilage contrasts with the low SI of the fat and bone marrow.

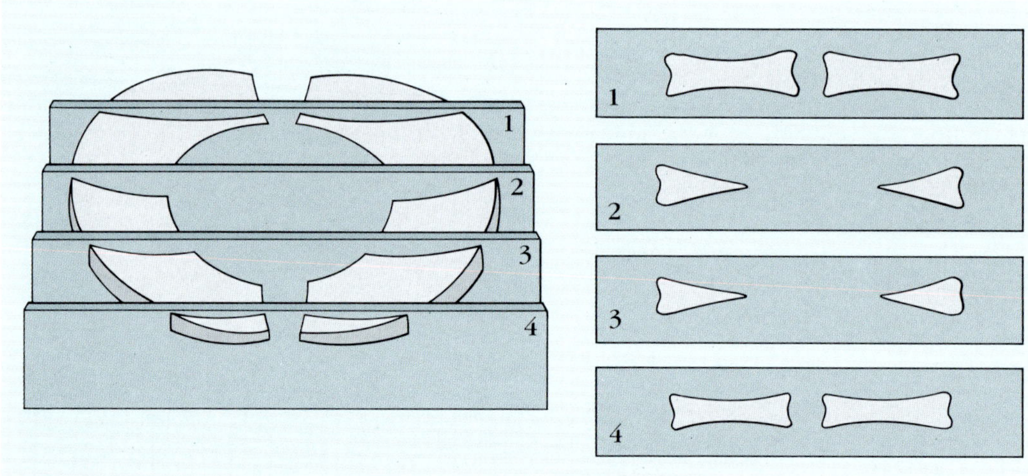

FIGURE 7.13. "Bow Tie" Configurations. Coronal sections through the anterior and posterior meniscal horns demonstrate a "bow tie" configuration.

The Femoropatellar Joint

The thick, relatively high SI articular hyaline cartilage of the patella is seen in sagittal sections using SE or fast imaging techniques (Fig. 7.12). A normal thinning of the patellar cartilage superiorly should not be confused with chondromalacia patellae. The patellar cartilage is thicker in the central region where it articulates with the hyaline cartilage of the femur in full extension.

The Tendons

Tendons, like menisci and ligaments, are hypointense in SE sequences (6). The quadriceps tendon inserts in the superior aspect of the patella (Fig. 7.12). The patellar tendon originates in the inferior margin of the patella and inserts in the tibial tuberosity. The patellar tendon should have straight, smooth margins and uniform thickness.

CORONAL PLANE

The Menisci

In coronal sections, the hypointense menisci are wider in the posterior and anterior horns and narrower in the midportions (Fig. 7.13). The same high SI line, secondary to the volume averaging seen in peripheral sagittal sections, can be seen in coronal views of the posterior horns simulating a horizontal tear.

The Ligaments

The lateral collateral ligament is formed by fibers of the fibular collateral ligament, iliotibial band, and arcuate ligament (15). The fibular collateral ligament, seen as a low SI band, extends from the lateral femoral condyle to the proximal fibula (Fig. 7.14) where it joins the biceps femoris tendon.

The superficial fibers of the medial (tibial) col-

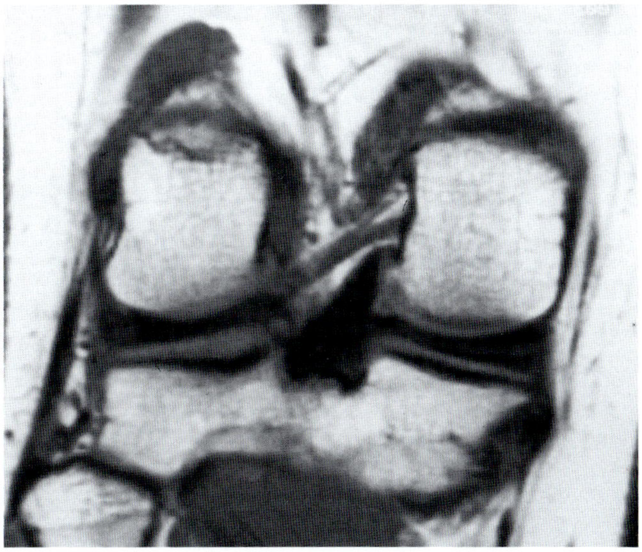

FIGURE 7.14. Normal Anterior Meniscofemoral Ligament. Coronal section [SE 2000/20]. The anterior meniscofemoral ligament extends from the posterior horn of the lateral meniscus to the inner surface of the medial femoral condyle. The posterior meniscal horns and the adjacent articular cartilage, as well as the other labeled structures, are well-defined.

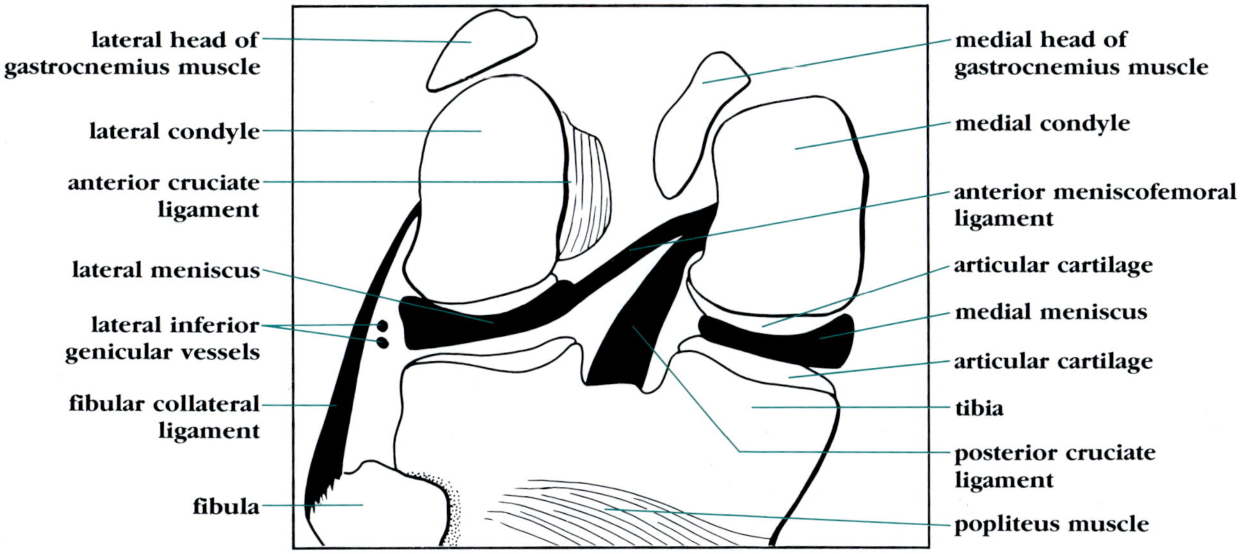

lateral ligament are seen as a thin, low SI band extending from the medial femoral condyle to the medial aspect of the proximal tibial epiphysis. The tibial collateral ligament is separated from the deep fibers or medial capsular ligament by an intraligamentous bursa (34) (Fig. 7.15).

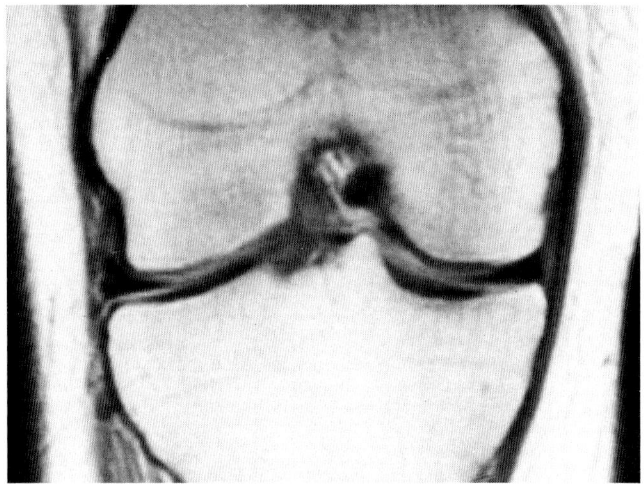

FIGURE 7.15. Normal Medial Collateral Ligament. The insertion of the posterior cruciate ligament in the inner aspect of the medial femoral condyle is demonstrated. Coronal section [SE 2000/20]. The menisci are seen as small triangles. The medial collateral ligament is well-defined in this section through the intercondylar notch.

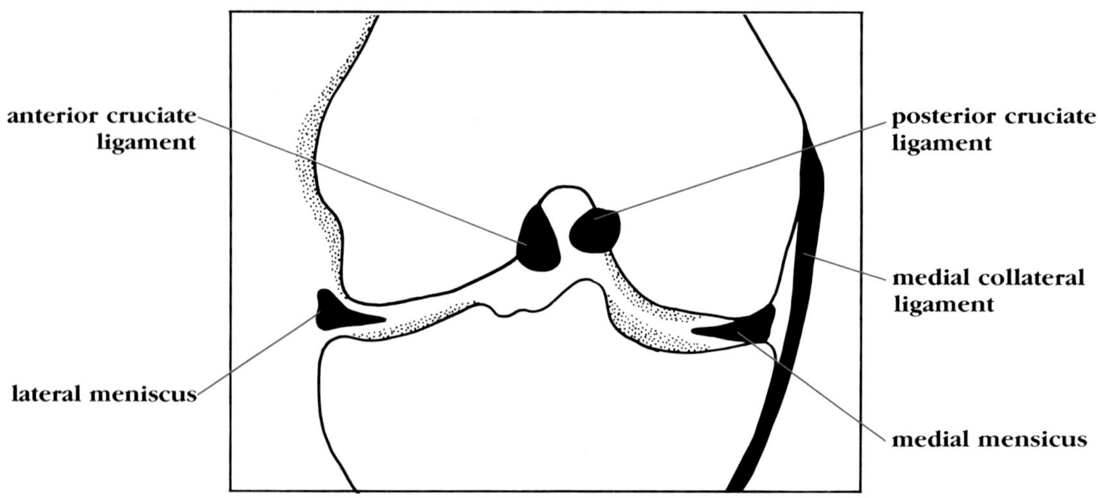

The iliotibial band originates from the fascia of the glutei muscles and extends caudad along the lateral aspect of the thigh, inserting in the tubercle of Gerdi of the proximal tibia (15,34) (Fig. 7.16).

The meniscofemoral ligaments (anterior and posterior) originate from the posterior horn of the lateral meniscus and blend with the fibers of the PCL (18,37) (Fig. 7.14).

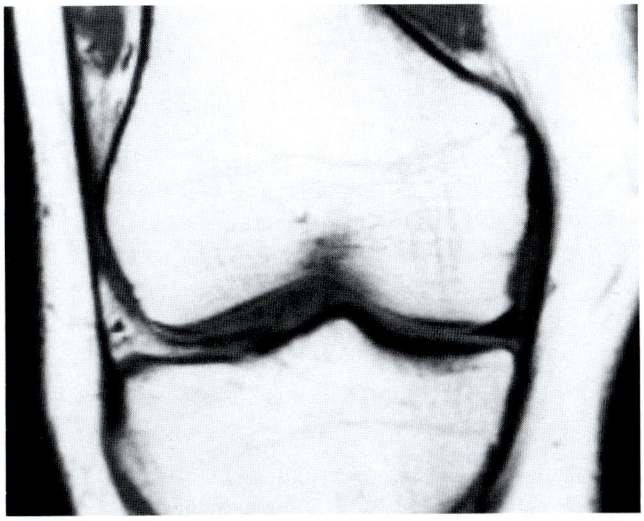

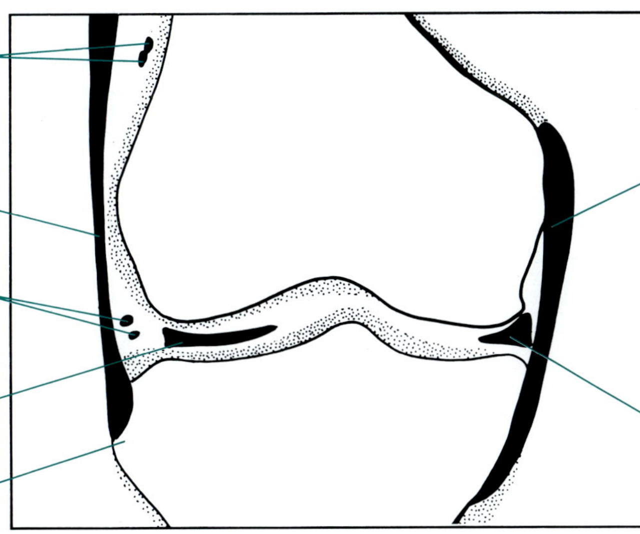

FIGURE 7.16. The Iliotibial Band Inserting in the Tubercle of Gerdi. Coronal section about 10 mm anterior to Fig. 7.15.

The transverse ligament is seen occasionally in the more anterior sections as a thin, transverse, hypointense band (Fig. 7.17).

The ACL is not well-visualized in coronal sections due to its oblique orientation. The PCL is well-identified from its origin in the medial femoral condyle to its insertion in the tibia (Fig. 7.14).

The Vessels

The popliteal vessels are seen midline in the most posterior sections. The superior genicular artery and vein (24,44) (medial and lateral) surround the distal femoral diaphysis (Fig. 7.16). The lateral inferior genicular vessels are seen adjacent to the lateral meniscus, under the fibular collateral ligament (Fig. 7.14) and iliotibial band (Fig. 7.16). The medial inferior genicular vessels can be seen more distally.

AXIAL PLANE

The Ligaments

The iliotibial band is identified in the lateral aspect of the knee as a thin line paralleling the lateral femoral condyle (Fig. 7.18). The fibular collateral ligament is seen as a rounded structure lateral to the insertion of the popliteus tendon in the lateral femoral condyle. It can be followed to its insertion in the fibula, joining the biceps femoris tendon (34,44) (Fig. 7.19).

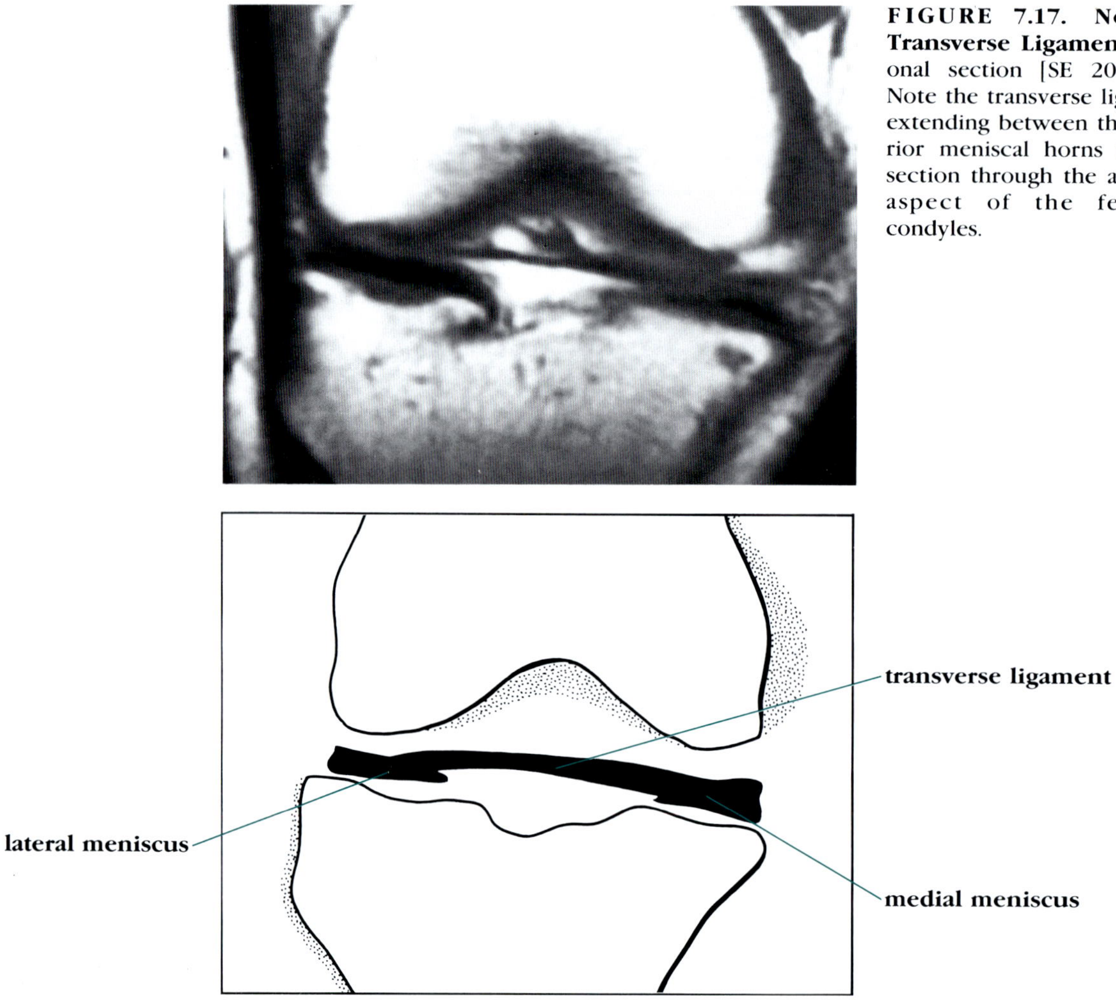

FIGURE 7.17. Normal Transverse Ligament. Coronal section [SE 2000/20]. Note the transverse ligament extending between the anterior meniscal horns in this section through the anterior aspect of the femoral condyles.

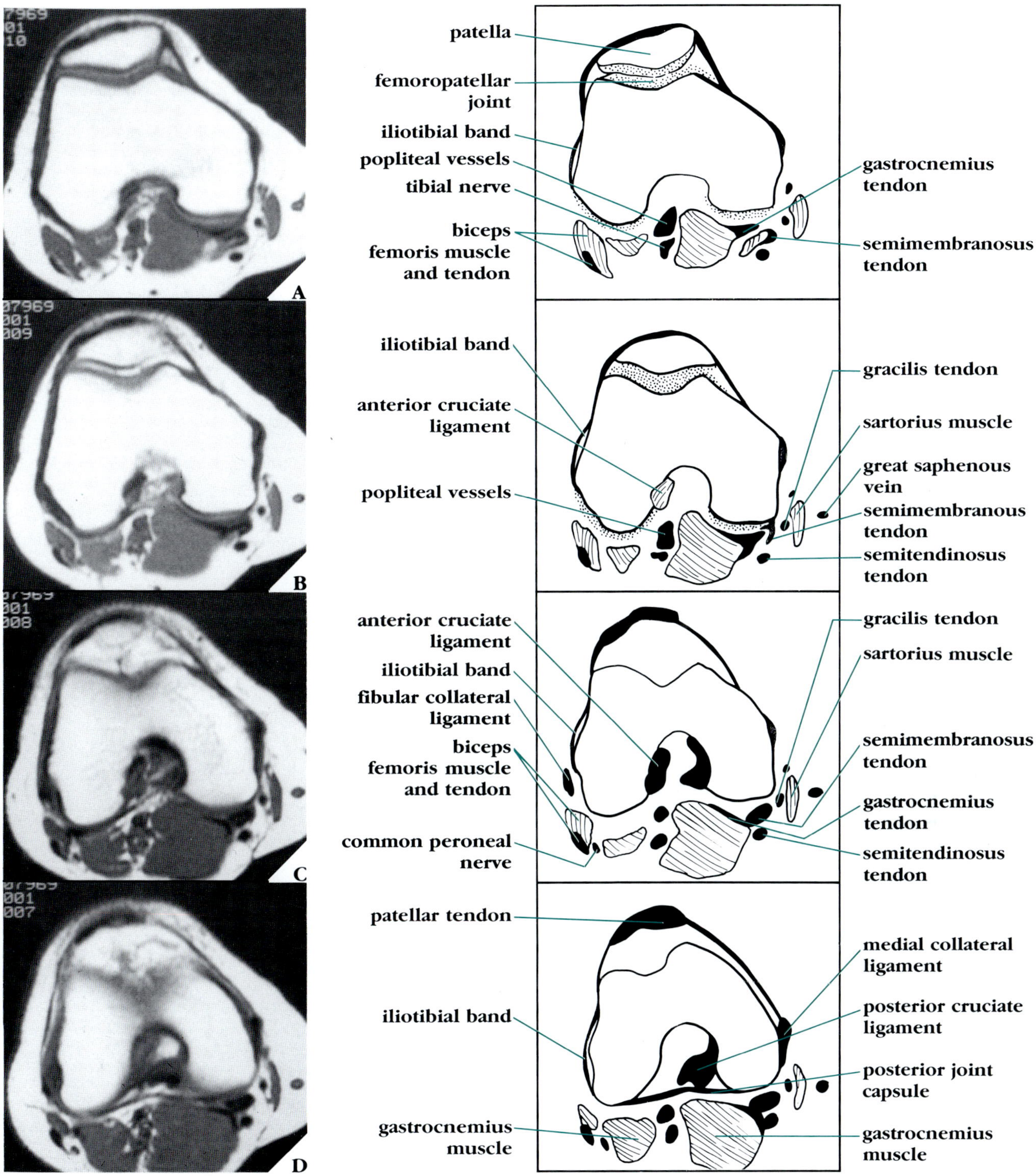

FIGURE 7.18. Femoropatellar Joint and Femoral Condyles. **A** and **B** lower half of the patella; **C** and **D** the femoral condyles. Axial sections [SE 2000/20]. The axial plane is useful for the evaluation of the anatomy of the intercondylar notch and the popliteal fossa.

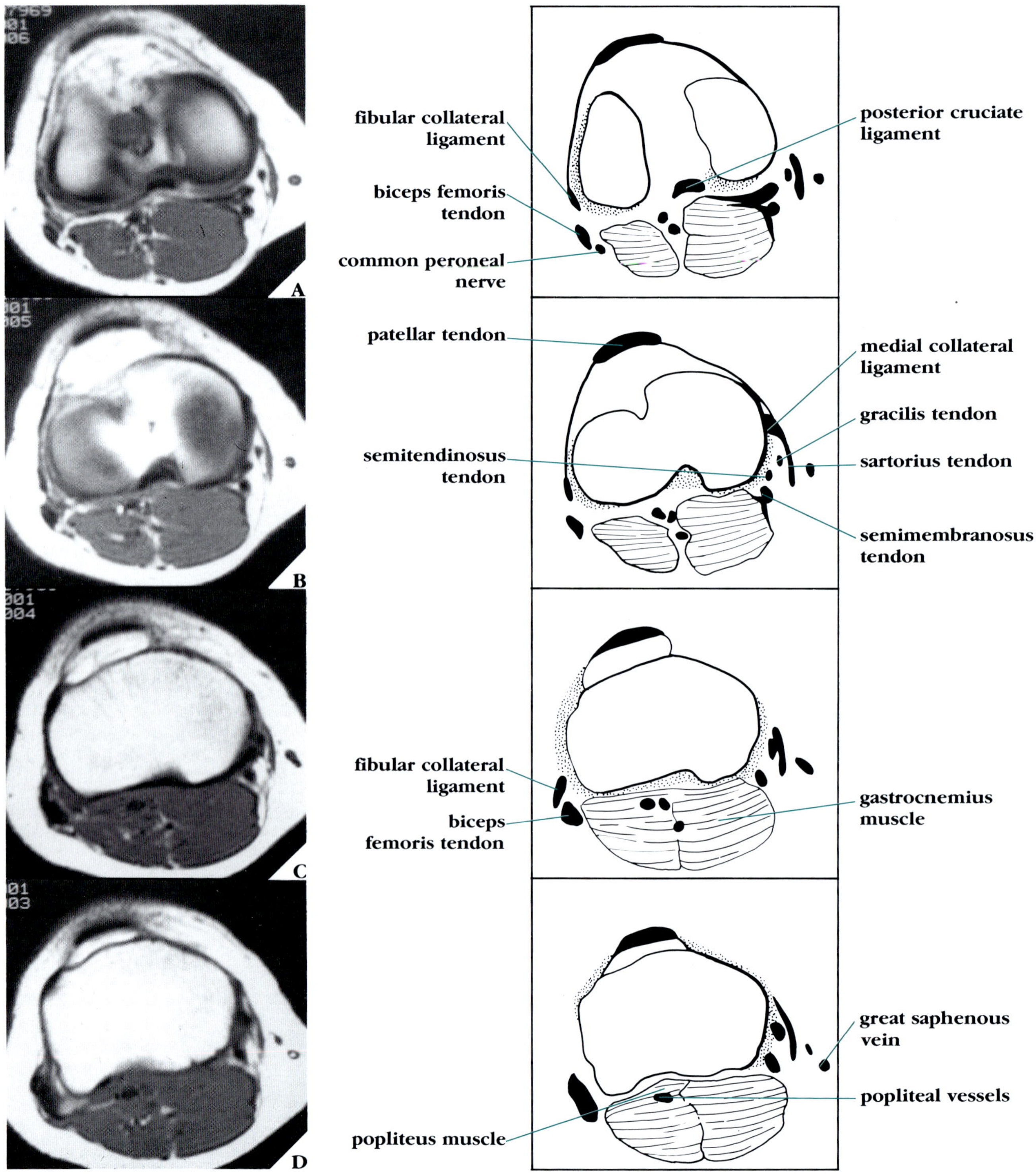

FIGURE 7.19. Normal Fibular Collateral Ligament and Biceps Femoris Tendon. **A** and **B** joint space; **C** and **D** proximal tibia. Axial sections [SE 2000/20]. The fibular collateral ligament and biceps femoris tendon can be followed to their insertion in the fibular head.

The ACL and PCL are easily identified in the intercondylar notch at their origin in the lateral and medial femoral condyles, and can be followed to their insertion in the tibial spine and posterior tibia (Figs. 7.18 and 7.19).

The medial collateral ligament is identified in the medial aspect of the knee as a fusiform, low SI structure adjacent to the medial femoral condyle (Figs. 7.18 and 7.19).

Occasionally, the transverse ligament is seen crossing the intercondylar notch anteriorly, behind the high SI infrapatellar fat (Fig. 7.20).

The Tendons

Anteriorly, inferior to the patella, the patellar tendon is seen as a thick, low SI structure extending from the inferior aspect of the patella to the tibial tuberosity (Figs. 7.18 and 7.19).

Laterally, the biceps femoris tendon is identified posterior to the fibular collateral ligament (Figs. 7.18 and 7.19).

Posteriorly and medially, the tendons of the semimembranosus and semitendinosus are seen as two round, hypointense dots medial to the tendon of the medial head of the gastrocnemius muscle (Fig.

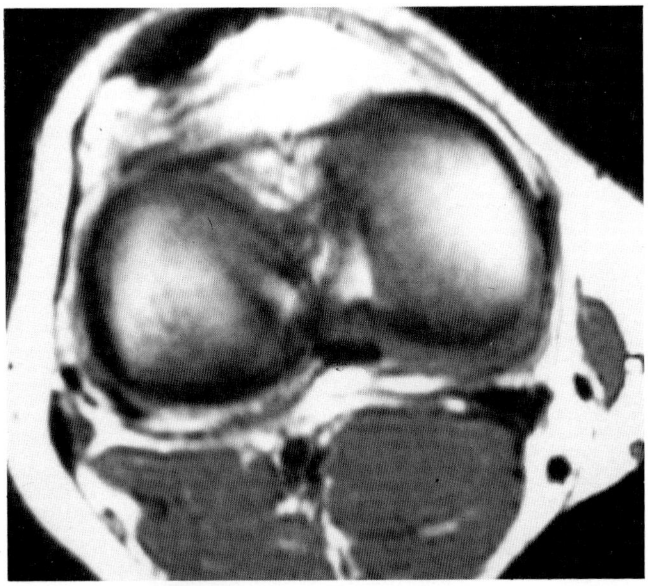

FIGURE 7.20. Intercondylar Notch with the Transverse Ligament. Axial section [SE 2000/20].

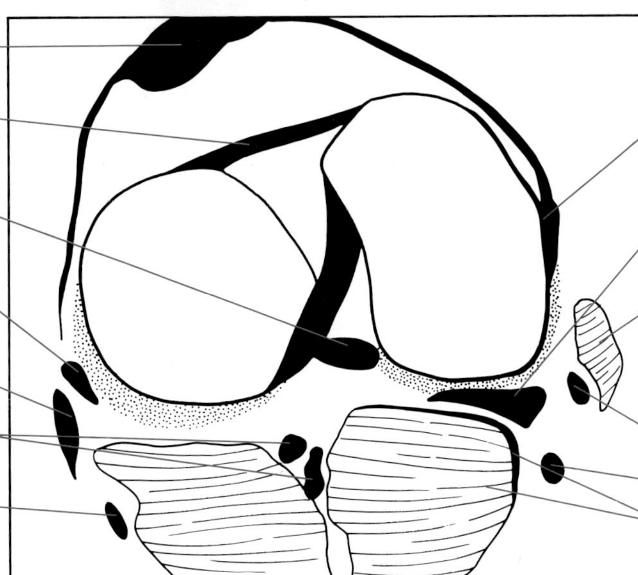

7.18). A star-shaped, intermediate signal structure between these tendons represents the gastrocnemius–semimembranosus bursa (Fig. 7.21). More peripherally, the gracilis tendon is seen as a third round dot adjacent to the sartorius muscle (Fig. 7.19). The semitendinosus, sartorius, and gracilis tendons are collectively called "pes anserinus" because of their resemblance to a goose foot on dissection (12,18,21,30,37).

The Neurovascular Bundle

The popliteal artery, vein, and tibial nerve are seen midline and posteriorly, between the medial and lateral heads of the gastrocnemius muscle, from anterior to posterior (Figs. 7.18 and 7.19). The artery is normally of low SI and the vein shows high SI due to slow flow. The common peroneal nerve is identified as a low SI dot, posterior to the laterally situated biceps femoris tendon (12) (Fig. 7.18).

MENISCAL LESIONS

Meniscal Tears

Normal menisci should have uniformly low SI. Degenerative or acute meniscal tears produce abnormal areas of intermediate or high SI within the meniscus, which may or may not extend to its articular surface. Based on the type of abnormality detected, meniscal lesions can be classified as follows (10,14,34,45) (Fig. 7.22):

Type I: A round area of high SI within the substance of the meniscus, not communicating with the articular margin (Fig. 7.23).

Type II: A linear area of high SI intensity within the meniscus, not communicating with the articular surface (Fig. 7.24).

Type III: A linear area of high SI extending to the superior or inferior articular surfaces of the meniscus (Figs. 7.25 and 7.26).

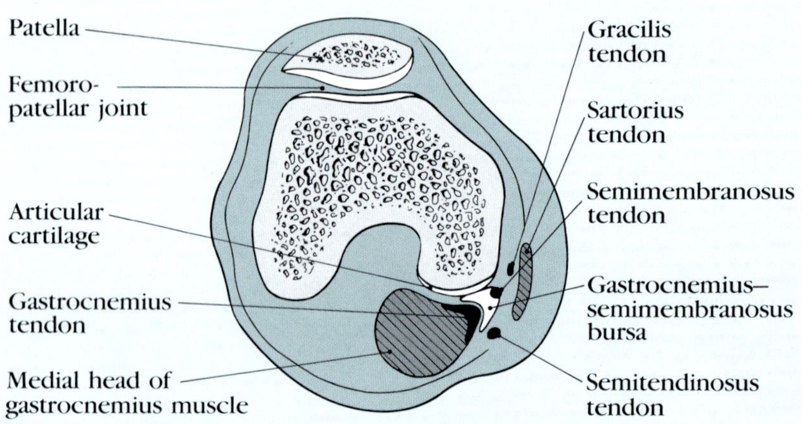

FIGURE 7.21. Recess of the Joint Capsule Between the Tendons of the Gastrocnemius and Semi-Membranosus. The synovium can herniate through this recess, producing a popliteal cyst.

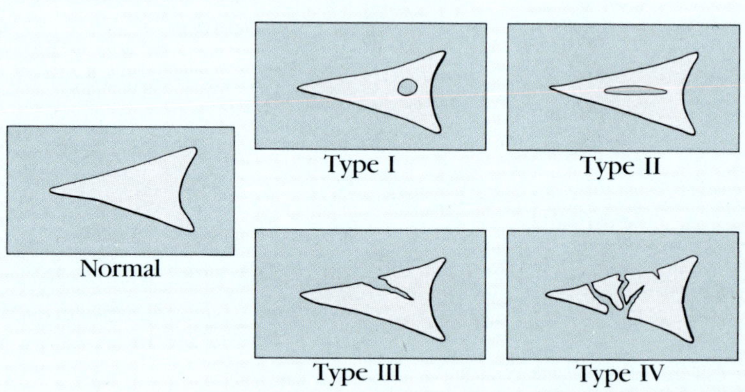

FIGURE 7.22. Types of Meniscal Lesions.

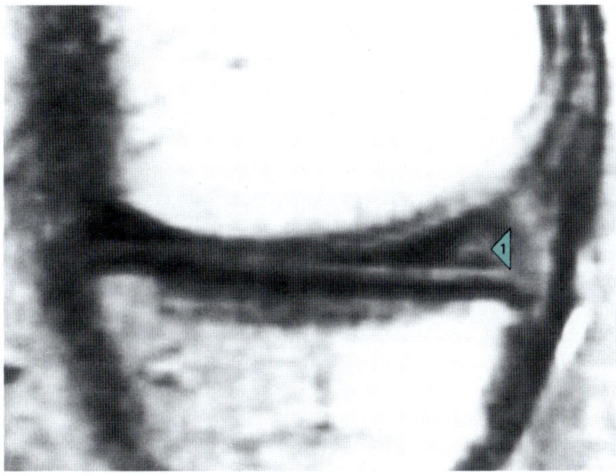

FIGURE 7.23. Type I Lesion of the Posterior Horn of the Medial Meniscus. Sagittal plane [SE 2000/20]. The intrameniscal lesion does not extend to the articular surface. Arrow **1** Type I lesion.

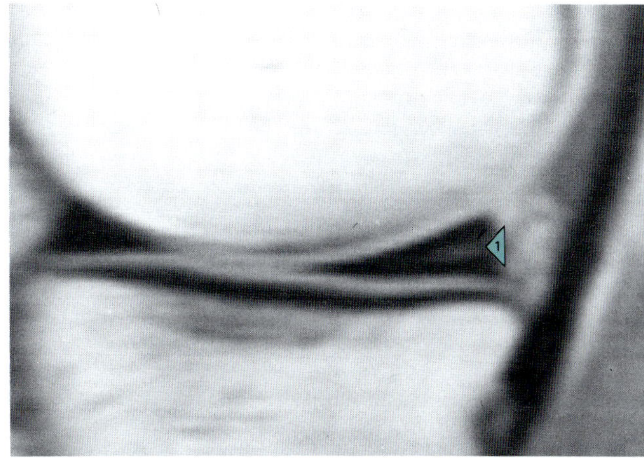

FIGURE 7.24. Type II Lesion of the Posterior Horn of the Medial Meniscus. Sagittal plane [SE 2000/20]. The configuration of this lesion is linear and does not communicate with the articular surface. Arrow **1** Type II lesion.

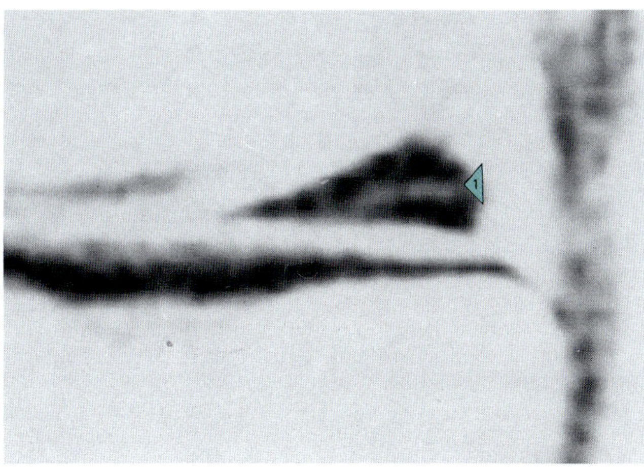

FIGURE 7.25. Type III Lesion of the Posterior Horn of the Medial Meniscus. Sagittal plane [SE 2000/20]. The window settings were adjusted to clearly illustrate the extension of the meniscal tear to the inferior articular surface. Arrow **1** Type III lesion.

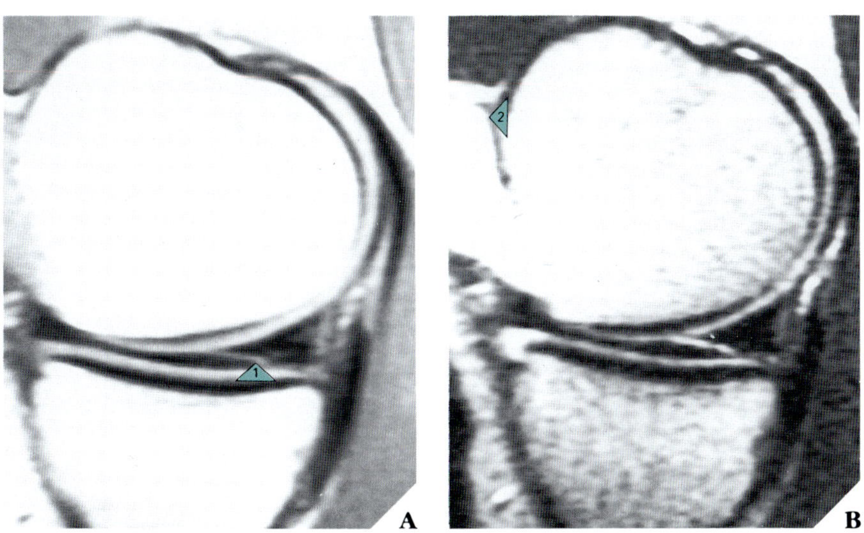

FIGURE 7.26. Type III Lesion of the Posterior Horn of the Medial Meniscus. A Sagittal plane [SE 2000/20]; **B** sagittal plane [SE 2000/80]. The meniscal tear extends to the superior and inferior articular surfaces of the meniscus. The lesion is more conspicuous on T2WI **(B)** because of the extension of joint fluid within the tear. Note the associated effusion anteriorly. Arrow **1** Type III lesion. Arrow **2** Joint effusion.

Type IV: A gross distortion of the SI or shape of the meniscus, which may or may not show fragmentation (Figs. 7.27–7.29).

Based on histologic examination of cadaver menisci and surgically removed menisci, it has been shown that type I and II lesions represent areas of hyaline or myxoid degeneration within the cartilage, found frequently in asymptomatic patients (48). Type II lesions probably represent a more advanced stage of degeneration more prone to evolve into frank tears. Type I and II lesions are not seen on arthroscopic examination, or may be seen as areas of discoloration of the meniscal surface.

As opposed to degenerative type I and II lesions, type III and IV lesions represent frank, acute or chronic, traumatic meniscal tears. Type III and IV lesions are typically symptomatic and in most cases require surgery.

Depending on their orientation, meniscal tears can be described as horizontal (Fig. 7.30), vertical or radial (Fig. 7.31), and oblique (Fig. 7.32).

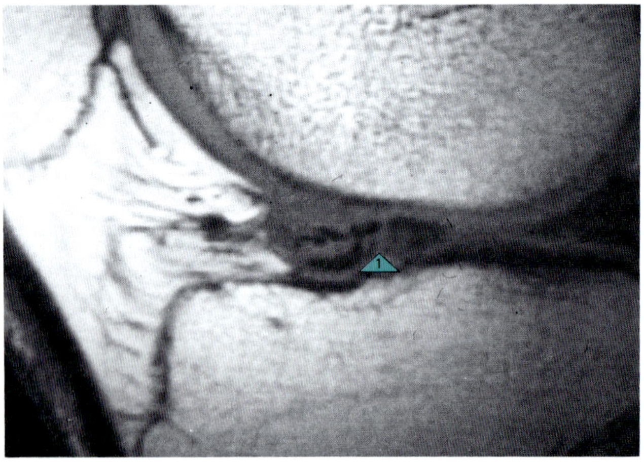

FIGURE 7.27. Type IV Lesion of the Anterior Horn of the Lateral Meniscus. Sagittal plane [SE 2000/20]. The entire anterior horn is fragmented. Arrow **1** Type IV lesion.

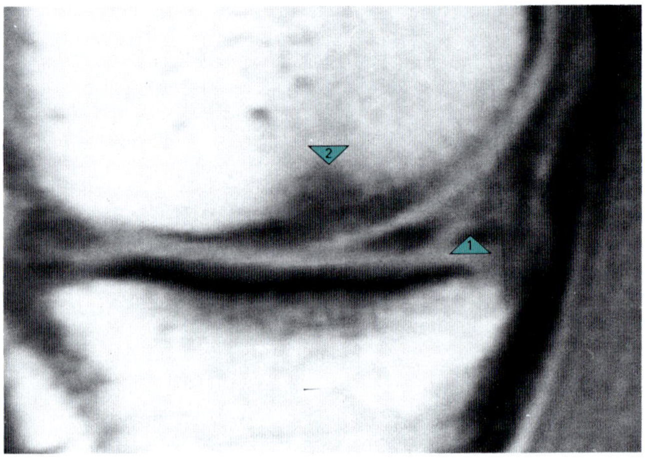

FIGURE 7.28. Type IV Lesion of the Posterior Horn of the Medial Meniscus. Sagittal plane [SE 2000/20]. The low SI lesion of the medial femoral condyle represents an area of bone contusion or an osteochondral fracture. Arrow **1** Type IV lesion. Arrow **2** Bone contusion.

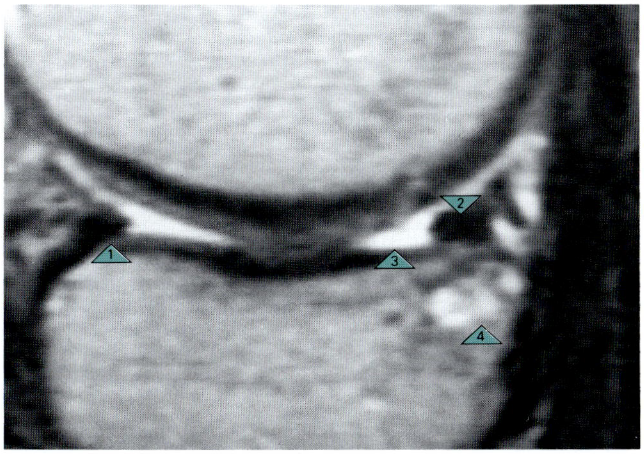

FIGURE 7.29. Type IV Lesion of the Posterior Horn of the Lateral Meniscus, with Amputation of the Free Edge of the Meniscus and Avulsion of the Semimembranosus Tendon. Sagittal plane [SE 2000/80]. The hyperintense signal in the joint space corresponds to an effusion. The periphery of the meniscus is very irregular. There is an area of high SI in the posterior tibial plateau corresponding to an avulsion of the semimembranosus tendon. Arrow **1** Normal anterior horn. Arrow **2** Amputated posterior horn. Arrow **3** Synovial fluid. Arrow **4** Semimembranosus tendon avulsion.

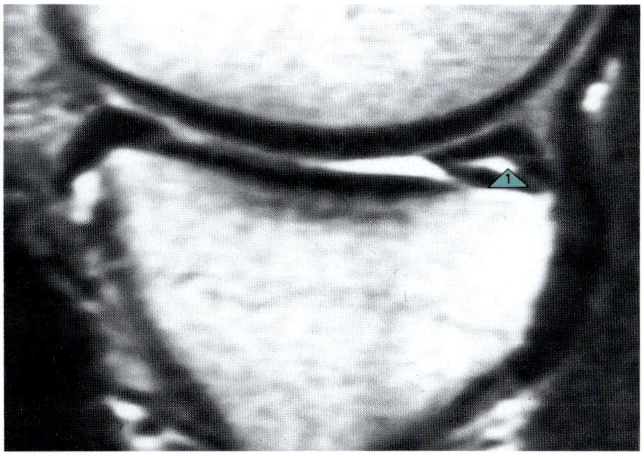

FIGURE 7.30. Horizontal Type III Tear of the Posterior Horn of the Medial Meniscus, with Small Joint Effusion. Sagittal plane [SE 2000/80]. Arrow **1** Horizontal tear.

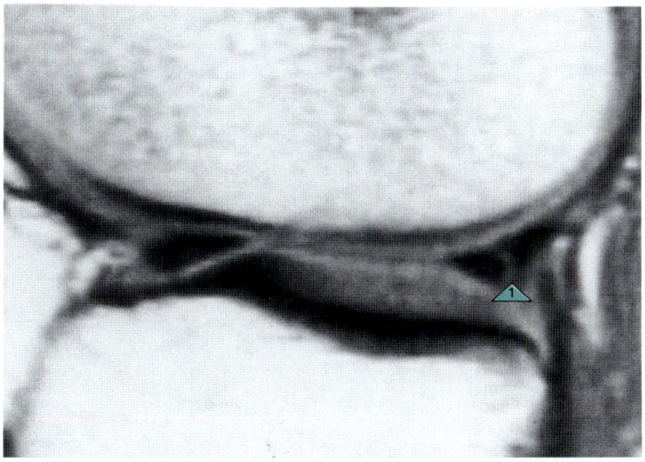

FIGURE 7.31. Vertical Type III Tear of the Posterior Horn of the Lateral Meniscus. Sagittal plane [SE 2000/20]. Arrow **1** Vertical tear.

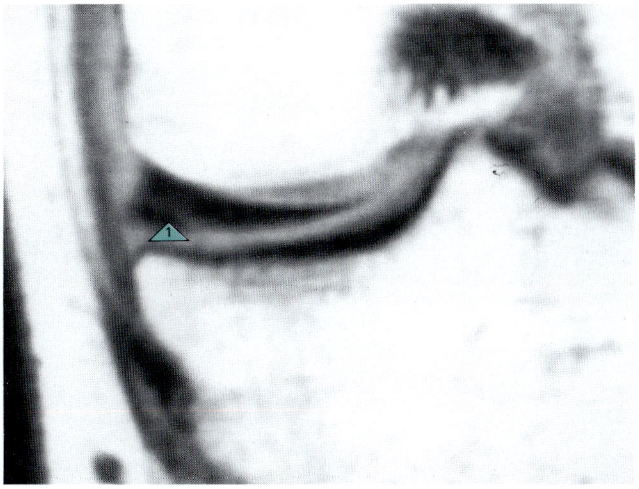

FIGURE 7.32. Oblique type III Tear of the Midportion of the Medial Meniscus, with Inferior Extension. Coronal plane [SE 2000/20]. Arrow **1** Oblique tear.

Bucket-handle tears (Figs. 7.33–7.35), parrot' beak tears (Fig. 7.36), and peripheral detachments (Fig. 7.37) are terms used to describe the configuration of distinctive tears seen on arthroscopic examination. These complex tears may be difficult to identify on MRI. The use of coronal and sagittal views as well as familiarity with the appearance of these tears should be helpful in avoiding false-negative interpretations (24).

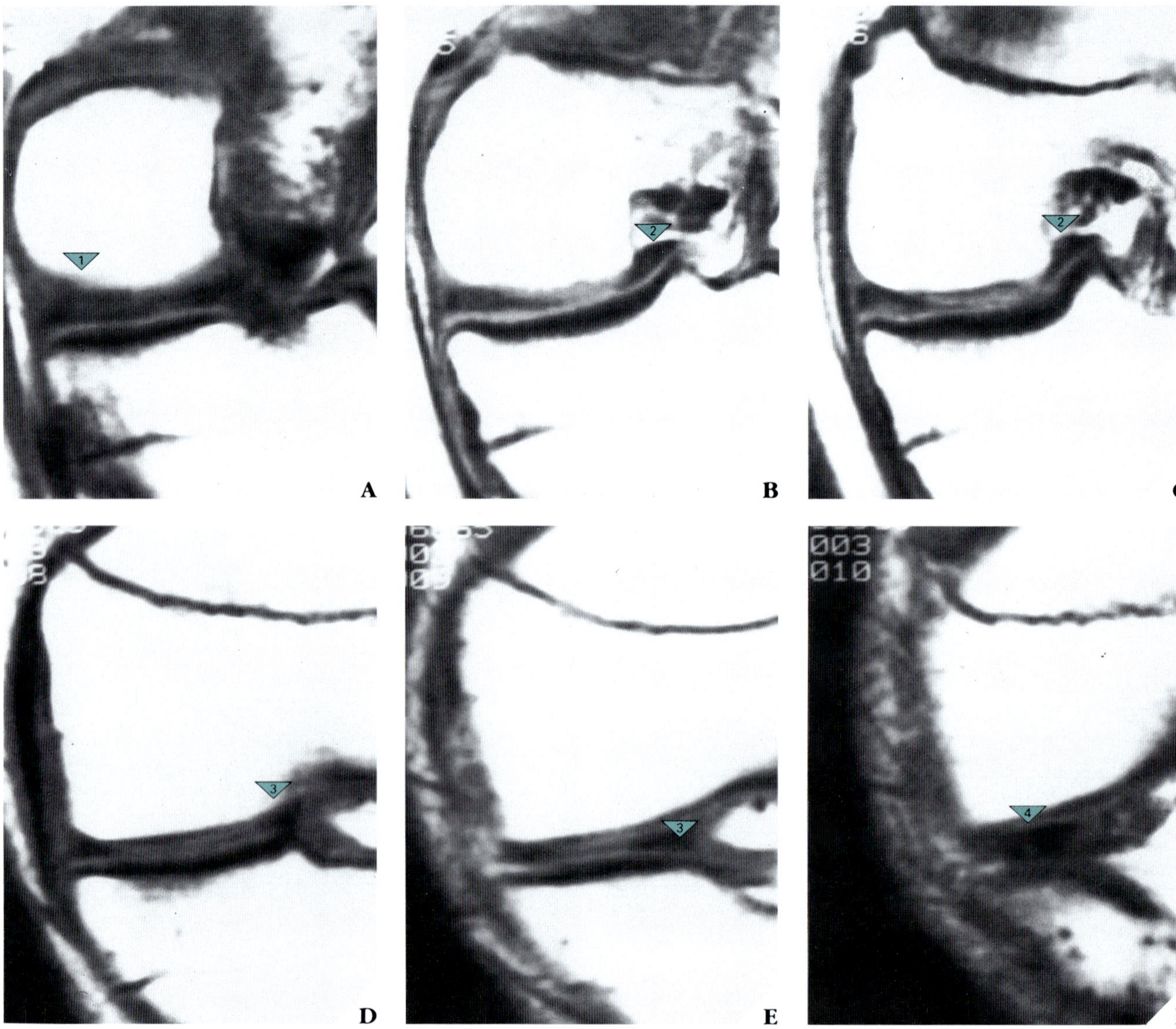

FIGURE 7.33. Bucket-Handle Tear of the Medial Meniscus. Coronal plane [SE 2000/20] through posterior (**A** and **B**), mid (**C** and **D**), and anterior (**E** and **F**) horns. The internally displaced meniscal fragment can be seen in the region of the intercondylar notch. The more posterior and anterior horns have a practically normal appearance. Arrow **1** Posterior horn. Arrows **2** Internally displaced meniscal fragment. Arrows **3** Meniscal fragment. Arrow **4** Anterior horn.

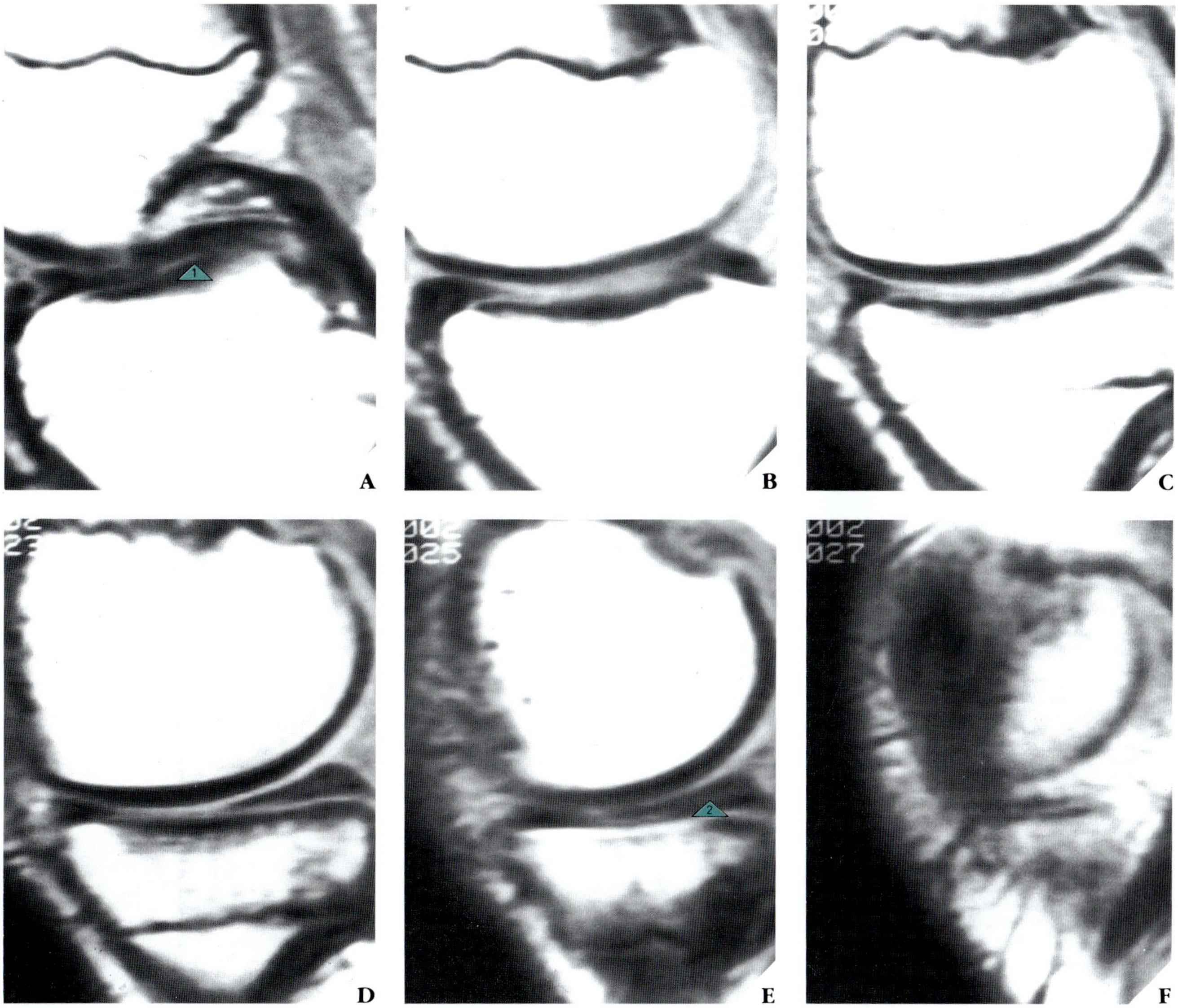

FIGURE 7.34. Bucket-Handle Tear of the Medial Meniscus. (Same patient as in Fig. 7.33.) Sagittal plane [SE 2000/20] through the intercondylar notch and meniscal attachments, **(A** and **B)**, the midportion of the medial meniscus, **(C** and **D)**, and the periphery of the medial meniscus (**E** and **F**). The internally displaced meniscal fragment is again noted **(A)**. A horizontal component of the tear is seen in peripheral sections. Arrow **1** Meniscal fragment. Arrow **2** Horizontal component of the tear.

Recurrent or persistent pain postmeniscectomy can be related to recurrent tear of the meniscal remnant (Fig. 7.38), lack of full excision of the tear, and the presence of free meniscal fragments. MRI may be helpful in the detection of these abnormalities following surgery.

Using state-of-the-art equipment, including dedicated surface coils, the sensitivity and specificity of MRI in the evaluation of meniscal lesions has been reported to be above 90 per cent, comparing favorably with arthrography and arthroscopy (34,35). Early publications reported an unacceptably high rate of false-positive MRI diagnoses. Most of these were related to type I and II lesions, not seen with arthroscopy. Another source of discrepancy between MRI and arthroscopy arises from the difficult visualization of the posterior horns of the menisci with arthroscopy. Tears located in the pos-

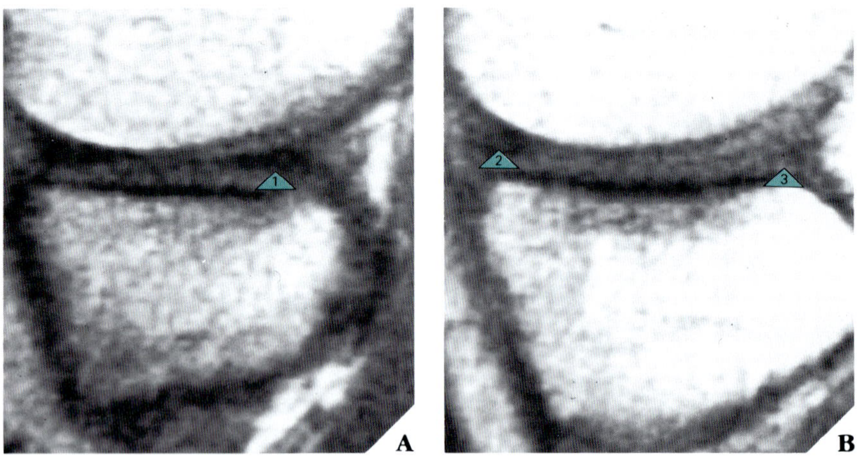

FIGURE 7.35. Bucket-Handle Tear with Migration of the Posterior Meniscal Fragment. Contiguous sections through the periphery of the medial meniscus [SE 2000/20]. The posterior horn is not seen in **B** due to internal fragment migration. Arrow **1** Normal posterior horn. Arrow **2** Anterior horn. Arrow **3** Missing posterior horn.

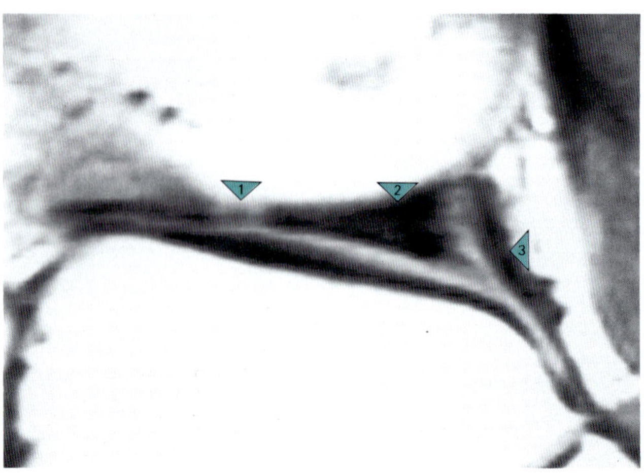

FIGURE 7.36. Parrot-Beak Tear of the Lateral Meniscus. Sagittal plane [SE 2000/20]. The tear is identified as a subtle gap in the midportion of the meniscus. Arrow **1** Parrot-beak tear. Arrow **2** Posterior horn. Arrow **3** Popliteus tendon.

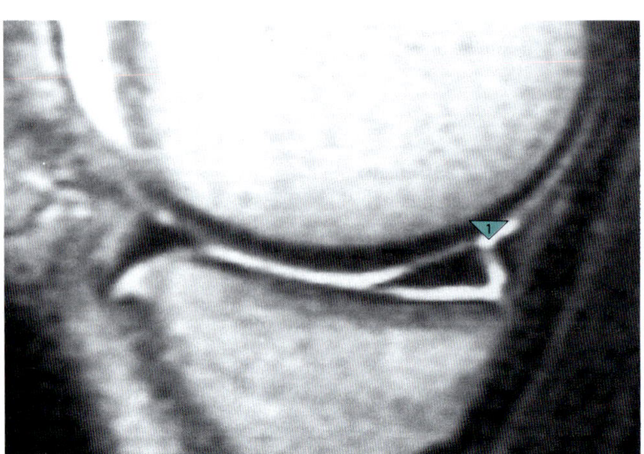

FIGURE 7.37. Peripheral Detachment of the Posterior Horn of the Medial Meniscus. Sagittal plane [SE 2000/80]. There is a vertical tear of the posterior horn of the medial meniscus with fluid extending into the tear. Arrow **1** Vertical peripheral detachment.

terior meniscal horns, especially those extending to the inferior surface, are very difficult to detect arthroscopically unless these areas are carefully probed. Probably the most important contribution of MRI in the investigation of meniscal tears is the high negative predictive value of the test, allowing for preoperative assessment of the presence, location, and type of lesion and selection of patients for surgery.

Figure 7.39 is a proposed algorithm reflecting the role of MRI in the evaluation of the painful knee with suspected meniscal tear. Arthrography or CT have not been included as diagnostic techniques for evaluation of the painful knee. Both techniques are invasive, and for practical purposes arthroscopy has replaced preoperative diagnostic evaluation in many institutions. Now MRI offers a highly sensitive, noninvasive method for detecting not only meniscal lesions but also ligament, tendon, synovial, and other soft-tissue lesions, with an almost 100 per cent negative predictive value (14). This makes MRI ideal for preoperative screening.

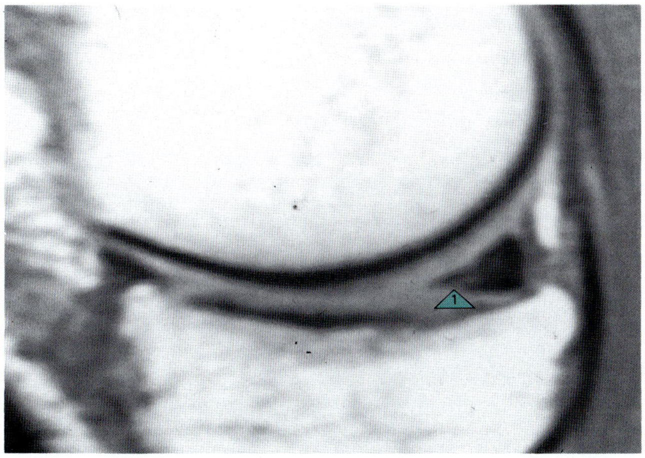

FIGURE 7.38. Postmeniscectomy. Tear of the meniscal remnant of the posterior horn of the medial meniscus. Sagittal plane [SE 2000/20]. Arrow **1** Meniscal remnant with tear.

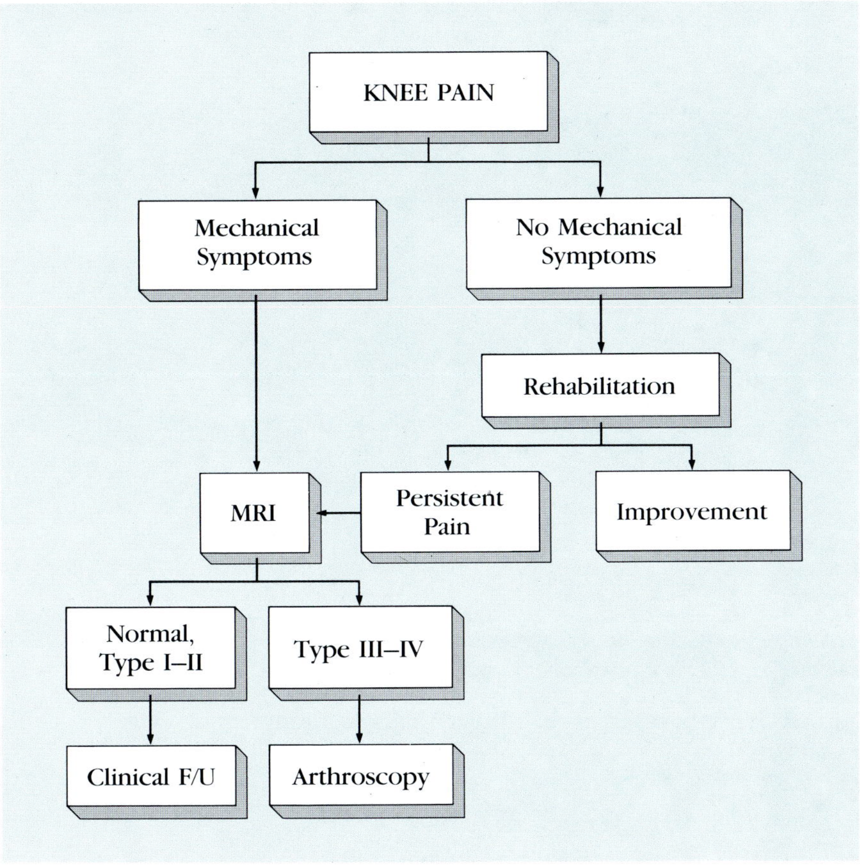

FIGURE 7.39. Algorithm for Evaluation of Patients Presenting with Knee Pain.

LIGAMENT AND TENDON LESIONS

Medial Collateral Ligament Tears

Acute tears of the medial collateral ligament produce disruption of the normal continuity of the ligament accompanied by soft-tissue edema and inflammation (31,34,51) (Fig. 7.40). Occasionally, associated impaction fractures are noted with intraosseous area of high SI on T2WI. Ligamentous sprain produces edema and local hemorrhage but the continuity of the ligament is preserved. Chronic tears may be seen as areas of localized thickening of the medial collateral ligament, without the MRI findings of associated acute edema or inflammation.

Lateral Collateral Ligament Tears

Similarly, acute lateral collateral ligament tears can be seen on MRI as disruption of the continuity of the ligament with an irregular contour and focal hyperintense areas on T2WI representing edema and/or hemorrhage (31,34,51) (Fig. 7.41). Associated avulsion fractures at the insertions of the iliotibial band (tubercle of Gerdy), the lateral collateral ligament (Segond fracture), and the fibular collateral ligament (fibular head fracture) can also be seen on MRI.

Anterior Cruciate Ligament Tears

Tears of the ACL are frequent. Lack of visualization of the ACL is highly suggestive of a ligamentous tear, although with standard 10–15-degree lateral rotation, the ACL is not seen in about 5 per cent of normal individuals. This necessitates confirmation with more oblique projections. Discontinuity of the fibers (Fig. 7.42), acute angulation (Fig. 7.43), and hyperintense areas within the ligament on T2WI are reliable signs of rupture. Buckling of the

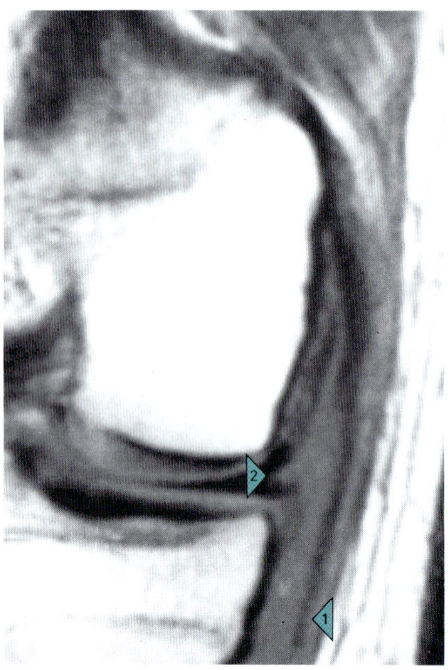

FIGURE 7.40. Acute Medial Collateral Ligament Rupture. Coronal plane [SE 2000/20]. The normally low SI ligament is detached at its tibial insertion. A peripheral tear of the medial meniscus is noted. There is edema and swelling about the ligament. Arrow **1** Separation of medial collateral ligament. Arrow **2** Meniscal tear.

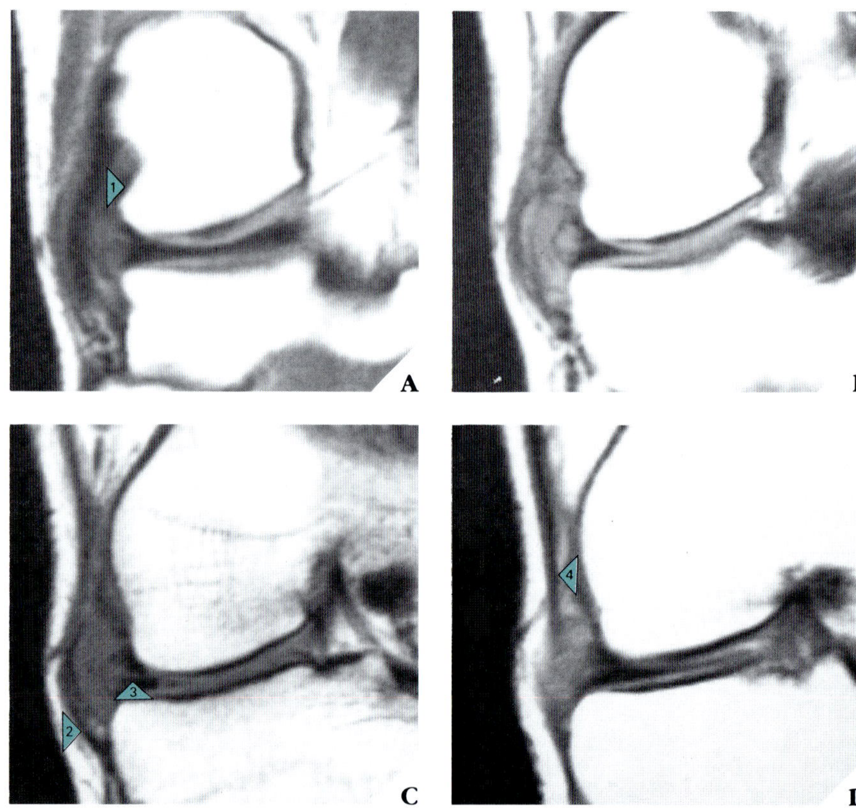

FIGURE 7.41. Lateral Collateral Ligament Tear. Four contiguous coronal sections [SE 2000/20] from posterior **(A)** to anterior **(D)**. The fibers of the tibial collateral ligament and the iliotibial band are disrupted. There is an underlying tear of the lateral meniscus. Arrow **1** Popliteus tendon. Arrow **2** Tibial collateral ligament. Arrow **3** Meniscal tear. Arrow **4** Iliotibial band.

posterior cruciate ligament is an indirect sign of ACL tear (Fig. 7.44). A focal, intermediate SI mass at the insertion of the ACL in the lateral femoral condyle is also a sign of ACL rupture, if volume averaging from the adjacent condyle is excluded (29,31,34,35,51).

Posterior Cruciate Ligament Tears

Similarly, rupture of the PCL can produce localized thickening, disrupture, and focal hyperintense areas on T2WI and hypointense areas on T1WI related to hematoma (34) (Fig. 7.45).

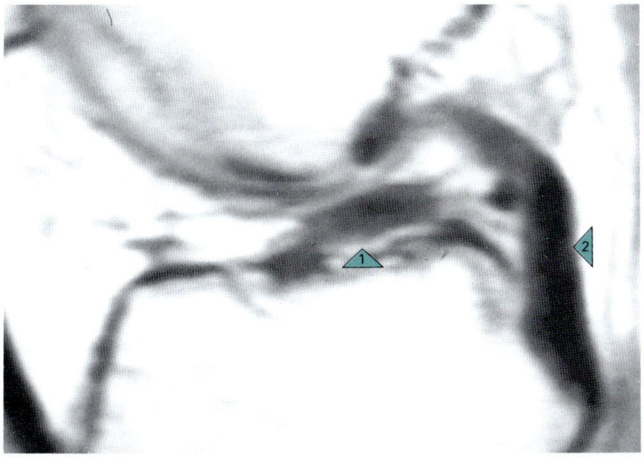

FIGURE 7.42. Complete Anterior Cruciate Ligament Tear. Sagittal plane [SE 2000/20]. Some fibers of the anterior cruciate ligament are seen in a horizontal position in the intercondylar notch area. Arrow **1** Anterior cruciate ligament. Arrow **2** Posterior cruciate ligament.

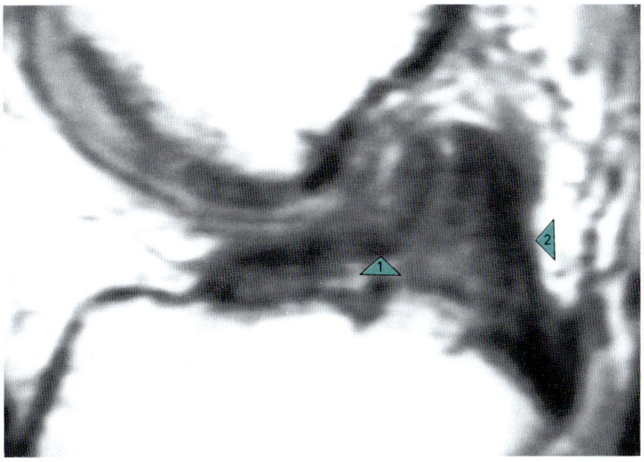

FIGURE 7.43. Incomplete Anterior Cruciate Ligament Tear. Sagittal plane [SE 2000/20]. There is acute angulation of the anterior cruciate ligament in the midportion of the ligament. Arrow **1** Anterior cruciate ligament. Arrow **2** Posterior cruciate ligament.

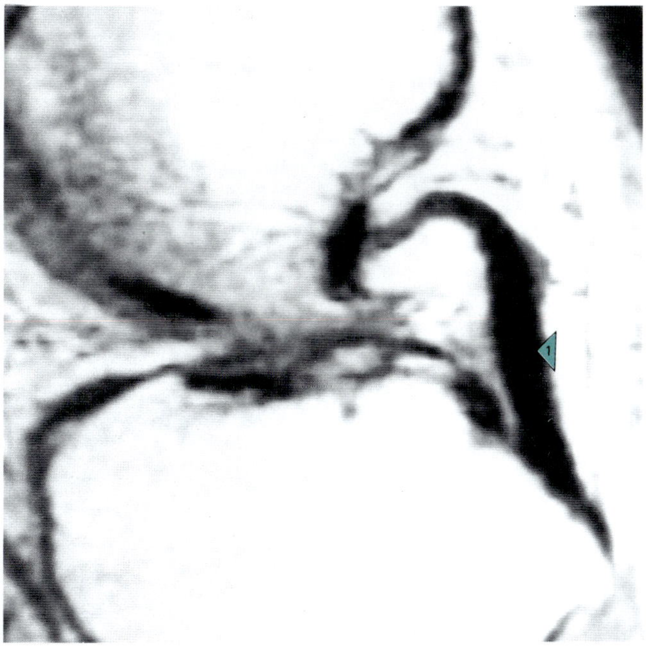

FIGURE 7.44. Buckling of the Posterior Cruciate Ligament. This is an indirect sign of anterior cruciate ligament tear. Sagittal plane [SE 2000/20]. Arrow **1** Posterior cruciate ligament.

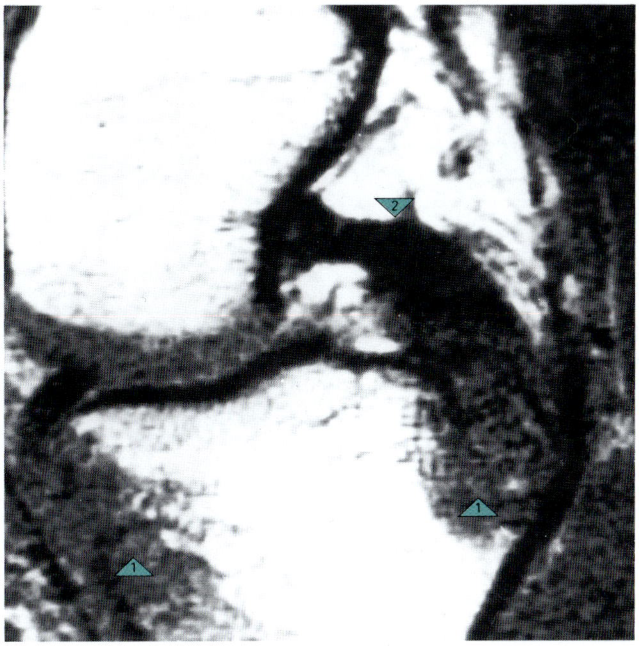

FIGURE 7.45. Avulsion of the Posterior Cruciate Ligament at the Tibial Insertion. Sagittal plane [SE 2000/20]. Low SI areas in the proximal tibia represent acute intraosseous hemorrhage related to fracture. Arrows **1** Fracture. Arrow **2** Posterior cruciate ligament.

Quadriceps and Patellar Tendon Ruptures

Tendon injuries in the region of the knee are relatively rare, but quadriceps and patellar tendon ruptures can occur (6). Incomplete tendon rupture can be seen on MRI as irregular thinning, localized thickening, and hyperintense areas within the tendon (Fig. 7.46). Disruption of the continuity of the tendon indicates complete rupture (Fig. 7.47). Edema and hemorrhage can be seen around the tendon rupture in the acute phase.

Chronic Patellar Tendinitis

Chronic patellar tendinitis occurs as a result of repeated stress on the patellar tendon. It is seen in "jumper's knee," and is also associated with Osgood–Schlatter disease. The condition is clinically manifested by patellar pain. Intratendinous scarring, degenerative changes, and focal necrosis of the tendon can be seen pathologically. Focal thickening and areas of high SI on long TR/TE images can be seen on MRI. These findings correspond to areas of chronic tears, necrosis, and inflammation within

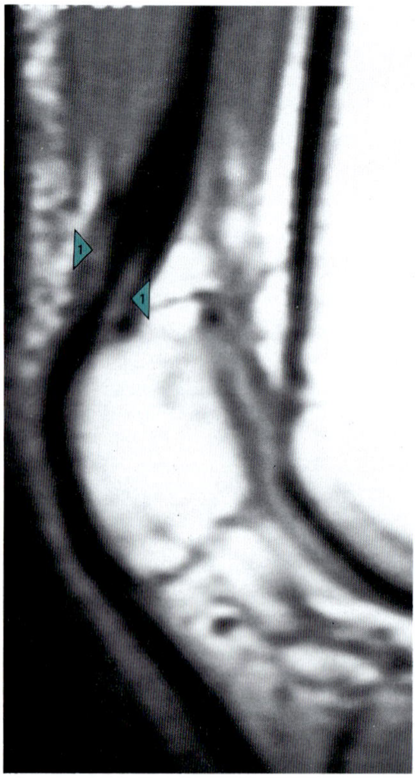

FIGURE 7.46. Partial Rupture of the Quadriceps Tendon. Sagittal plane [SE 2000/20]. Intermediate SI lines within the tendon represent edema or hemorrhage between the tendon fibers. Arrow **1** Partial quadriceps tendon rupture.

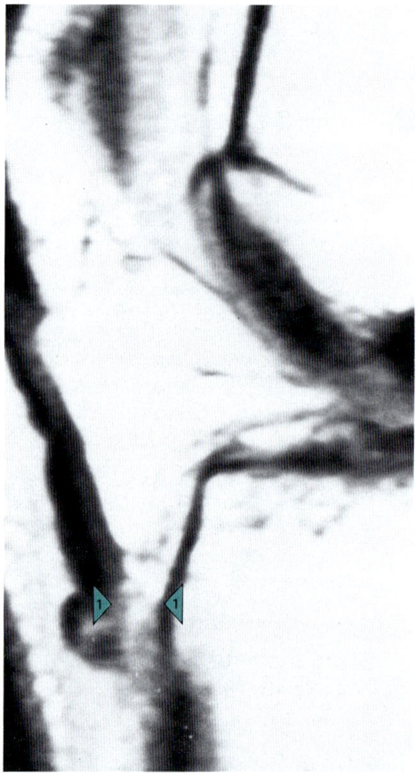

FIGURE 7.47. Avulsion of the Patellar Tendon at the Insertion in the Tibial Tuberosity. Sagittal plane [SE 2000/20]. Arrow **1** Patellar tendon rupture.

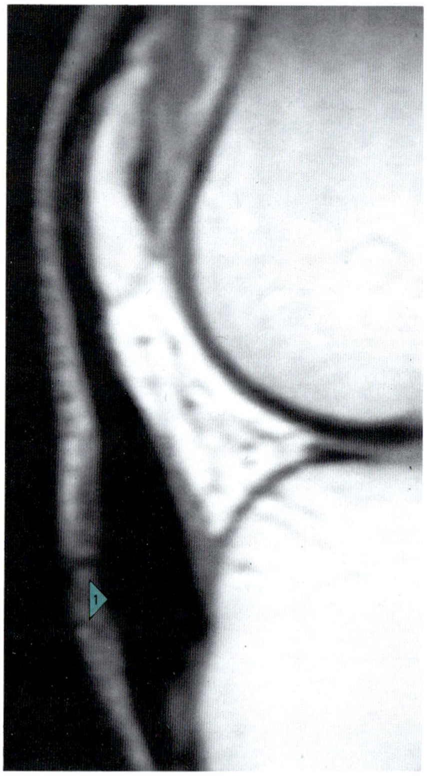

FIGURE 7.48. Chronic Patellar Tendinitis. Sagittal plane [SE 2000/20]. There is thickening of the distal patellar tendon (arrow **1**).

the tendon substance (8) (Fig. 7.48). Enlargement of the distal patellar tendon supports the diagnosis of Osgood–Schlatter disease, whereas enlargement of the tendon just inferior to the patella is more indicative of "jumper's knee."

ARTICULAR CARTILAGE AND BONE LESIONS

Articular or hyaline cartilage is seen on MRI as intermediate SI tissue on T1WI and T2WI, covering the articular surfaces of the patella, femoral condyles, and tibial plateau (12). The SI of the articular cartilage is higher than that of the fibrocartilage of the menisci, owing to the higher content of water-binding proteoglycan molecules (19,26).

Osteochondral Lesions

Articular cartilage lesions can be related to acute or repetitive trauma (osteochondral fracture, osteochondritis dissecans, and chondromalacia patellae), degenerative osteoarthritis, and inflammatory diseases. In osteochondral lesions and osteochondritis dissecans, a small segment of subchondral bone becomes hypointense on T1WI and T2WI (23,33) (Figs. 7.49 and 7.50). This is probably related to bone contusion and subsequent fibrous healing or development of osteonecrosis. The overlying articular cartilage can either be sep-

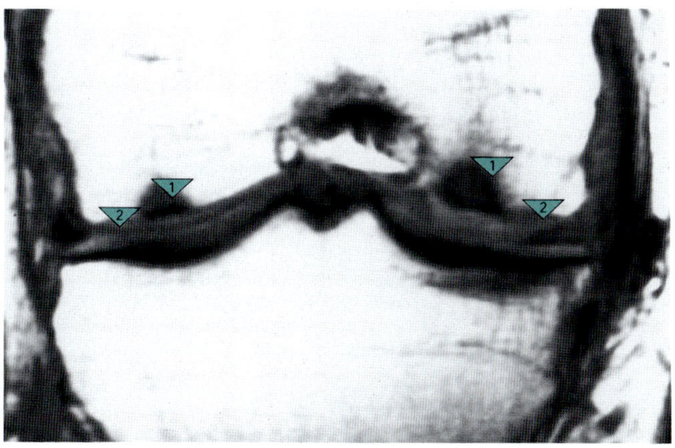

FIGURE 7.49. Osteochondral Lesions of Both Femoral Condyles. Coronal plane [SE 2000/20]. The low SI areas of both femoral condyles represent bone trauma (contusion). In this case the overlying articular cartilage is preserved. Arrows **1** Bone contusion. Arrows **2** Articular cartilage.

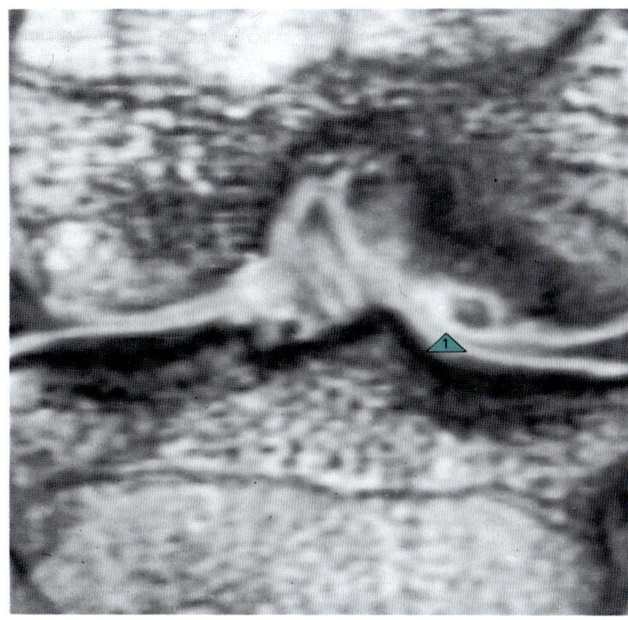

FIGURE 7.50. Osteochondritis Dissecans Involving the Medial Femoral Condyle. Coronal plane [3DFT GRASS 50/20, 40 degrees]. There is an osteochondral fragment partially detached from the medial condyle. The cartilaginous portion is bulging inferiorly. Arrow **1** Bulging cartilage.

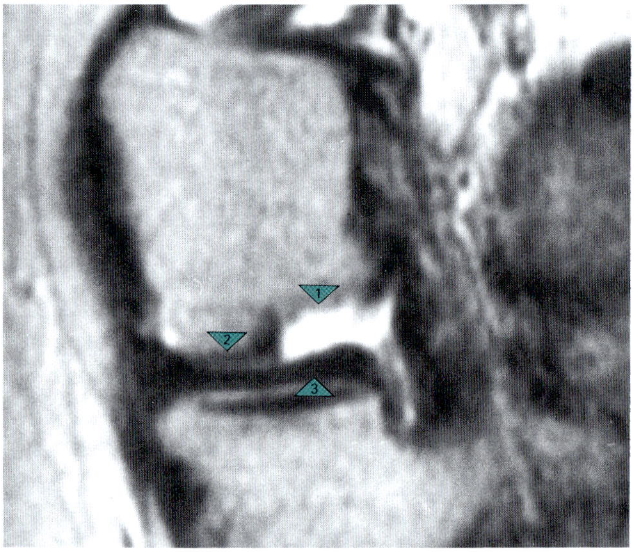

FIGURE 7.51. Osteochondral Fracture of the Medial Femoral Condyle. Coronal plane through posterior knee [SE 2000/80]. There is a missing segment of bone and overlying articular cartilage in the medial femoral condyle. The area of the osteochondral fracture is filled with fluid from a synovial effusion. Arrow **1** Osteochondral fracture. Arrow **2** Articular cartilage. Arrow **3** Posterior horn of the medial meniscus.

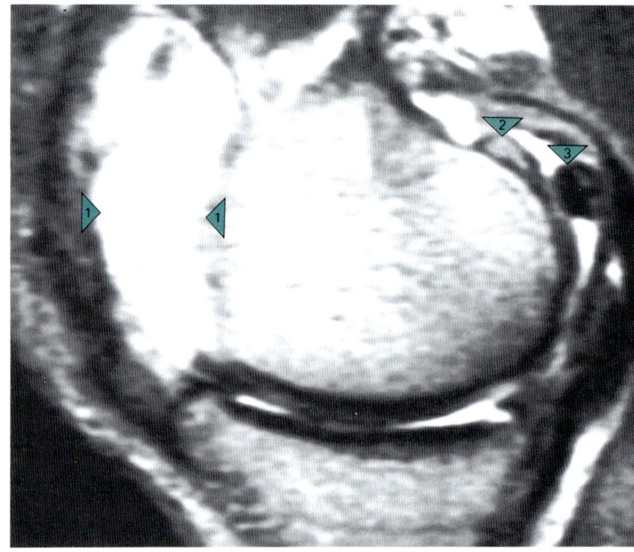

FIGURE 7.52. Osteocartilaginous Loose Body. Sagittal plane [SE 2000/80]. The articular cartilage of the posterior aspect of the medial femoral condyle is irregular in thickness. A low SI loose body can be seen posteriorly and a high SI joint effusion can be readily identified. Arrow **1** Joint effusion. Arrow **2** Articular cartilage. Arrow **3** Loose body.

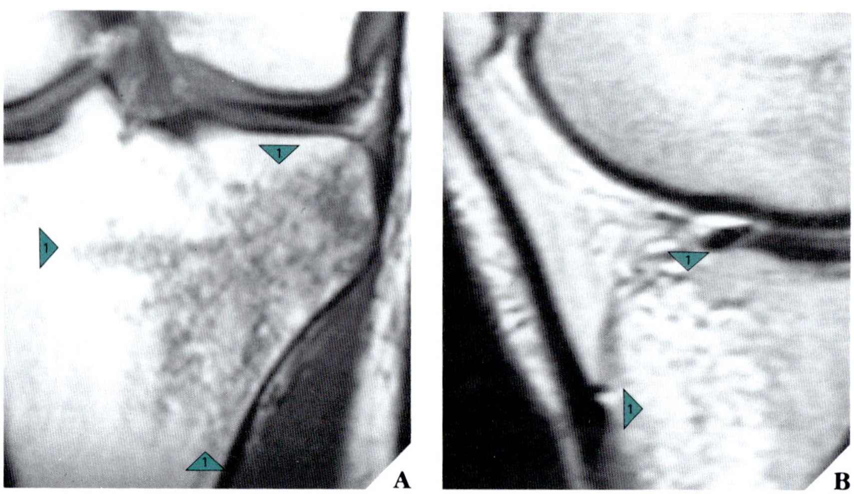

FIGURE 7.53. Occult Intraosseous Fracture of the Proximal Tibia. A Coronal plane [SE 800/20]; **B** sagittal plane [SE 2000/80]. There is low SI on T1WI **(A)** and high SI on T2WI **(B)** involving the proximal metaphysis and epiphysis of the tibia, and reflecting the acute intraosseous bleeding. Arrows **1** Intraosseous hemorrhage.

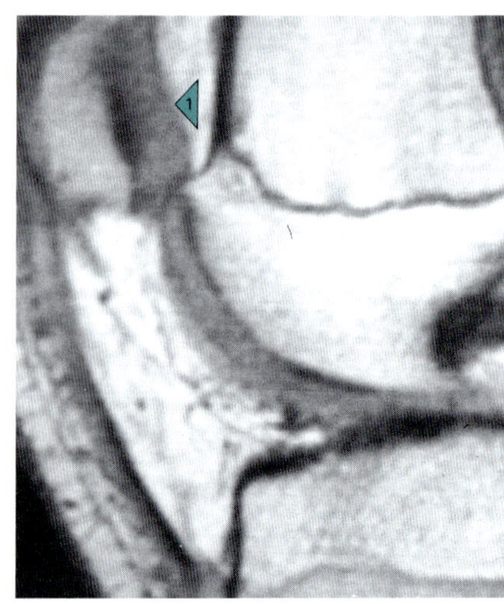

FIGURE 7.54. Chondromalacia Patellae, Early Stage. Sagittal plane [SE 2000/80]. The patellar articular cartilage is thickened. Arrow **1** Swollen articular cartilage.

arated from the surrounding cartilage (Fig. 7.51) or remain attached to it. In some cases the osteochondral fragment can become a loose intraarticular body (Fig. 7.52). This can calcify, showing low SI on MRI in all pulse sequences, or ossify, showing a hyperintense signal on T1WI due to the presence of fatty marrow.

Occult Intraosseous Fractures

A more severe manifestation of bone trauma, occult intraosseous fracture, or bone contusion, can be demonstrated by MRI, in the absence of plain-film findings. This entity has been described in the femur and tibia (Fig. 7.53). Irregular areas of low SI on T1WI, and high SI on T2WI, can be seen within the bone marrow space, probably representing a microscopic compression fracture of the trabecular bone, with associated edema and hemorrhage. A follow-up MRI will demonstrate resolution of the bone marrow changes.

Chondromalacia Patellae

In chondromalacia patellae, the articular cartilage of the patella is traumatized. During the early stages it becomes swollen or blistered (57) (Fig. 7.54); subsequently, irregularity and focal thinning develop (Fig. 7.55), and finally, the subchondral bone is exposed due to complete loss of the hyaline cartilage (Fig. 7.56).

Osteoarthritis

In early osteoarthritis, focal hypointense areas within the cartilage or cartilage thinning without joint space narrowing can be seen on MRI (26) (Fig.

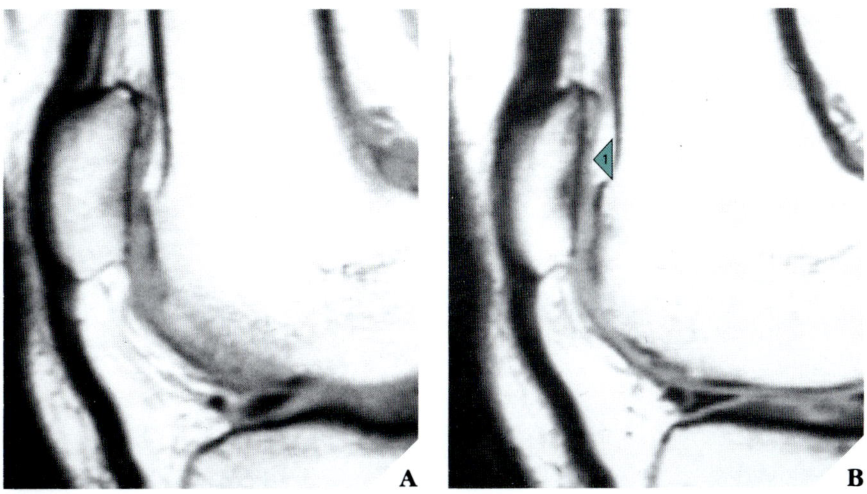

FIGURE 7.55. Chondromalacia Patellae, Intermediate Stage. Two contiguous sagittal sections [SE 2000/20]. The patellar cartilage shows a central area of focal thinning. Arrow **1** Focal cartilage thinning.

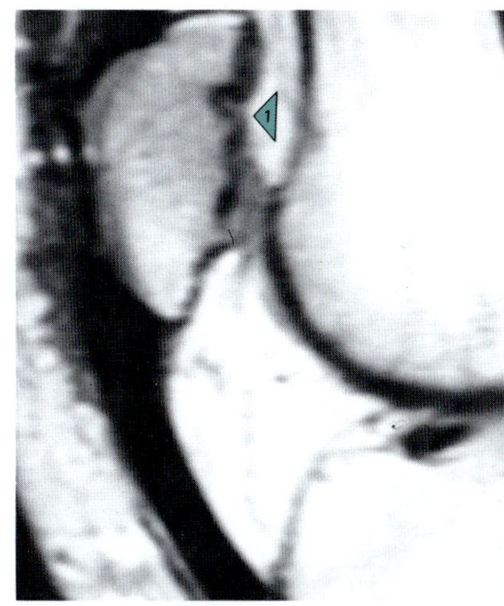

FIGURE 7.56. Chondromalacia Patellae, Late Stage. Sagittal plane [SE 2000/20]. The patellar articular cartilage shows deep ulcerations, to the level of the subchondral cortex, which also shows marked irregularity. Arrow **1** Loss of articular cartilage.

7.57). In advanced degenerative disease, osteophytes, joint space narrowing, and diffuse, irregular cartilage loss are seen (Fig. 7.58).

SYNOVIAL LESIONS

Synovial Effusions

Synovial effusions of the knee are seen on MRI as low SI collections on T1WI, and high SI on T2WI following the anatomic boundaries of the joint capsule and suprapatellar recess. There are no SI characteristics to help distinguish between inflammatory and infectious joint effusion (5,10).

Rheumatoid Arthritis

Rheumatoid arthritis affects the knee joint in about 10 per cent of cases. The classic stigmata of rheumatoid arthritis including synovial proliferation (pannus), erosions, subluxations, effusions, popliteal cysts, and cartilage thinning are well depicted by MRI (3,58) (Figs. 7.59 and 7.60). Differentiation between joint effusion and pannus formation may be difficult unless hemosiderin is deposited in the inflammatory pannus. In these cases the low SI of the pannus, which is seen in all pulse sequences (Fig. 7.60), is due to the paramagnetic effect of hemosiderin.

Pigmented Villonodular Synovitis

Pigmented villonodular synovitis (PVNS) is believed to be an inflammatory condition of the synovium, most frequently affecting the knee joint. It can present as a diffuse form (Fig. 7.61) or nodular form

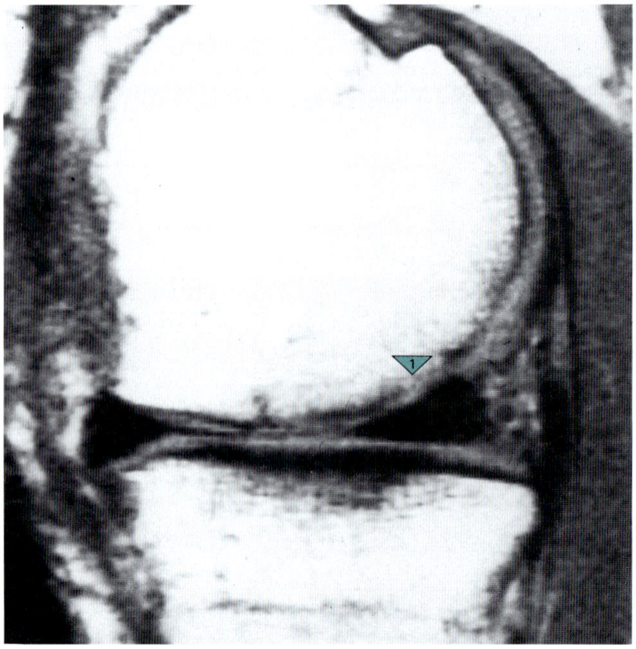

FIGURE 7.57. Early Osteoarthritis. Sagittal plane [SE 2000/20]. There is a focal loss of articular cartilage thickness in the medial femoral condyle, with underlying cortical irregularity. Arrow **1** Articular cartilage thinning.

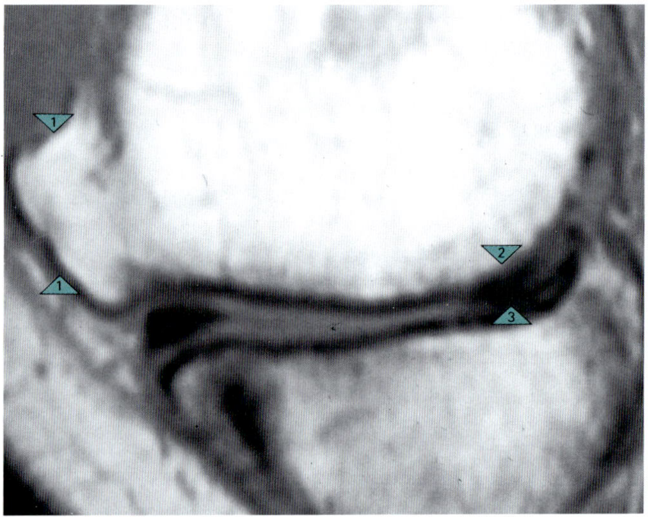

FIGURE 7.58. Advanced Osteoarthritis. Sagittal plane [SE 2000/20]. The cartilage of the posterior aspect of the femoral condyle is practically lost, and the subchondral cortex is in contact with the posterior horn of the medial meniscus. There is a large anterior osteophyte. Arrow **1** Osteophyte. Arrow **2** Loss of articular cartilage. Arrow **3** Posterior horn of medial meniscus.

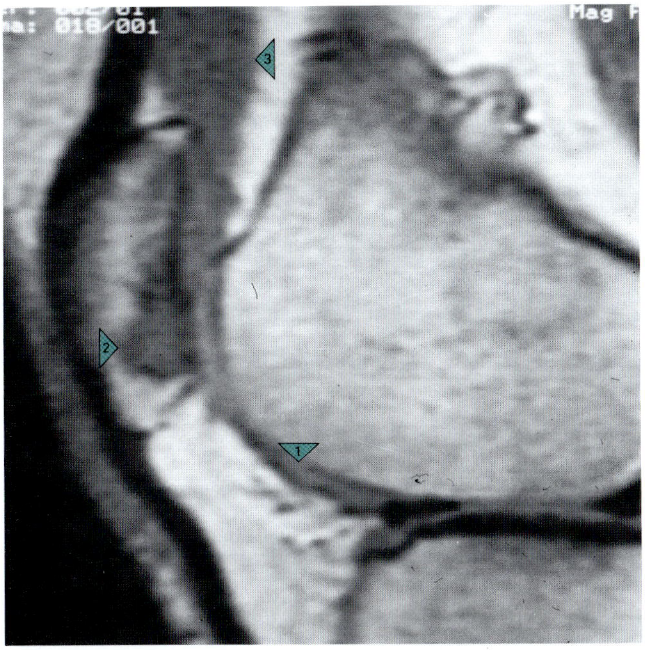

FIGURE 7.59. Rheumatoid Arthritis. Sagittal plane [SE 2000/20]. The anterior articular cartilage of the lateral femoral condyle is irregular and abnormally thin. There is an erosion of the posterior aspect of the patella and a joint effusion with distension of the suprapatellar recess. Arrow **1** Articular cartilage thinning. Arrow **2** Bone erosion. Arrow **3** Joint effusion.

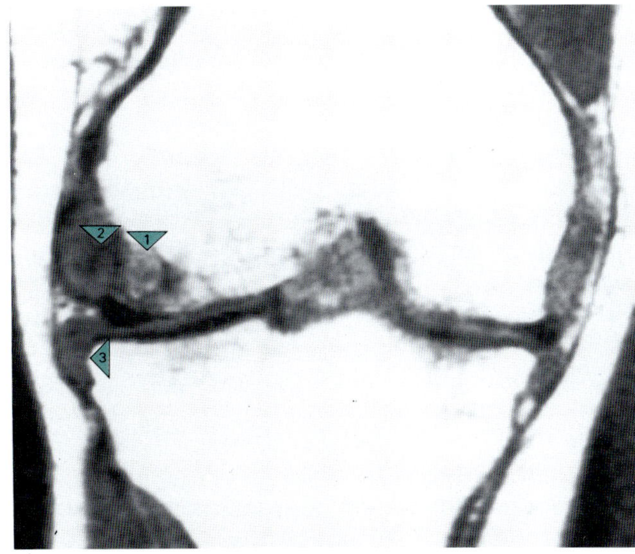

FIGURE 7.60. Rheumatoid Arthritis. Coronal plane [SE 2000/20]. There is a marginal erosion of the lateral femoral condyle. Overlying low SI tissue represents hypertrophied pannus, surrounded by a small amount of joint effusion. Arrow **1** Bone erosion. Arrow **2** Pannus. Arrow **3** Joint effusion.

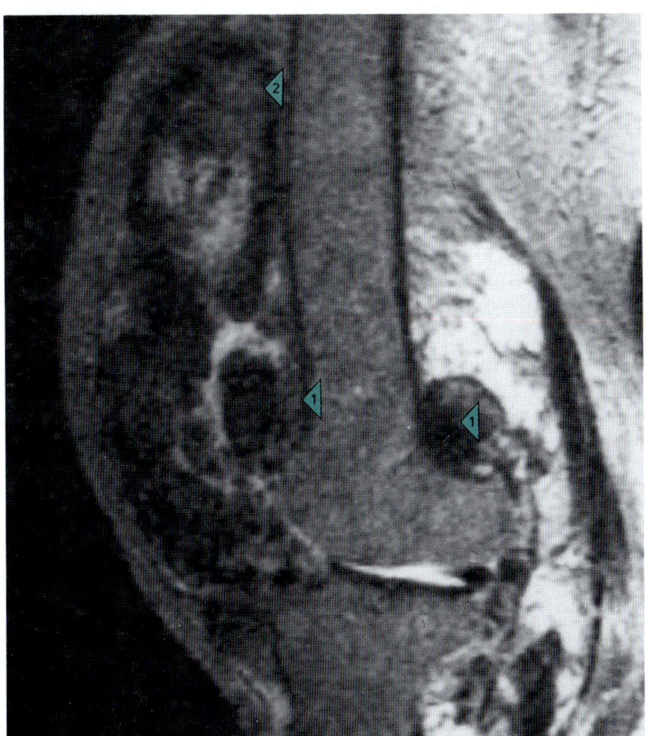

FIGURE 7.61. Diffuse Form of Pigmented Villonodular Synovitis. Sagittal plane [SE 2000/80]. Multiple soft-tissue masses are present within the distended joint capsule. The low SI seen in multiple areas within the mass represents regions rich in hemosiderin-laden macrophages. The paramagnetic effect of the hemosiderin produces this low SI. The synovium in the suprapatellar recess of the joint is also thickened and hypointense, due to a combination of fibrosis and hemosiderin deposition. Arrows **1** Masses of pigmented villonodular synovitis. Arrow **2** Thick joint capsule.

(Fig. 7.62). In the diffuse form, low SI, irregular, intrasynovial nodules are seen associated with bone erosions. Localized masses can be identified on arthrography in the focal form. Plain-film diagnosis may be difficult in the focal form, due to the absence of calcification and erosions. In both forms hemosiderin-laden macrophages, histiocytes, and multinucleated giant cells are present pathologically. The paramagnetic effect of the hemosiderin is responsible for the low SI seen on T1WI and T2WI in the diffuse and focal forms of PVNS (23,27,45).

Hemophilic Arthropathy

Hemophilic arthropathy develops as a consequence of repeated episodes of hemarthrosis. Specific pathologic findings include inflammatory synovial reaction, hemosiderin deposition, periarticular fibrosis, and also secondary degenerative changes with articular cartilage thinning, subchondral cysts, and subarticular sclerosis. A characteristic MRI finding is diffuse thickening of the synovium, often with low SI on T1WI and T2WI, due to hemosiderin deposition and periarticular fibrosis ("black synovium" sign) (Fig. 7.63). Changes of seconary degenerative osteoarthritis, including articular cartilage thinning and joint space narrowing, can also be detected on MRI (28,45,55).

Synovial Osteochondromatosis

Synovial osteochondromatosis occurs as a consequence of metaplasia of synovial cells into cartila-

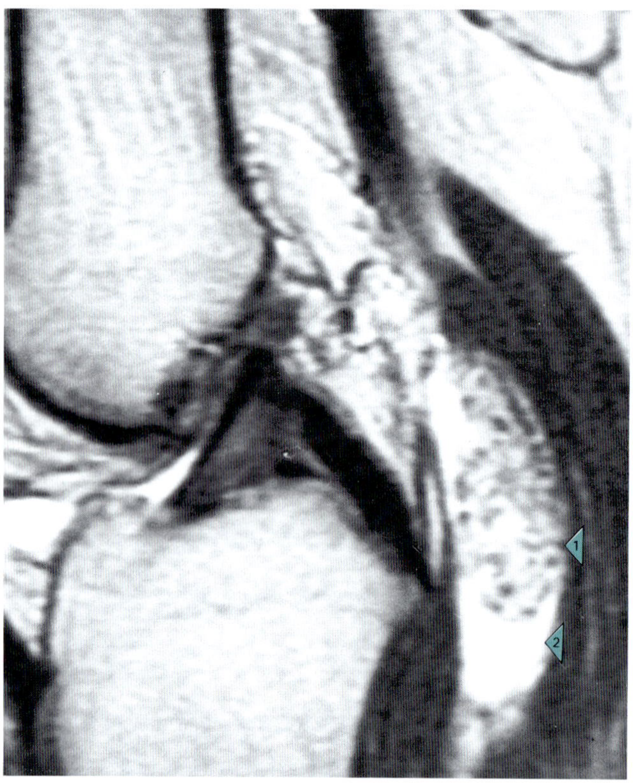

FIGURE 7.62. Nodular Form of Pigmented Villonodular Synovitis. Sagittal plane [SE 2000/80]. There is a small joint effusion and a popliteal cyst. Within the cyst, there is an irregular mass surrounded by hyperintense fluid. The areas of low SI represent regions of hemosiderin deposition. Arrow **1** Focal pigmented villonodular synovitis. Arrow **2** Popliteal cyst.

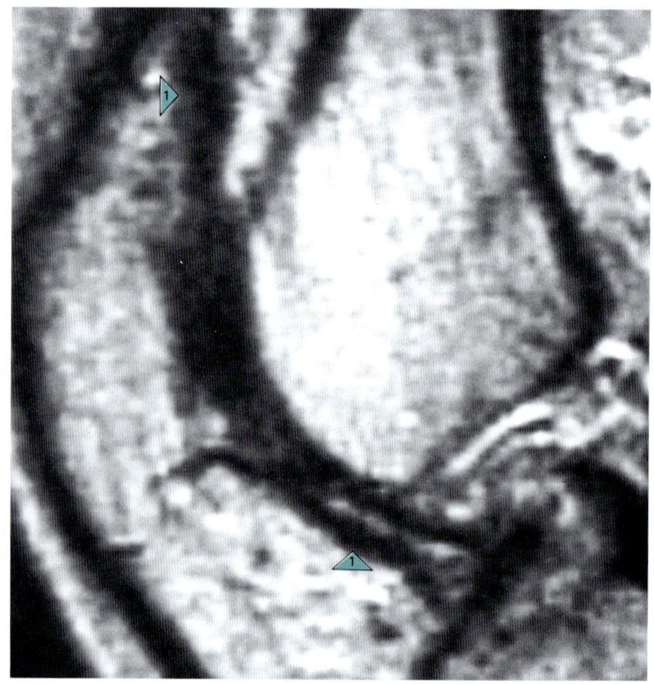

FIGURE 7.63. Hemophilic Arthropathy. Sagittal plane [SE 2000/80]. There is no joint effusion, but the synovium is hypointense and thickened due to fibrosis and hemosiderin deposition, both caused by repeated hemarthrosis. Arrows **1** Thick synovium.

ginous cells. This entity presents clinically with mechanical symptoms and pain due to the presence of intraarticular loose bodies, often detectable on plain films because of their characteristic calcification. In typical cases, the MRI appearance is similar to that of multiple post-traumatic loose bodies. These show varying degrees of low SI on both T1WI and T2WI, depending on the amount of calcification. If ossification with bone marrow formation is present, the SI of the synovial osteochondromas may be similar to that of the bone marrow (Fig. 7.64). Occasionally, the cartilaginous component of synovial osteochondromatosis may be seen as a hyperintense area on T2WI (13). Joint effusion is frequently associated with this condition.

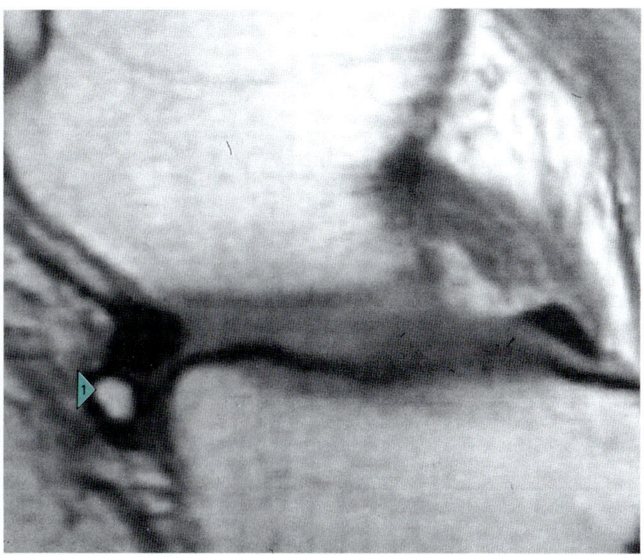

FIGURE 7.64. Synovial Osteochondromatosis. Sagittal plane [SE 2000/20]. There is a hyperintense, anteriorly located intraarticular loose body. The SI is similar to that of the bone marrow of the distal femur. Arrow **1** Synovial osteochondromatosis.

NONTUMORAL MASSES

Palpable, nontumoral masses around the knee include popliteal cysts, bursitis, meniscal cysts, ganglion cysts, hematomas, and aneurysms. Infectious and tumoral conditions are described in detail in Chapters 10 and 11.

Popliteal Cysts

Popliteal or Baker cysts are frequently found when performing MRI examinations for meniscal lesions. They are usually associated with joint effusions and often associated with meniscal tears (3,23,34,49). Popliteal cysts are identified as fluid collections in the popliteal space (Fig. 7.65). Their relationship

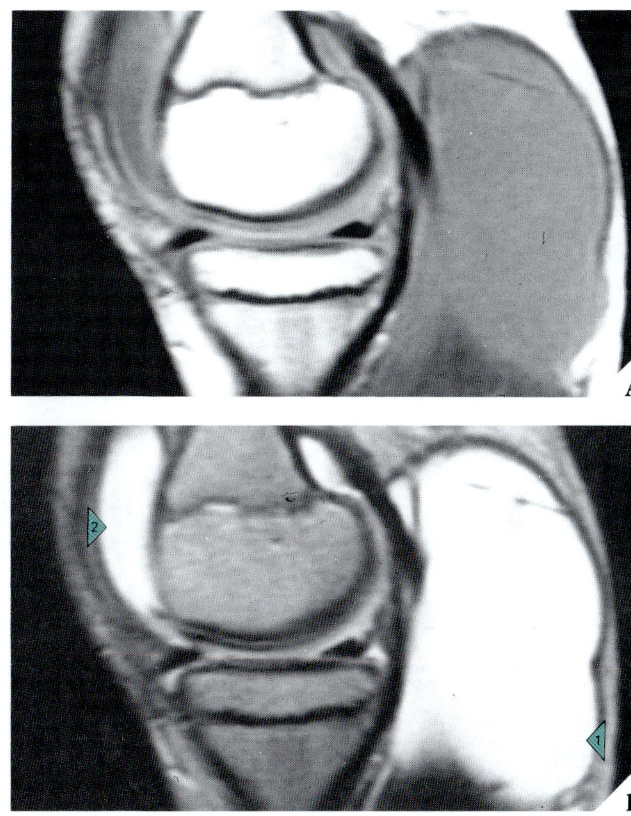

FIGURE 7.65. Popliteal Cyst. A Sagittal plane [SE 2000/20]; **B** sagittal plane [SE 2000/80]. Characteristic SI changes of fluid within this large popliteal cyst are seen on PDWI **(A)** and T2WI **(B).** Arrow **1** Popliteal cyst. Arrow **2** Joint effusion.

with the gastrocnemius–semimembranosus bursa is better appreciated in the axial plane (30) (Fig. 7.66).

Bursitis

Bursitis can be defined as a noninfectious inflammation of a bursa frequently related to trauma. The MRI findings reflect those of a fluid collection including low SI on T1WI and PDWI, and high SI on T2WI (34). In fact, differentiation between bursitis and soft-tissue abscess by MRI may be difficult without clinical correlation and knowledge of the anatomic locations of the bursae. The prepatellar (Fig. 7.67), infrapatellar (Fig. 7.68), and anserine bursa (Fig. 7.69) can be identified in their expected

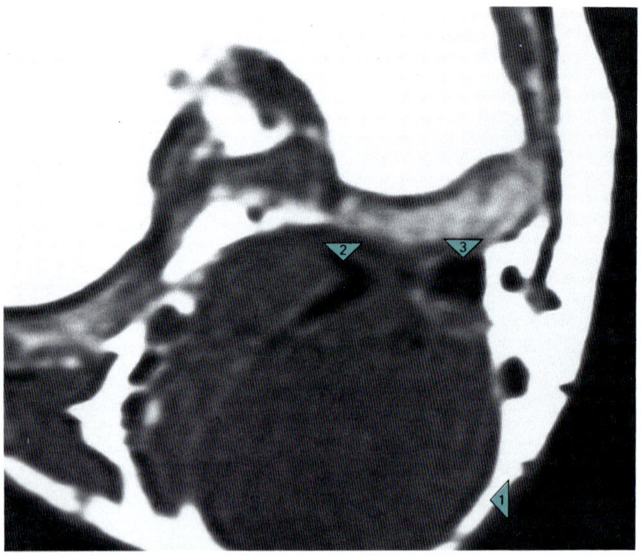

FIGURE 7.66. Popliteal Cyst. Axial plane [SE 800/20]. The cyst herniates between the gastrocnemius and semimembranosus tendons. Arrow **1** Popliteal cyst. Arrow **2** Gastrocnemius tendon. Arrow **3** Semimembranosus tendon.

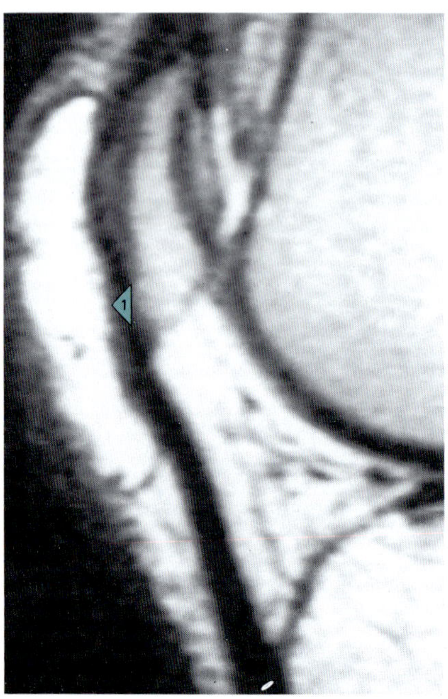

FIGURE 7.67. Prepatellar Bursitis. Sagittal plane [SE 2000/20]. The low SI region anterior to the patella corresponds to the inflamed bursa. Arrow **1** Prepatellar bursitis.

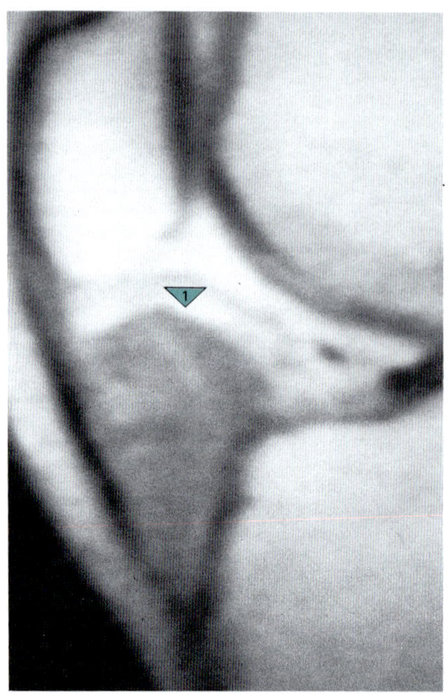

FIGURE 7.68. Infrapatellar Bursitis. Sagittal plane [SE 800/25]. The low SI mass behind the patellar tendon represents the bursa swollen with fluid. Arrow **1** Infrapatellar bursitis.

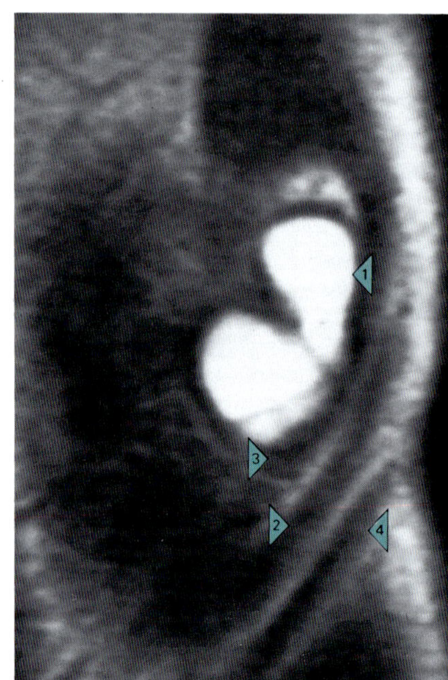

FIGURE 7.69. Anserine Bursitis. Sagittal plane [SE 2000/80] through the superficial medial aspect of the knee. The high SI anserine bursa protrudes anterior to the tendons of the pes anserinus. Arrow **1** Anserine bursitis. Arrow **2** Sartorius tendon. Arrow **3** Gracilis tendon. Arrow **4** Semitendinosus tendon.

anatomic location when fluid accumulates within their cavity. The anserine bursa is located between the tendons of the sartorius, gracilis, and semitendinosus muscles (pes anserinus), in the medial aspect of the knee, and should not be confused with a popliteal cyst (21). Popliteal cysts are located more posteriorly, herniating between the gastrocnemius and semimembranosus tendons. A joint effusion is practically always present with popliteal cysts, but not in anserine bursitis.

Meniscal Cysts

Meniscal cysts (Fig. 7.70) are probably traumatic in origin and are often associated with meniscal tears, although developmental and inflammatory etiologies have also been proposed. Meniscal cysts are more frequently located in the lateral meniscus, at the joint line level. In most cases these cysts present clinically as asymptomatic palpable masses; however, they may occasionally be symptomatic to the point of requiring surgical excision (9,21).

Ganglion Cysts

Ganglion cysts (Fig. 7.71) are lesions containing jelly-like viscous fluid occurring around the knee. They present clinically as palpable masses, in atypical locations for popliteal cysts. Ganglion cysts are similar in SI to meniscal cysts, or other fluid-containing lesions (hypointense on T1WI and hyperintense on T2WI). Unlike ganglion cysts, meniscal cysts are associated with meniscal tears and are located adjacent to the menisci (9). Although separated from the menisci, ganglion cysts can have connections with the joint capsule. A synovial lining may be present in both meniscal cysts and ganglion cysts.

Hematomas

Soft-tissue hematomas involving the calf muscles and knee region are often confused with deep venous thrombosis or ruptured popliteal cysts, since they are often accompanied by pain and swelling. Early diagnosis is important to prevent development of posterior compartment syndrome and extensive muscle damage. The SI of a hematoma changes over time. As a general rule, the SI increases with time on T1WI, and, on T2WI, is hyperintense in all stages of the evolution. Occasionally, heterogeneous areas of lower SI on T1WI and T2WI are seen, associated with fluid levels. These areas have

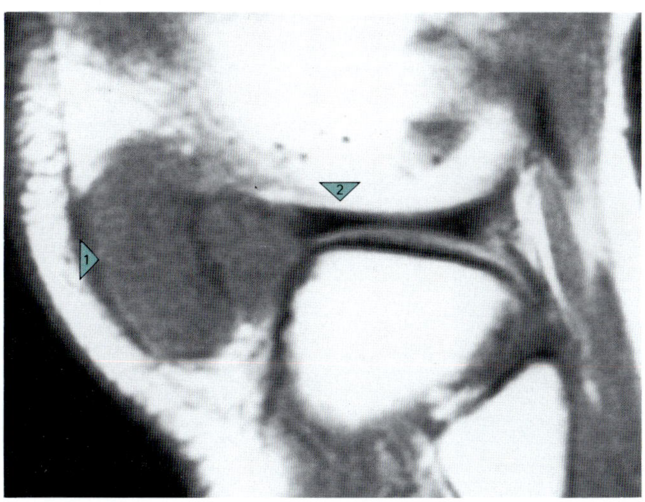

FIGURE 7.70. Meniscal Cyst. Sagittal plane [SE 2000/20]. Large, lobulated, hypointense mass represents a cyst arising from the anterior horn of the lateral meniscus, and extending to the infrapatellar fat pad. Arrow **1** Meniscal cyst. Arrow **2** Lateral meniscus.

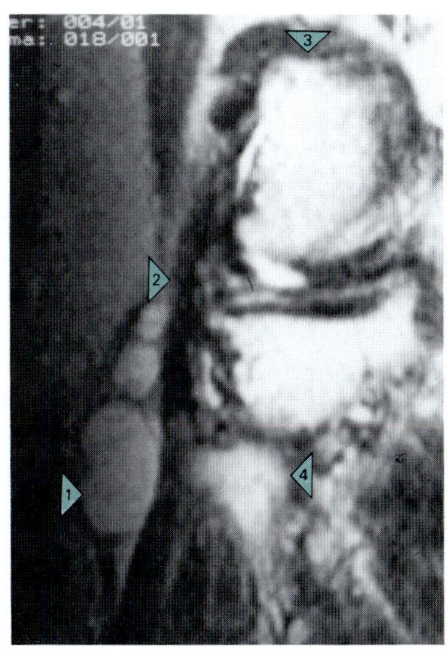

FIGURE 7.71. Ganglion Cyst. Coronal plane [SE 2000/80]. A polylobulated, fluid-filled mass arising from the tibiofibular joint is seen in the lateral aspect of the knee, lateral to the fibular collateral ligament. Arrow **1** Ganglion cyst. Arrow **2** Fibular collateral ligament. Arrow **3** Lateral condyle. Arrow **4** Fibula.

been found to correspond to fibrotic clots at surgery (Fig. 7.72), and the heterogeneity of the SI is due to the paramagnetic effects of the hemoglobin breakdown (39,45,50,53). Occasionally, low SI rims are seen in the periphery of the hematomas, in all pulse sequences. These are believed to represent regions of hemosiderin-laden macrophages.

Aneurysms

Popliteal artery aneurysms (Fig. 7.73) are most often diagnosed clinically and confirmed by ultrasound and angiography. Occasionally, these patients are referred for MRI evaluation because the characteristic pulsations are not identified and soft-tissue tumors are suspected. MRI shows a flow artifact,

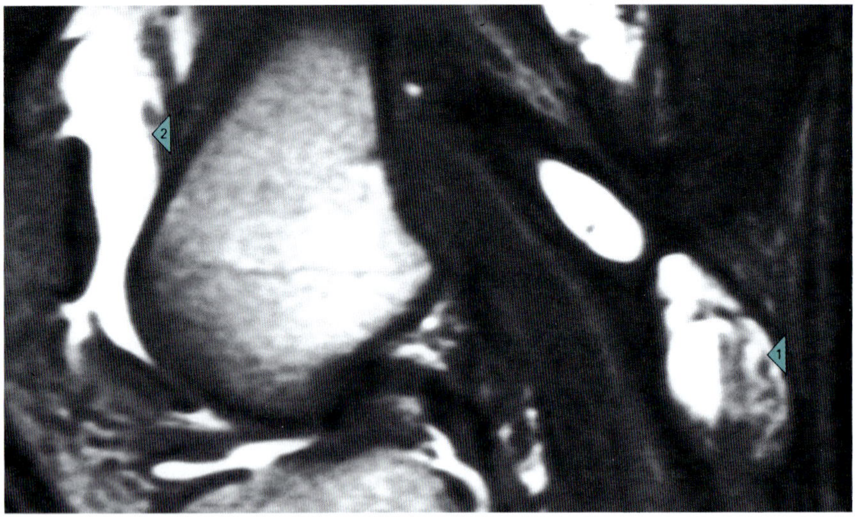

FIGURE 7.72. Hematoma of the Gastrocnemius Muscle. Sagittal plane [SE 2000/80]. There are two hyperintense fluid collections in the lateral head of the gastrocnemius muscle, with a fluid level in the inferior one. (MRI obtained about 2 weeks after trauma). Arrow **1** Hematoma. Arrow **2** Joint effusion.

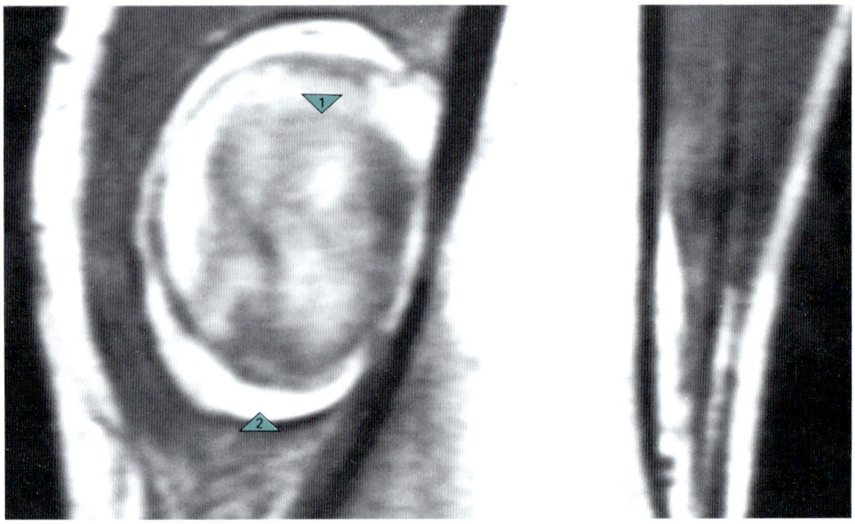

FIGURE 7.73. Popliteal Artery Aneurysm. Coronal plane [SE 2000/20]. Large mass in the medial aspect of the distal thigh. Peripheral rim of low SI represents the calcified aneurysm wall, seen on plain films. Mural thrombus is seen peripherally as a high SI band inside the aneurysm. The slowly flowing blood inside the aneurysm shows low SI. Arrow **1** Lumen of aneurysm. Arrow **2** Mural thrombus.

peripheral mural thrombus with a hyperintense signal, and central areas of lower SI representing the flowing blood (34,59). The calcified wall may be seen as a low SI rim at the periphery of the aneurysm. Patency of the popliteal vein can be assessed by MRI.

AVASCULAR NECROSIS

Avascular necrosis (osteonecrosis, bone infarction, aseptic necrosis) of the knee can occur either as a spontaneous event or as a consequence of a known predisposing condition. Such conditions include steroid use, systemic lupus erythematosus, pancreatitis, alcoholism, sickle cell anemia, and Gaucher's disease, among others.

Spontaneous Osteonecrosis

Spontaneous osteonecrosis of the knee is more frequently seen in the medial femoral condyle, although it has been described in the lateral femoral condyle and tibial plateau. The syndrome is characterized by intense knee pain, not related to trauma, and occurs more frequently in middle-aged and elderly females (23,34,41). Spontaneous osteonecrosis has been associated with meniscal tears in at least one report (38). MRI can demonstrate a well-demarcated area of low SI on T1WI and T2WI in the subchondral bone (Fig. 7.74), occasionally surrounded by a hyperintense halo on T2WI.

Secondary Osteonecrosis

Secondary osteonecrosis can occur in the subchondral bone or in the medullary space and can involve the femur, the tibia, or both. Characteristic MRI findings include hypo- or isointense areas surrounded by a halo of high SI on T2WI, and a peripheral band of low SI on both T1WI and T2WI (double-line sign) (2,4,36) (Fig. 7.75). The hyperintense area corresponds to granulation tissue and

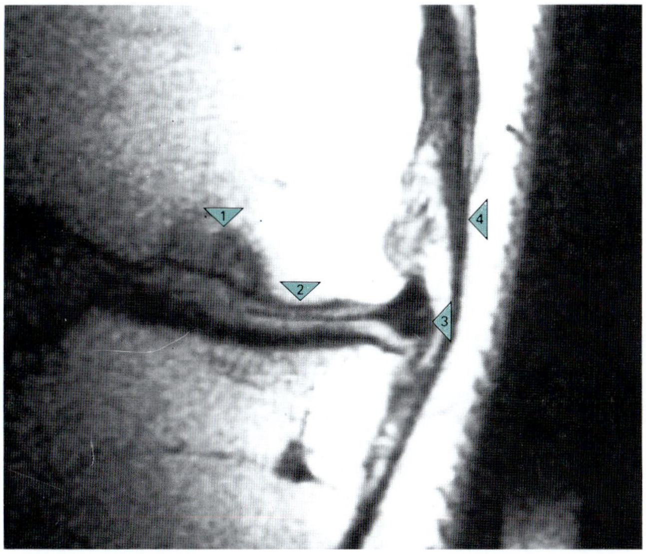

FIGURE 7.74. Spontaneous Osteonecrosis of the Medial Femoral Condyle. Coronal plane [SE 600/20]. The well-demarcated area of low SI is associated with depression and flattening of the subchondral cortex. The articular cartilage is normal. Arrow **1** Osteonecrosis. Arrow **2** Articular cartilage. Arrow **3** Medial meniscus. Arrow **4** Medial collateral ligament.

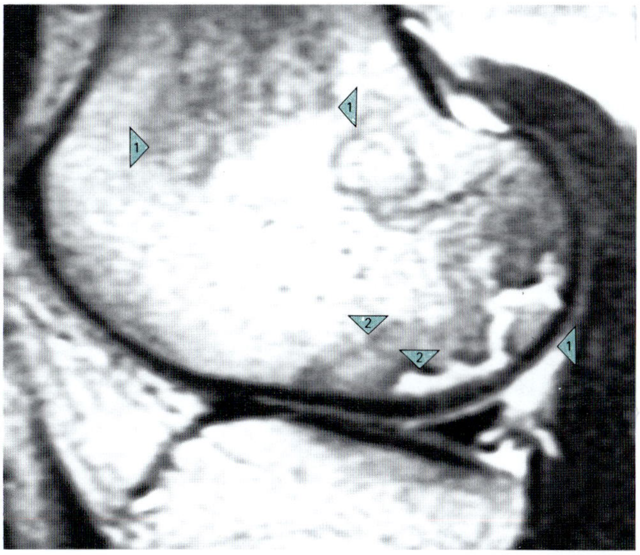

FIGURE 7.75. Avascular Necrosis of the Distal Femur. Sagittal plane [SE 2000/80]. Both the metaphysis and subarticular area of the medial femoral condyle are involved. The small area of infarction of the femoral condyle is surrounded by hyperintense and hypointense bands representing the reactive interface ("double-line" sign). Arrows **1** Avascular necrosis. Arrow **2** "Double-line" sign.

hypervascularity at the reactive interface. The hypointense area corresponds to the region of osteoblastic activity and bone sclerosis. In the medullary space, undulating hypointense lines demarcating the bone infarction can be seen (Fig. 7.76). In the late stages of avascular necrosis, subchondral fracture can develop in a manner similar to that of the femoral head (subchondral crescent sign). This subchondral fracture can be seen on MRI as a crescent-shaped area of high SI on T2WI (Fig. 7.77).

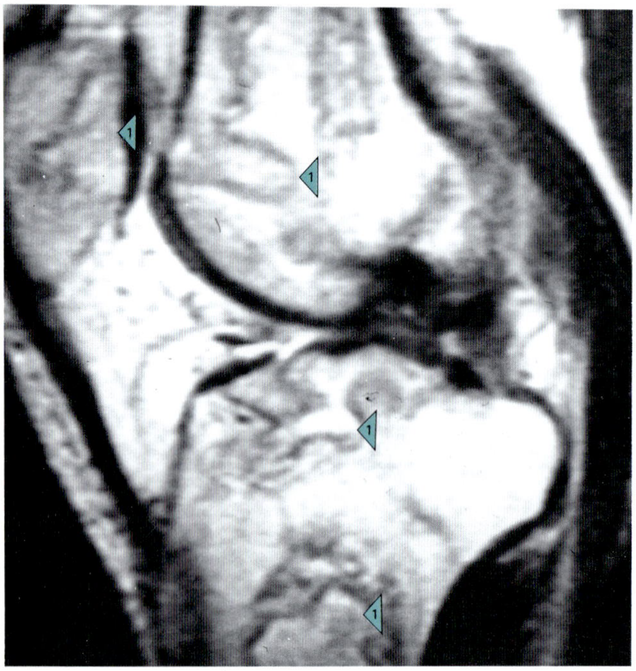

FIGURE 7.76. Avascular Necrosis Secondary to Steriod Use. Sagittal plane [SE 2000/80]. There are multiple areas of medullary infarction involving the femur, tibia, and patella in this patient with systemic lupus erythematosus. The infarctions are demarcated by a low SI line, probably representing the zone of osteoblastic activity of the reactive interface. Arrows **1** Avascular necrosis.

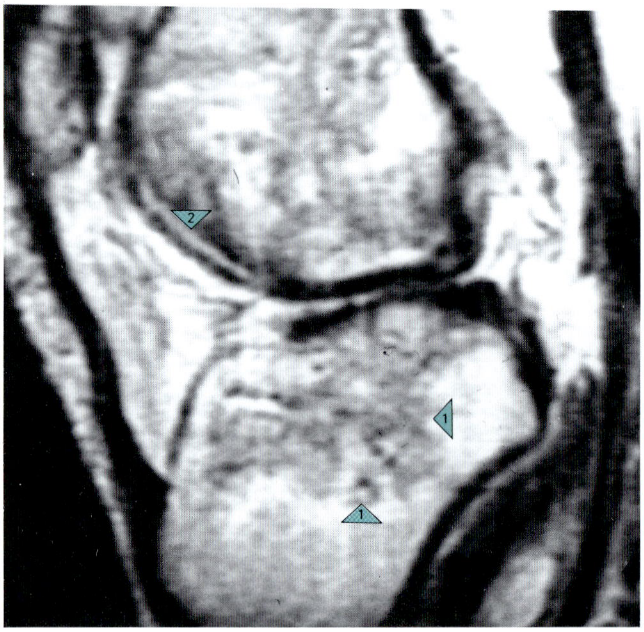

FIGURE 7.77. Avascular Necrosis of the Femur and Tibia. Sagittal plane [SE 2000/80]. There is a subchondral fracture (plain-film "crescent sign") in the anterior aspect of the lateral femoral condyle. Sagittal plane [SE 2000/80]. The patient was treated with steroids for a renal transplant. The medullary infarctions are seen as inhomogeneous areas of low and high SI in the distal femur and proximal tibia. The subchondral fracture is seen anteriorly as a hyperintense crescent area in the subchondral region. Arrow **1** Avascular necrosis. Arrow **2** Subchondral fracture.

REFERENCES

1. Bellon EM, Haacke EM, Coleman PE, et al: MR artifacts: a review. AJR 1986;147:1271–1281.
2. Beltran J, Burk JM, Herman LJ, et al: Avascular necrosis of the femoral head: early MRI detection and radiological correlation. Magn Reson Imaging 1987;5:431–442.
3. Beltran J, Caudill JL, Herman LJ, et al: Rheumatoid arthritis: MR imaging manifestations. Radiology 1987;165:153–157.
4. Beltran J, Herman LJ, Burk JM, et al: Femoral head avascular necrosis: MR imaging with clinical-pathologic correlation. Radiology 1988;166:215–220.
5. Beltran J, Noto AM, Herman LJ, et al: Joint effusions: MR imaging. Radiology 1986;158:133–137.
6. Beltran J, Noto AM, Herman LJ, Lubbers LM: Tendons: high-field-strength, surface coil MR imaging. Radiology 1987;162:735–740.
7. Beltran J, Noto AM, Mosure JC, et al: The knee: surface-coil MR imaging at 1.5 Tesla. Radiology 1986;159:747–751.
8. Bodne D, Quinn SF, Murray WT, et al: Magnetic resonance imaging of chronic patellar tendinitis. Skeletal Radiol 1988;17:24–28.

9. Burk DL, Dalinka MK, Kanal E, et al: Meniscal and ganglion cysts of the knee: MR evaluation. AJR 1988;150:331–336.
10. Burk DL, Kanal E, Brunberg JA, et al: 1.5T surface-coil MRI of the knee. AJR 1986;147:293–300.
11. Burk DW, Dalinka MK, Schiebler ML, et al: Strategies for musculoskeletal magnetic resonance imaging. Radiol Clin North Am 1988;26:653–672.
12. Christoforidis AJ: Atlas of Axial, Sagittal and Coronal Anatomy. Philadelphia, WB Saunders, 1988.
13. Cohen EK, Kressel HY, Frank TS, et al: Hyaline cartilage-origin bone and soft-tissue neoplasms: MR appearance and histologic correlation. Radiology 1988;167:477–481.
14. Crues JV III, Mink JH, Levy TL, et al: Meniscal tears of the knee: accuracy of MR imaging. Radiology 1987;164:445–448.
15. Dietz GW, Wilcox DM, Montgomery JB: Segond tibial condyle fracture: lateral capsular ligament avulsion. Radiology 1986;159:467–469.
16. Felmlee JP, Ehman RL: Spatial presaturation: a method for supressing flow artifacts and improving depiction of vascular anatomy in MR imaging. Radiology 1987;164:559–564.
17. Frahm J, Haase A, Mathaei D: Rapid three-dimensional NMR imaging using the FLASH technique. J Comput Tomogr 1986;10:363–368.
18. Gray H: Anatomy of the Human Body, ed 30, Clemente CD (ed). Philadelphia, Lea & Febiger, 1985.
19. Gylys-Morin VM, Hajek PC, Sartoris DJ, Resnick D: Articular cartilage defects. Detectability in cadaver knees with MR. AJR 1987;148:1153–1157.
20. Haggar AM, Froelich JW, Hearshen DO, Sadasivan K: Meniscal abnormalities of the knee: 3DFT fast-scan GRASS MR imaging. AJR 1988;150:1341–1344.
21. Hall FM, Joffe N: Case report. CT imaging of anserine bursa. AJR 1988;150:1107–1108.
22. Harms SE, Muschler G: Three-dimensional MR imaging of the knee using surface coils. J Comput Assist Tomogr 1986;10:773–777.
23. Hartzman S, Reicher MA, Bassett LW, et al: MR imaging of the knee. Part II. Chronic disorders. Radiology 1987;162:553–557.
24. Herman LJ, Beltran J: Pitfalls in MR imaging of the knee. Radiology 1988;167:775–781.
25. Kneeland JB, Jesmanowicz A, Froncisz W, et al: High-resolution MR imaging using loop-gap resonators. Radiology 1986; 158:247–250.
26. Konig H, Sauter R, Deimling M, Vogt M: Cartilage disorders: comparison of spin-echo, CHESS, and FLASH sequence MR images. Radiology 1987;164:753–758.
27. Kottal RA, Vogler JB III, Matamoros A, et al: Pigmented villonodular synovitis: report of MR imaging in two cases. Radiology 1987;163:551–553.
28. Kulkarni MV, Drolshagen LF, Kaye JJ, et al: MR imaging of hemophilic arthropathy. J Comput Assist Tomogr 1986;10:445–449.
29. Lee JK, Yao L, Phelps CT, et al: Anterior cruciate ligament tears: MR imaging compared with arthroscopy and clinical tests. Radiology 1988;166:861–864.
30. Lee KR, Cox GC, Neff JR: Cystic masses of the knee: arthrographic and CT evaluation. AJR 1987;148:329–334.
31. Li DKB, Adams ME, McConkey JP: Magnetic resonance imaging of the ligaments and menisci of the knee. Radiol Clin North Am 1986;24:209–227.
32. Lufkin RB, Vortruba J, Reicher M, et al: Solenoid surface coils in magnetic resonance imaging. AJR 1986:146:409–412.
33. Mesgarzadeh M, Sapega AA, Bonakdarpour A, et al: Osteochondritis dissecans: analysis of mechanical stability with radiography, scintigraphy and MR imaging. Radiology 1987; 165:775–780.
34. Mink JH, Reicher MA, Crues JV: Magnetic Resonance Imaging of the Knee. New York, Raven Press, 1987.
35. Mink JH, Levy T, Crues JV III: Tears of the anterior cruciate ligament and menisci of the knee: MR imaging evaluation. Radiology 1988;167:769–774.
36. Mitchell DG, Rao V, Dalinka MK, et al: Femoral head avascular necrosis: correlation of MR imaging, radiographic staging, radionuclide imaging, and clinical findings. Radiology 1987;162:709–715.
37. Netter FH: The CIBA Collection of Medical Illustrations. Musculoskeletal System. Summit, NJ, CIBA-GEIGY Corporation, 1987.
38. Norman A, Baker ND: Spontaneous osteonecrosis of the knee and medial meniscal tears. Radiology 1978;129:653–656.
39. Pakter RL, Fishman EK, Zerhouni EA: Calf hematoma: computed tomographic and magnetic resonance findings. Skeletal Radiol 1987;16:393–396.

40. Patton JA, Kulkarni MV, Craig JK, et al: Techniques, pitfalls and artifacts in magnetic resonance imaging. Radiographics 1987;7:505–519.
41. Pollack MS, Dalinka MK, Kressel HY, et al: Magnetic resonance imaging in the evaluation of suspected osteonecrosis of the knee. Skeletal Radiol 1987;16:121–127.
42. Porter BA, Hastrup W, Richardson ML, et al: Classification and investigation of artifacts in magnetic resonance imaging. Radiographics 1987;7:271–287.
43. Pusey E, Lufkin RB, Brown RJK, et al: Magnetic resonance imaging artifacts: mechanism and clinical significance. Radiographics 1986; 6:891–911.
44. Reicher MA, Rauschning W, Gold RH, et al: High-resolution magnetic resonance imaging of the knee joint: normal anatomy. AJR 1985;145:895–902.
45. Reicher MA, Hartzman S, Basett LW, et al: MR imaging of the knee. Part II. Traumatic disorders. Radiology 1987;162:547–551.
46. Rubin JI, Gomori JM, Grossman RJ, et al: High field MR imaging of extracranial hematomas. AJR 1987;148:813–817.
47. Spritzer CE, Vogler JB, Martinez S, et al: MR imaging of the knee: preliminary results with 3DFT GRASS pulse sequence. AJR 1988; 150:597–603.
48. Stoller DW, Martin C, Crues JV, et al: MR imaging—pathologic correlation of meniscal tears. Radiology 1987;163:731–735.
49. Sundaram M, McGuire MH, Fletcher J, et al: Magnetic resonance imaging of lesions of synovial origin. Skeletal Radiol 1986; 15:110–116.
50. Terrier F, Revel D, Pajannen H, et al: MR imaging of fluid collections. J Comput Assist Tomogr 1986;10:953–962.
51. Turner DA, Prodromos CC, Petasnick JP, et al: Acute injury of the ligaments of the knee: Magnetic resonance evaluation. Radiology 1985;154:717–722.
52. Tyrrell RL, Gluckert K, Pathria M, Modic MT: Fast three-dimensional MR imaging of the knee: comparison with arthroscopy. Radiology 1988;166:865–872.
53. Unger EC, Glazer HS, Lee JKT, et al: MRI of extracranial hematomas: preliminary observations. AJR 1986;146:403–408.
54. Wehrli FW: Introduction to Fast-Scan Magnetic Resonance, publication 7229, Milwaukee, WI, General Electric Medical Systems, 1986, pp 1–12.
55. Wilson DA, Prince JR: MR imaging of hemophilic pseudotumors. AJR 1988;150:349–350.
56. Yao L, Lee JK. Occult intraosseous fractures: detection by MR imaging. Radiology 1988;167:749–54.
57. Yulish BS, Montanez J, Goodfellow DB, et al: Chondromalacia patellae: assessment with MR imaging. Radiology 1987;164:763–766.
58. Yulish BS, Lieberman JM, Newman AJ, et al: Juvenile rheumatoid arthritis: assessment with MR imaging. Radiology 1987;165:149–152.
59. Zeitler E, Kaiser W, Schuierer G, et al: Magnetic resonance imaging of aneurysms and thrombi. Cardiovasc Intervent Radiol 1986; 8:321–328.

FIGURE CREDITS

Figures 7.27 and 7.45. From Beltran J, Noto AM, Mosure JC, et al: The knee: surface-coil MR imaging at 1.5 Tesla. Radiology 1986; 159: 747–751.

Figure 7.35. From Herman LJ, Beltran J: Pitfalls in MR imaging of the knee. Radiology 1988;167; 775–781.

Figure 7.47. Courtesy of L. Burk, MD.

Figure 7.68. Courtesy of B. Flanigan, MD.

CHAPTER EIGHT

Javier Beltran

The Ankle and Foot

TECHNICAL CONSIDERATIONS

IMAGING PLANES AND PULSE SEQUENCES

Coronal, axial, and sagittal planes are obtained for routine imaging of the ankle and subtalar joints (Fig. 8.1). Oblique planes perpendicular and parallel to the long axes of the metatarsal bones can be obtained for imaging of the forefoot (Fig. 8.2). Axial and sagittal images may be sufficient for evaluation of tendon lesions (6). Sagittal images with the foot in plantar flexion and dorsiflexion are useful for evaluation of Achilles tendon ruptures (see Tendon Lesions).

Traumatic, inflammatory, or tumoral conditions should be evaluated with T1WI and T2WI in at least two planes. T1WI alone may be sufficient for assessment of tendon ruptures.

FIELD OF VIEW

An FOV of 16 to 24 cm is adequate for ankle imaging. Smaller FOV will produce aliasing artifacts and will decrease the S/N by decreasing both the

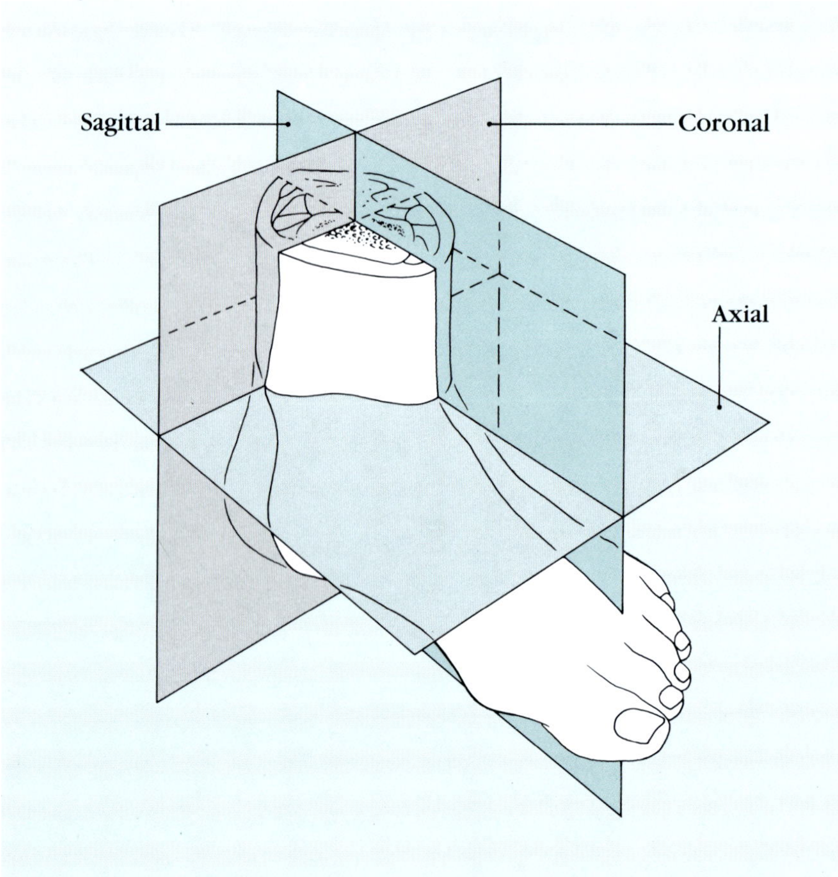

FIGURE 8.1. Routine planes of section for ankle imaging.

pixel size and the number of protons per voxel. However, if an FOV smaller than 16 cm must be used, the S/N can be increased by increasing slice thickness or NEX. Pseudo-filtering methods can be used to suppress aliasing artifacts. Presaturation with a 90-degree RF pulse should prevent pulsation artifacts.

▼▼▼ *POSITIONING*

The patient is positioned feet first, supine. Occasionally the lateral decubitus position may be indicated due to the patient's physical condition. In these cases, planes of section oriented obliquely to the magnetic field may be necessary to obtain images in the orthogonal planes of the foot and ankle.

▼▼▼ *ACQUISITION MATRIX*

In most cases, a matrix of 256 × 128 pixels is adequate for ankle and foot imaging.

▼▼▼ *SLICE THICKNESS*

A slice thickness of 5 mm with a 1-mm interslice gap can be routinely used. For the evaluation of small parts, such as the tendons or ligaments, a slice

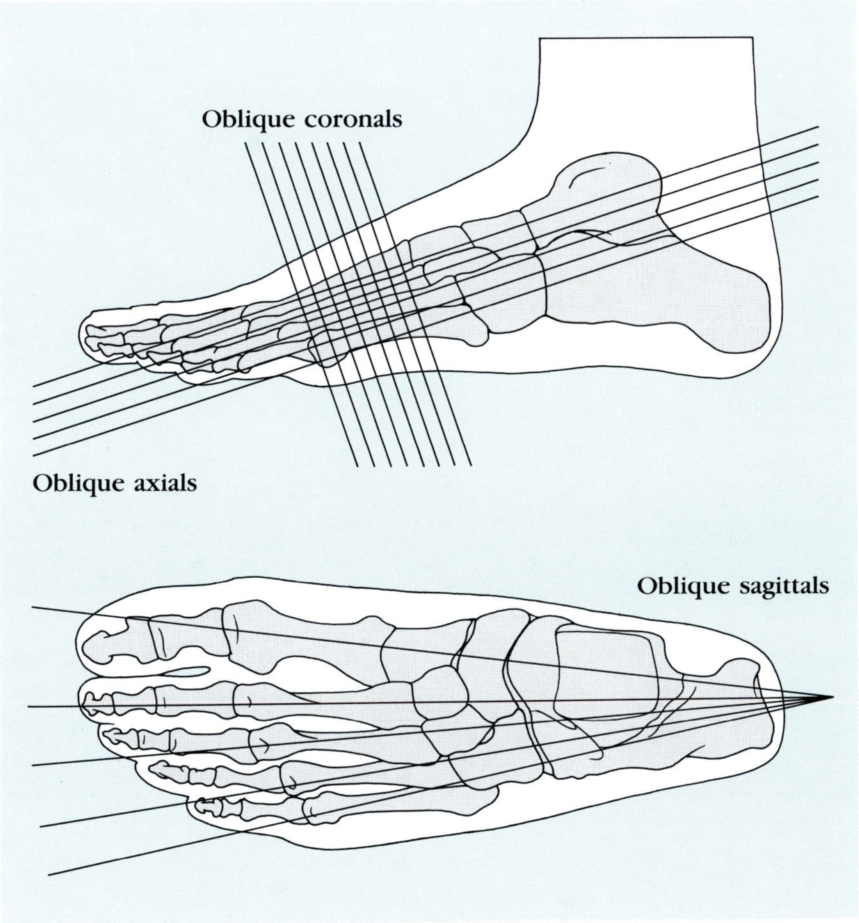

FIGURE 8.2. Routine planes of section for foot imaging.

thickness of 3 mm can be used alternatively. Thin slices decrease both C/N and S/N. This can be compensated for by increasing NEX.

▼▼▼ *SURFACE COILS*

As in other joints, surface coils are essential for high-resolution MR imaging. A transmit–receive wrap-around extremity coil (Fig. 8.3) or a 10-cm diameter receive-only loop-gap coil (Fig. 8.4) provide excellent S/N for imaging the entire ankle or foot. Smaller transmit–receive coils of 4-cm diameter have been used with an FOV of 8 cm for detailed imaging of the small ligaments and tendons of the ankle and foot (12).

NORMAL ANATOMY

▼▼▼ *AXIAL PLANE*

The Ligaments

The anterior and posterior tibiofibular ligaments are seen anterior and posterior to the tibiofibular syndesmosis in sections obtained above the tibiotalar joint (Fig. 8.5).

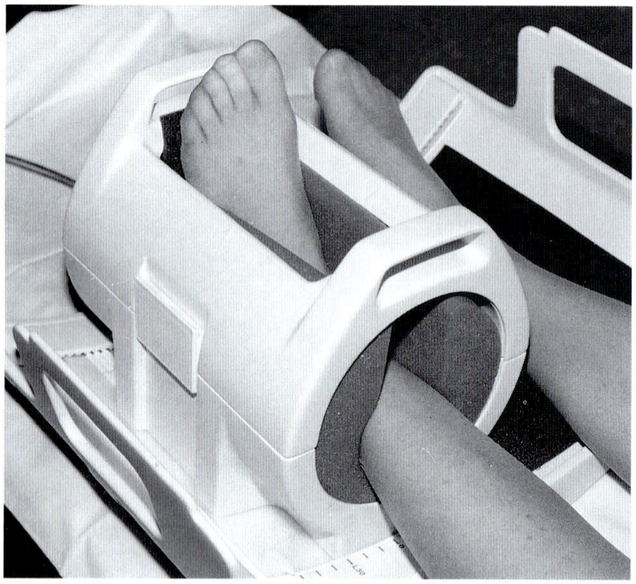

FIGURE 8.3. Extremity coil positioned for ankle imaging.

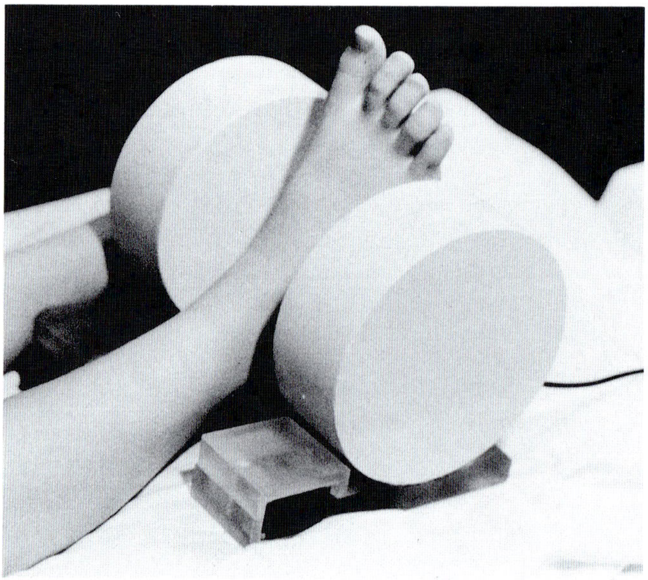

FIGURE 8.4. Loop-gap resonator coil positioned for ankle imaging.

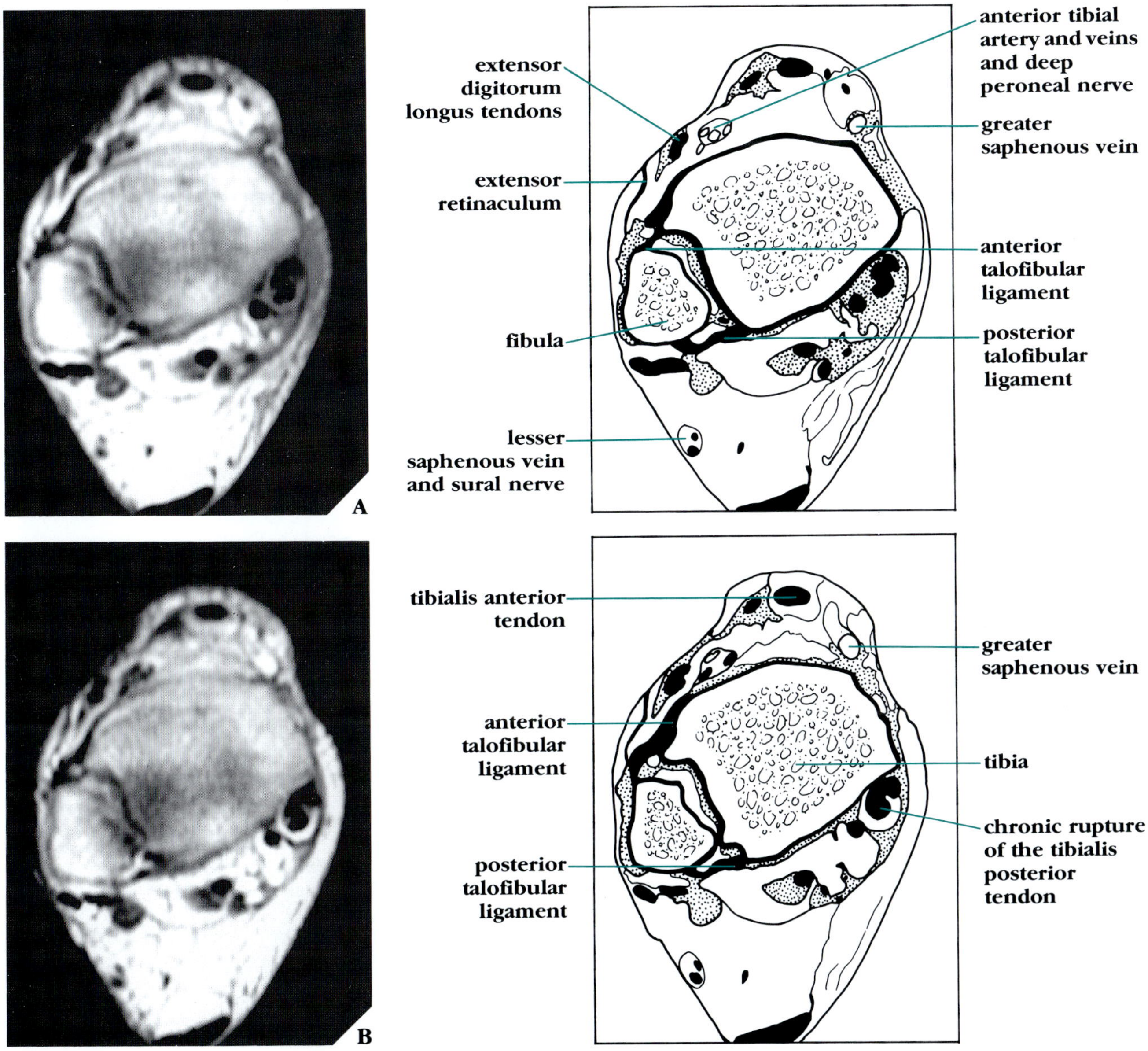

FIGURE 8.5. Normal Anterior and Posterior Tibiofibular Ligaments. A,B Two consecutive axial sections through the distal tibiofibular syndesmosis [SE 2000/20] demonstrating the anterior and posterior tibiofibular ligaments. This patient also had a chronic rupture of the tibialis posterior tendon (also see Fig. 8.29).

The anterior and posterior talofibular ligaments are seen in more distal sections. The anterior talofibular ligament is more consistently identifiable than the posterior ligament (Fig. 8.6).

The Tendons

All tendons should show uniformly low SI in all SE pulse sequences, with the exception of the Achilles and tibialis posterior tendons. On long TR techniques, these tendons occasionally show small areas of intermediate SI within the tendon substance, near their insertions in the calcaneal tuberosity and the navicular bone, respectively (19). The synovial sheaths and the small amount of synovial fluid around the tendons of the ankle and foot are not seen with MRI unless there is an inflammatory process.

In the lateral aspect of the ankle, the peroneus brevis and longus tendons are identified (2,4,12,20,23) (Fig. 8.7). The peroneus brevis tendon is located superior and anterior to the peroneus longus tendon. The peroneus tendons pass beneath the superior and inferior peroneal retinacula and insert into the base of the fifth metatarsal (peroneus

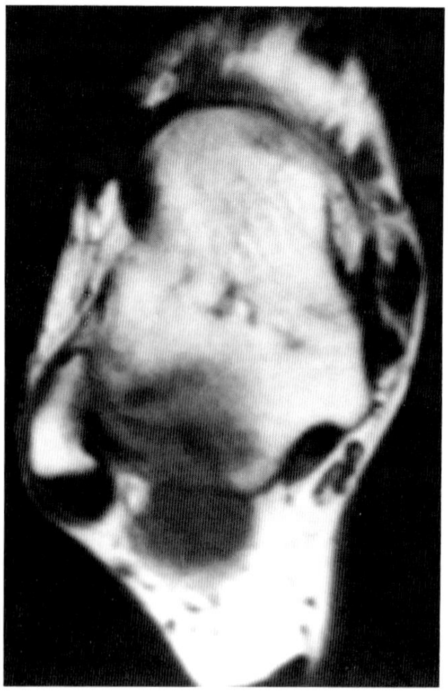

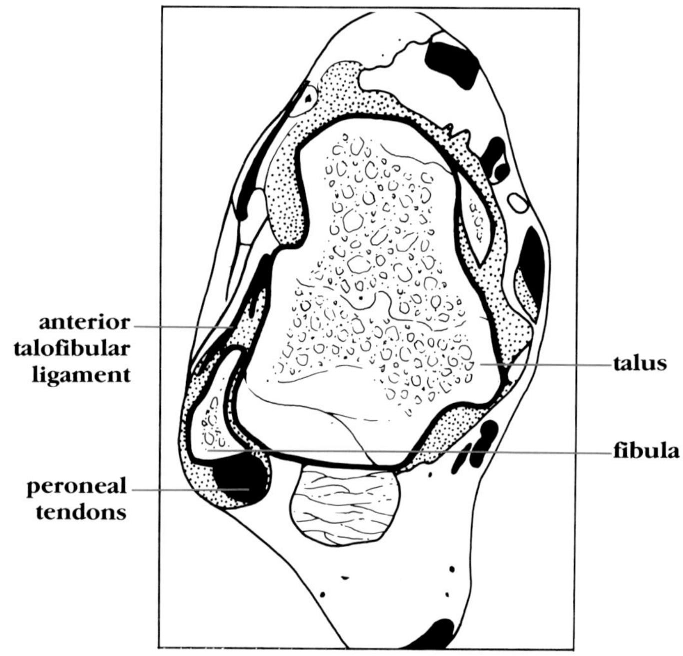

FIGURE 8.6. Normal Anterior Talofibular Ligament. Axial section [SE 2000/20] through the lateral malleolus and talus.

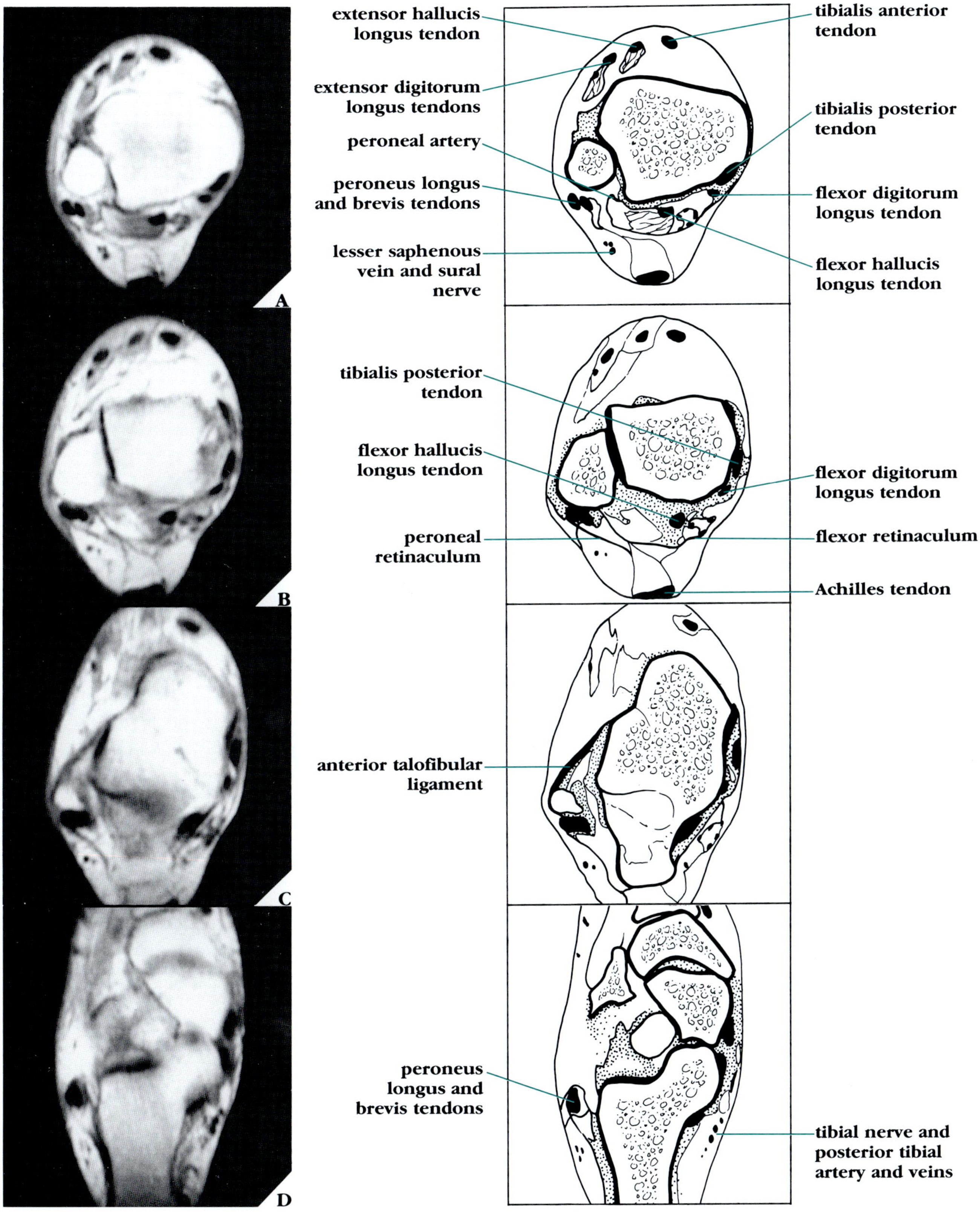

FIGURE 8.7. Normal Ankle Tendon Anatomy. Four consecutive axial sections [SE 800/20] from the distal tibiofibular syndesmosis **(A)** to the calcaneus **(D)**.

brevis) and the first metatarsal (peroneus longus) (Fig. 8.8). In some cases it is difficult to separate these tendons in the region of the posterior malleolus and lateral aspect of the calcaneus.

Anteriorly, the tibialis anterior, extensor hallucis longus, extensor digitorum longus, and peroneus tertius tendons are seen beneath the extensor retinaculum (Fig. 8.5). The tibialis anterior is the

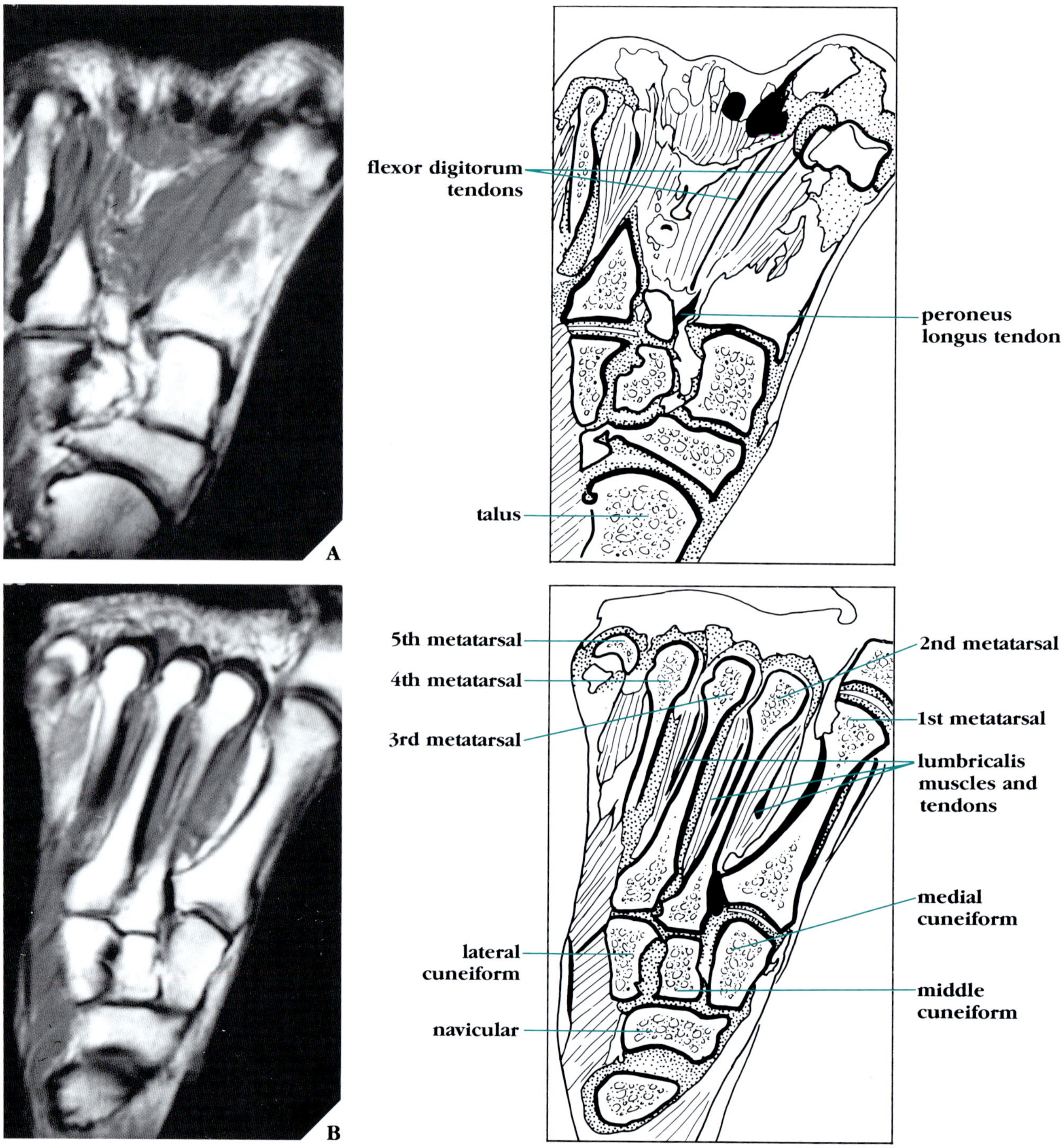

FIGURE 8.8. Normal Foot Anatomy. Two contiguous oblique axial sections [SE 800/20] through the metatarsal area. The plane of section is angled along the longitudinal axis of the metatarsals.

thickest of the anterior tendons. The peroneus tertius is the thinnest and often the most difficult to identify.

Medially, the tibialis posterior, flexor digitorum longus, and flexor hallucis longus tendons are identified from anterior to posterior, beneath the flexor retinaculum (Fig. 8.7). The tendons of the medial and lateral aspects of the ankle are similar to one another in thickness, with the exception of the tibialis posterior tendon, which is slightly thicker and becomes somewhat less homogeneous in SI as it approaches its insertion in the navicular bone.

Posteriorly, the Achilles tendon is seen inserting in the calcaneal tuberosity (2,4,17) (Fig. 8.7). The Achilles tendon should measure no more than 1 cm in the anterior–posterior dimension. Superiorly, it narrows to about 1–2 cm in the transverse diameter and, inferiorly, it widens at the level of its insertion in the calcaneal tuberosity.

Axial sections through the foot and along the longitudinal axis of the metatarsals show the flexor tendons (Fig. 8.8).

The Neurovascular Structures

The peroneal artery can be seen as a small, low SI dot adjacent to the posterior tibiofibular ligament (Fig. 8.7). The lesser saphenous vein and sural nerve are seen as black dots posterior to the peroneal tendons, outside the peroneal retinaculum. The vein is the largest of the dots (12).

Anteriorly, the anterior tibial artery and veins and the deep peroneal nerve can be seen as a series of small dots located posterior to the extensor hallucis longus tendon. The artery is the larger dot (Fig. 8.5). Because of their small size, these structures can be difficult to identify unless specialized surface coils and small FOV are used (12).

The posterior neurovascular bundle is formed by the tibial nerve and the posterior tibial artery and veins. The tibial nerve divides into the lateral and medial plantar nerves. These structures can be seen as a series of small dots posterior to the tendons of the ankle's medial aspect, inside the flexor retinaculum (Fig. 8.7).

The greater saphenous vein is seen in the anterior medial aspect of the ankle as a rounded structure of intermediate or high SI on PDWI and T2WI (Fig. 8.5).

▼▼▼ *SAGITTAL PLANE*

The Ligaments

The only ligament that can be well-identified in the sagittal plane is the interosseous or talocalcaneal ligament. This ligament is oriented obliquely in the sinus tarsi from medial to lateral and from superior to inferior. Midline sagittal sections demonstrate the interosseous ligament as a thin, black line surrounded by hyperintense fat on T1WI (4, 12) (Fig. 8.9). The fan-like fibers of the interosseous ligament can be seen in more medial sections (Fig. 8.10).

The Tendons

Medially, the tibialis posterior, flexor digitorum longus, and flexor hallucis longus tendons are well-identified. Occasionally, the tibialis posterior ten-

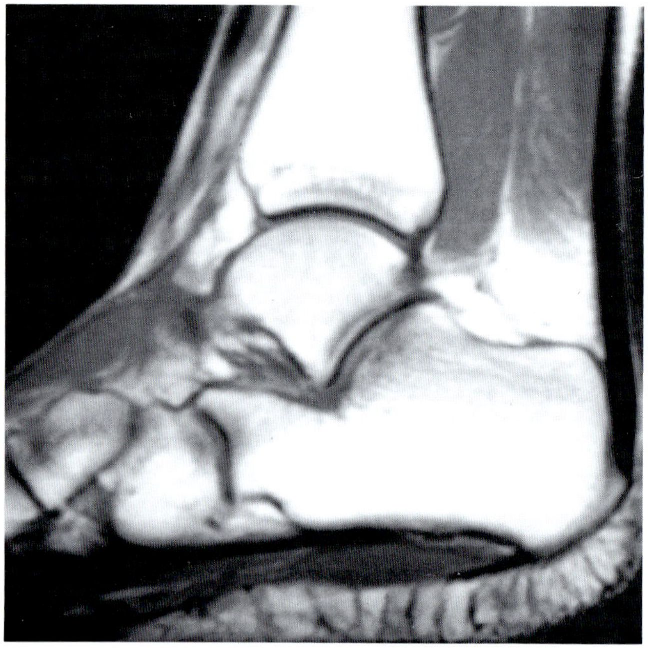

FIGURE 8.9. **Normal Achilles Tendon.** Midline sagittal section [SE 800/20] demonstrating the Achilles tendon. Note the uniformly low SI of the tendon contrasting with the high SI of the anterior fat pad.

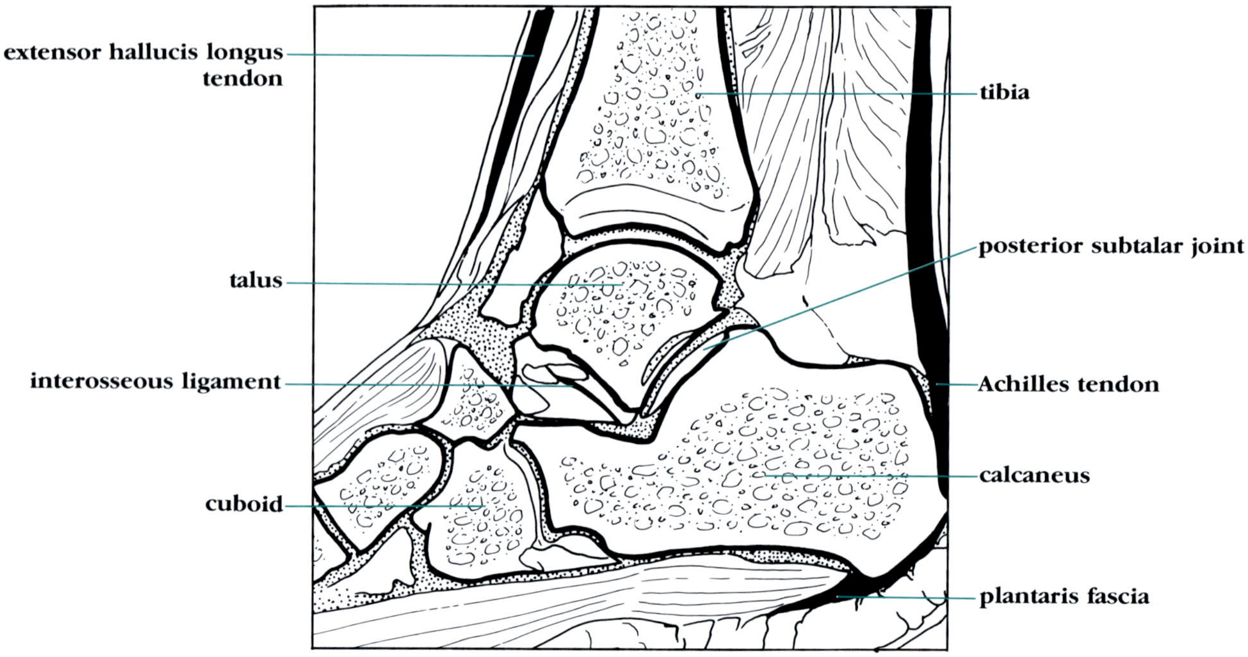

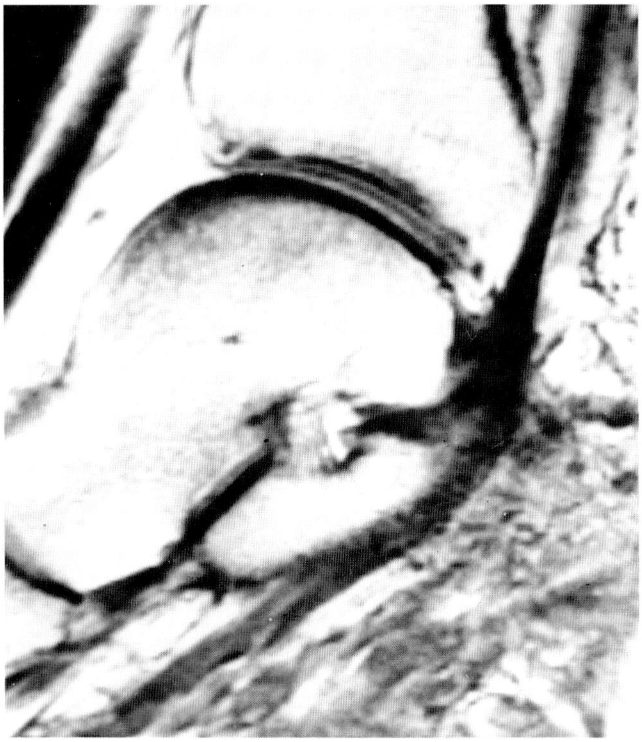

FIGURE 8.10. Normal Flexor Hallucis Longus Tendon. Sagittal section [SE 600/20] through the sustentaculum tali. The flexor hallucis longus tendon is located posterior to the tibia, and curves beneath the sustentaculum tali. Note the fibers of the interosseous ligament in the sinus tarsi area and the articular cartilage of the tibiotalar joint.

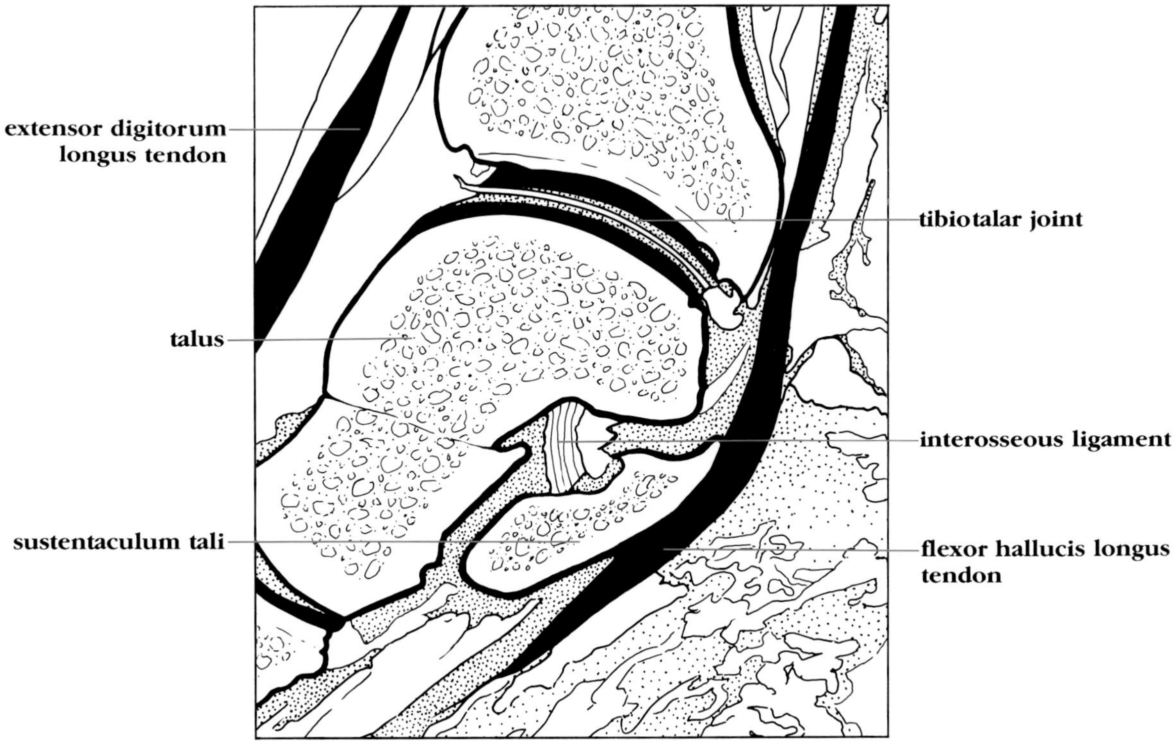

don can be followed to its insertion in the posterior aspect of the navicular bone (4,12) (Fig. 8.11). The flexor digitorum longus is seen posterior and medial to the tibialis posterior (Fig. 8.11). In more medial sections, the flexor hallucis longus is identified beneath the sustentaculum tali (Fig. 8.10). Sagittal sections through the foot, along the long axis of the metatarsal, show the flexor tendons (Fig. 8.12).

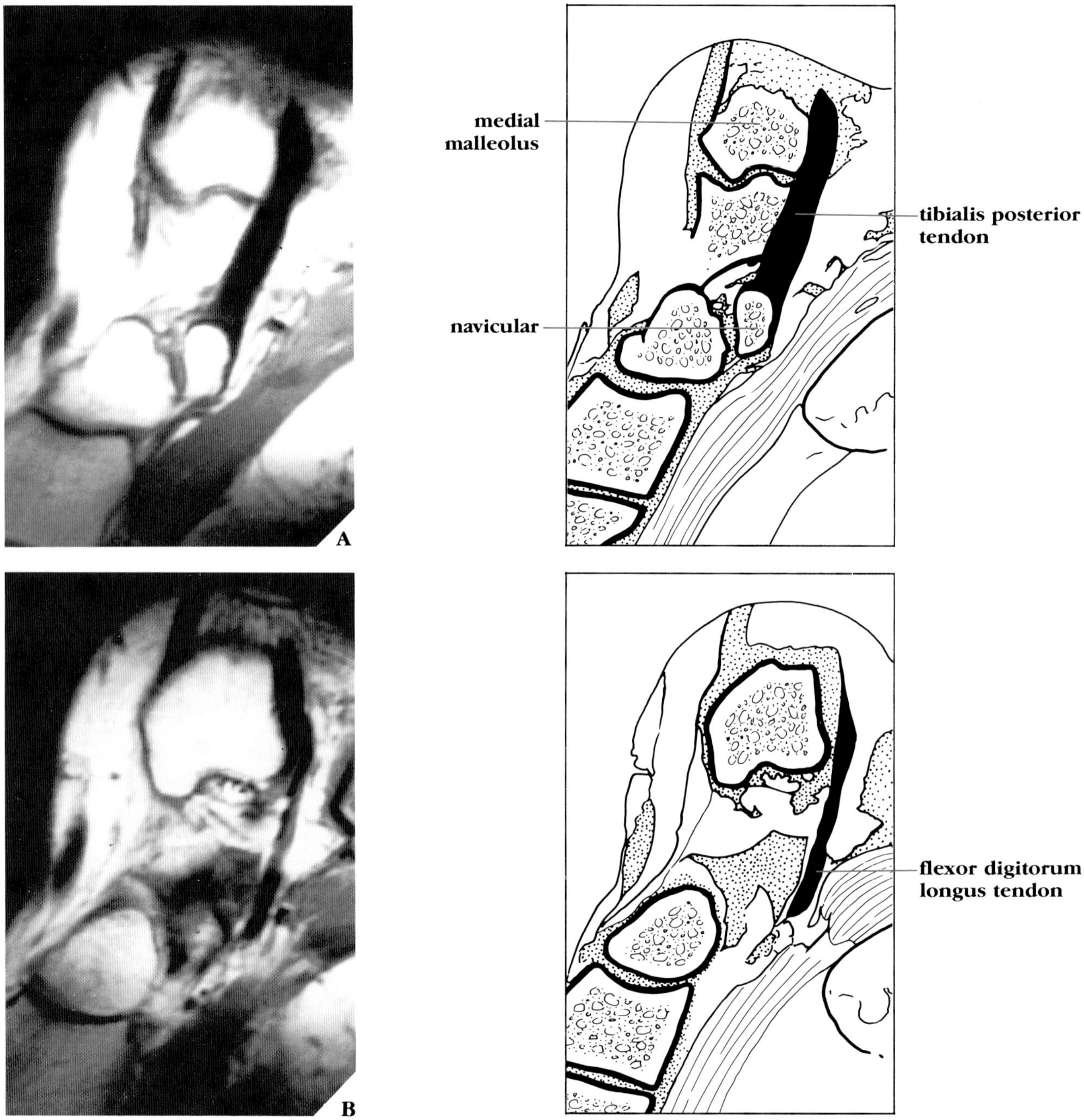

FIGURE 8.11. Normal Tibialis Posterior and Flexor Digitorum Longus Tendons. Two consecutive sagittal sections [SE 800/20] through the medial malleolus. The thick tibialis posterior tendon curves around the posterior aspect of the tibia and inserts in the navicular bone **(A)**. A more lateral section **(B)** shows the flexor digitorum longus tendon.

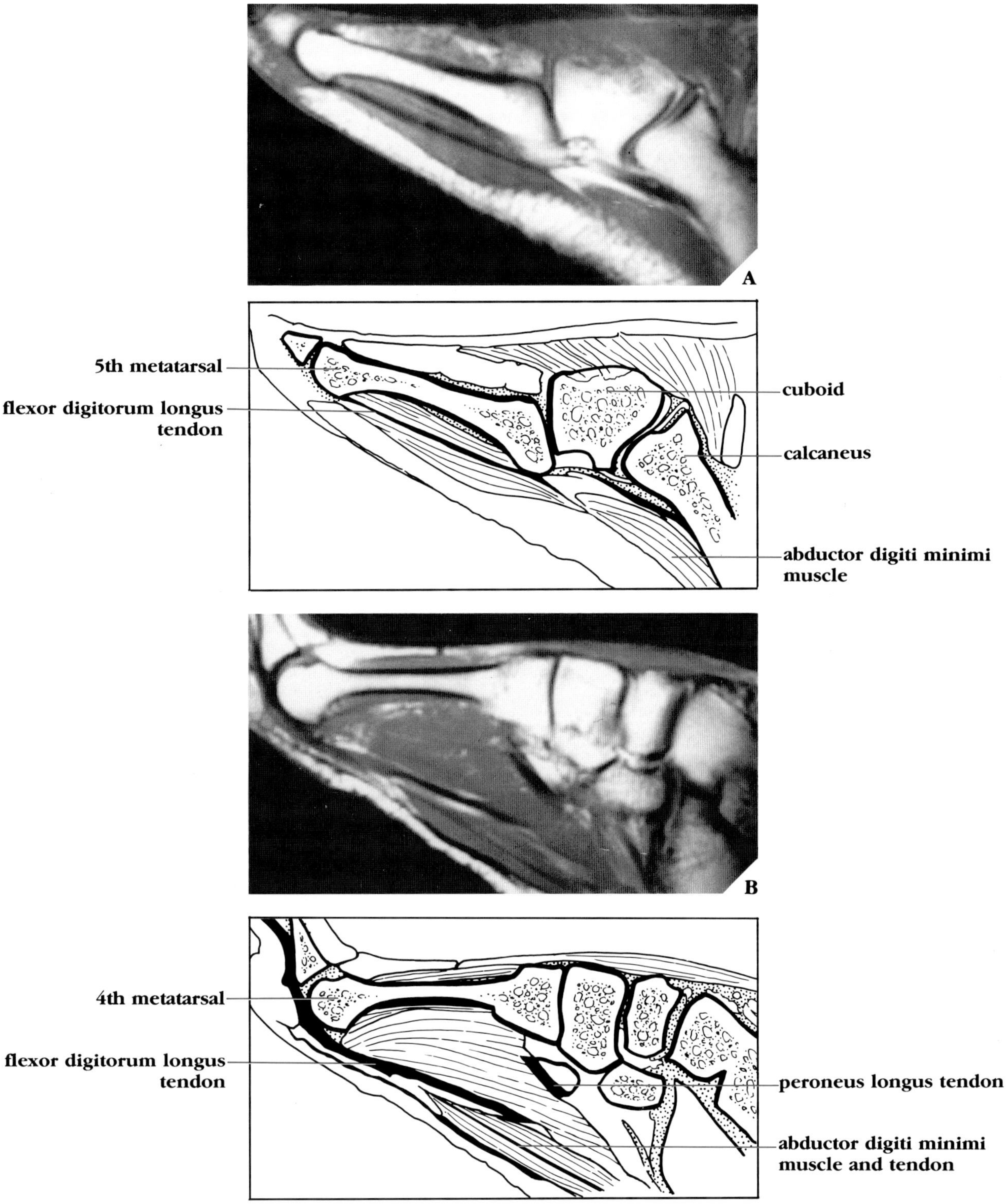

FIGURE 8.12. Normal Sagittal Foot Anatomy. Two oblique sagittal sections [SE 800/20] through the fifth **(A)** and the fourth metatarsals **(B)**. The plane of section is oriented along the longitudinal axis of the metatarsals. Note the flexor tendons in the plantar region of the foot.

Laterally, the peroneus longus and brevis tendons are seen (Fig. 8.13). Occasionally, the insertion of the peroneus brevis at the base of the fifth metatarsal can be identified (Fig. 8.13).

Posteriorly, the Achilles tendon inserting in the calcaneal tuberosity is identified (4,17) (Fig. 8.9). Anteriorly, the tibialis anterior, extensor hallucis longus, and extensor digitorum longus can be identified in consecutive sections (Figs. 8.9, 8.10, and 8.13).

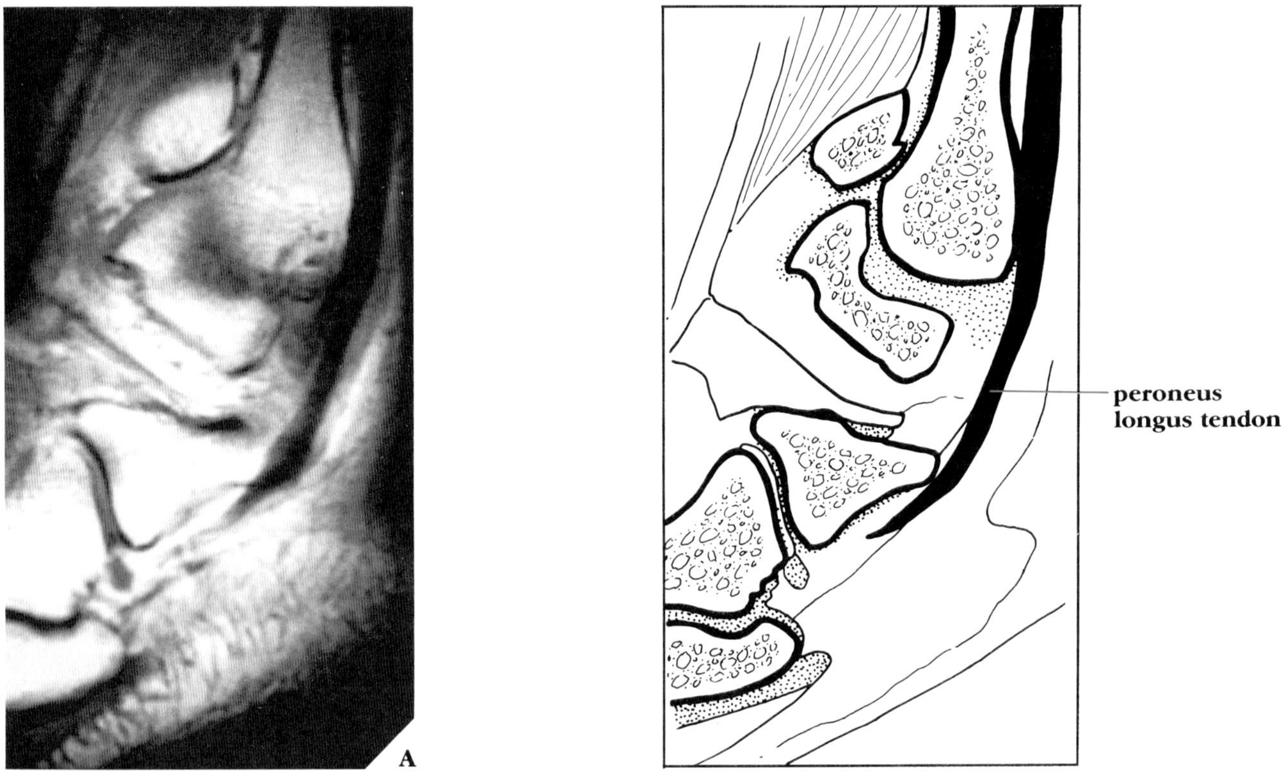

FIGURE 8.13. Normal Peroneus Brevis and Longus Tendons. Sagittal sections [SE 800/20] through lateral malleolus showing the peroneus longus **(A)** and the peroneus brevis **(B)** tendons as they curve around the lateral malleolus. Note the insertion of the peroneus brevis tendon in the base of the fifth metatarsal **(B).**

▼▼▼ *CORONAL PLANE*

The Ligaments

The posterior sections demonstrate the posterior talofibular and calcaneofibular ligaments laterally (Fig. 8.14). The medial talocalcaneal ligament is seen medially (12) (Fig. 8.14).

More anterior sections through the midtalar area demonstrate the deltoid ligament with its tibiotalar and tibiocalcaneal components (Fig. 8.15). The interosseous ligament is well-identified in the region of the sinus tarsi, surrounded by fat (Fig. 8.15). Using small-diameter surface coils and small FOV, high-resolution images of these ligaments can be obtained (12) (Figs. 8.16 and 8.17).

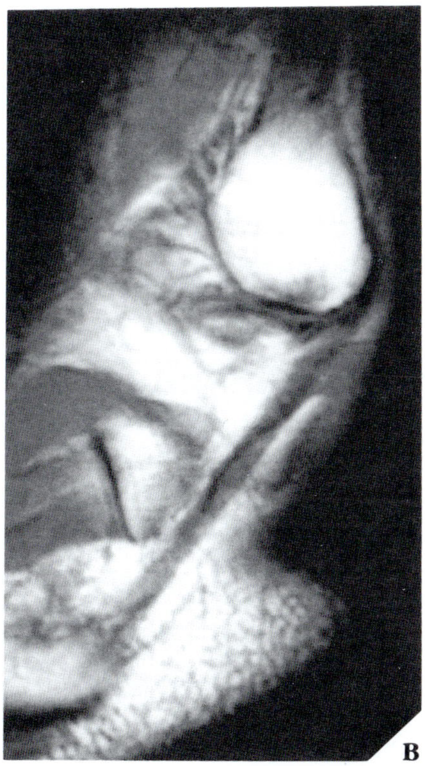
B

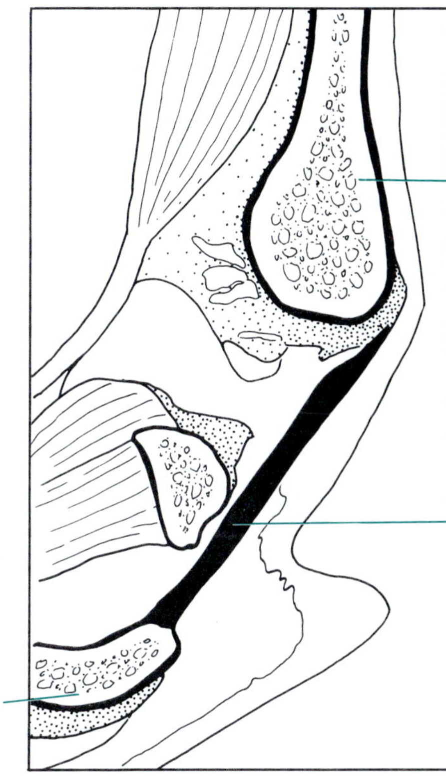

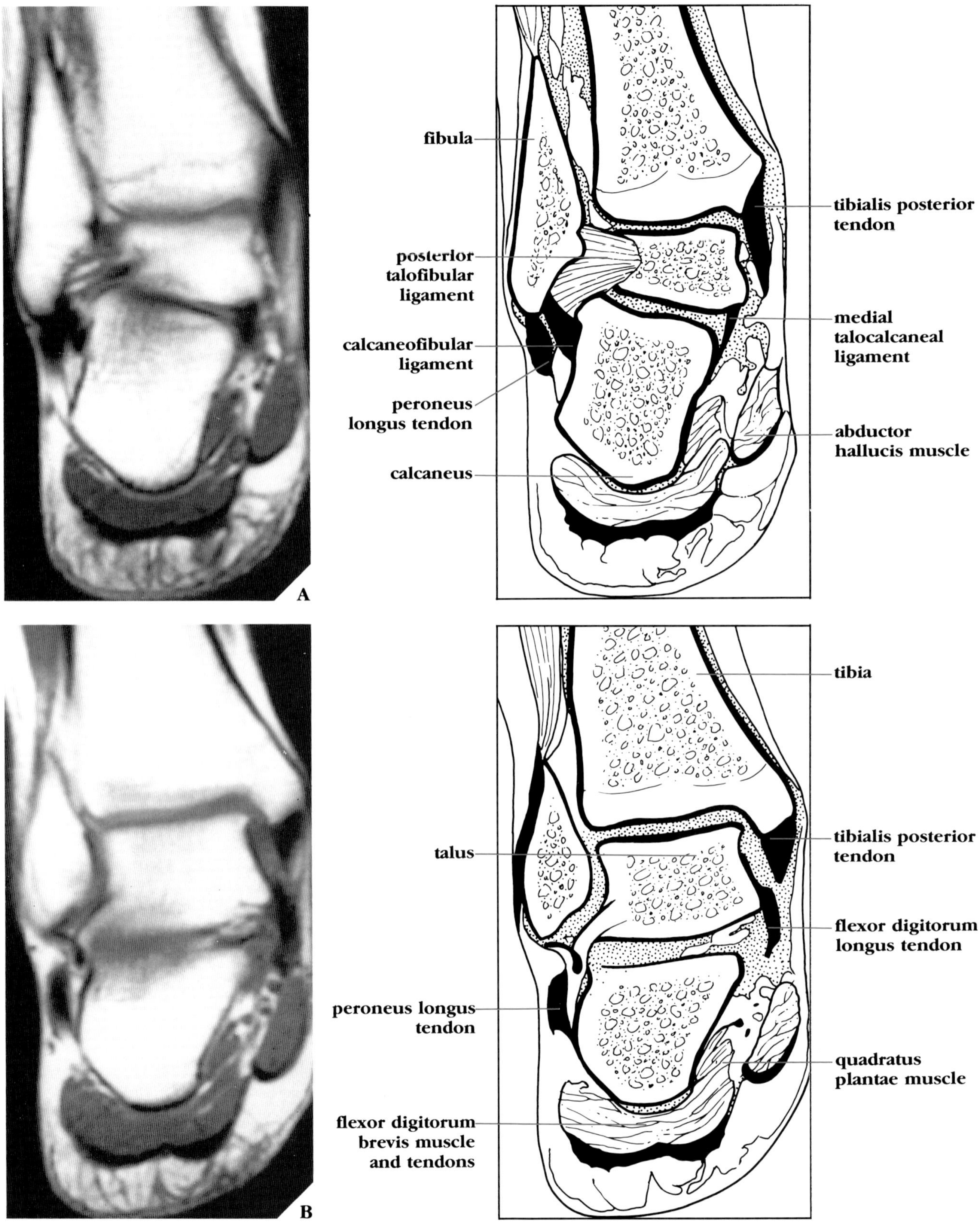

FIGURE 8.14. Normal Posterior Talofibular and Calcaneofibular Ligaments. A,B Two consecutive coronal sections [SE 800/20] through posterior aspect of the ankle demonstrating the posterior ligaments and tendons.

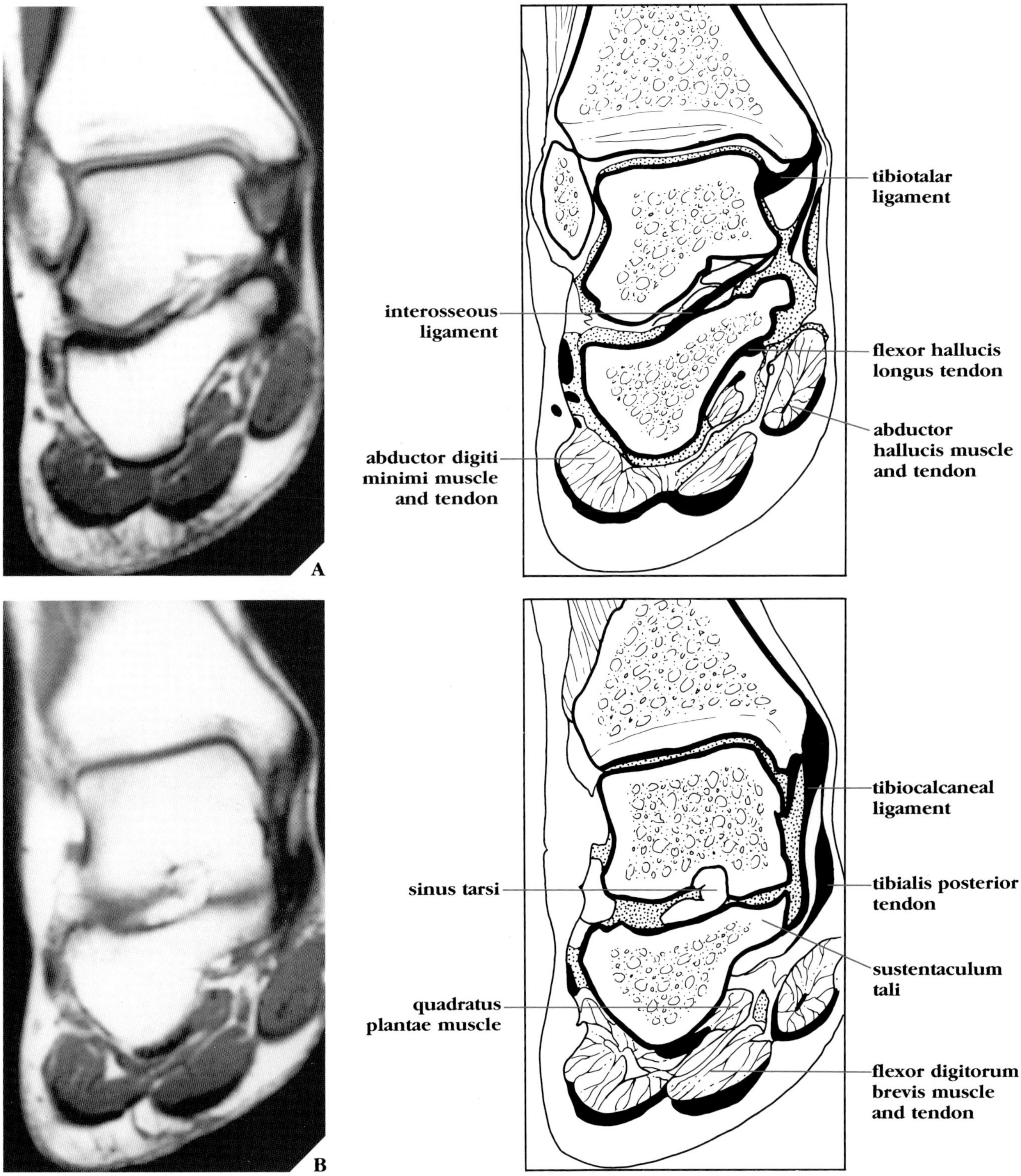

FIGURE 8.15. Normal Deltoid Ligament. A,B Two consecutive coronal sections [SE 800/20] anterior to those shown in Figure 8.14, through the tibiotalar joint and subtalar joint. Note the different components of the deltoid ligament in the medial aspect of the ankle. The interosseous ligament is shown in **A**. Also note the muscles and tendons of the plantar aspect of the foot.

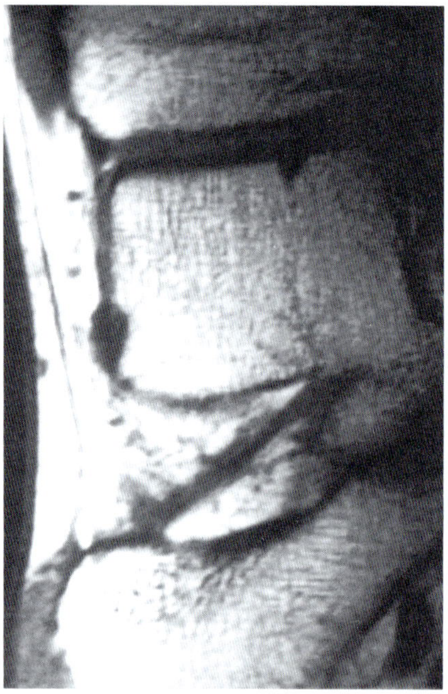

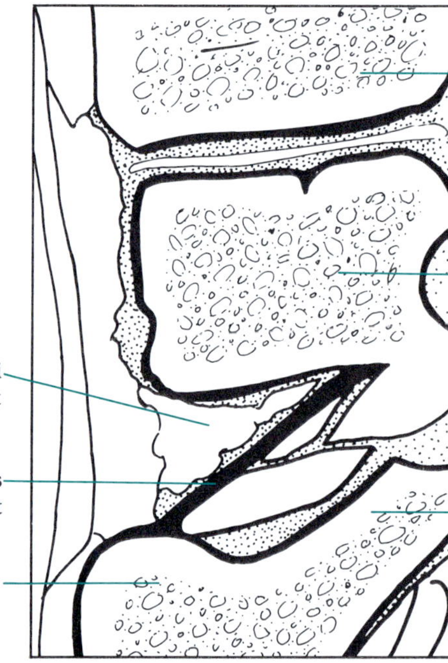

FIGURE 8.16. Normal Interosseous Ligament. Coronal section obtained with a high resolution transmit–receive surface coil over the lateral aspect of the ankle [SE 500/20]. The interosseous ligament surrounded by fat in the sinus tarsi is well demonstrated.

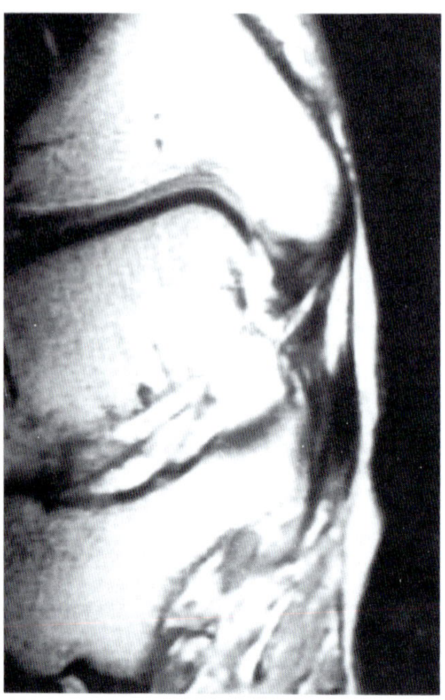

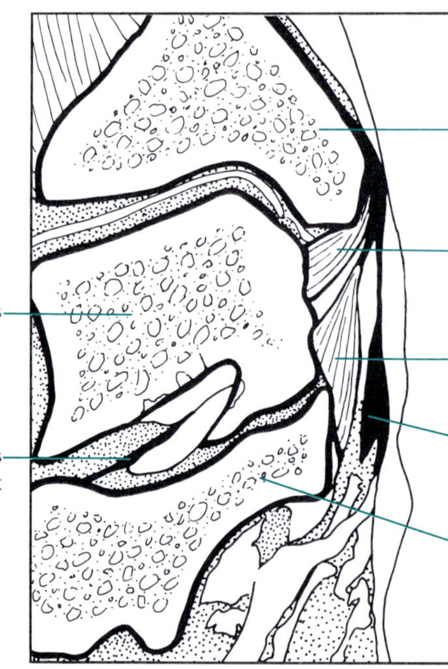

FIGURE 8.17. Normal Deltoid Ligament. Coronal section obtained with a high resolution transmit–receive surface coil placed in the medial aspect of the ankle. Note the tibiotalar and tibiocalcaneal ligaments.

The Tendons

In coronal sections, the peroneus longus and brevis tendons are identified laterally, and the tibialis posterior, flexor digitorum longus, and flexor hallucis longus are identified medially (Figs. 8.14 and 8.15). The flexor hallucis longus is easily identified due to its location beneath the sustentaculum tali (4) (Fig. 8.15).

In the plantar aspect of the heel, from lateral to medial, the following muscles are seen: abductor digiti minimi, flexor digitorum brevis, abductor hallucis, and quadratus plantae (Figs. 8.14 and 8.15).

Coronal sections through the foot, in a direction perpendicular to the longitudinal axes of the metatarsals, demonstrate the extensor tendons and the tendons of the plantar aspect of the foot (Fig. 8.18).

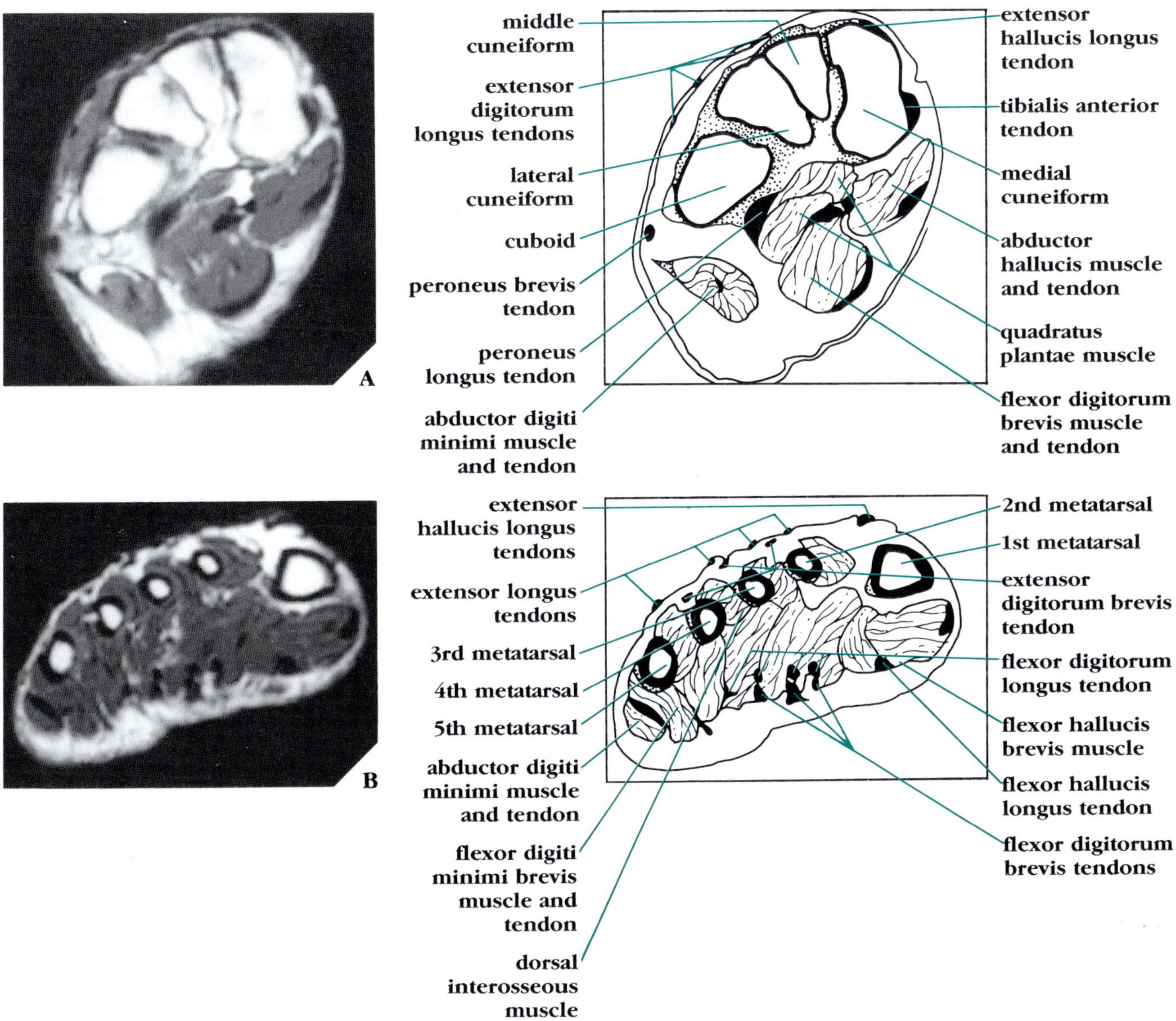

FIGURE 8.18. Normal Foot Anatomy. Two coronal sections through the distal tarsal row **(A)** and the midmetatarsal area **(B)** [SE 800/20]. The extensor and flexor muscles and tendons are demonstrated.

LIGAMENTOUS LESIONS

Collateral Ligament Tears

The anterior talofibular ligament is the most commonly injured ligament of the ankle (5). As an isolated injury, it occurs when the foot is forced into internal rotation during plantar flexion and inversion. Axial plane MRI most successfully demonstrates anterior talofibular ligament ruptures, because the ligament runs in an almost horizontal direction (Fig. 8.19). The main MRI findings are absent ligament and inflammatory changes in the region anterior to the fibula.

Deltoid ligament tears frequently occur in association with lateral ligament injuries and bony lesions. Coronal MRI demonstrates absent ligaments in the tibiotalar area with inflammatory changes (Fig. 8.20).

Calcaneofibular ligament tears are rare as isolated injuries. They are more often associated with lesions of the anterior talofibular ligament. Calcaneofibular ligament ruptures occur when the foot is dorsiflexed and inverted. MRI diagnosis is again based on lack of visualization of the normal calcaneofibular ligament in the coronal plane and secondary inflammatory changes (Fig. 8.21).

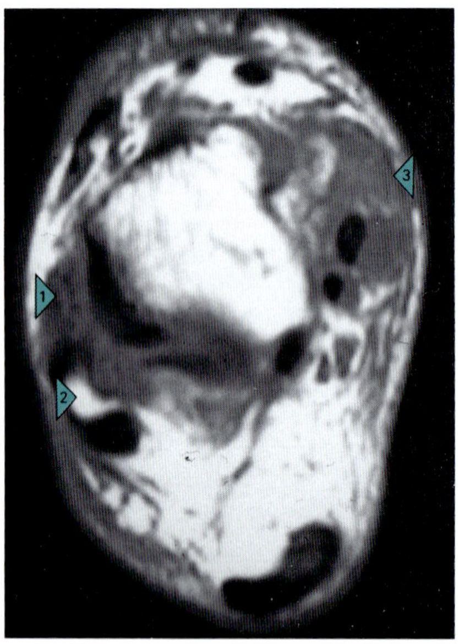

FIGURE 8.19. Anterior Talofibular Ligament Tear. Axial section [SE 2000/20] through the distal lateral malleolus. The anterior talofibular ligament is absent. Low SI inflammatory changes are seen in the medial and lateral aspects of the ankle. This patient also suffered a deltoid ligament injury. Arrow **1** Anterior talofibular ligament tear. Arrow **2** Lateral malleolus. Arrow **3** Inflammatory tissue.

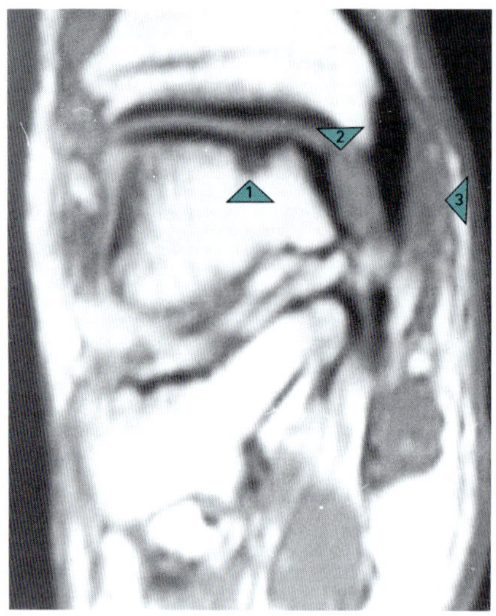

FIGURE 8.20. Deltoid Ligament Tear. Coronal section through tibiotalar and subtalar joints [SE 2000/20]. The tibiotalar and tibiocalcaneal ligaments are absent. Inflammatory changes around the lateral malleolus are seen. Arrow **1** Posttraumatic talar cyst. Arrow **2** Deltoid ligament tear. Arrow **3** Inflammatory tissue.

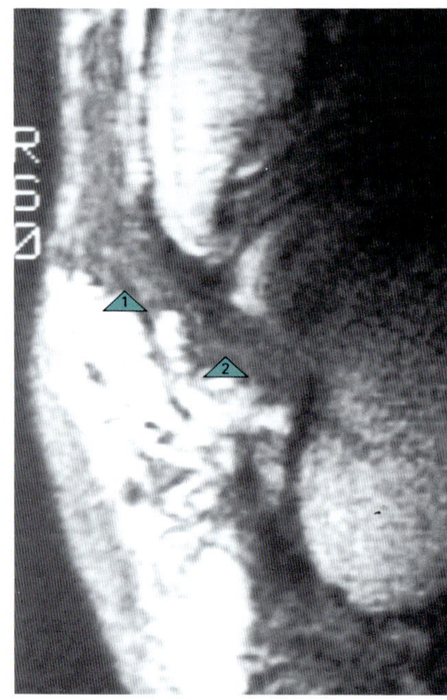

FIGURE 8.21. Calcaneofibular Ligament Tear. Coronal section though the lateral malleolus [SE 500/20] obtained with a small diameter transmit–receive coil. The image was obtained shortly after external reduction and casting for dislocation of the talus. The calcaneofibular ligament is absent and the lateral malleolar area is infiltrated with edema and hemorrhage. Arrow **1** Inflammatory tissue. Arrow **2** Calcaneofibular ligament tear.

Sinus Tarsi Syndrome

Interosseous or talocalcaneal ligament rupture occurs as a result of single or repeated ankle sprains with foot inversion but without dorsiflexion or plantarflexion (26). The calcaneofibular ligament and lateral talocalcaneal ligament frequently rupture before the interosseous ligament is injured. Clinically, patients present with pain in the lateral aspect of the ankle, increased by pressure over the lateral opening of the sinus tarsi, and instability of the hindfoot, hence the name sinus tarsi syndrome (16). Local injections of anesthetic into the sinus tarsi abolish the pain. During the acute stage, surgical exploration demonstrates ligament tears or bleeding within the sinus tarsi. Synovial hyperplasia obliteration of the synovial recesses of the subtalar joint, and healing of the ligaments with scar tissue are found in chronic sinus tarsi syndrome.

Coronal MRI of the subtalar joint in patients with sinus tarsi syndrome demonstrates absent interosseous ligament and inflammatory changes in the sinus tarsi area; normal fat is replaced by tissue of low SI on T1WI and high SI on T2WI (Fig. 8.22). Subcortical cysts at the calcaneal insertion of the interosseous ligament can also be seen.

TENDON LESIONS

Posttraumatic tendon injuries, tendon entrapment, and acute or chronic tenosynovitis are relatively frequent conditions involving the tendons around the ankle and foot. Ultrasonography, CT, tenography, and more recently MRI, have been used for evaluation of these conditions. CT is somewhat limited to the axial plane (12,22,23,24). Sagittal CT reconstructions from thin section axial images have relatively poor spatial resolution (2). Ultrasonography is an inexpensive, noninvasive, and readily available method for tendon evaluation. The disadvantages of US are its limitation to superficially located structures and low spatial resolution (15). Tenography is an accurate but invasive technique (8,28). In the last few years, MRI has been tested for evaluation of tendon injuries (2,12) and dramatic depiction of normal and abnormal anatomy of the tendons has been achieved. However, the high cost of MRI prevents its use as a routine diagnostic tool for uncomplicated tendon and ligamentous injuries.

Tendon ruptures can be easily diagnosed clinically. However, such data as the exact location and type of tendon rupture, and the size of the gap be-

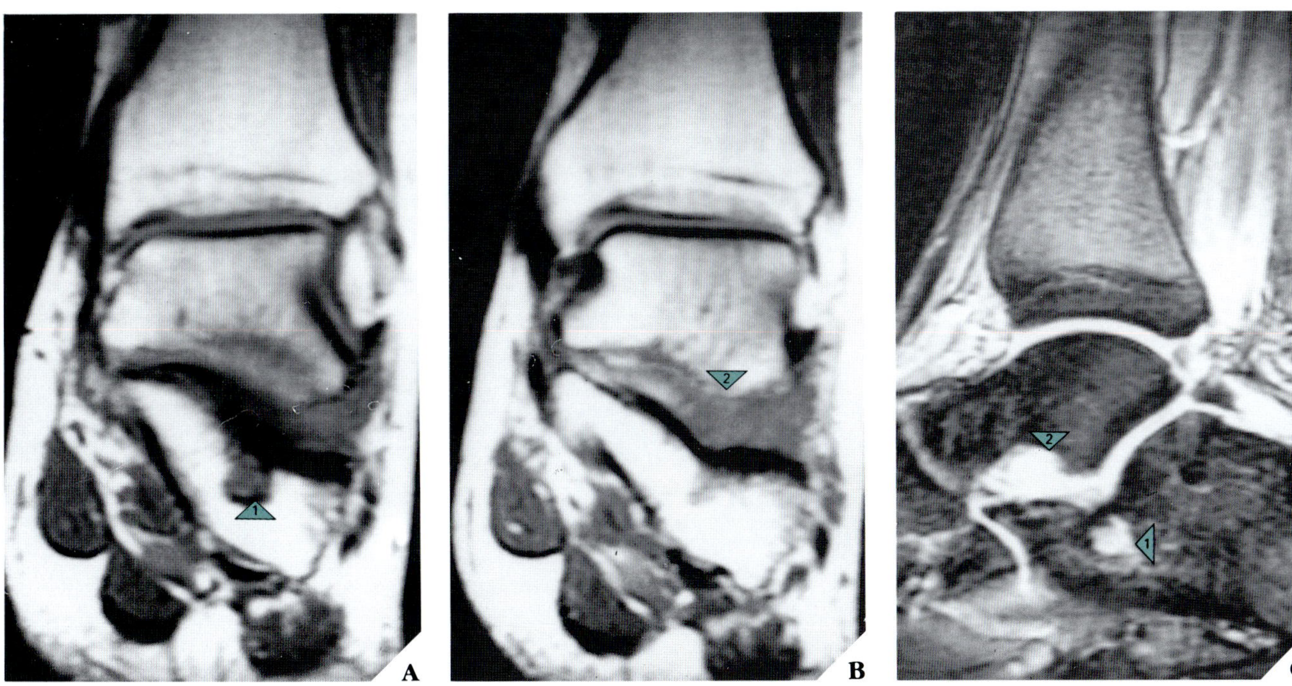

FIGURE 8.22. Sinus Tarsi Syndrome with Interosseous Ligament Rupture. A and **B** Two consecutive coronal sections [SE 2000/20] through the subtalar joint showing absent interosseous ligament in the sinus tarsi, and replacement of the fat by low SI inflammatory tissue. **C** Sagittal GRASS image demonstrates high SI in the sinus tarsi area. Note the presence of a subcortical calcaneal cyst in **A** and **C**. Proliferation of the synovium and inflammatory changes with absent interosseous ligament were found surgically. Arrows **1** Calcaneal cyst. Arrows **2** inflammatory tissue.

tween the ends of the complete tear, may be difficult to assess clinically. Tendons can rupture in several different ways (Fig. 8.23). Preoperative knowledge of the type of tendon injury will help surgeons in planning the appropriate treatment.

Achilles Tendon Rupture

Avulsion of the Achilles tendon from its calcaneal insertion is easily diagnosed with plain films by demonstrating the bony fragment.

Achilles tendon rupture at the musculotendinous interface is seen on MRI as a discontinuity of the tendon in an area where the gastrocnemius muscle is present anteriorly (Fig. 8.24). Conservative treatment is preferred over surgical treatment in this type of rupture.

Incomplete Achilles tendon rupture is the MRI diagnosis when the continuity of the tendon is preserved, but diffuse thinning or localized areas of thickening, with regions of high SI on PDWI and

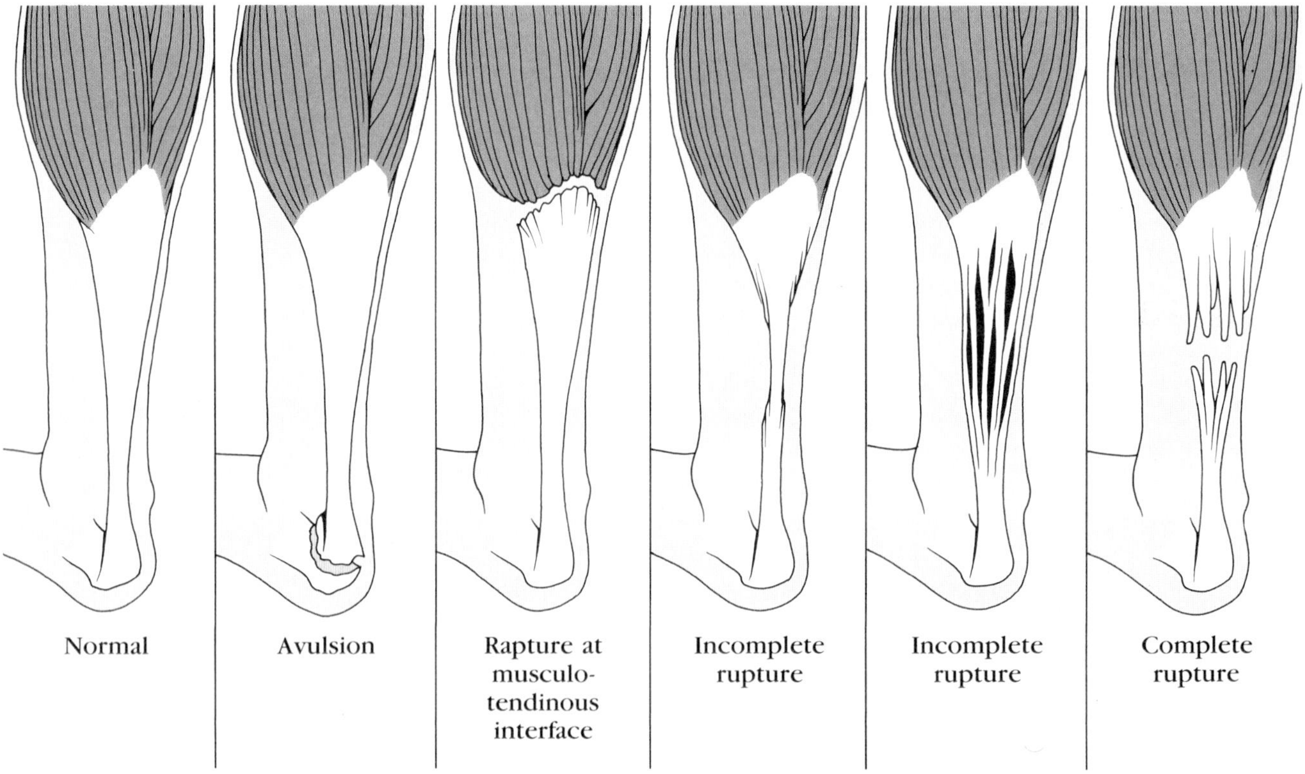

FIGURE 8.23. Types of tendon rupture.

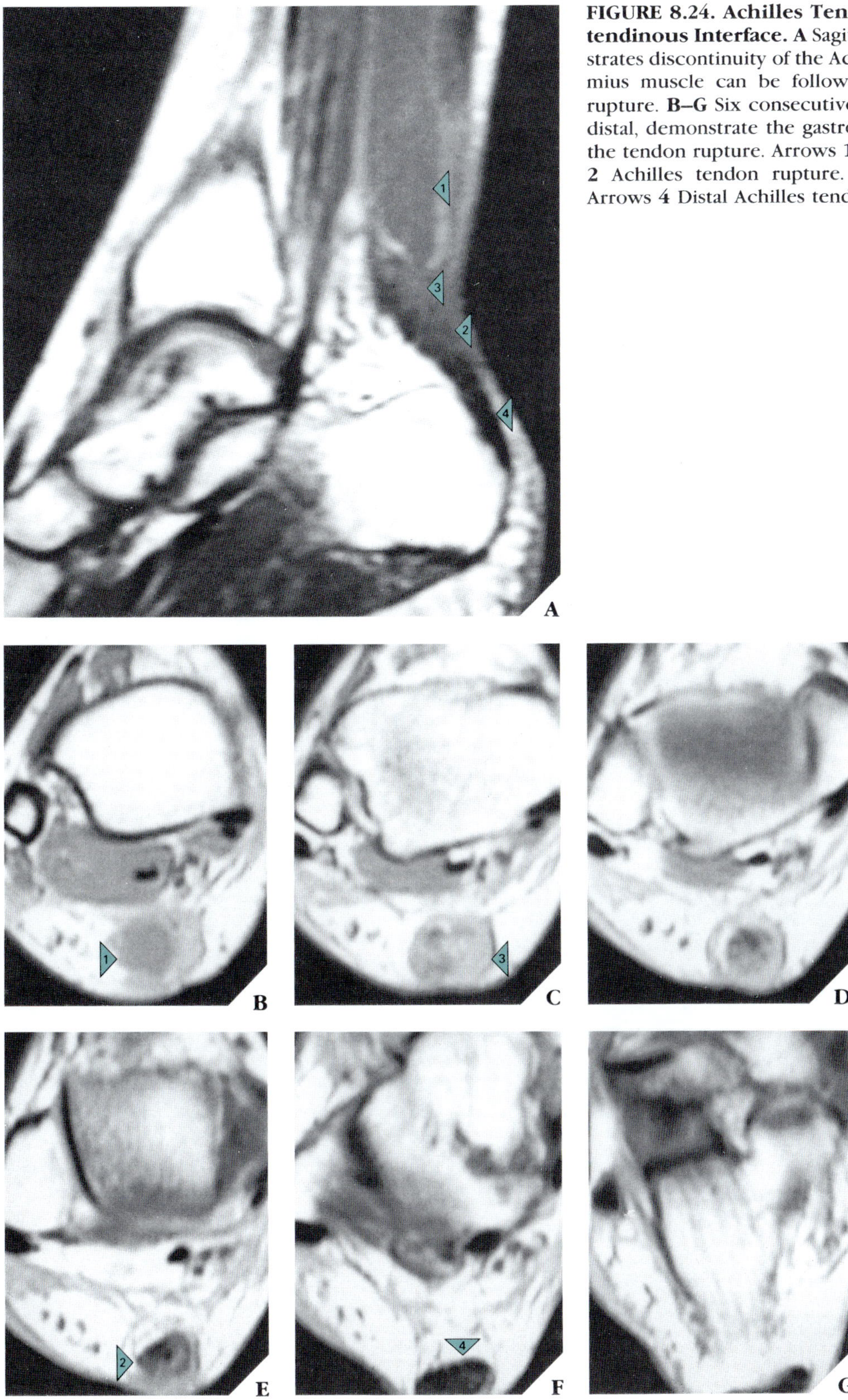

FIGURE 8.24. Achilles Tendon Rupture at the Musculotendinous Interface. A Sagittal section [SE 2000/20] demonstrates discontinuity of the Achilles tendon, but the gastrocnemius muscle can be followed to the level of the tendon rupture. **B–G** Six consecutive axial images, from proximal to distal, demonstrate the gastrocnemius muscle to the level of the tendon rupture. Arrows **1** Gastrocnemius muscle. Arrows **2** Achilles tendon rupture. Arrow **3** Inflammatory tissue. Arrows **4** Distal Achilles tendon.

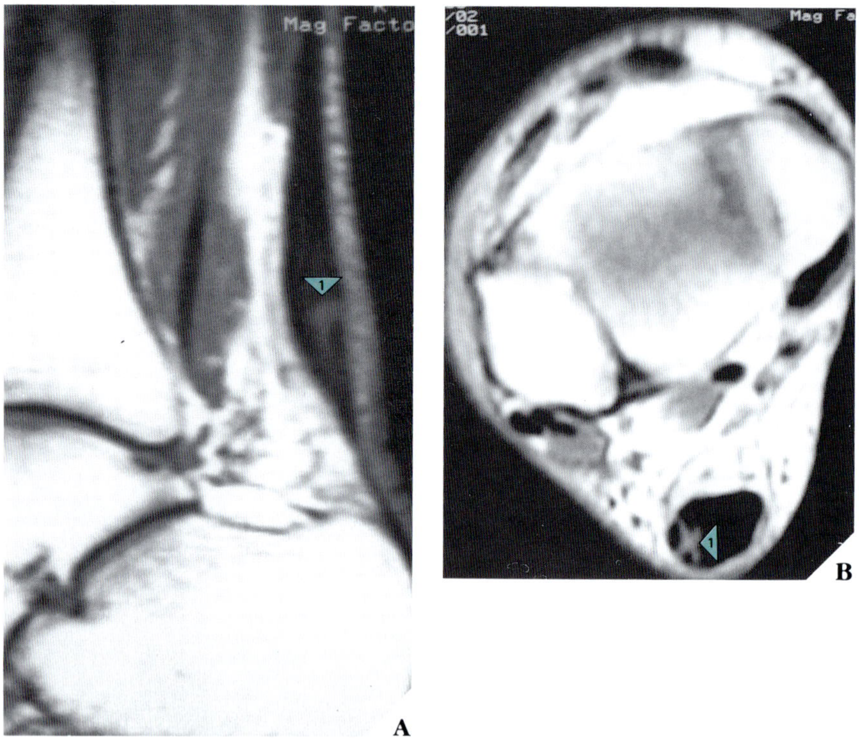

A

B

FIGURE 8.25. Incomplete Achilles Tendon Rupture. A Sagittal image [SE 2000/20] demonstrates localized thickening of the Achilles tendon and a small area of increased SI representing an incomplete rupture. **B** the same findings can be seen in the axial section [SE 500/20]. Arrows **1** Incomplete Achilles tendon rupture.

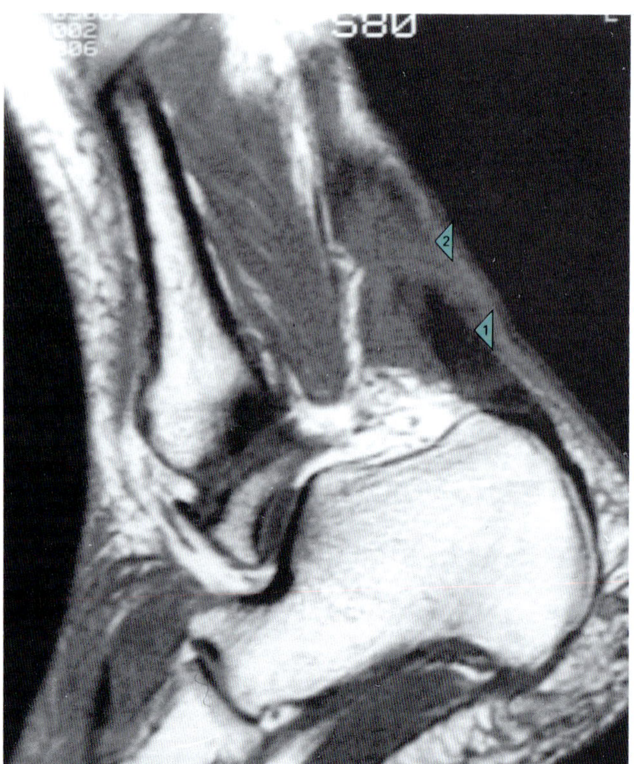

FIGURE 8.26. Complete Achilles Tendon Rupture. Sagittal section [SE 2000/20] showing discontinuity of the Achilles tendon near its insertion. Inflammatory tissue is seen at the rupture site. Arrow **1** Complete Achilles tendon rupture. Arrow **2** Inflammatory tissue.

T2WI, are seen within the tendon substance (17) (Fig. 8.25). Conservative treatment is also preferred in this type of tear.

Complete Achilles tendon rupture is seen on MRI as a complete discontinuity of the tendon, with irregularity and separation of the tendon ends. This space is filled with edema and hemorrhage, showing low SI on T1WI and high SI on T2WI (2,17) (Fig. 8.26). Complete ruptures can be treated conservatively with cast immobilization of the ankle in plantar flexion. However, if the tendon ends are not well-approximated, conservative treatment may fail and surgical management may then be indicated. MRI can be used to assess the size of the gap once the ankle has been casted in plantarflexion (Fig. 8.27).

Peroneal Tendon Rupture

Rupture of the peroneal tendons occurs as a result of trauma and laceration at the lateral aspect of the ankle (21,24). Irregularity and lack of continuity of the tendons with inflammatory changes are seen

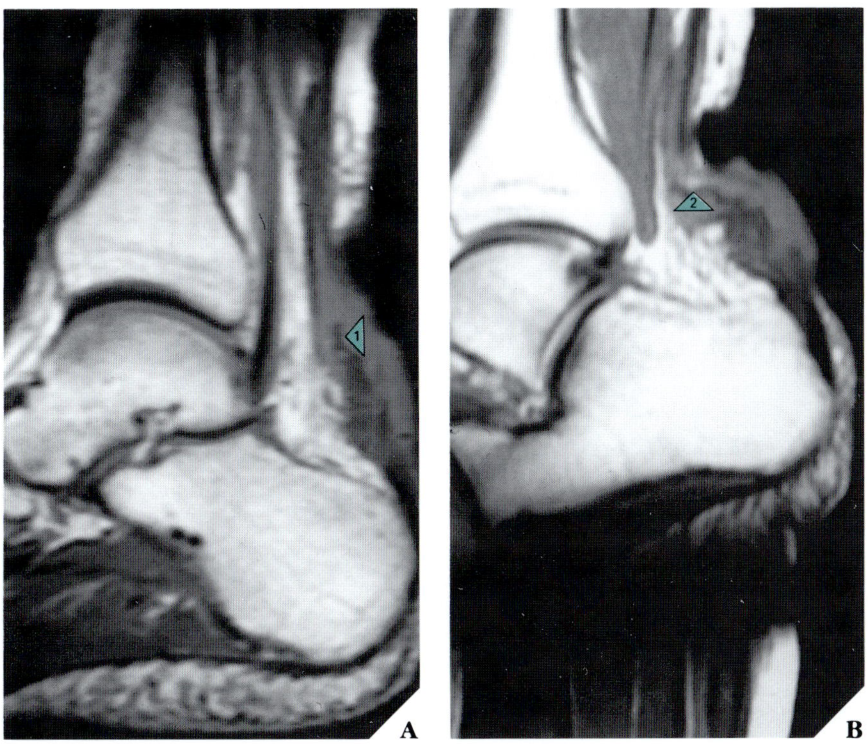

A B

FIGURE 8.27. Complete Achilles Tendon Rupture. Sagittal sections in **A** dorsiflexion and **B** plantarflexion [SE 800/20]. The ends of the ruptured Achilles tendon are separated from each other by a 1-cm gap. Healing can be expected without surgical repair. Arrow **1** Achilles tendon rupture in dorsiflexion. Arrow **2** Achilles tendon rupture in plantar flexion.

on MRI (Fig. 8.28). Peroneal tendon subluxation or dislocation is seen in patients with a shallow peroneal canal in the posterior aspect of the lateral malleolus (21,25). Axial MRI can depict the abnormal position of the tendons.

Tibialis Posterior Tendon Rupture

Chronic tibialis posterior tendon rupture is a well-defined clinical entity, occurring in middle-aged females with no history of trauma. The clinical presentation is pain and swelling of the medial aspect of the ankle, and the onset of flat-foot deformity (7,19,22). The tibialis posterior tendon probably ruptures due to chronic stress and tension in the region of the medial malleolus. Rheumatoid arthritis and seronegative arthritides may predispose the patient to tendon rupture. Surgery is the treatment of choice, because conservative treatment, including immobilization and antiinflammatory medication, is most often unsuccessful. Incomplete ruptures can be repaired by anastomosis, but triple arthrodesis may be required in complete ruptures with large gap (9,10). MRI demonstrates irregular SI, and thickening or thinning of the tibialis posterior tendon in incomplete tendon ruptures (Fig. 8.29). Discontinuity of the tendon with associated inflammatory changes is seen in complete ruptures (19).

Tendon Entrapment

Tendon entrapment following comminuted fracture of the calcaneus can involve the peroneal tendons (24) or the tendons of the medial aspect of the ankle. MRI shows bony fragments, inflammatory changes, and scar tissue surrounding the entrapped tendons (Fig. 8.30).

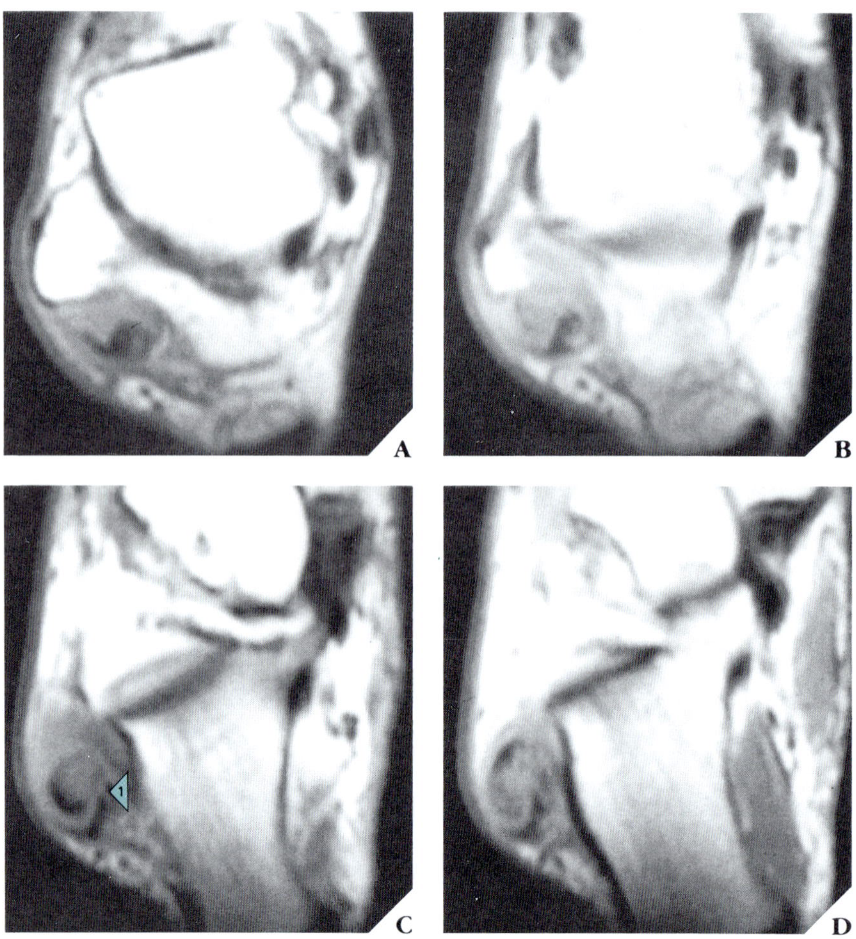

FIGURE 8.28. Peroneal Tendon Rupture. Four consecutive axial images from **A** proximal to **D** distal [SE 800/20]. Peroneus brevis and longus tendons show marked irregularity and surrounding inflammatory changes. Tendon rupture resulted from a laceration in the lateral aspect of the joint. Arrow **1** Peroneal tendon rupture.

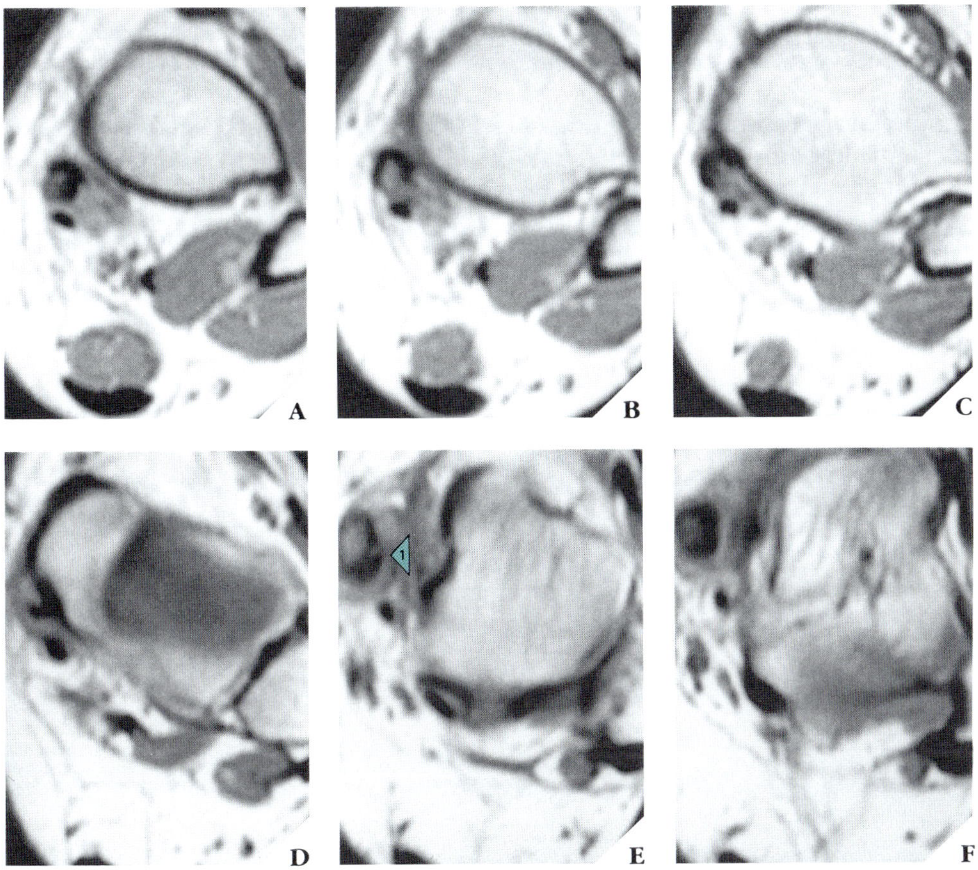

FIGURE 8.29. Chronic Rupture of the Tibialis Posterior Tendon. Six consecutive axial sections through the ankle from **A** proximal to **F** distal [SE 800/20]. The tibialis posterior tendon shows irregular SI and inflammatory changes around the rupture. Arrow **1** Chronic tibialis posterior tendon rupture.

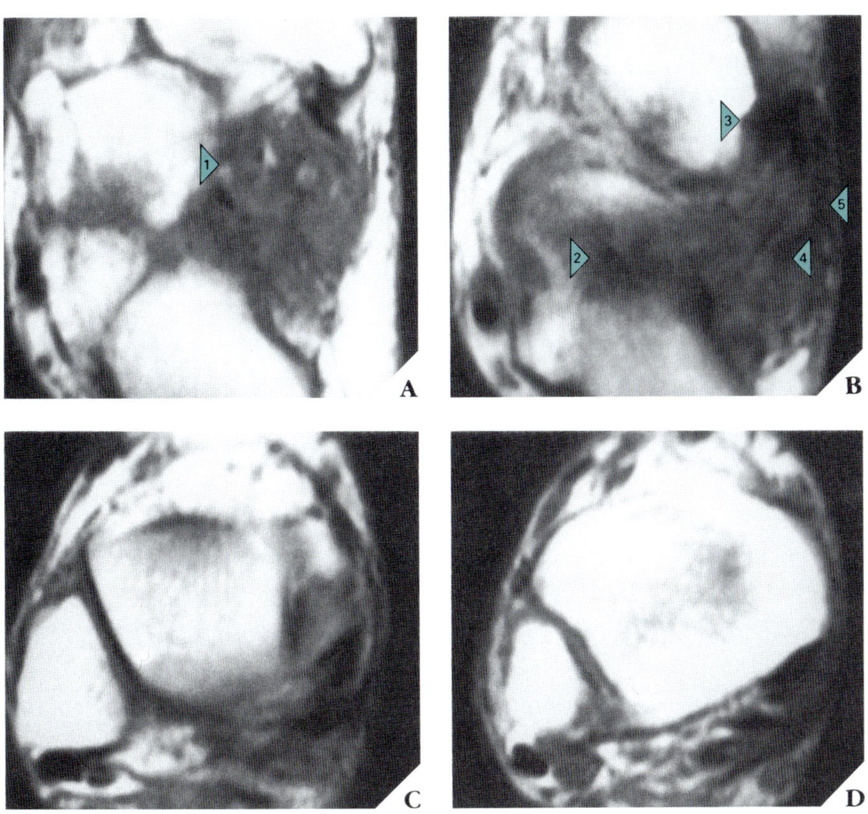

FIGURE 8.30. Tendon Entrapment. Four consecutive axial sections from **A** distal to **D** proximal [SE 800/20]. There is a fracture of the calcaneus with inflammatory changes medially, surrounding the tibialis posterior, flexor digitorum longus, and flexor hallucis longus tendons. In fact, the flexor tendons are absent in the more distal section **(A)** due to rupture. Arrow **1** Inflammatory tissue. Arrow **2** Calcaneal fracture. Arrow **3** Tibialis posterior tendon. Arrow **4** Flexor hallucis longus tendon. Arrow **5** Flexor digitorum longus tendon.

Tenosynovitis

Tenosynovitis is an inflammatory or infectious condition of the tendons and synovial sheaths (Fig. 8.31). The synovial sheath and synovial fluid are not normally seen on MRI. Distension of the synovial sheath by fluid can be seen in acute tenosynovitis, with a relatively normal tendon (2) (Fig. 8.32). In chronic tenosynovitis the tendon and synovial sheath are thickened and fluid surrounds the tendon (Fig. 8.33). If no synovial sheath is present, only tendon thickening is demonstrated on MRI (Fig. 8.34).

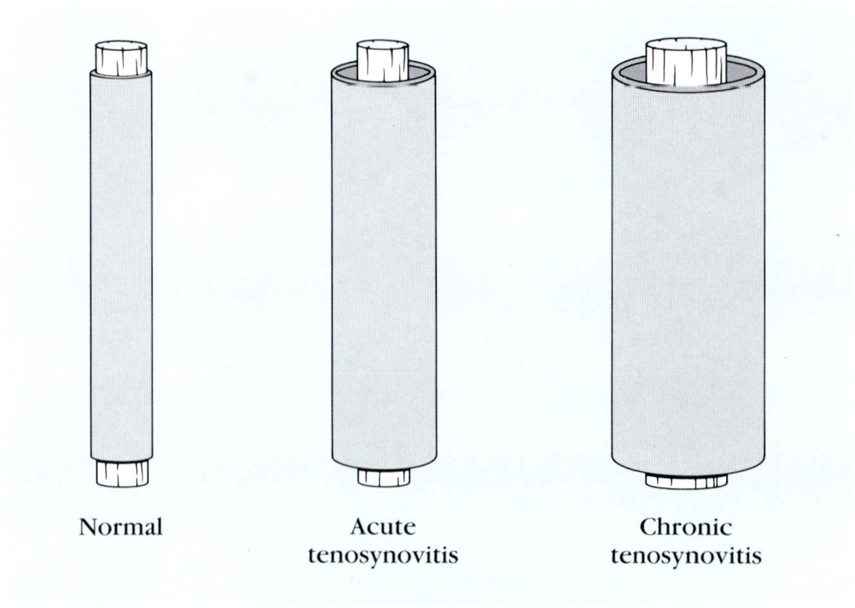

FIGURE 8.31. Changes seen in tenosynovitis.

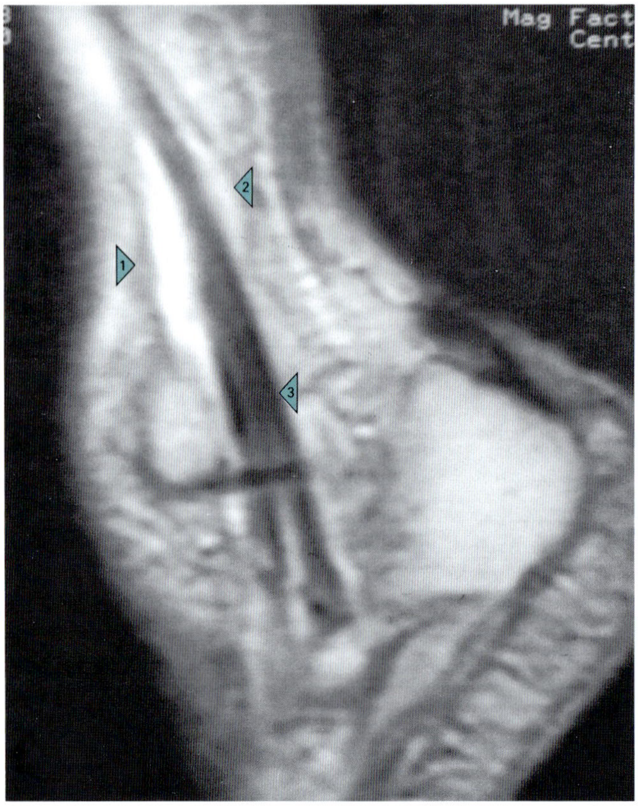

FIGURE 8.32. Acute Tenosynovitis of the Peroneal Tendons. Sagittal section [SE 2000/80] through the peroneal tendons demonstrating high SI fluid distending the synovial sheath. The tendons are normal in size. Arrow **1** Synovial sheath. Arrow **2** Fluid. Arrow **3** Peroneal tendons.

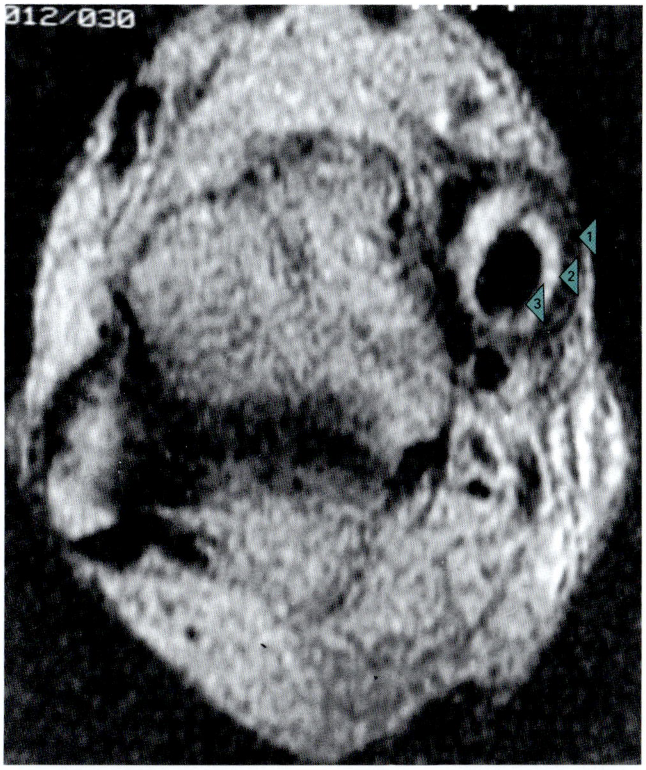

FIGURE 8.33. Chronic Tenosynovitis of the Tibialis Posterior Tendon. Axial section [SE 2000/80] demonstrating thickening of the tendon and synovial sheath, with a high SI synovial effusion. Arrow **1** Synovial sheath. Arrow **2** Fluid. Arrow **3** Tibialis posterior.

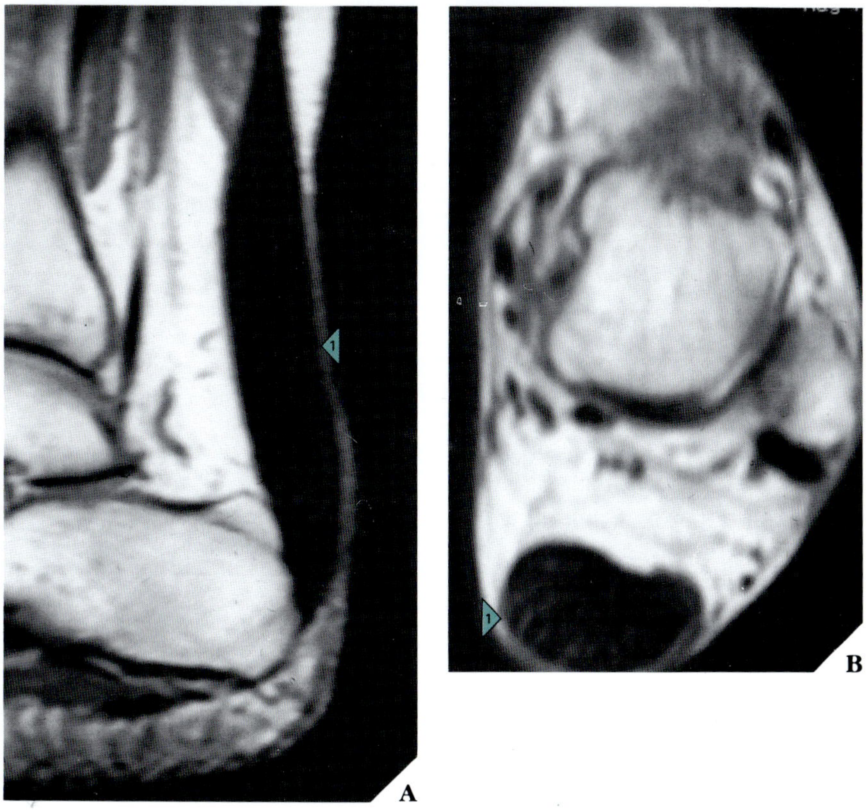

FIGURE 8.34. Chronic Achilles Tendinitis. A Sagittal and **B** axial sections [SE 2000/20] demonstrating marked thickening of the Achilles tendon, but no tendon rupture. Arrows **1** Chronic Achilles tendinitis.

ARTICULAR LESIONS

Joint Effusion

The presence and extent of inflammatory and traumatic injuries involving the ankle and foot can be assessed with MRI (4). As in other joints, joint effusions are seen on MRI as low SI fluid collections on T1WI and high SI on T2WI within the confines of the joint space (1) (Fig. 8.35). Normally, there is only a very small amount of fluid in the joints of the ankle and foot. Occasionally, this fluid is seen on MRI as a thin hyperintense line in the joint space on T2WI. Infection, trauma, and inflammation can induce joint effusions. The etiology of the effusion cannot be determined based on the MRI characteristics (1).

Synovial Cysts

Synovial cysts are relatively rare in the ankle and foot. MRI demonstrates rounded fluid collections in the vicinity of a joint, with low SI on T1WI and high SI on T2WI (Fig. 8.36).

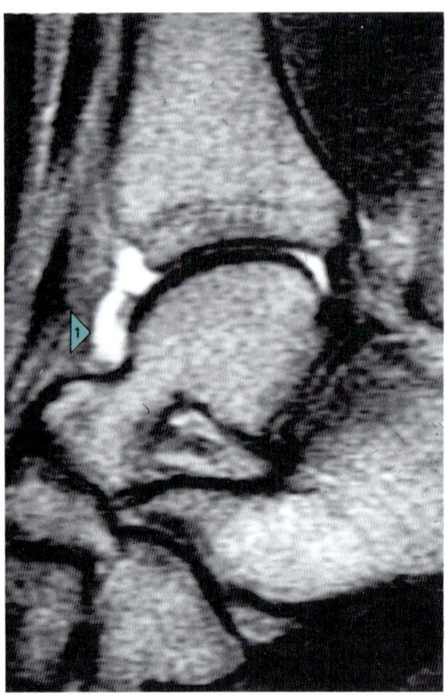

FIGURE 8.35. Joint Effusion. Sagittal section [SE 2000/80] showing high SI fluid distending the joint capsule anteriorly. Arrow **1** Joint effusion.

Rheumatoid Arthritis

Rheumatoid arthritis involves the ankle joint less frequently than other larger joints (30). The metatarsophalangeal joints are more often involved. Bony erosions, joint effusions, synovial proliferation, and subluxations can be seen on MRI but plain films can also demonstrate most of these findings.

Hemophilic Arthropathy

Hemophilic arthropathy frequently involves the tibiotalar joint. The main pathologic findings are synovial proliferation and fibrosis due to repeated episodes of hemarthrosis, followed by secondary degenerative osteoarthritis, and joint space narrowing. MRI can depict these abnormalities, showing thickening of the synovium and joint capsule with low SI in all pulse sequences due to the paramagnetic effect of hemosiderin (4,13) (Fig. 8.37).

Osteochondral Lesions

Acute trauma to the ankle and foot can induce ab-

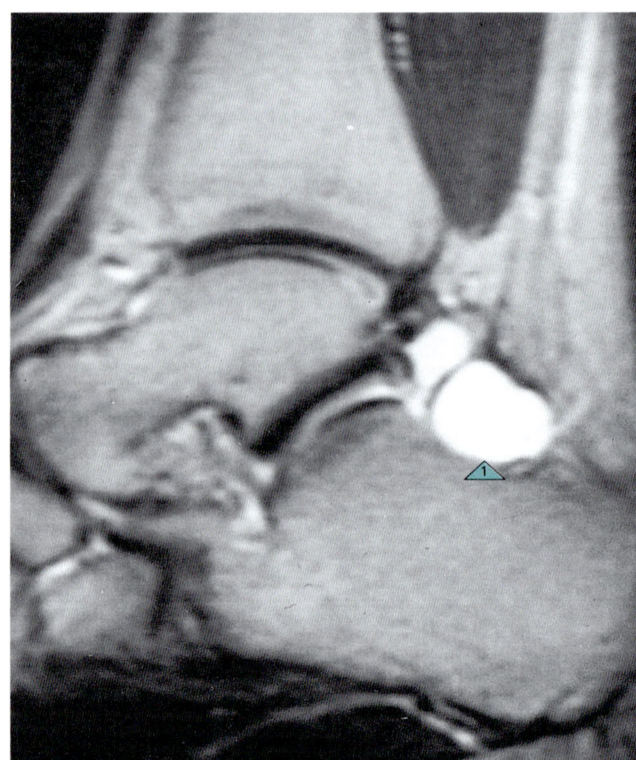

FIGURE 8.36. Synovial Cyst. Sagittal section [SE 2000/80] showing a well-demarcated, high SI fluid collection adjacent to the posterior aspect of the subtalar joint. No associated joint effusion is seen. Arrow **1** Synovial cyst.

normalities often not seen on plain films. Osteochondral fractures or osteochondritis dissecans occur in the talar dome, frequently involving the middle third of the lateral border and the posterior third of the medial border, depending on the mechanism of injury (Fig. 8.38). MRI is useful for assessing the cartilaginous portion of the fracture (4) (Fig. 8.39). Surgical removal of the fragment may be necessary if the lesion produces joint incongruity.

Fractures

Stress fractures and occult intraosseous fractures may not be detected with plain films during the acute and subacute stages, and bone scintigraphy is often required to provide the diagnosis (14,29). Stress fractures occur in bones under unusual stress and are more frequent in the metaphysis and diaphysis of the bone. Occult intraosseous fractures occur following a single traumatic event and are more often located in the epiphysis. MRI findings

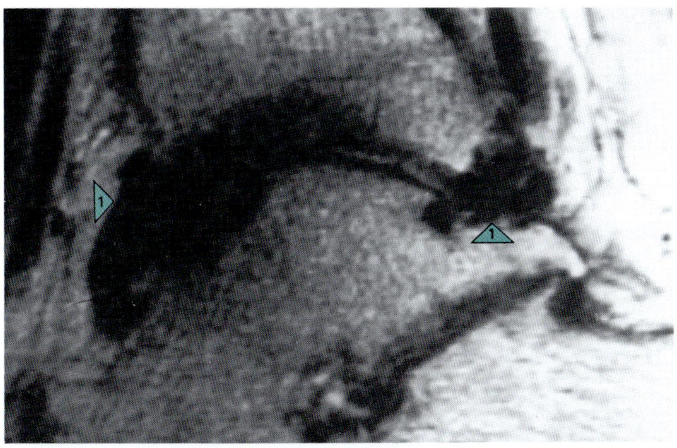

FIGURE 8.37. Hemophilic Arthropathy. Sagittal section [SE 1500/80] showing low SI proliferative synovium in the anterior and posterior aspects of the tibiotalar joint. Arrows **1** Proliferative synovium.

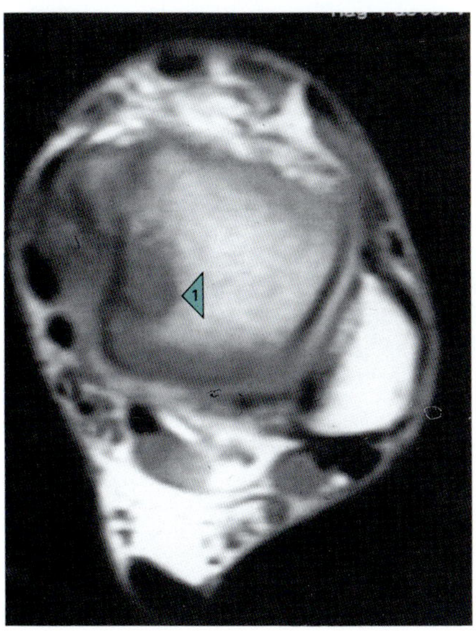

FIGURE 8.38. Osteochondral Fracture. Axial section [SE 500/20] showing a small low SI area in the dome of the talus, near its medial side. Arrow **1** Osteochondral fracture.

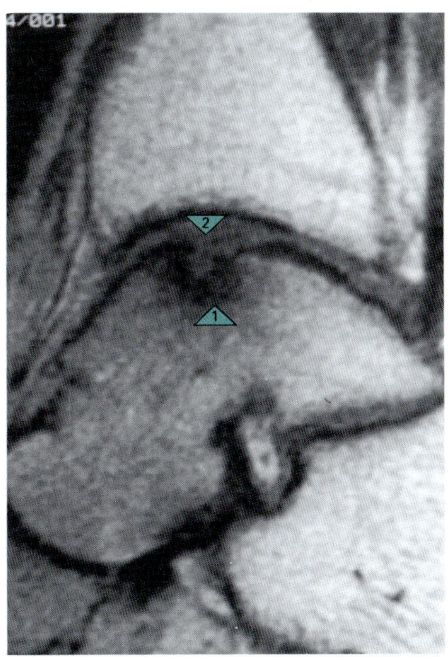

FIGURE 8.39. Osteochondral Fracture. Sagittal section [SE 800/20] showing a defect in the talar dome, with intermediate SI articular cartilage filling. The lesion is surrounded by a rim of low SI reactive bone. Arrow **1** Osteochondral fracture. Arrow **2** Articular cartilage.

in stress fractures differ from those in occult intraosseous fractures (14). In stress fractures, one can see a low SI band on T1WI and T2WI, contiguous with the cortex. A high SI periosteal reaction may also be seen on T2WI. On the other hand, occult intraosseous fractures demonstrate ill-defined areas of low SI on T1WI and high SI on T2WI, without periosteal reaction (29) (Fig. 8.40).

Posttraumatic complications include secondary osteoarthritis and loose intraarticular bodies. In posttraumatic osteoarthritis there is narrowing of the joint space, irregularity of the articular cartilage (Fig. 8.41), sclerosis, osteophytes, and subchondral cysts. Sclerosis is seen on MRI as low SI areas in all pulse sequences involving the subchondral bone (Fig. 8.42). Intraarticular loose bodies can be seen

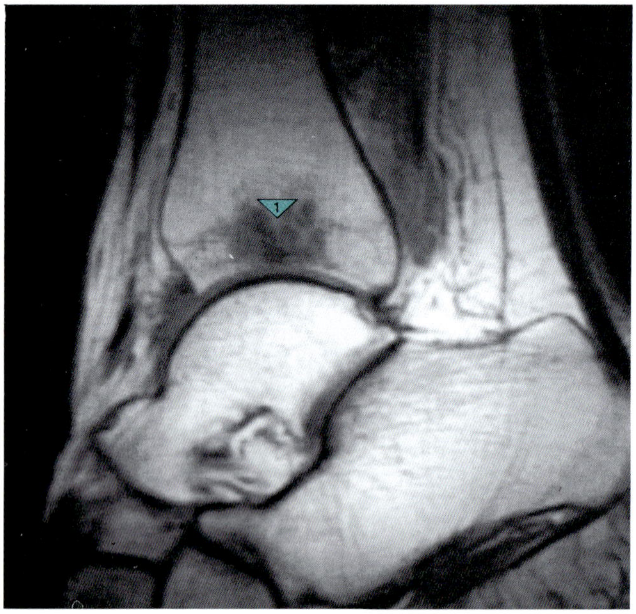

FIGURE 8.40. Occult Intraosseous Fracture. Sagittal section [SE 500/20] showing an ill-defined area of low SI in the distal tibial epiphysis. Arrow **1** Occult intraosseous fracture.

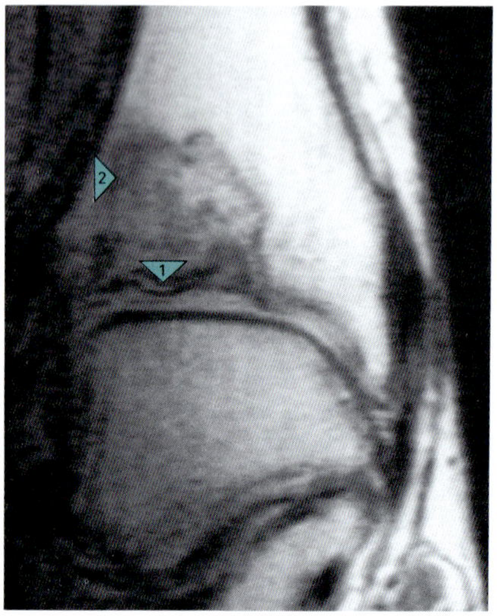

FIGURE 8.41. Posttraumatic Osteoarthritis. Coronal section [SE 500/20]. There is an irregular, low SI area in the distal tibial probably representing the changes related to fracture healing. Observe thinning of the articular tibial cartilage with irregular subchondral cortex and normal articular cartilage of the talus. Arrow **1** Thinning of the articular cartilage. Arrow **2** Posttraumatic lesion.

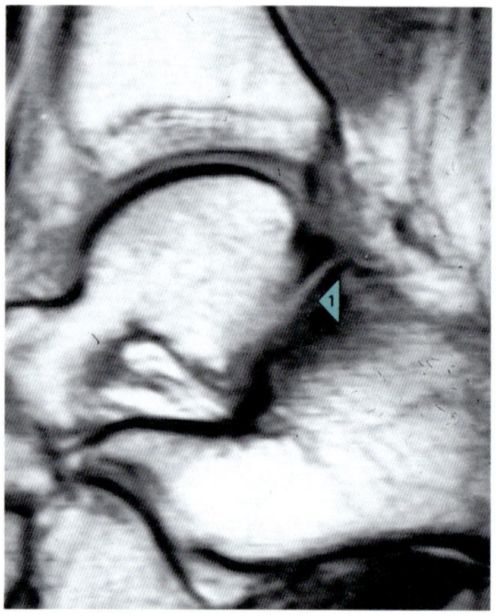

FIGURE 8.42. Posttraumatic Osteoarthritis of the Subtalar Joint. Sagittal section [SE 2000/20] showing irregular, low SI subchondral cortex of the posterior subtalar joint, with narrowing of the joint space. Arrow **1** Osteoarthritis of posterior subtalar joint.

on MRI as high SI fragments on T1WI and low SI on T2WI, similar to fat, or low SI in all pulse sequences, depending on the amount of fatty marrow and calcification within the loose bodies (Fig. 8.43).

Avascular Necrosis

Avascular necrosis of the talus occurs most frequently in the body of the talus, secondary to neck fractures associated with dislocations of the subtalar joint, ankle joint, or both. The radiographic diagnosis of avascular necrosis of the body of the talus is based on the absence of a thin rim of subchondral osteoporosis about 4 weeks following the injury (Hawkins sign) (18). The presence of osteoporosis indicates an intact vascular supply. MRI can also demonstrate the presence or absence of avascular necrosis (Figs. 8.44 and 8.45) and assess its extent (Fig. 8.46).

Neuroarthropathy

Neuroarthropathy of the foot and ankle is frequently related to diabetes mellitus. The classical radiographic signs of neuroparthropathy include extensive bone fragmentation and joint dislocation, without associated osteoporosis. It frequently involves the phalanges and metatarsophalangeal joints, followed by the tarsometatarsal joint, and the tibiotalar joint. The differential diagnosis between neuroarthropathy and osteomyelitis may be

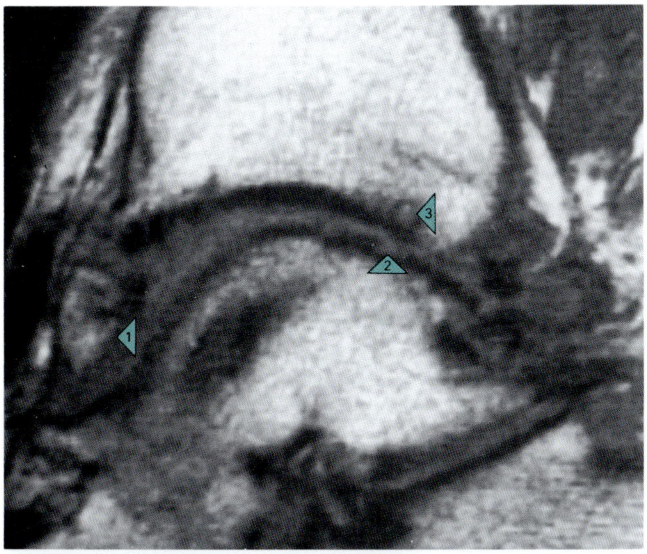

FIGURE 8.43. Posttraumatic Osteoarthritis and Loose Body. Sagittal section [SE 500/20]. There is a large loose body in the anterior recess of the tibiotalar joint. Note narrowing of the posterior aspect of the joint space and a small subchondral cyst in the tibial side. Arrow **1** Loose body. Arrow **2** Tibiotalar joint narrowing. Arrow **3** Subchondral cyst.

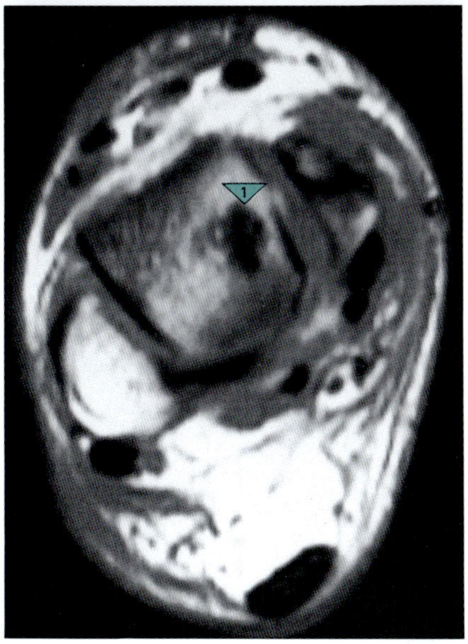

FIGURE 8.44. Posttraumatic Avascular Necrosis of the Talus. Axial section [SE 2000/20] through the dome of the talus. This patient had a fracture dislocation of the talus four weeks prior to the MRI examination. A small area of osteonecrosis is seen in the talar dome. Arrow **1** Avascular necrosis.

difficult during the early stages of the disease. In neuroarthropathy, MRI demonstrates bone destruction and low SI in all pulse sequences within the medullary space (Fig. 8.47). In osteomyelitis, high SI areas on T2WI involving the medullary space are seen (3.27).

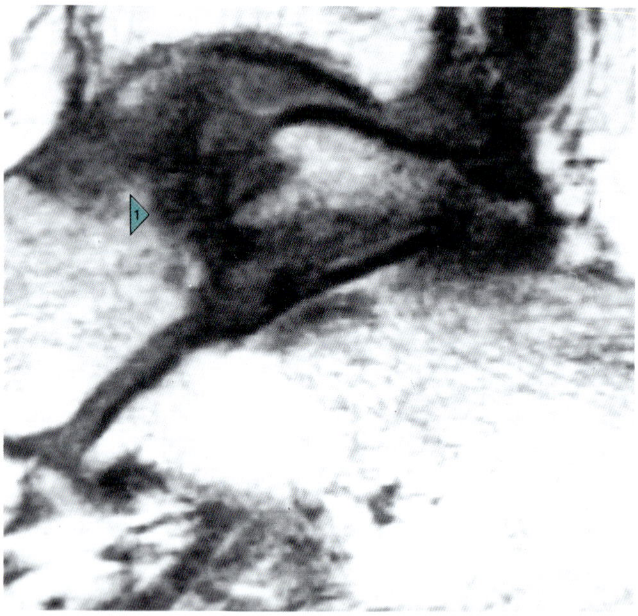

FIGURE 8.45. Talar Fracture Without Avascular Necrosis. Sagittal section [SE 800/20]. Note the vertical fracture through the body of the talus. There is high SI in the medullary space of the anterior and posterior fragments similar to that of the calcaneus and distal tibia, indicating viable bone. Arrow **1** Talar fracture.

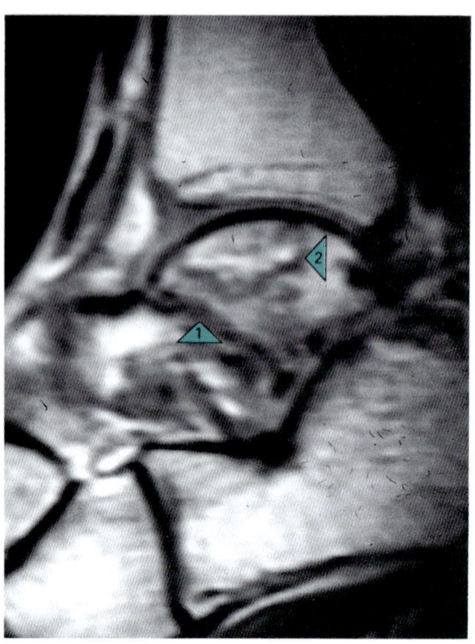

FIGURE 8.46. Posttraumatic Avascular Necrosis. Sagittal section [SE 2000/80]. Note an oblique fracture through the neck of the talus and a large area of avascular necrosis of the body of the talus, involving most of the articular surface. Arrow **1** Talar fracture. Arrow **2** Avascular necrosis.

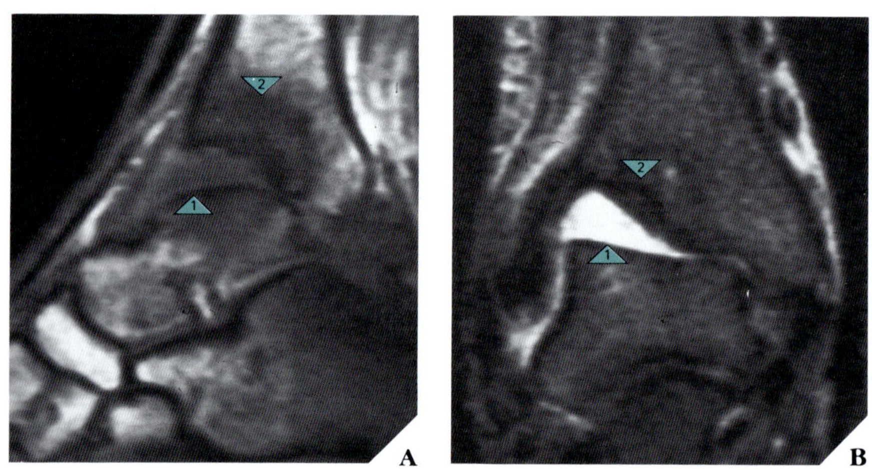

FIGURE 8.47. Neuroarthropathy of the Tibiotalar Joint. A Sagittal [SE 2000/20] and **B** coronal [SE 2000/80] sections. There is destruction of the anterior half of the distal tibia with low SI of the medullary space on T1WI **(A)** and T2WI **(B),** in this diabetic patient. A joint effusion is also present. Arrows **1** Joint effusion. Arrows **2** Neuroarthropathy.

REFERENCES

1. Beltran J, Noto AM, Herman LJ, et al: Joint effusions: MR imaging. Radiology 1986;158:133–137.
2. Beltran J, Noto AM, Herman LJ, Lubbers LM: Tendons: high-field-strength, surface coil MR imaging. Radiology 1987;162:735–740.
3. Beltran J, Noto AM, McGhee RB, et al: Infections of the musculoskeletal system: high-field-strength MR imaging. Radiology 1987; 164:449–454.
4. Beltran J, Noto AM, Mosure JC, et al: Ankle: surface coil MR imaging at 1.5 T. Radiology 1986;161:203–209.
5. Bostrom L: Sprained ankles. 3. Clinical observations in recent ligament ruptures. Acta Chir Scand 1965;11:560–569.
6. Burk DL, Dalinka MK, Schiebler ML, et al: Strategies for musculoskeletal magnetic resonance imaging. Radiol Clin North Am 1988;26:653–672.
7. Funk DA, Cass RJ, Johnson KA: Acquired adult flat foot secondary to posterior tibial tendon pathology. J Bone Joint Surg (Am) 1986;68:95–102.
8. Gilula LA, Oloff L, Caputi DPM, et al: Ankle tenography: a key to unexplained symptomatology. Radiology 1984;151:581–587.
9. Jahss MH: Spontaneous rupture of the tibialis posterior tendon: clinical findings, tenographic studies, and a new technique of repair. Foot Ankle 1982;3:158–166.
10. Johnson KA: Tibialis posterior tendon rupture. Clin Orthop 1983;177:140–147.
11. Keysler CK, Gilula LA, Hardy DC, et al: Soft-tissue abnormalities of the foot and ankle: CT diagnosis. AJR 1988;150:845–850.
12. Kneeland JB, Macrandar S, Middleton WD, et al: MR imaging of the normal ankle: correlation with anatomic sections. AJR 1988; 151:117–123.
13. Kulkarni MV, Drolshagen LF, Kaye JJ, et al: MR imaging of hemophiliac arthropathy. J Comput Assist Tomogr 1986;10:445–449.
14. Lee JK, Yao L: Stress fractures: MR imaging. Radiology 1988;169:217–220.
15. Mathieson JR, Connell DG, Cooperberg PL, Lloyd-Smith DR: Sonography of the Achilles tendon and adjacent bursae. AJR 1988;151: 127–131.
16. O'Connor D: Sinus tarsi syndrome. A clinical entity. J Bone Joint Surg (Am) 1958;40: 720.
17. Quinn SF, Murray WT, Clark RA, Cochran CF: Achilles tendon: MR imaging at 1.5T. Radiology 1987;164:767–770.
18. Resnick D, Niwayama G: Diagnosis of Bone and Joint Diseases, ed. 2. Philadelphia, W.B. Saunders Co, 1988.
19. Rosenberg ZS, Cheung Y, Jahss MH, et al: Rupture of the posterior tibial tendon: CT and MR imaging with surgical correlation. Radiology 1988;169:229–235.
20. Rosenberg: ZS, Feldman F, Singson RD, Kane R: Ankle tendons: evaluation with CT. Radiology 1988:166:221–226.
21. Rosenberg ZS, Feldman F, Singson RD, Price GJ: Peroneal tendon injury associated with calcaneal fractures: CT findings. AJR 1987; 149:125–129.
22. Rosenberg ZS, Jahss MH, Noto AM, et al: Rupture of the posterior tibial tendon: CT and surgical findings. Radiology 1988;176: 489–493.
23. Solomon MA, Gilula LA, Oloff LM, et al: CT scanning of the foot and ankle: 1. Normal anatomy. AJR 1986;146:1192–1203.
24. Solomon MA, Gilula LA, Oloff LM, Oloff J: CT scanning of the foot and ankle: 2. Clinical applications and review of the literature. AJR 1986;146:1204–1214.
25. Szczukowski M, St Pierre RK, Fleming LL, Somogyi J: Computerized tomography in the evaluation of peroneal tendon dislocations: a report of two cases. Am J Sports Med 1983:11:444–447.
26. Taillard W, Meyer JM, Garcia J, Blanc Y: The sinus tarsi syndrome. Int Orthop 1981;5: 117–130.
27. Tang JSH, Gold RH, Bassett LW, Seeger LL: Musculoskeletal infection of the extremities: evaluation with MR imaging. Radiology 1988;166:205–209.
28. Teng MMH, Destouet JM, Gilula LA, et al: Ankle tenography: a key to unexplained symptomatology. Part I: normal tenographic anatomy. Radiology 1984;151:575–580.
29. Yao L, Lee JK: Occult intraosseous fracture: detection with MR imaging. Radiology 1988;167:749–751.
30. Yulish BS, Lieberman JM, Newman AJ, et al: Juvenile rheumatoid arthritis: assessment with MR imaging. Radiology 1987;165:149–152.

FIGURE CREDITS

Figures 8.10, 8.16, 8.17, 8.21, 8.37, 8.39, 8.41: From Beltran J, Noto AM, McGhee RB et al: Ankle: surface coil MR imaging at 1.5T. Radiology 1986;161:203–209. Reprinted with permission.

Figures 8.26, 8.32: From Beltran J, Noto AM, Herman LJ, Lubbers LM: Tendons: high-field-strength, surface coil MR imaging. Radiology 1987;162:735–740.

CHAPTER NINE

Donald W. Chakeres
Andrew J. Curtin
Robert V. Bulas

The Spine

TECHNICAL CONSIDERATIONS

SURFACE COILS

Most imagers have a body receive coil built into the housing of the scanner. Whole-body images of the spine can be made but are not of ideal quality because the body coil is distant from the spine. Surface coils are usually receive-only antennae that can be placed in close approximation to the body part being studied (7). The type of surface coil varies based on the type of imaging unit and its geometry.

A dedicated surface coil is mandatory for high quality spine imaging. Surface coils provide improved S/N and exclude signals from unwanted regions (such as the abdomen or chest), thus diminishing motion artifacts. A commonly used surface coil is a 10 × 28-cm rectangular "license-plate"-shaped receive-only coil. This type of coil allows for a large region of visualization on both sagittal and coronal images and maintains adequate signal for axial imaging. Smaller circular coils (10–15 cm) could be used for better resolution. However, the smaller volume of tissue studied may exclude some areas of pathology and therefore they are not recommended for survey exams. Visualization of the deep structures within the spine can be a problem with smaller coils and larger patients.

PATIENT POSITIONING

Patients are studied in a supine neutral position, with the spine centered over an underlying surface coil. Some surface coils are designed to wrap around the cervical spine. Occasionally, decubitus or prone positioning may be necessary when the patient cannot maintain a supine position. Change in the positioning of the surface coil may also be necessary. Flexion or extension positions can be used to provide functional information about stability and subluxation, particularly for cervical disorders such as rheumatoid arthritis.

PLANES OF SECTION

Both the axial and sagittal planes are routinely used for MR imaging of the spine. Coronal images are obtained less commonly, but they can be very helpful in cases of scoliosis or tumors that bridge into the chest, retroperitoneum, or abdomen. MRI also has the capacity to obtain sections at any angle of obliquity (6). Sections angled parallel to the vertebral endplates or oriented toward specific structures may help delineate pathology in difficult cases.

SLICE THICKNESS

Sagittal and axial 5.0-mm thick sections with a 1-mm gap between slices are used routinely for evaluation of the thoracic and lumbar spine. Small gaps between slices reduce the artifact of signal "cross talk" from adjacent slices. In the cervical region, 3-mm thick sagittal sections and 5.0-mm thick axial sections are obtained. The thinner sections limit the regions studied and usually require extra averages to maintain an adequate S/N. Thicker sections (10 mm) are less desirable since they lead to more partial volume imaging problems. Partial volume averaging occurs when two different structures only partially fill a voxel, which leads to averaging of their signal intensities. This averaging produces diminished spatial resolution and distorts small anatomic structures (5). Partial volume averaging can best be avoided by using thin sections.

SIZE OF ACQUISITION MATRIX

The matrix of the image is important in image resolution. The greater the number of pixels the higher the possible resolution. Since it requires more time to obtain a higher pixel matrix image, it is not always used. Also, since the voxels are smaller with a higher number of pixels, the noise may increase, actually decreasing the image quality. Most MR images use 256 points for the frequency-encoded gradient axis. The phase-encoded gradient axis requires a change in acquisition time, since each point requires a unique measurement. Most images have a minimum of 128 points. A 256 × 256 matrix requires twice the acquisition time. Some units have a choice of a 192 × 256 matrix as well.

For sagittal images, a 256 × 256 matrix is best. For axial images, 128 × 256 pixels with double averages are best because the FOV is usually smaller for the axial than for the sagittal images and noise suppression is more important.

PULSE SEQUENCE DESIGN

The pulse sequence design is very important to the quality of the exam because it affects the image contrast (29). To better characterize the tissues, two sets of images are usually made at the same anatomic level using different pulse sequences. Short TR/TE (SE 800/20) T1WI are used for nearly every study because the acquisition time is relatively short (3.5–7 min) and the images are of high resolution.

A T2WI or PDWI is essential to complement the T1-weighted series. PDWI or T2WI can be obtained using a number of different techniques.

Proton-density- or T2-weighted images are made at the same anatomic locations as the T1WI. A heavily PDWI can be completed in axial and sagittal planes using a gradient refocused at a steady state (GRASS) technique and a low flip angle. Some gradient refocused techniques such as TR/TE100/12 msec and 30-degree flip angle gradient echo images only produce single slices, but in a very short time. Some manufacturers have released software for a PDWI GRASS technique that generates multiple sequential images that appear very similar to multislice spin echo T2WI, but in a fraction of the time (4 min). As compared with routine SE images, the PD images are of higher, more consistent quality and are free of most motion artifacts. An alternate pulse sequence for T2WI is a long TR (1500–3000) multiple TE (20–100) SE technique, but the acquisition time may be long (10–13 minutes).

FIELD OF VIEW

The size of the FOV effects the total region visualized and the image quality. The S/N changes in proportion to the square of the FOV; thus, decreasing the FOV by one half generates only one fourth as much signal. Sagittal images are made with an FOV of 32 cm to include a large span of the spine whenever a survey study is needed, as in metastatic disease. An FOV of 24 cm is used when the region in question is smaller, as in cervical spine radiculopathy studies. A smaller FOV leads to wrap-around artifacts, a decrease in the anatomic levels studied, and a loss of S/N. Axial images can be made with a 20-cm FOV since a smaller region is needed than for sagittal and coronal images. FOV's of less than 20 cm are of poor quality unless a number of averages are obtained. The sagittal images will be adequate with a single average (two excitations). The axial images require two averages (four excitations).

PHASE AND FREQUENCY GRADIENT AXES

The localization of the pixel intensity for each voxel is defined by both phase and frequency information. The orientation of these axes to the patient is operator-dependent and can be very important to image quality. The choice of phase and frequency gradient axes is important to avoid artifacts from wrap-around, aliasing, and motion (19). A large

FOV is the simplest solution to wrap-around artifact, but resolution is reduced. Wrap-around artifact effects both phase and frequency directions but is worse in the phase-encoded axis. By aligning the phase gradient with the shortest axial dimension of the spine, wrap-around artifacts are reduced. In sagittal and coronal images the frequency gradient parallels the long axis of the spine. In axial images the frequency gradient runs side to side. Special software pulse sequences are also available to reduce wrap-around artifact.

Since the phase-encoded axis is most affected by motion artifacts (such as flowing blood), assignment of the axes so that the artifacts do not cross the most crucial segment of the images is very important. Pulse sequences with extra pulses that saturate the spins of the tissues outside the FOV can also be helpful. This saturation technique limits cerebral spinal fluid (CSF) and blood-flow artifacts from entering the imaging slice.

▼▼▼ *RESPIRATORY GATING, CARDIAC GATING, AND MOTION SUPPRESSION*

Programs that help suppress motion artifacts from respiration by measuring movement with bellows should be used if available, particularly for the thoracic and lumbar spine: Strict respiratory gating is not used, but a software program which uses the order of the phase gradients to help limit the ghost artifacts is very helpful. There is no time lost with this technique. Cardiac gating is not routinely used because it limits the choice of the TR, increases the acquisition and set-up time, and may not give ideal images (24). Motion suppression techniques, such as null moment gradient programs, can resolve many of the pulsatile motion artifacts at high fields ($>$ 1 Tesla). This technique lengthens the TE and lowers the total number of sections.

▼▼▼ *PARAMAGNETIC CONTRAST AGENTS*

The paramagnetic contrast agent Gd-DTPA can be helpful in certain clinical situations (25). Contrast agents are important in the evaluation of spinal drop metastases, carcinomatosis meningitis, septic meningitis, disc protrusion versus surgical scar, and intramedullary spinal tumors. The Food and Drug Administration has not specifically approved Gd-DTPA for dedicated indications outside of the head, but this does not limit its use to just the brain. Numerous reports indicate that Gd-DTPA is useful in spine imaging. A routine series of T1WI and PDWI is usually obtained first, followed by contrast injection and T1WI only.

▼▼▼ *THREE-DIMENSIONAL VOLUME DATA ACQUISITION*

Three-dimensional volume data acquisition may be important in the future because of its shorter acquisition times. The use of the more rapid gradient-refocused pulse sequences makes it applicable in the routine clinical setting. One of the main advantages is that very thin sections (1 mm) can be obtained. Greater manipulation of the MR images, similar to reconstructed CT, is possible in any plane.

▼▼▼ *METAL ARTIFACTS*

Some artifacts cannot be overcome (12). Certain metal implants will warp the magnetic field so much that the signal will be lost, producing large low-signal regions (Fig. 9.1). Occasionally, regions near metal wires or rods can be studied, but the quality of the study may be so compromised that they are of marginal clinical value. These types of artifacts are particularly troublesome on gradient echo and T2WI SE studies. If possible, early scanning of patients before implantation or use of other metal structures is recommended (as in the case of cervical trauma and support halos). MR safe halos and spinal traction devices are commercially available.

▼▼▼ *SEDATION*

For children sedation is essential in many situations. A combination of oral or rectal chloral hydrate and sleep deprivation works well for the majority of patients. This simple technique is safe and very reliable. Stronger sedatives such as Valium or hypnotics are occasionally necessary. General anesthesia is rarely needed at our institution, yet it is common in some centers. A carbon dioxide apnea monitor is used to monitor sedated patients. Cardiac monitoring is usually not used because the signals are distorted during scanning.

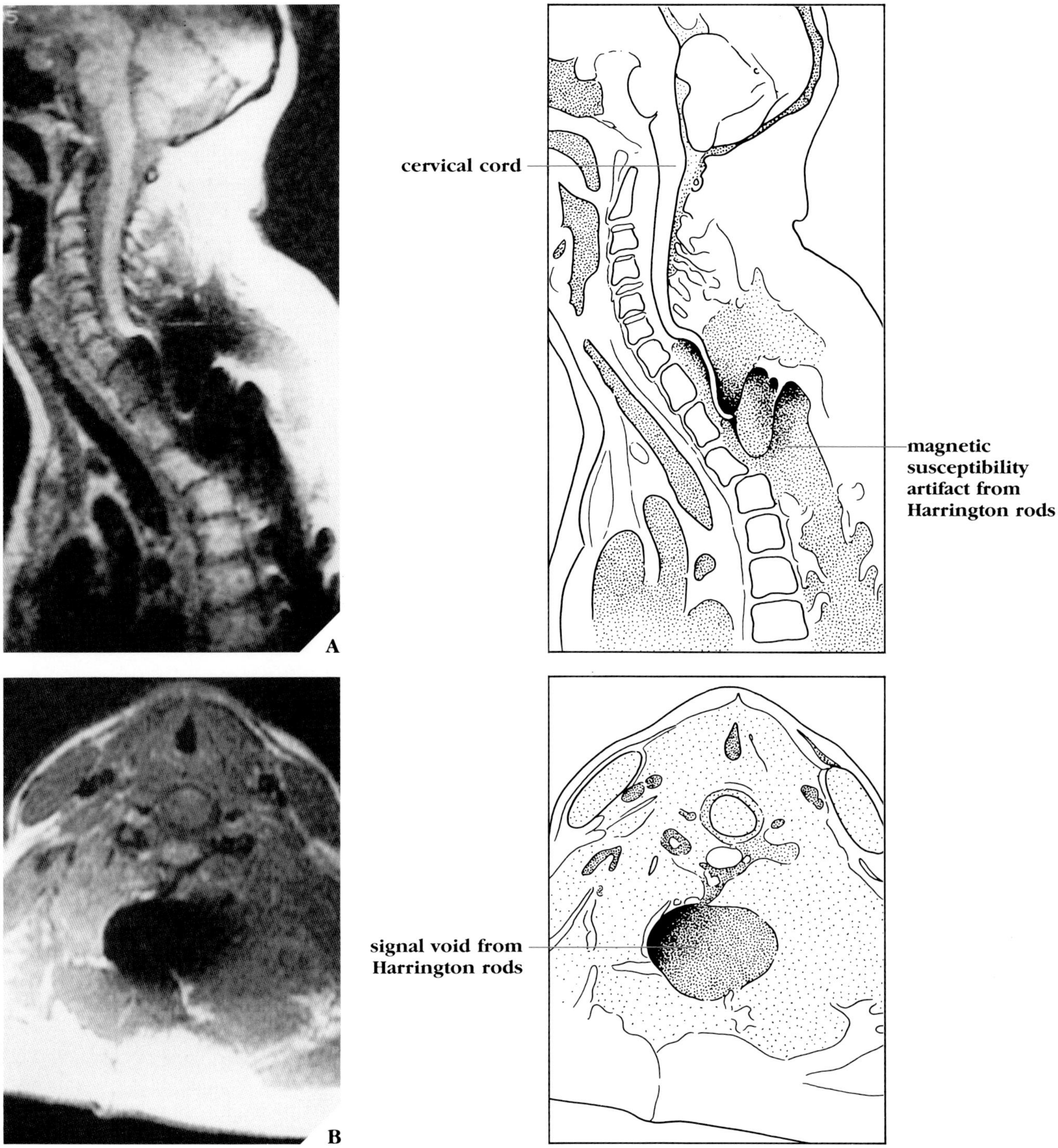

FIGURE 9.1. Upper Thoracic Metal-Wire Artifacts. A Sagittal section [SE 800/20]. Thoracic Harrington rods placed in this young trauma patient produce large signal void areas. This is due to the distortion of local magnetic fields by the metallic rods. Although the thoracic cord is not visualized, enough of the cervical area can be evaluated to rule out the presence of a syrinx. **B** Axial [SE 800/20]. Axial scans show severe magnetic field irregularities which are quite localized and do not streak across the image (as in CT) marring the whole scan. The lower cervical cord is well-identified and normal despite the adjacent artifact.

NORMAL ANATOMY

For a simple noninvasive exam, MRI provides quite remarkable visualization of the normal anatomy of the spine. Other than in the evaluation of the cortical bone structures, MRI is superior to CT and plain films for spinal imaging. The major components of the spine include the cortical bone, disc interspaces, marrow spaces, epidural spaces, dura, CSF, nerves, epidural veins, and the spinal cord (Figs. 9.2 and 9.3).

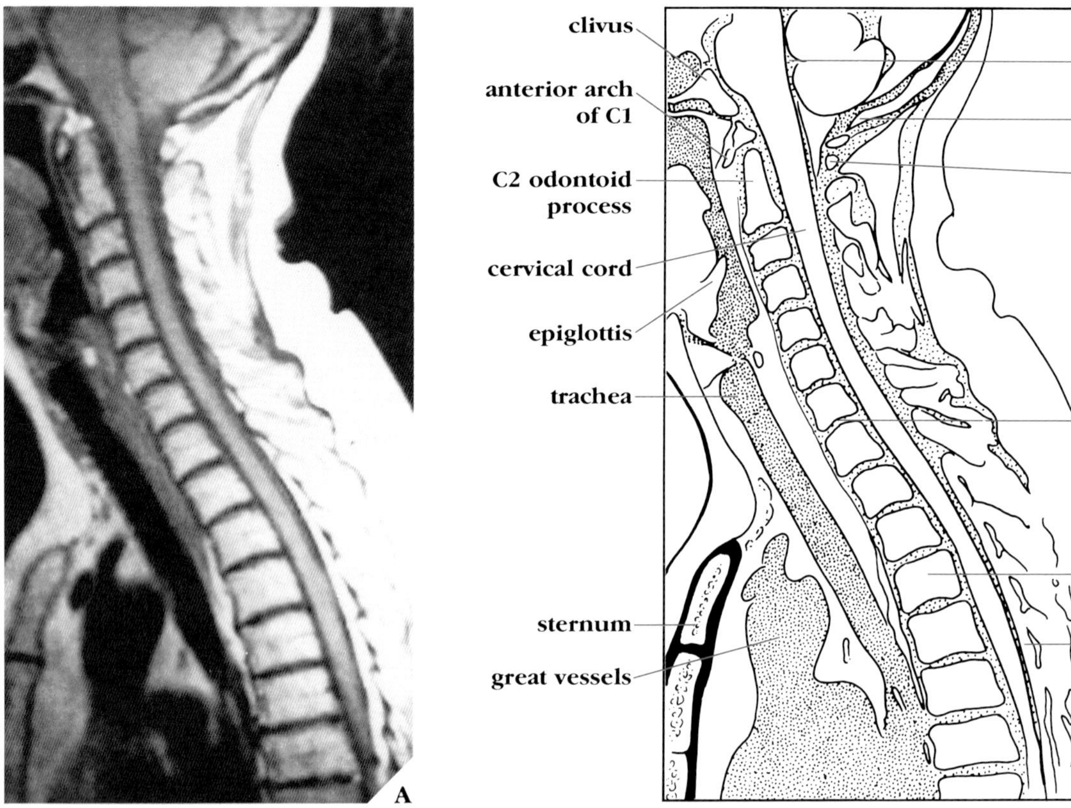

FIGURE 9.2. Normal Cervical Spine. A Sagittal midline section [SE 800/20]. This 32-cm FOV T1WI spans from the posterior fossa to the midthoracic cord. The craniocervical junction is well-outlined. The foramen magnum is defined by the fat within the occipital bone and clivus. The anterior and posterior arches of C1 appear as small oval marrow-containing structures at the upper cervical spine. The spinal cord is of an intermediate signal intensity outlined by lower SI CSF. The cervical and thoracic intervertebral discs are relatively low SI structures. The sternum, portions of the mediastinum, and the

On short TR/TE T1WI the cortical bone, dura, and fibrous-ligamentous structures are low signal due to the low water content or to the highly structured chemical form of the protons (Fig. 9.4). The marrow spaces are not quite as intense as the brighter epidural and other fat spaces. The triglycerides of the fatty spinal structures have short T1 and T2 relaxation times which account for the high signal.

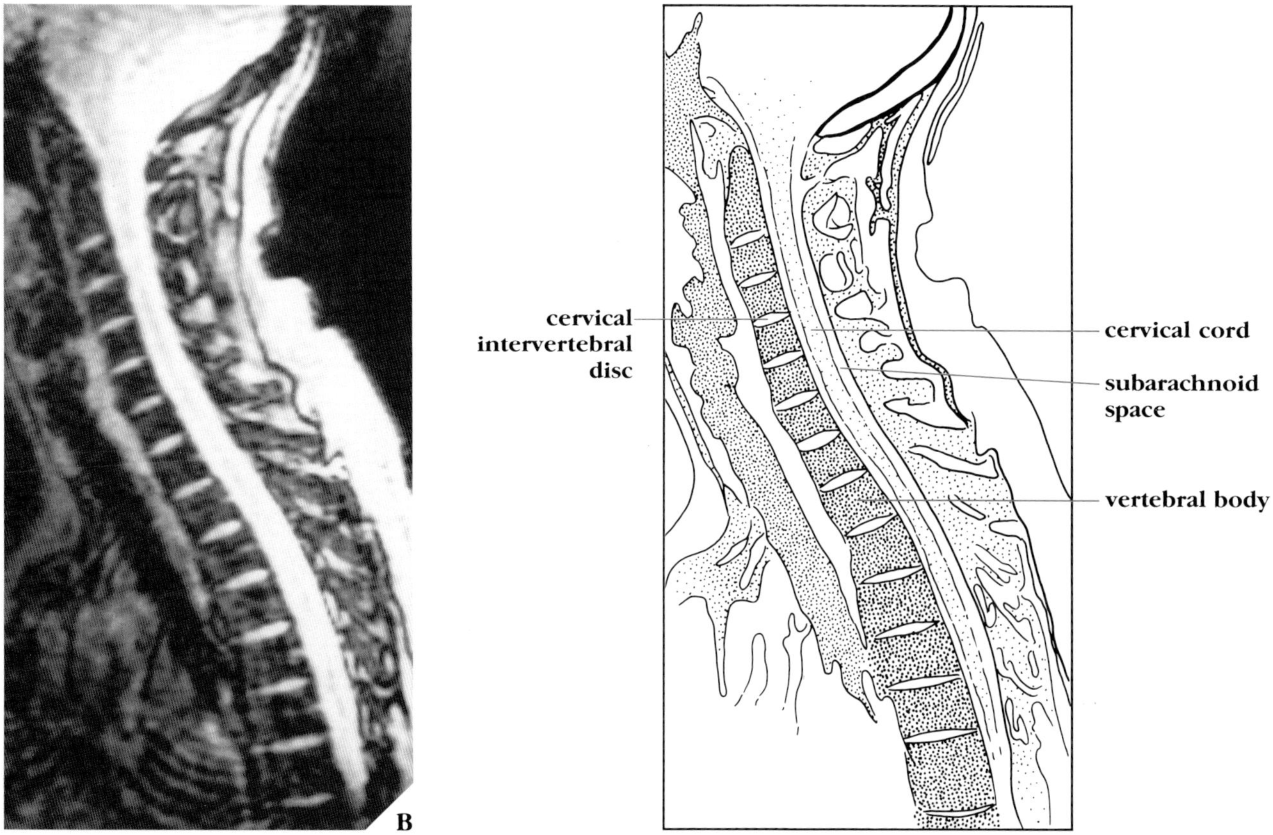

great vessels are commonly seen, as are the trachea and hypopharynx. **B** Sagittal midline section [GRASS 1000/12, flip angle 22.5 degrees]. On this PDWI, the high water content of the intervertebral discs produces very high SI similar to that of CSF. The cervical cord is moderately clear within the much higher intensity CSF spaces. The vertebral bodies exhibit a more uniformly low SI. Note that the signal is more intense posteriorly because it is closer to the surface coil.

The CSF is low signal due to a long T1 and motion effects. The apparent signal of the nerve root is dependent upon its location. In the CSF, the nerve roots appear relatively bright, while in the epidural spaces they are of relatively low signal (Figs. 9.5 and 9.6). The spinal cord is intermediate in signal, and some of the internal anatomy of the gray and white matter structures can be seen on axial images (Fig. 9.3).

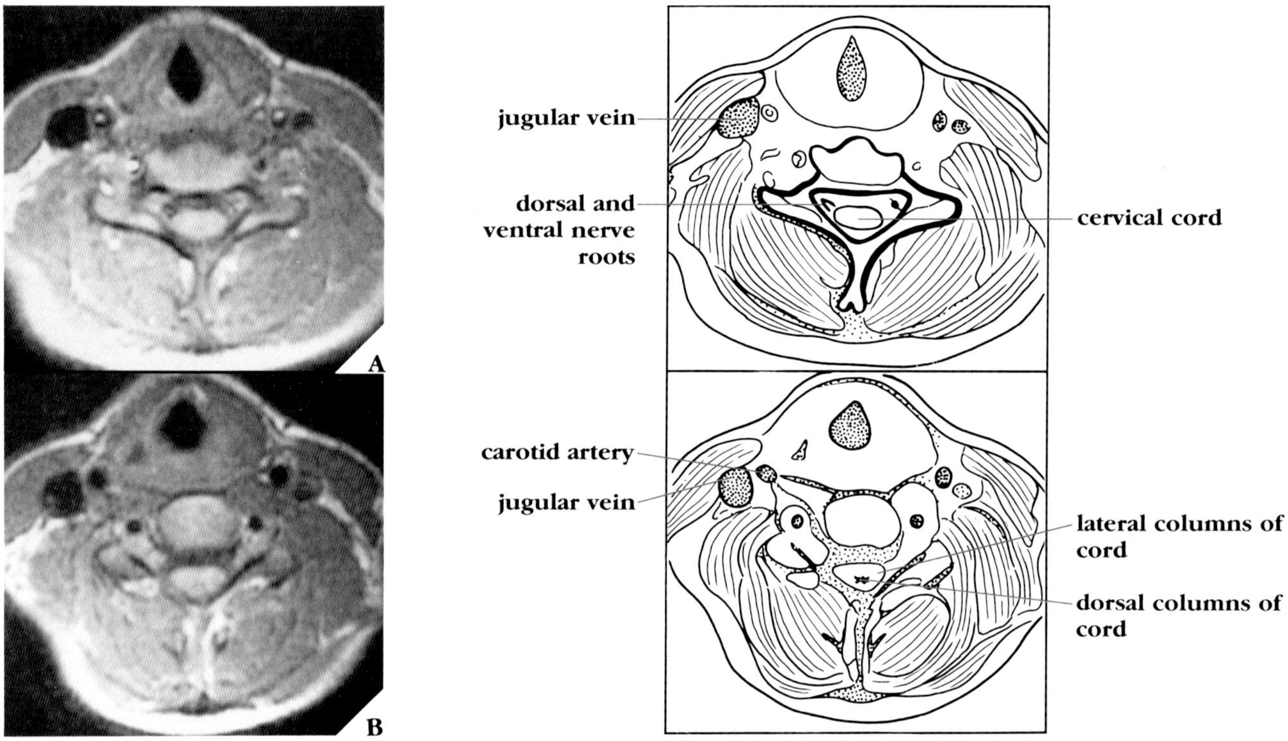

FIGURE 9.3. Normal Cervical Spine. A Axial section [SE 800/20]. This T1WI cervical image at the level of the pedicles demonstrates some of the definable internal morphology of the spinal cord. A small triangular low SI region located posteriorly is related to the posterior columns. A thin ring of low SI CSF surrounds the cord. Small ventral and dorsal nerve roots are often identified as slightly higher signals in the CSF. The facet joints are not as well-evaluated in the cervical region due to their horizontal orientation and small size. Usually vascular structures appear as signal voids related to rapidly flowing blood; however high signal can occasionally be seen, as in the right vertebral artery. **B** Axial section [SE 800/20]. This T1WI at the level of the foramina demonstrates the presence of some fat in the neural exit foramina, but these foramina are not as well-evaluated as those in the lumbosacral region. The large dominant right jugular vein and bilateral carotid arteries are

On T2WI and PDWI the cortical bone, dura and fibrous structures are low signal (Fig. 9.4B). The marrow spaces are usually low signal due to the short T2 and lower PD of fat. The epidural spaces are of low signal as well. The CSF has the highest signal due to its high PD and long T2. The nerve roots and spinal cord are low SI compared with the CSF (Figs. 9.3 and 9.6).

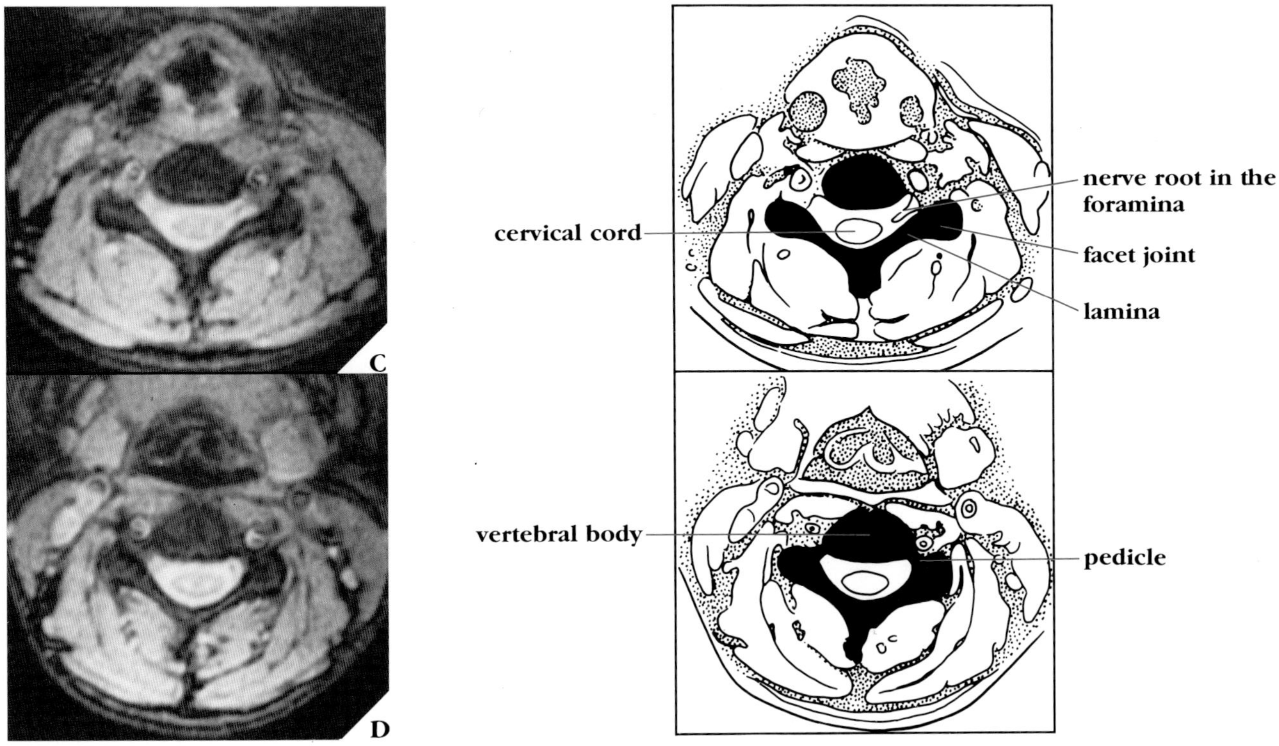

seen in the posterior triangles of the neck. Signal void of the air-containing glottis is present. **C** Axial section [GRASS 1000/ 12, flip angle 22.5 degrees]. The neural exit foramina, as well as some of the nerve roots, are better appreciated on this PD "myelographic technique" image. The cervical cord in transverse section is again well-visualized with some consistently definable internal morphology. **D** Axial section [GRASS 1000/ 12, flip angle 22.5 degrees]. This PDWI is made at the level of the cervical pedicles and demonstrates the low SI vertebral body and high SI CSF around the spinal cord. Note that on this pulse sequence, the blood vessels are not signal voids. The two gray matter tracts of the spinal cord are small paramidline high signals.

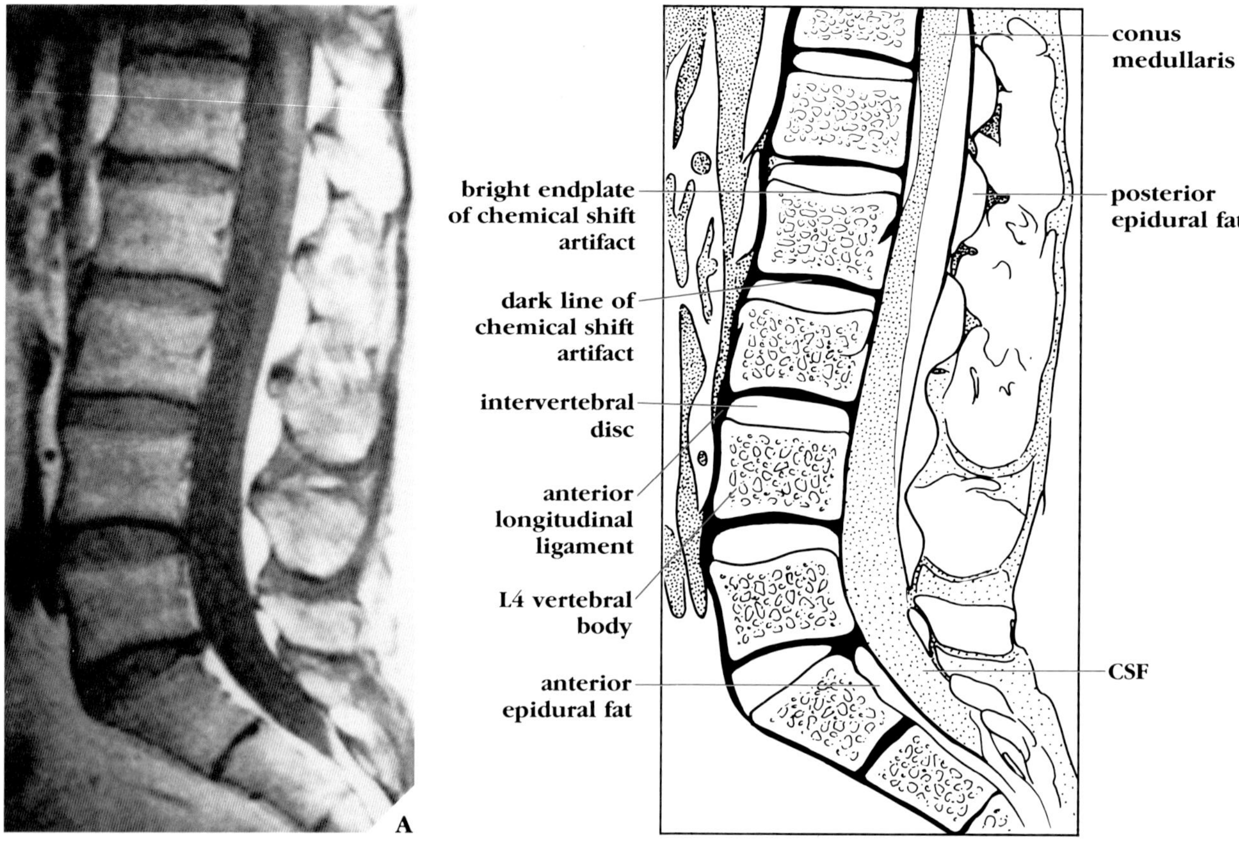

FIGURE 9.4. Normal Lumbar Spine. A Sagittal midline section [SE 800/20]. On this T1WI the tip of the conus medullaris is identified at T12-L1 in the posterior portion of low SI CSF of the thecal sac. Epidural fat is of a very high SI, most clearly seen posteriorly, but also seen anteriorly at the lumbosacral junction. Intervertebral discs are of a somewhat low SI due to their high water content. The low SI lines along the ventral and dorsal vertebral body are related to the anterior longitudinal ligament and cortical bone of the vertebral bodies. The ligament also spans and covers the anterior and posterior portions of the discs. The thin black line along the inferior endplate and the bright line at the superior portion of each vertebral body are due to a chemical shift artifact. The protons of the fatty marrow precess at approximately one pixel, a fre-

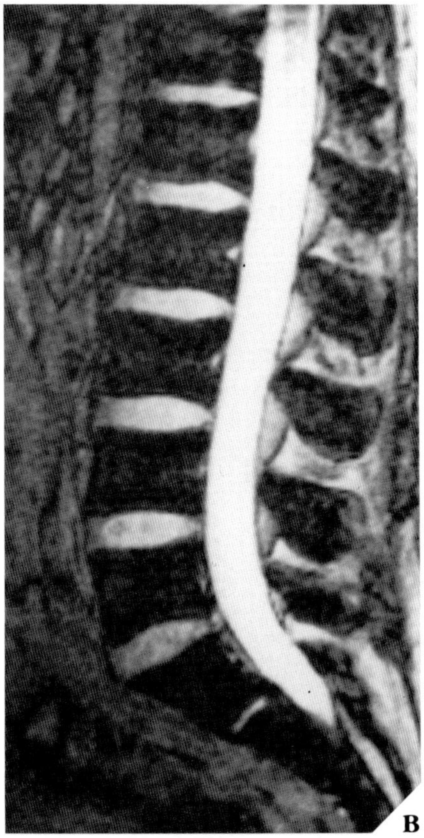

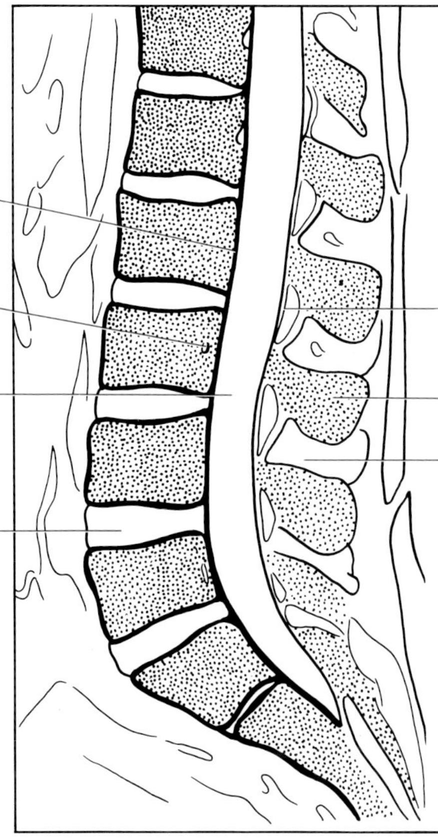

quency different than that of the water protons. **B** Sagittal midline section [GRASS 1000/12, flip angle 22.5 degrees]. This PDWI has a similar appearance to the myelographic technique, due to the very high gray scale contrast. There is clear delineation of the dural margins of the thecal sac filled with high SI CSF. The posterior longitudinal ligament and dura are silhouetted against the high water signal of the CSF and the intravertebral discs. The epidural fat is of more intermediate low signal and vertebral bodies are of a very low SI. Interspinous process ligaments are well defined. A high SI midposterior cleft in the vertebral bodies is related to the basivertebral veins. Note that the conus medullaris is lower in SI and is somewhat obscured by the high CSF signal, as compared with the T1WI.

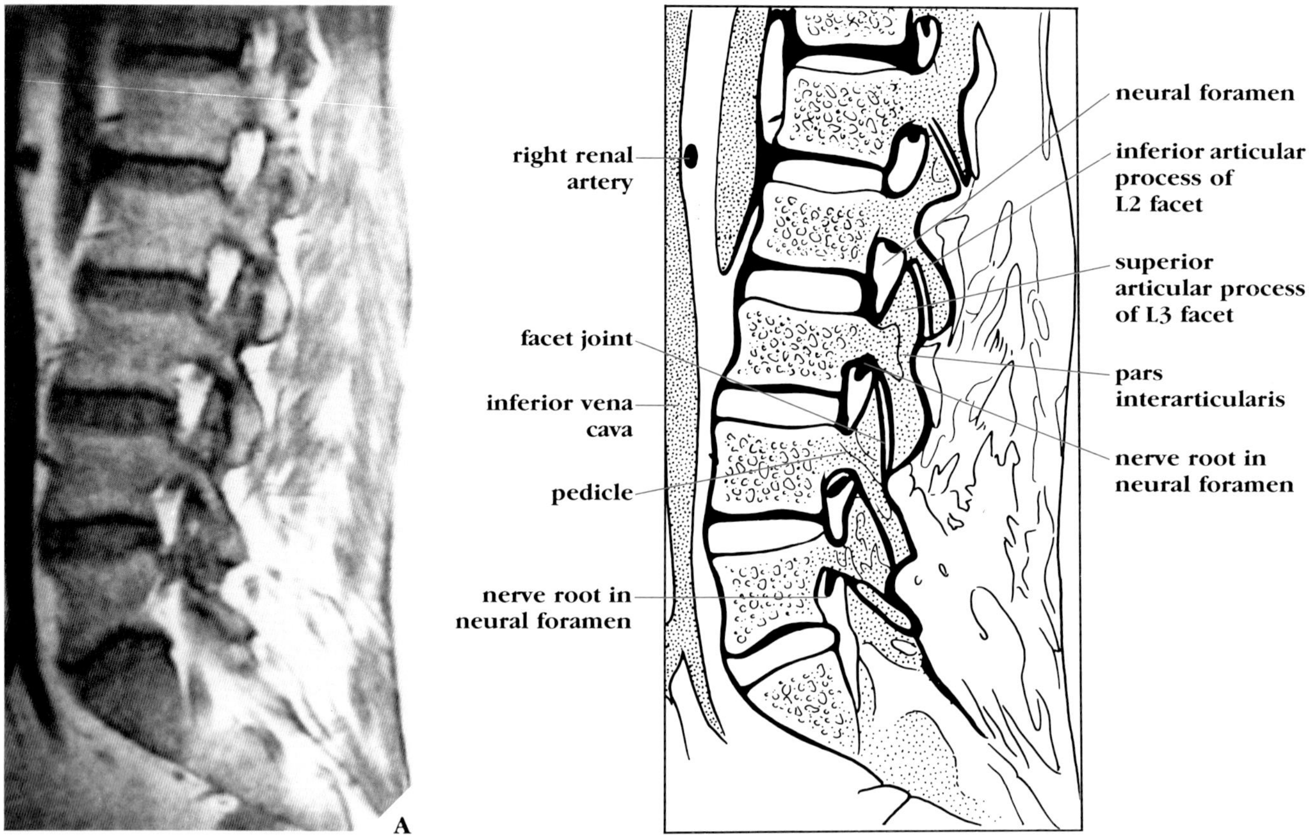

FIGURE 9.5. Normal Lumbar Spine. A Parasagittal section [SE 100/20]. This T1WI, made off the midline at the level of the pedicles, demonstrates the epidural fat, which fills the medial neural-exit foramina. The lower SI nerve roots course through the superior aspect of the foramen just inferior to the pedicles. Note that the medial segments of the nerves are well superior to the intervertebral disc. The superior and inferior articular processes of the facets are clearly demonstrated as is

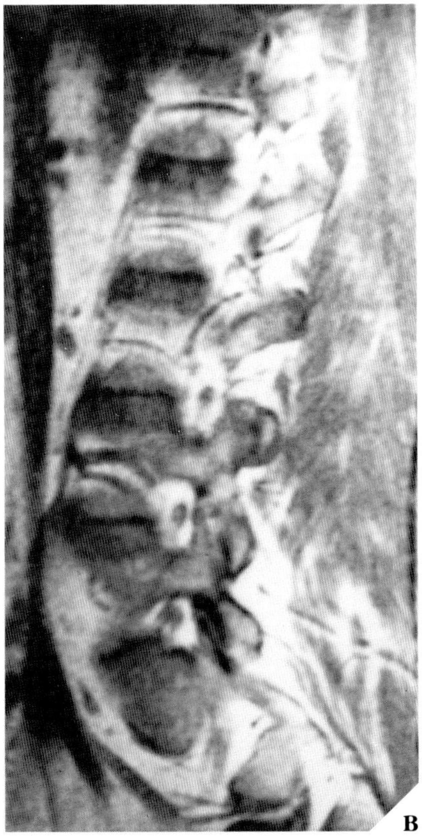

B

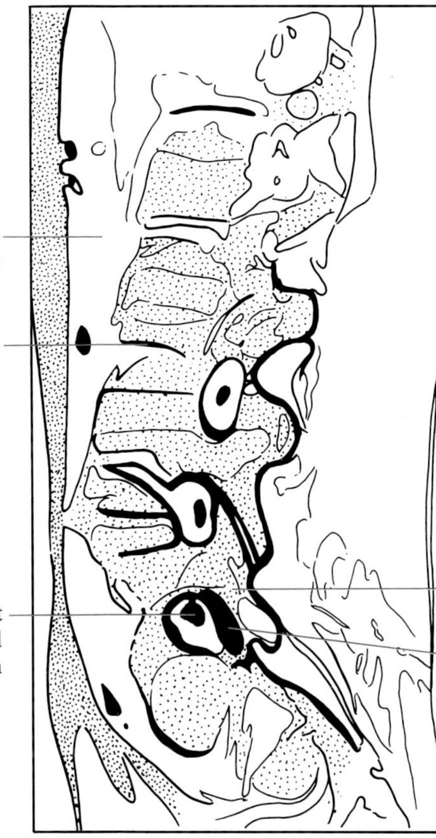

each pars interarticularis. **B** Parasagittal section [SE 800/20]. This section is also T1WI, but is made slightly more laterally in the neural foramina. The nerve roots are completely surrounded by high SI epidural fat in rather spacious exit foramen and are centered at the level of the disc. Lumbar arteries are commonly seen in the paraspinal fat, forming a ring around the vertebral bodies.

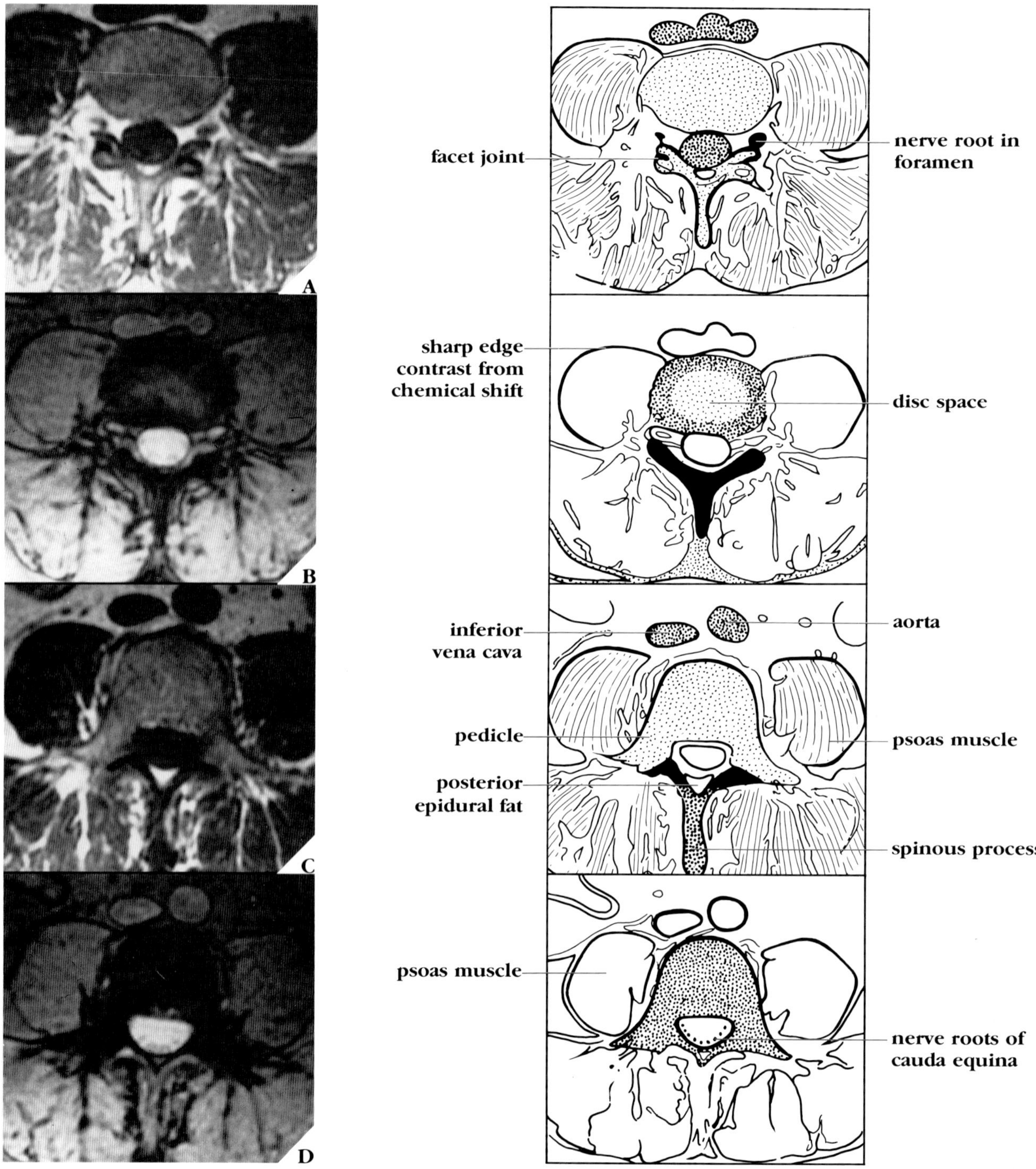

FIGURE 9.6. Normal Lumbar Spine. A Axial section [SE 800/20]. The nerve roots are surrounded by high SI fat in the neural foramen on this T1WI. The ventral margin of the low SI thecal sac at the disc level is convex outward and the canal is ample in size. The facet joints are well seen as the two low SI arcs of the cortical bone, which face the articular surfaces. **B** Axial section [GRASS 1000/12 flip angle 22.5 degrees]. This PDWI demonstrates low signal nerve roots of the cauda equina surrounded by high SI CSF. The anterior margin of the thecal sac is well-delineated. The individual nerve-root sheaths in the foramen also appear at a somewhat high SI. Some signal is visible from the disc interspace. **C** Axial section [SE 800/20]. This T1WI is made at the level of the pedicles, which are of normal length and flare slightly outward. A triangular region of high SI fat is seen in the posterior epidural space. **D** Axial section [GRASS 1000/12 flip angle 22.5 degrees]. This PDWI demonstrates the high SI CSF within the thecal sac. Note the low SI chemical shift artifact at the interface between the fat in the retroperitoneum and the water of the psoas muscles. This is not dependent on the phase or frequency axes. This artifact acts like edge enhancement.

NORMAL VARIANTS

Disc Interspaces

Children have a very high signal from the disc interspaces on PDWI and T2WI because their water content is very high (Fig. 9.2). As patients age, there is a normal desiccation of the discs. By the third decade, many of the discs begin to lose signal, height, and water. By the sixth to seventh decade, most of the discs are dry and have low signal. Thus, due to these normal changes, it is very important to know the patient's age before interpretation.

Bone Marrow

The distribution of the fat-containing marrow is different in children than in adults (27). The amount of fatty marrow increases with age. In very small children, the marrow spaces of the spine contain less fat so the signal on T1WI is not as high as that of adults. In adults this finding could suggest that there was replacement of the fatty marrow, but the appearance is normal in children and should not be confused with pathology. Slight heterogeneity of the marrow spaces is not uncommon and is, in fact, particularly common in patients who are on steroids or who have renal failure. Fatty marrow inclusions are common in elderly patients and appear as regions of high signal on T1WI. Fatty marrow inclusions are relatively low signal on PDWI and T2WI compared with CSF. After radiation therapy, the fatty marrow replaces the hematopoietic water component. This leads to a high signal from post-radiated bone on T1WI (20).

Vertebral Hemangioma

Incidental vertebral hemangiomas (Fig. 9.7) may be encountered on MRI (23). Characteristically, hemangiomas appear on T1WI as regions of bright signal fatty infiltration of a vertebral body. Many

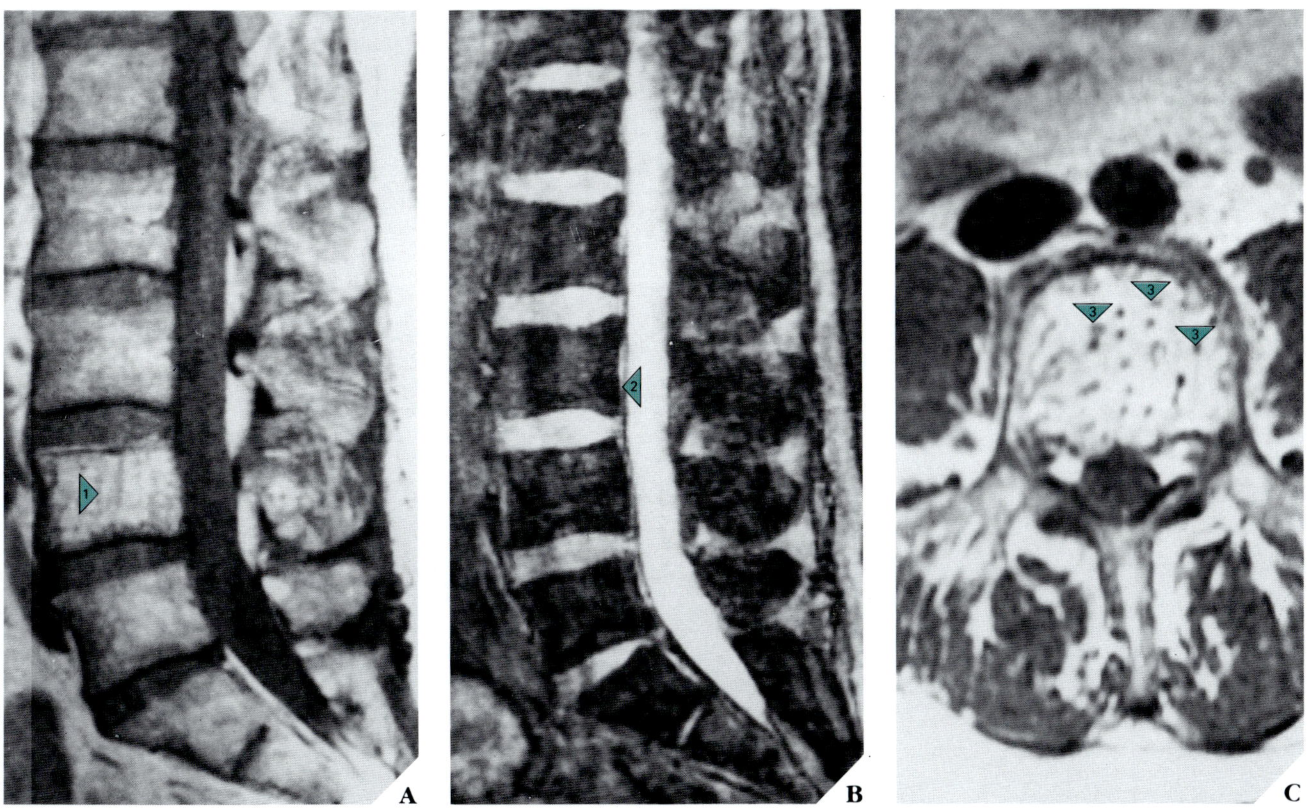

FIGURE 9.7. Lumbar Vertebral Hemangioma. A Sagittal section [SE 800/20]. An incidental fourth lumbar vertebral body hemangioma is seen on this T1WI. The hemangioma has a slightly higher SI than the vertebral bodies due to its high fat content. Note the low SI vertical striations of dense bone struts in the vertebra. The vertebral body is not compressed or destroyed. **B** Sagittal section [GRASS 1000/12, flip angle 22.5 degrees]. On this PDWI the hemangioma is not seen. There is no expansion or distortion of this normal-appearing vertebral body. Also note the well-delineated posterior longitudinal ligament. **C** Axial section [SE 800/20]. On this T1WI, the dense struts of the hemangioma generate no signal and appear as small dots within the higher fat signal of the vertebral hemangioma. Arrow **1** Hemangioma. Arrow **2** Posterior longitudinal ligament. Arrows **3** Bone struts.

small low signal foci are evident due to dense cortical bone. There is no bony destruction or expansion.

Perineural Cysts

Tarlov or perineural subarachnoid cysts are simple regions of focal dilatation of the axillary sheaths of the nerve roots in or near the neural foramen (Fig. 9.8). They may become quite large (> 1.5 cm) and mimic tumors, with enlargement of their respective foramina. Preservation of adjacent epidural fat, multiple cysts, and lack of symptoms are common findings for Tarlov cysts. Differentiation of a Tarlov cyst from a tumor, particularly a cystic neurofibroma, may be impossible on MRI and myelography may thus be required.

Conjoint Nerve Root

Conjoint nerve roots, which occur when two roots exit through one enlarged root sheath, are common normal variations (Fig. 9.9). They are by far most common in the lumbar region. Characteristically, the dural sac and axillary root sheaths are bulbous at the lateral recess. The visualization of double roots within the sheath and the preservation of the epidural fat are the most helpful findings to exclude the presence of disc protrusion.

Residual Pantopaque

Since Pantopaque (iophendylate, myelogram contrast) has a fatty, oil base component, it has an SI similar to subcutaneous fat, with a high signal on T1WI (Fig. 9.10) (13). The distribution is characteristic as well, with visualization of a fluid–fluid level or multiple small droplets. Chemical shift artifacts can help confirm the presence of fatty Pantopaque in the dural sac. Trapped Pantopaque which does not fall to the dependent portion of the canal suggests arachnoiditis.

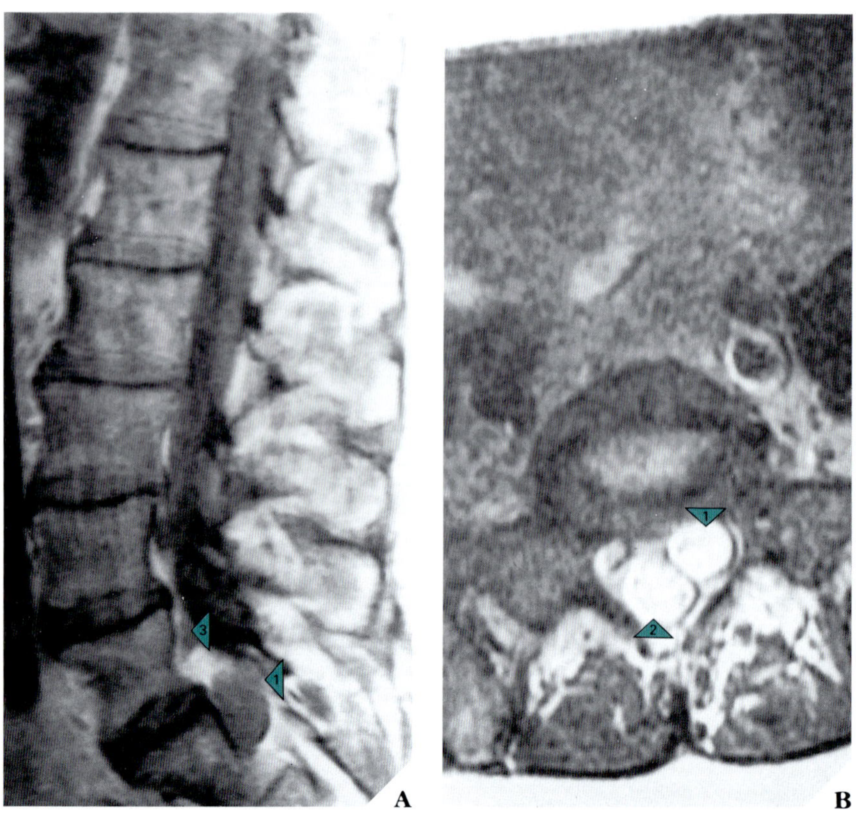

FIGURE 9.8. Sacral Tarlov Root Cyst. **A** Parasagittal [SE 1500/20]. This T1WI of the lumbar spine shows a large Tarlov cyst of CSF signal surrounded by epidural fat. The nerve root directly superior is normal in size. **B** Axial section [SE 1500/80]. On this more T2-weighted image, the perineural Tarlov cyst is of high water signal similar to that of the caudal thecal sac. Also note that the chemical shift artifacts in the left-to-right frequency-encoded direction produce asymmetric signal margins. Arrows **1** Tarlov cyst. Arrow **2** Dural sac. Arrow **3** Normal nerve root.

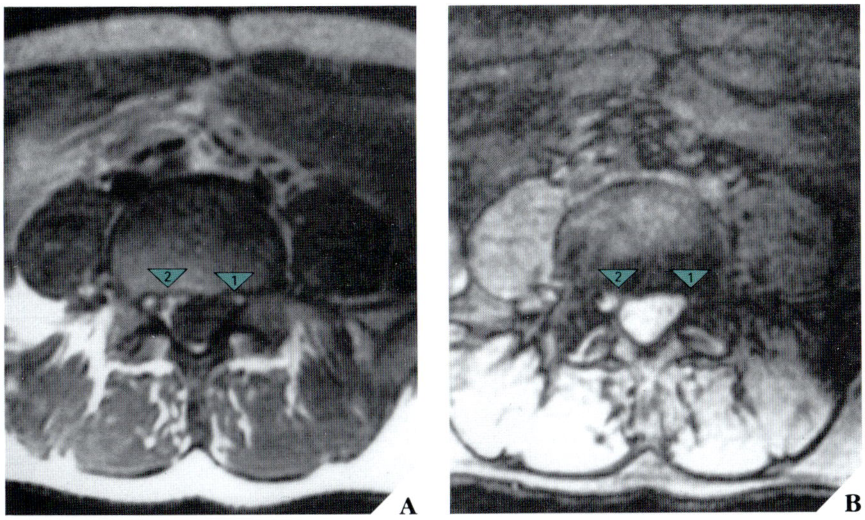

FIGURE 9.9. Lumbar Conjoint Nerve Root Sheath. A Axial section [SE 800/20]. A conjoint nerve-root sheath is seen (on the left) on this T1WI. The conjoint root sheath is bulbous in comparison to the normal sheath (on the right). The absence of fat surrounding the nerve root on the left should not be confused with a disc herniation. **B** Axial section [GRASS 1000/12, flip angle 22.5 degrees]. This PDWI confirms the CSF nature of the asymmetric dural sac and nerve-root sheaths, secondary to a conjoint root (on the left). Arrows **1** Conjoint nerve root. Arrows **2** Normal nerve root.

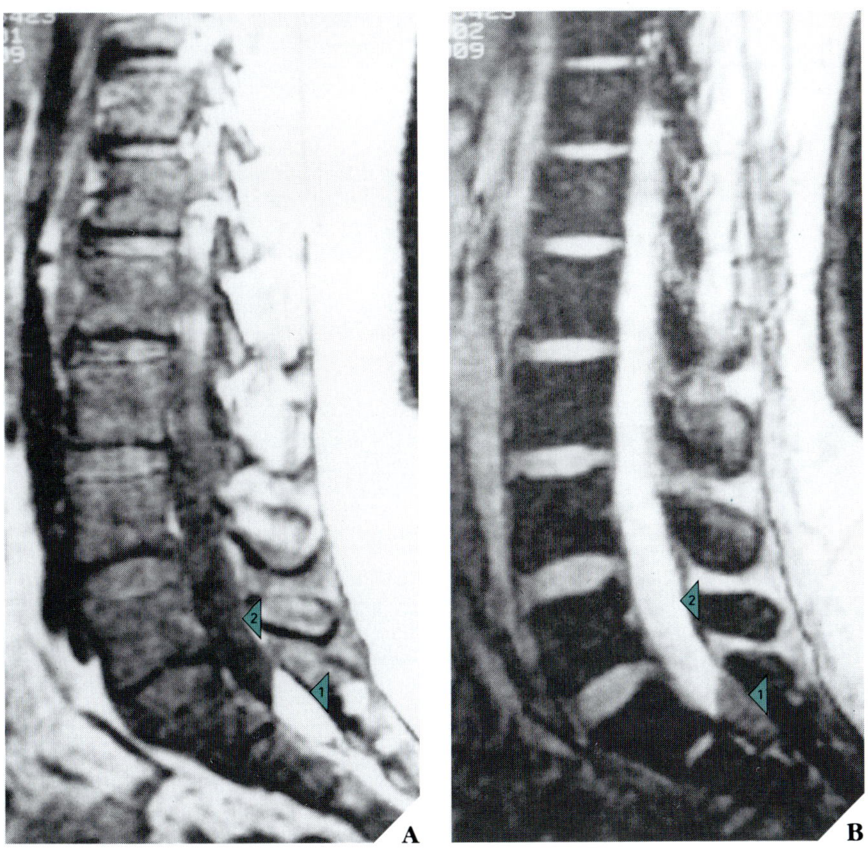

FIGURE 9.10. Pantopaque in Lumbar Spine. A Sagittal section [SE 800/20]. In this supine patient, Pantopaque (iophendylate) myelographic contrast layers are seen in the caudal sac. The fatty oil-based contrast has a high SI similar to that of the subcutaneous fat on this T1WI. The CSF is dark. Notice that a chemical shift artifact is present, with a low SI superior margin as seen in the vertebral bodies; this finding confirms that this is not hemorrhage but fat. **B** Sagittal section [GRASS 1000/12, flip angle 22.5 degrees]. On this PDWI the CSF is high SI whereas the oil-based contrast is of lower SI, similar to that of the retroperitoneal fat. The signal of the subcutaneous fat is higher than the Pantopaque because of its close proximity to the coil. Arrows **1** Pantopaque. Arrows **2** CSF.

CONGENITAL AND DEVELOPMENTAL ANOMALIES

MRI excels in the evaluation of congenital and developmental anomalies of the spine since all of the components of the spine are shown simultaneously. The MRI evaluation is mainly an assessment of neuroanatomic development and therefore only T1WI are usually performed.

Butterfly and Hemivertebrae

Butterfly and hemivertebrae are anomalous triangle-shaped vertebrae caused by congenital segmentation aberrations (Fig. 9.11A). Butterfly and hemivertebrae can be seen as isolated incidental anomalies of little clinical significance, but are more commonly associated with other findings including dysraphic defects, scoliosis, kyphosis, diastematomyelia, dysplastic fused ribs, tethered cord, or a neuroenteric cyst.

Caudal Regression, Sacral Agenesis

Caudal regression or sacral agenesis is the absence or hypoplasia of the sacrum and lower lumbar spine (Fig. 9.11B). This anomaly is usually quite serious and may be accompanied by a host of other organ anomalies, including rectum and bladder deformities.

Block Vertebrae

Block vertebrae are two or more fused vertebral

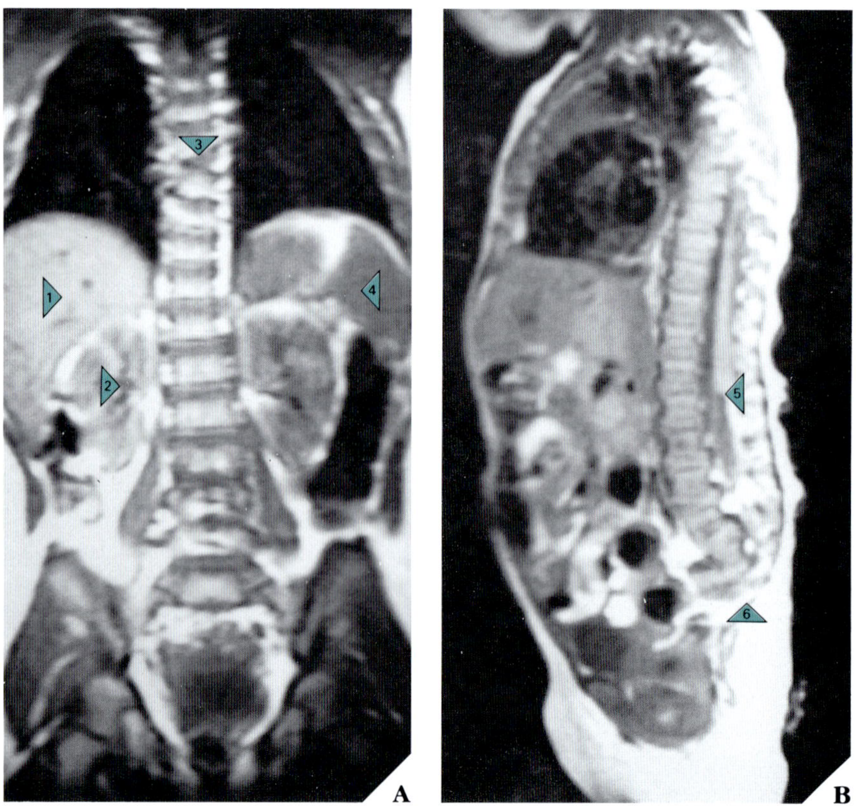

FIGURE 9.11. A Thoracic Hemivertebra. Coronal section [SE 800/20]. A defect in the right cartilaginous growth of a midthoracic vertebra has resulted in a triangle-shaped hemivertebra in this young child. On this T1WI the kidneys and other retroperitoneal structures can also be evaluated for associated anomalies and are outlined by high SI fat. **B Sacral Agenesis.** Sagittal section [SE 800/20]. In another infant, the sacrum has failed to develop and ends abruptly on this T1WI. There is some widening of the caudal thecal sac and a tethered cord to the low lumbar segment. Arrow **1** Liver. Arrow **2** Kidney. Arrow **3** Hemivertebra. Arrow **4** Spleen. Arrow **5** Tethered cord. Arrow **6** Absent sacral segment.

bodies without interspaces. The posterior elements may also be fused but the neural foramina usually remain patent. If the interspace is narrowed in the anterior to posterior dimension, the fusion is likely to be congenital or to have occurred early in development (as in juvenile rheumatoid arthritis). Block vertebrae acquired from surgery, trauma, and infection usually will not have all of these characteristics. Many block vertebrae are incidental. They can however be associated with other anomalies, including Arnold–Chiari malformation, Klippel–Feil syndrome, and Sprengel's deformity.

Anomalies of the Odontoid

Anomalies of the odontoid include agenesis, hypoplasia (Fig. 9.12), ligamentous laxity, or os odontoideum. These anomalies are relatively common. The most significant aspect of these anomalies is cervical instability, with canal encroachment on the upper spinal cord with flexion. These anomalies are common in Morquio's syndrome and Trisomy 21. Ligamentous laxity is particularly common in Down's syndrome. Os odontoideum is a congenital or early acquired abnormality in which the odontoid does not fuse with the second vertebral body. Confusion with an acute or non-union odontoid fracture is common. Enlargement of the os is common and may also lead to spinal stenosis. Visualization of a sharp cortical margin around the os and the second vertebral body suggests that the separation is chronic and is not secondary to acute trauma.

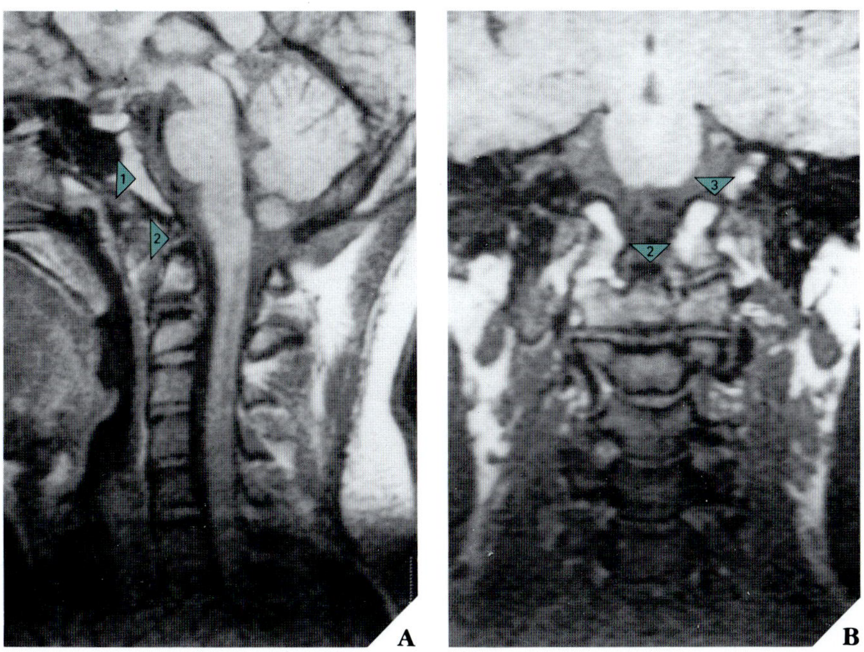

FIGURE 9.12. Hypoplastic Odontoid. **A** Sagittal section [SE 800/20]. The superior growth center of the odontoid has failed to develop and only a hypoplastic dens arises from a normal second vertebral body. The anterior arch of the first cervical vertebra is not visualized because of fusion to the occiput. Note that there is no significant spinal stenosis. **B** Coronal section [SE 800/20]. The second cervical vertebral body is normal but only a rudimentary odontoid process has formed. The atlas has fused with the occiput so that there are no occipital condyles. Arrow **1** Clivus. Arrows **2** Hypoplastic odontoid. Arrow **3** Fused lateral masses of C1 to occipital condyle.

Pars Interarticularis Defects

One of the more common abnormalities of the lumbar spine is a pars interarticularis defect (Fig. 9.13). These are not true congenital abnormalities, since they are not present at birth and may be stress fractures. Spondylolytic defects of the posterior elements between the superior and inferior facets may be associated with spondylolisthesis (anterior subluxation). However, these defects are commonly associated with disc degeneration and back pain, which may mimic disc protrusions. CT and plain radiographs are more accurate than MRI in defining the pars defect and possible bony sclerosis, but MRI better demonstrates many other aspects, including disc degeneration, foraminal distortion, and possible spinal stenosis or enlargement.

Idiopathic Scoliosis

Idiopathic scoliosis is commonly seen in young females. There is no underlying, causal neurologic or congenital disorder such as is present with neurofibromatosis or a butterfly vertebrae. Classically,

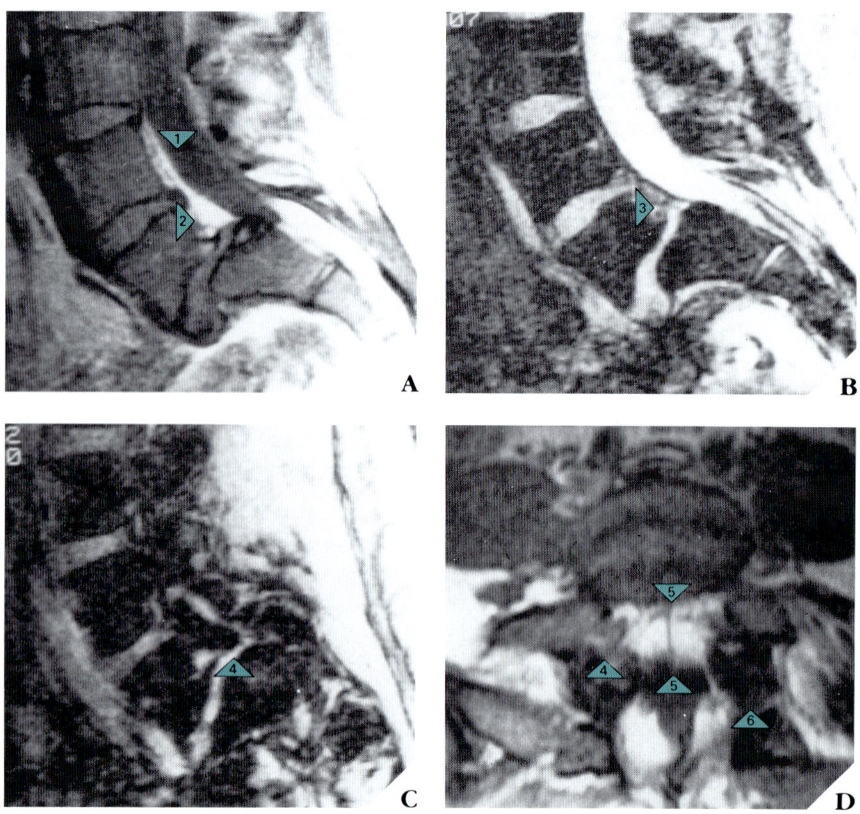

FIGURE 9.13. Lumbosacral Spondylolisthesis. A Sagittal section [SE 800/20]. This T1WI demonstrates anterior subluxation of the fifth lumbar vertebrae on the first sacral body. The thecal sac remains unaffected. An expanded anterior epidural fat space fills the space between the anterior margin of the thecal sac and the posterior margin of the vertebral body. Due to subluxation, there are portions of the disc which are not covered by the endplates, causing the illusion of a disc protrusion. **B** Sagittal section [GRASS 1000/12, flip angle 22.5 degrees]. On this PDWI, the high SI thecal sac remains unaffected by the spondylolisthesis. The high signal from the intervertebral discs indicates no significant degeneration of the disc, despite some narrowing at the lumbosacral junction. **C** Parasagittal section [GRASS 1000/12, flip angle 22,5 degrees]. The pars interarticularis defect is identified as a high SI cleft between segments of the facet articular processes. This PDWI was made off-midline at the level of the facets. This is the etiology for the spondylolisthesis. **D** Axial section [SE 800/20]. This T1WI demonstrates the defects in the pars interarticularis as low SI sclerotic regions anterior to the facet joints. Elongation of the anterior-to-posterior dimension of the spinal canal has occurred because the vertebral body has slid forward and the posterior elements have remained dorsal. Arrow **1** Thecal sac. Arrow **2** Enlarged epidural space. Arrow **3** Spondylolisthesis. Arrows **4** Pars interarticularis defect. Arrows **5** Widened AP dimension of canal. Arrow **6** Facet joint.

the patients have sigmoid scoliosis of the thoracic spine with a correcting curve so that the alignment of the head is not affected. MRI can help exclude the presence some unexpected disorders such as syrinx, but in general MRI has not made a significant impact on the evaluation of this disorder.

Spondylolisthesis

Spondylolisthesis (Fig. 9.13) occurs when a superior vertebral body is shifted anteriorly or posteriorly (reverse spondylolisthesis) in relation to the adjacent inferior body. This subluxation can be secondary to acquired causes such as pars interarticularis defects, trauma, fractures, destructive tumors, or severe degeneration of the facets or discs. The congenital cases are frequently quite severe with marked subluxation. These congenital cases occur most commonly at the lumbosacral junction. Absent or dysplastic facets allow severe anterior subluxation, to the point that the lumbar vertebrae slide into the pelvis.

In patients with bilateral pars interarticularis defects and spondylolisthesis, there will be elongation of the anterior to posterior dimension of the spinal canal. This occurs because the posterior elements remain fixed while the vertebral body slides forward and opens the canal. Spondylolisthesis secondary to degeneration of the discs and the facets results in spinal stenosis. In this case the posterior elements slide forward with the vertebral body and the hypertrophic facet changes. This results in dorsal as well as ventral narrowing of the spinal canal.

Facet Anomalies

Significant isolated facet anomalies are uncommon. There are usually multiple associated findings including unilateral hypoplasia of one facet and pedicle. The contralateral facet may show rotation, hypertrophy, and deformity. Those patients with severe vertebral body anomalies frequently have associated posterior element anomalies as well.

Spinal Dysraphism

Spinal dysraphism is due to nonclosure of the primary neural tube and most often affects the posterior elements of the spine (2). The exact etiology for this failure of closure is not known. MRI, US, and serum testing for elevated fetal alpha-protein in the mother can be used as prenatal screening exams. At this time, the FDA has not approved MRI as a routine exam during pregnancy. The mildest form of spinal dysraphism is an incidental common variation in the lumbosacral region occurring when there is spina bifida without any other abnormality.

Meningoceles are extensions of the dural sac through a defect in the spinal canal (Fig. 9.14). They may be covered by skin, subcutaneous fat, or mus-

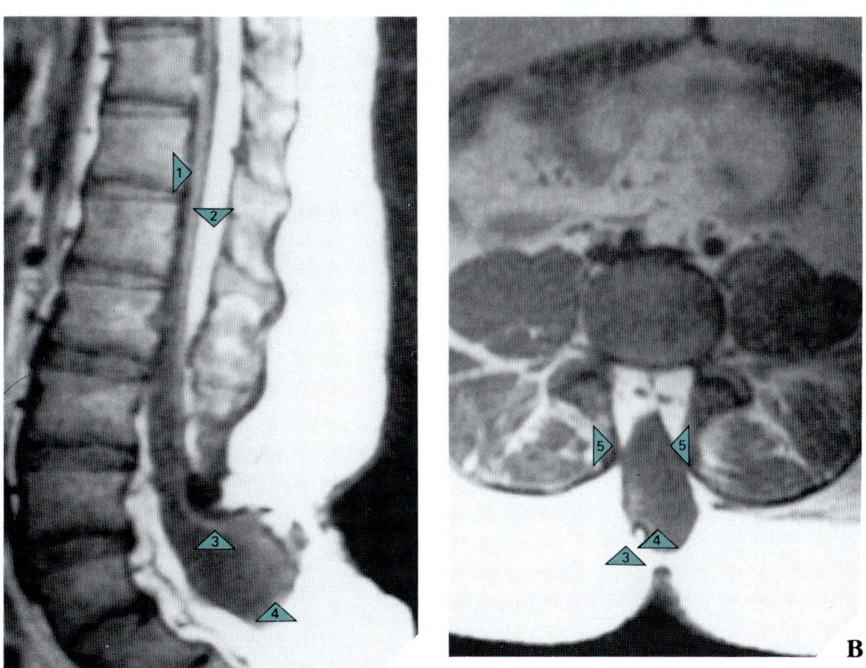

A B

FIGURE 9.14. Lumbosacral Spinal Dysraphism. A Sagittal section [SE 800/20]. On this T1WI from a pediatric patient, a meningomyelocele is seen with a large low SI CSF sac containing a thin segment of tethered cord, which extends into the posterior spine. The dysraphic segment, or posterior element defect, extends approximately from the fourth lumbar vertebrae through the sacrum. The widened lumbar canal and the large amount of epidural fat in the posterior thoracic canal are also part of this congenital anomaly. Note that the lower thoracic and lumbar spinal cord is dysplastic and small. **B** Axial section [SE 800/20]. The defect in the posterior elements is well-identified in the axial plane. The spinal canal is large with anterior epidural fat. The meningocele extends posteriorly. A small lipoma in the tethered cord appears as a small high SI region within the posterior aspect of the sac. Arrow **1** Thin dysplastic cord. Arrow **2** Intracanicular lipoma. Arrows **3** Tethered cord to dysraphic segment. Arrows **4** Meningomyelocele. Arrows **5** Spinal dysraphism.

cle. If there is no covering, rupture and secondary infection are more common. They may vary in size from a few millimeters to huge masses which can be almost as large as the child. If central neural elements such as the spinal cord enter the sac, then it is called a meningomyelocele. Most meningomyeloceles are in the lumbosacral region (Fig. 9.14). The spinal canal is usually enlarged, with significant spina bifida. The vertebral bodies may be dysplastic. The posterior elements may be absent, hypoplastic, or anomalous. In the vast majority of meningoceles the dural sac extends through a posterior defect, but one must be alert to the possibility of an anterior meningomyelocele. Intradural and extradural lipomas are common. The lipoma is normal fat in an unusual distribution. When found inside the spinal canal it can act as a mass and lead to cord compression. This is, however, uncommon. There is a high association between syrinx and tethered cord.

Diastematomyelia

Diastematomyelia is a congenital, longitudinal fissure of the spinal cord (Fig. 9.15) (10). Usually the cord is divided into two symmetric segments with a normal number of roots. Diplomyelia is true duplication of the spinal cord and is much less common. Diastematomyelia is usually associated with some anomaly of the bony spine including butterfly vertebrae, hemivertebrae, dysraphism or kyphoscoliosis. There is an anomalous bony, cartilaginous, or fibrous ridge dividing both the spinal canal and cord. Diastematomyelia is commonly seen in patients with meningomyeloceles and in such cases the cord may also be tethered.

Neuroenteric Cysts

Neuroenteric cysts are developmental abnormalities of the notochord with a persistent connection or fibrous band from the gut and the dorsal skin (8). They are much less common than meningo-

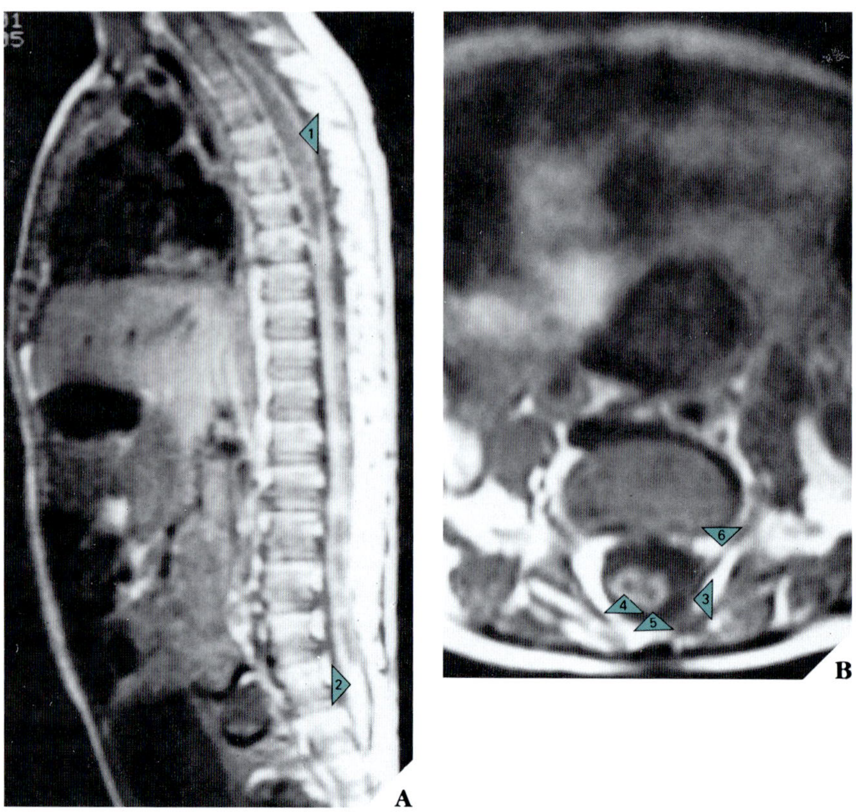

FIGURE 9.15. Thoracic Syrinx and Lumbar Diastematomyelia. A Sagittal section [SE 800/20]. In this young patient, the spinal cord is tethered to the low sacrum. There is also an upper thoracic cord syrinx seen as a low SI channel within the expanded spinal cord. **B** Axial section (SE 800/20). The split cord, diastematomyelia, with a small ventral cleft is well seen on this axial T1WI. The fibrosis division of the cord is not well seen as a separate entity. Each portion of the split cord also contains a small low SI syrinx cavity. Because the cord is tethered, individual dorsal and ventral roots exit laterally rather than descending as normal. The spinal canal is enlarged. Also, incomplete fusion of the posterior elements suggests a mild dysraphic defect. Arrow **1** Syrinx. Arrow **2** Tethered cord. Arrow **3** Dysraphism. Arrow **4** Syrinx cavities. Arrow **5** Diastematomyelia. Arrow **6** Horizontal orientation of nerve roots.

celes. These complex deformities are seen most commonly in the thoracic region and involve the anterior portion of the vertebral bodies. The spinal cord may be tractioned forward into an enlarged defect of the posterior portion of the vertebral body. A fibrous or CSF tract through the vertebrae into the paraspinal space is common. An isolated intradural cyst may be seen but is uncommon. An associated paraspinal cyst or anterior meningocele may mimic a lung mass on chest films. The trachea, esophagus, and other viscera may have deformities as well, and therefore extensive gastrointestinal, respiratory, and neural evaluation is important.

Developmental Spinal Stenosis

Many patients have developmentally narrow or small spinal canals (Fig. 9.16). Most cases of developmental spinal stenosis have no distinct underlying osseous syndrome. They are commonly seen in large powerfully built patients. The spinal canal should be a minimum of 10 mm in anterior to posterior dimension, at essentially every level. Most patients can be diagnosed on plain radiographs when the anterior to posterior dimension (from the posterior margin of the vertebral body to the spinelaminar line) is small, but MRI more accurately stages the severity of the disorder. With spinal stenosis of the lumbar region, the interpedicular distance may cause progressive narrowing from superior to inferior in contrast to the normal widening. Many patients with developmental spinal stenosis develop symptoms at a relatively young age. The development of minor hypertrophic degenerative changes may lead to neural compression. These patients are also more prone to significant neurologic injury following minor trauma. It is important to obtain PDWI or T2WI because the degree of stenosis may be underestimated by T1WI. This occurs because the cortical bone, ligaments, and CSF may all have low SI.

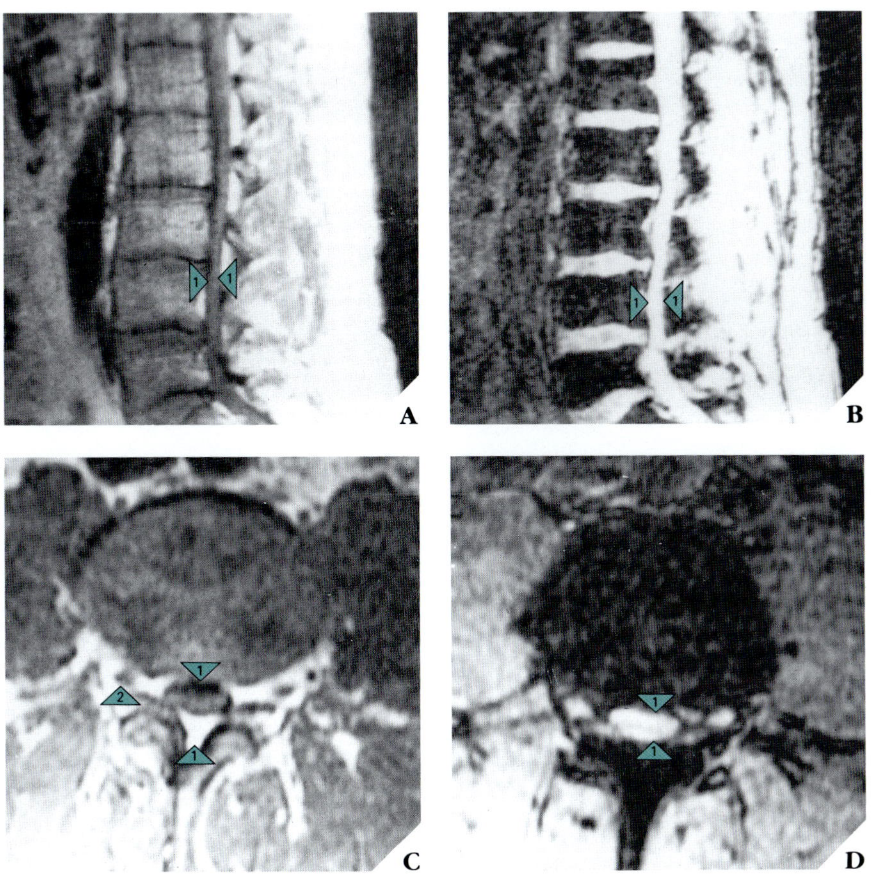

FIGURE 9.16. Developmental Lumbar Spinal Stenosis. A Sagittal section [SE 800/20]. On this T1WI the lumbar canal is very narrow congenitally and only measures 5-mm, in anterior-to-posterior dimension. There is little in the way of bulging annuli. The thecal sac is not as low SI as normal because the canal is crowded by the nerve roots of the cauda equina, with only a small amount of CSF present. **B** Sagittal section [GRASS 1000/20, flip angle 22.5 degrees]. The disc spaces are well maintained with only minor disc bulges. The diffuse narrowing of the spinal canal is developmental in etiology. However, minor degenerative changes can have a significant effect. **C** Axial section [SE 800/20]. The thecal sac is flattened and very narrow in the anterior is posterior dimension. The posterior portion of the dural sac is packed with nerve roots. The facet joints maintain a more coronal orientation due to short, broad pedicles. **D** Axial section [GRASS 1000/12, flip angle 22.5 degrees]. The small, flat spinal canal is confirmed on this PDWI. Arrows **1** Small canal. Arrow **2** Nerve root.

Achondroplasia

Achondroplasia (Fig. 9.17) is one hereditary etiology for developmental spinal stenosis. Patients with this disorder have a multitude of bony findings, only one of which is related to the spine (28). Severe spinal canal stenosis is very serious since many patients are disabled not only by pain, but also by severe neurologic compromise. Myelography is frequently difficult due to the size of the canal. Thus, MRI offers a true advantage. The vertebral bodies are frequently dysmorphic in shape and acquired degenerative changes are common. These patients demonstrate very narrow spinal canals at any or all levels including the foramen magnum. The spinal canal is small due to the short, thick pedicles and a narrow interpedicular distance, which leads to spinal cord and root compression. There may be associated hypertrophic degenerative changes that accentuate the stenosis.

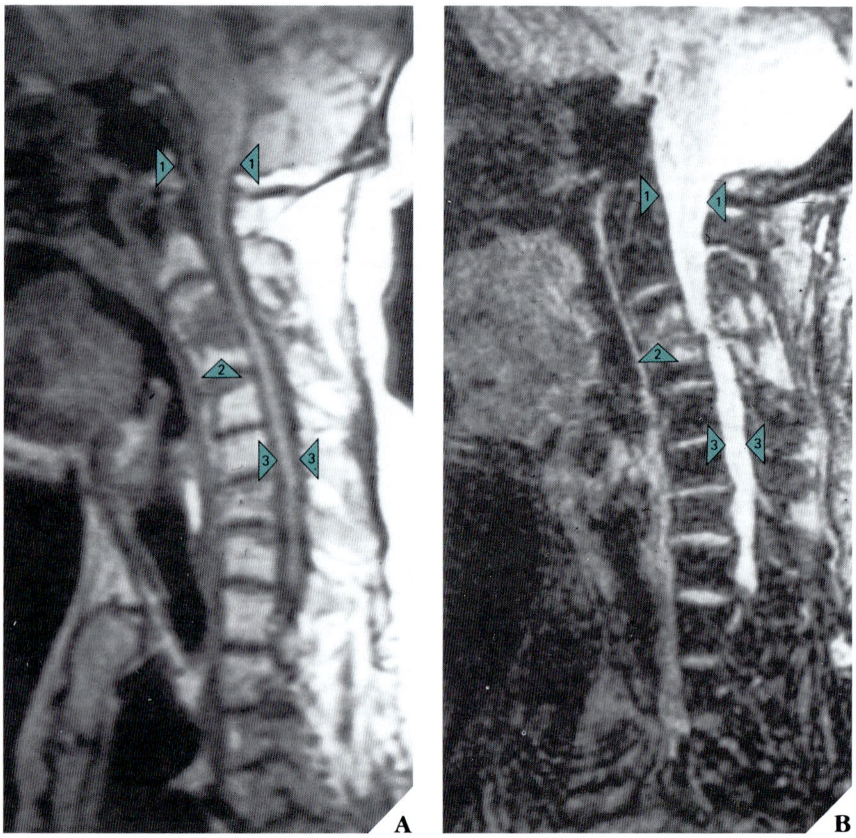

FIGURE 9.17. Achondroplasia. A Sagittal section [SE 800/20]. On this T1WI there is crowding of the craniocervical junction by a congenitally small foramen magnum. The thecal sac is narrowed, particularly at the third–fourth cervical level, where there is anterior subluxation. The low signal changes in the adjacent endplate are secondary to replacement of the normal fatty bone marrow by reactive changes. **B** Sagittal section [GRASS 1000/12, flip angle 22.5 degrees]. There is reversal of the normal lordotic curve. At the narrowed third–fourth cervical interspace there is anterior subluxation and replacement of the normal marrow spaces, secondary to reactive endplate changes. On this PDWI the anterior to posterior spinal canal dimension is very small. The spinal cord is narrowed and there is crowding of the craniocervical junction at a congenitally small foramen magnum. Arrows **1** Narrow foramen magnum. Arrows **2** Reactive changes of endplates. Arrows **3** Congenital canal stenosis.

DEGENERATIVE DISEASE

Degenerative joint disease of the spine is a form of noninflammatory osteoarthritis (9,15,16). It encompasses a wide range of hypertrophic changes which occur with normal aging. Their occurrence can be accelerated by nearly any insult to the spine such as trauma, arthritis, or infection. The specific findings of osteoarthritis will be reviewed separately though simultaneous multiple findings are usually present (Fig. 9.18).

Degenerative changes of the spine are the most common indication for evaluation of the spine by MRI. MRI has had a profound impact on the identification of the degree and type of degenerative findings. MRI frequently provides more than the combined information of myelography, CT, and discography. Overall, a significant cost savings can be incurred if MRI is used as the primary exam, at the exclusion of multiple other exams. The area in which MRI is limited is the evaluation of calcified, air containing, or densely fibrotic changes. On MRI these types of degenerative findings can be inferred from low SI regions; however, they are less well characterized than on CT. Degenerative changes have a propensity for certain levels such as the lower lumbosacral spine and the midcervical region. The individual changes, such as osteophyte development and disc protrusion, are essentially identical at all levels.

Disc Disease

Prior to MRI, discography was the best morphologic exam for the evaluation of degeneration of the disc interspaces. MRI has largely replaced the invasive and painful discogram. MRI allows for a direct evaluation of the internal morphology of the disc. There

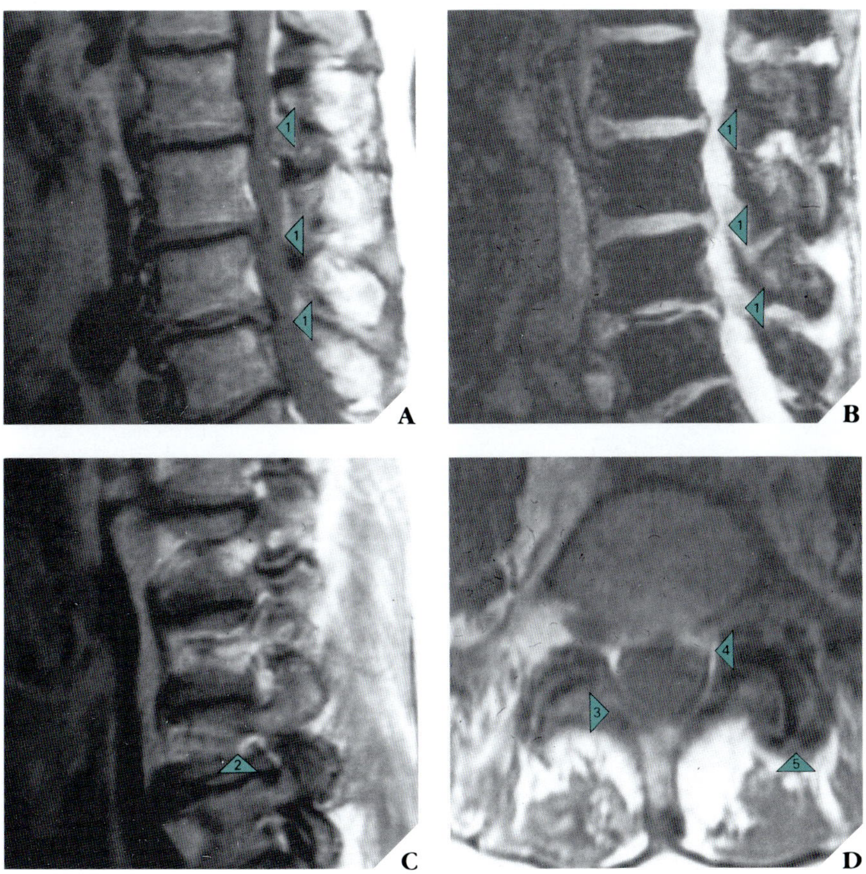

FIGURE 9.18. Acquired Lumbar Spinal Stenosis. A Sagittal section [SE 800/20]. On this T1WI, the intervertebral disc spaces are narrow and there are multilevel anterior and posterior bulging annuli. **B** Sagittal section [GRASS 1000/12, flip angle 22.5 degrees]. On this PDWI there are bulging annuli compressing the high SI thecal sac anteriorly. There are also dorsal impressions on the sac due to ligamentous hypertrophy. The combined hypertrophic changes lead to spinal stenosis. **C** Parasagittal section [SE 800/20]. The neural foramina are shorter in cephalocaudad dimension due to narrowing of the disc space and a narrowed anterior to posterior dimension, secondary to hypertrophy of the facet joints. **D** Axial section [SE 800/20]. This axial T1WI demonstrates hypertrophic degeneration of the facet joints with expansion and irregularity of the joints. There is less epidural fat in the lateral recesses and neural foramina, secondary to the hypertrophic changes. The ligamentum flavum, adjacent to the lamina, is only slightly thickened at this level. Arrows **1** Bulging annuli. Arrow **2** Lateral recess foraminal stenosis. Arrow **3** Ligamentum flavum hypertrophy. Arrow **4** Lateral recess narrowing. Arrow **5** Hypertrophic facet degeneration.

is a loss of signal on PDWI and T2WI from the discs associated with degeneration of the disc. Comparison of the relative SI between the discs and the CSF is important since all of the discs may be degenerated. The loss or absence of MRI signal from the interspace, and the loss of height are best seen on PDWI and T2WI. These findings correlate quite accurately with degeneration on discography (Fig. 9.19). There is a normal low SI central horizontal cleft within the disc, due to dense collagen fibers, which should not be confused with disease. Total loss of signal in an irregular pattern suggests either a vacuum disc phenomenon or dense calcification. These two cannot be differentiated on MRI.

Loss of disc height, vacuum phenomenon, and sclerotic, irregular endplates are indirect findings of degeneration on CT and plain films. Bulging or disc protrusion may or may not be present with severe disc degeneration.

A patient who has a single narrowed low SI degenerated disc may have no other radiographic findings. This limited finding may account for all of the classic findings of sciatica or disc protrusion short of muscle weakness. Thus, recognition of a

FIGURE 9.19. Lumbar Disc Degeneration. A Sagittal section [SE 1500/20]. There are mild bulging annuli at each lumbar level, with a small central disc protrusion at the lumbosacral junction interspace. The disc protrusion has a fusiform configuration of the disc material, with a focal deformity. The disc space heights are well maintained. The SI from the lumbosacral disc is slightly lower than the other but it is not obvious. Note that the delineation of the CSF and bone marrow is not as clear on this longer TR SE image as on the more T1WI with a shorter TR. **B** Sagittal section [SE 1500/80]. Degeneration of the lumbosacral disc is more clearly demonstrated on this more T2WI. The low SI of the interspace is related to dessication of the disc, secondary to degeneration. Note that the small focal protruding disc is also of low SI, allowing it to stand out sharply against the high SI CSF. There are no reactive endplate changes at this level. Note the normal, low SI cleft within the upper lumbar discs. Arrow **1** Focal disc protusion. Arrow **2** Normal internuclear disc cleft. Arrow **3** Decreased signal from dry disc.

dry degenerated disc on MRI can be quite significant. There may be associated reactive changes of the endplates (Fig. 9.20). High or low signal changes on either the T1WI or PDWI and T2WI paralleling the endplates may occur. If the signal is always decreased there is sclerosis. There may be fatty infiltration of the marrow. There may also be replacement of the fatty marrow spaces by tissue, which would suggest inflammation. These findings may mimic those of septic discitis, except that with aseptic discitis the disc should be low SI due to degeneration (14).

Disc Bulges, Protrusion, and Extrusions

Disc bulges are very common and may be found with or without disc degeneration (Fig. 9.18). Bulging discs probably represent early disc-degenerative changes with lax annular ligaments. The findings may be quite impressive but of little clinical significance. The most characteristic finding of a disc bulge is that the disc extends in a symmetric fashion beyond the endplates in all directions. Frequently there is compression of the dural sac but not of the nerve roots (since the bulge is at the level

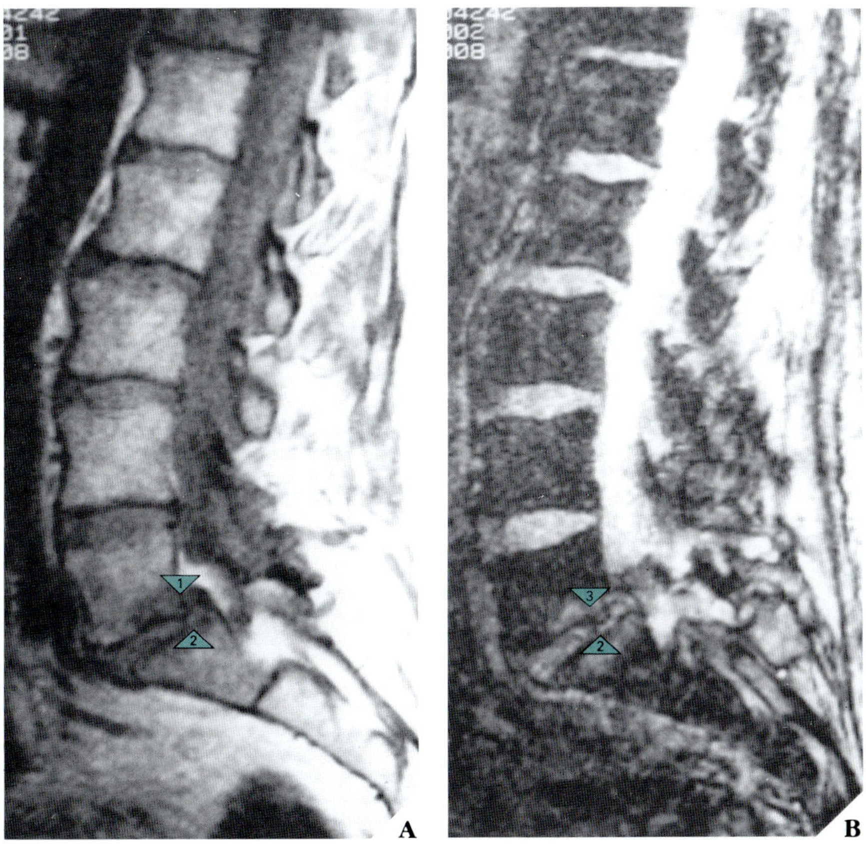

FIGURE 9.20. Lumbar Disc Degeneration and Reactive Changes of the Endplates. A Sagittal section [SE 800/20]. There is slight anterior spondylolisthesis of the fifth lumbar vertebral body on the first sacral body due to both degeneration of the intervertebral disc and to facet disease. The lumbosacral interspace is narrowed and demonstrates decreased signal intensity. The juxtaposed endplates show lower SI than the remaining fatty marrow spaces, suggesting replacement or bony sclerosis. **B** Sagittal section [GRASS 1000/12, flip angle 22.5 degrees]. The normal disc interspaces demonstrate high SI on this PDWI. The lumbosacral intervertebral disc is clearly lower in SI than the other discs, consistent with degeneration. The endplates have high SI due to increased water content from reactive changes. Arrow **1** Spondylolisthesis. Arrow **2** Endplate reactive changes. Arrow **3** Desiccated disc.

of the interspace where the roots have already exited the neural foramen). It is sometimes difficult to make the distinction between a large bulging disc and a central disc protrusion.

Disc herniations are usually asymmetric or focal protrusions of disc material beyond the limits of the endplate (Figs. 9.21 and 9.22). Their exact location denotes specific associated clinical symptoms and consequences. If the disc protrusion is far lateral, it may extend into the retroperitoneum and have no symptoms. If it occurs in the neural foramina, there may be root compression and obliteration of the normal epidural fat (Fig. 9.23). If the protrusion is centered at the lateral recess, there may be both root and dural sac deformity. Central disc protrusions may compress only the anterior dural sac. The SI of disc protrusions is quite variable. If the disc is not degenerated, the SI parallels the normal disc. Thus, it may be more visible on T1WI than on PDWI and T2WI (Fig. 9.24). If the disc material is dry with a low SI, it will be difficult to see on T1WI, but will be visible as a low signal defect on PDWI and T2WI (Fig. 9.25). Because of this signal variance, both pulse sequences are essential to diagnose a protrusion.

Disc extrusions are disc protrusions where the material migrates a significant distance from the interspace. Disc extrusions may mimic epidural tumors such as metastases. The findings of disc extrusions are similar to those seen with disc protrusions except for the unusual location of the epidural mass.

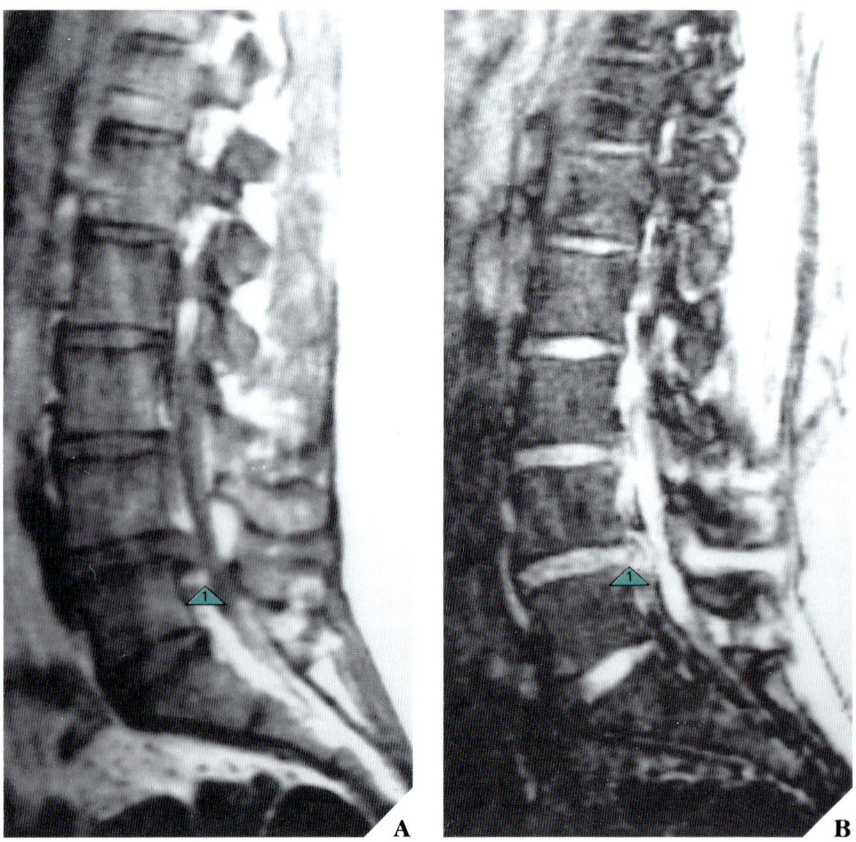

FIGURE 9.21. Lumbar Disc Herniation and Degeneration. A Parasagittal section [SE 1500/20]. On this T1WI, a large central focal disc herniation is seen at the fourth–fifth interspace. The herniation displaces the epidural fat and comes in contact with the thecal sac. **B** Parasagittal section [SE 1500/80]. The fourth–fifth interspace disc protrusion is not as well-identified on this PDWI because the protruding disc which is not degenerated has a high SI. A very thin line of dura can be seen delineating the posterior margin of the disc protrusion. Arrows **1** L4–5 disc protrusion.

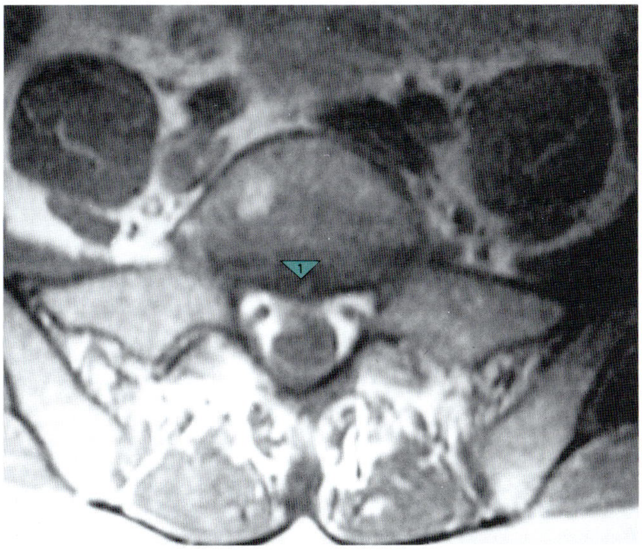

FIGURE 9.22. Central Lumbar Disc Protrusion. Axial section [SE 800/20]. A large central disc protrusion is seen on this T1WI obliterating the epidural fat, anterior to the dural sac. The SI of the disc is slightly different than that of the CSF. The dura is slightly compressed on the right. The fat about the nerve roots remains intact. Arrow **1** Central disc protrusion.

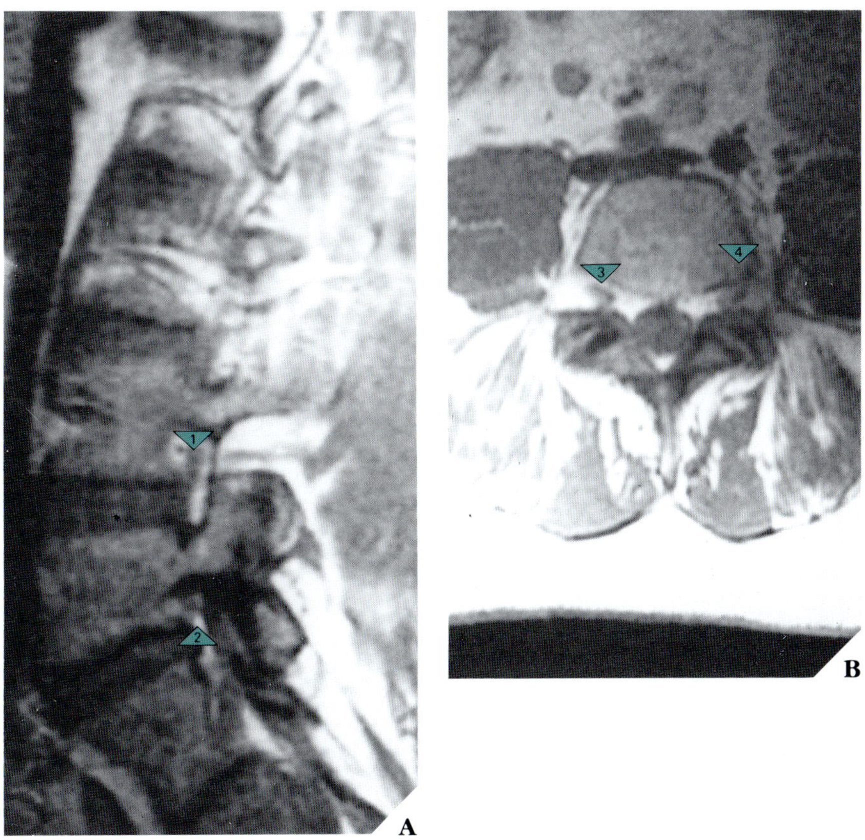

FIGURE 9.23 Lateral Lumbar Disc Herniation. A Parasagittal section [SE 800/20]. On this T1WI, the normal epidural fat is not seen in the fifth lumbar–first sacral neural foramen. This absence of normal fat is secondary to a lateral disc protrusion, which produces foraminal stenosis. At the fourth–fifth lumbar neural foramen a normal nerve root and a small spinal artery are surrounded by high SI fat, though the foramina is somewhat small. **B** Axial section [SE 800/20]. On the left there is obliteration of the fat in the neural-exit foramen due to a lateral disc protrusion. On the right, there is a normal amount of epidural fat surrounding the anterior and posterior margins of the nerve root in the right neural foramen. Arrow **1** Nerve root in neural foramen. Arrow **2** Obliteration of fat. Arrow **3** Nerve root surrounded by fat. Arrow **4** Lateral disc protrusion.

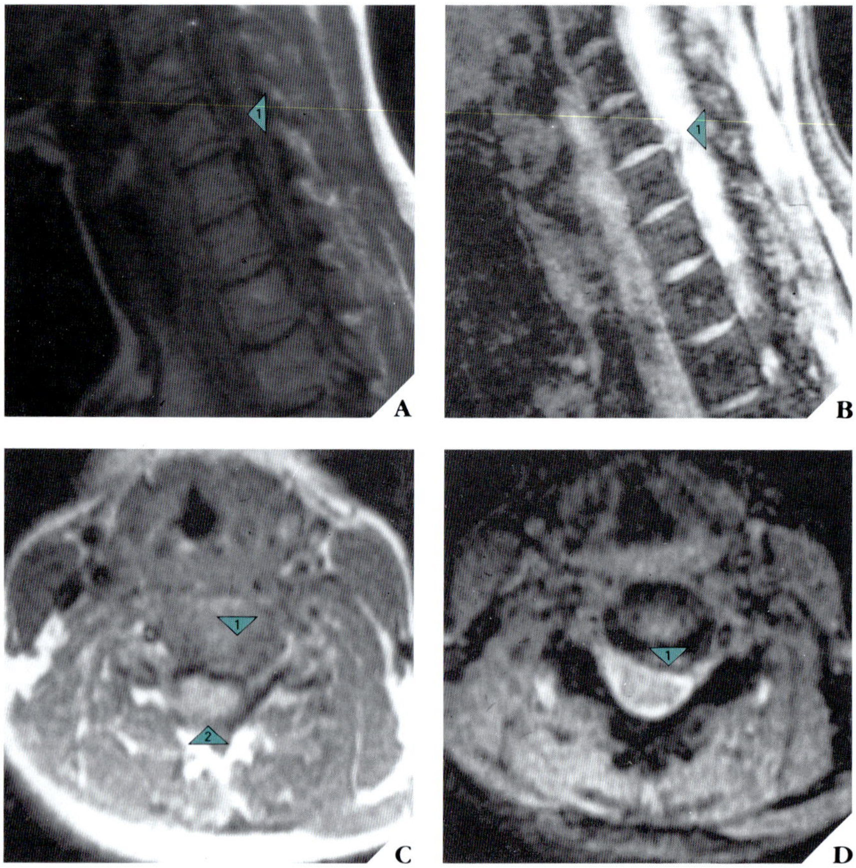

FIGURE 9.24. Cervical Disc Protrusion. A Sagittal section [SE 800/20]. This T1WI is a paramidline view demonstrating a small epidural mass which extends from the interspace, secondary to a lateral disc protrusion. The spinal cord is not well seen. Some straightening of the normal cervical lordosis is present but the intervertebral disc heights are well-maintained. **B** Sagittal section [GRASS 1000/12 flip angle 22.5 degrees]. On this PDWI the small, focal disc protrusion is visualized as a high SI region, elevating the low SI dura. **C** Axial section [SE 800/20]. The spinal cord is slightly rotated (clockwise) by the left lateral disc protrusion. The neural foramen is slightly smaller on the left as well. The disc protrusion is practically isointense with the spinal cord and is therefore not well seen. **D** Axial section [GRASS 1000/12, flip angle 22.5 degrees]. The left lateral disc protrusion is well seen since the protrusion still contains water. The disc posteriorly displaces the dura and the thecal sac. The thin dark band represents the dura. The disc still maintains some high SI due to its high water content. Arrows **1** Focal disc protrusion. Arrow **2** Rotated spinal cord.

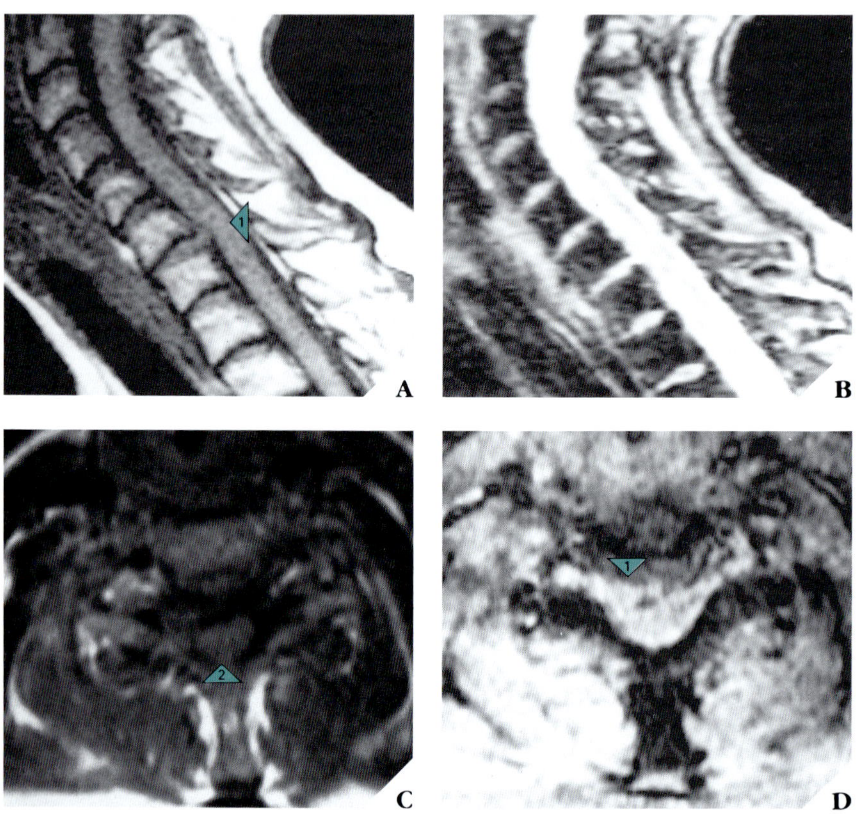

FIGURE 9.25. Cervical Disc Protrusion. A Sagittal section [SE 800/20]. On this T1WI a small focal, intermediate SI disc protrusion can be easily seen lifting the low SI dura. There is slight impression on the cervical cord. **B** Sagittal section [GRASS 1000/12, flip angle 22.5 degrees]. On this PDWI, the disc protrusion is invisible because the high SI disc protrusion is isointense with the CSF. To avoid such false-negative exams, multiple pulse sequences are essential. **C** Axial section [SE 800/20]. The small right lateral disc protrusion is seen as an intermediately intense mass that fills the neural foramen and slightly indents the cord. **D** Axial section [GRASS 1000/12, flip angle 22.5 degrees]. Once again, the disc protrusion is poorly seen on this PDWI. A small segment of the dura is displaced just to the right of midline. Arrows **1** Focal disc protrusion. Arrow **2** Slightly rotated cervical cord.

Scheuermann's Disease

Scheuermann's disease is an idiopathic disorder in children and young adults, in which there is a premature degeneration of the interspaces (Fig. 9.26). It may be related to abnormal development of the endplates. The findings mimic those of degenerative aging changes but occur in a young age group. Similar findings can be seen in children who have had spinal radiation. MRI findings include extensive multiple-level low signal changes of the degenerated discs, Schmorl node deformities, scoliosis, and kyphosis. Schmorl nodes are essentially disc protrusions into the endplates. Disc material is seen inside small sclerotic cavities adjacent to the endplates.

Facet Degeneration

The facets degenerate (Fig. 9.18) just as any other joint, with narrowing of the cartilaginous surfaces, sclerosis, hypertrophy, vacuum joint phenomenon, and irregularity. Most of these findings are visible on MRI. Encroachment on the lateral spinal canal and the lateral recess may produce neurologic compression.

Ligamentous Hypertrophy

The ligamentum flavum frequently enlarges with degenerative disease (Fig. 9.18). Hypertrophy may be seen in conjunction with many other degenerative findings; it is rarely seen in isolation. The ligaments, due to their fibrous nature, are seen as low signal regions bridging one lamina to the next. Thickening and calcification may lead to significant spinal stenosis. The first finding may be obliteration of the epidural fat in the posterior triangle of the spinal canal. Focal enlargement, with exostosis-like growths from the lamina, can be seen in the lower thoracic level causing posterior spinal cord compression.

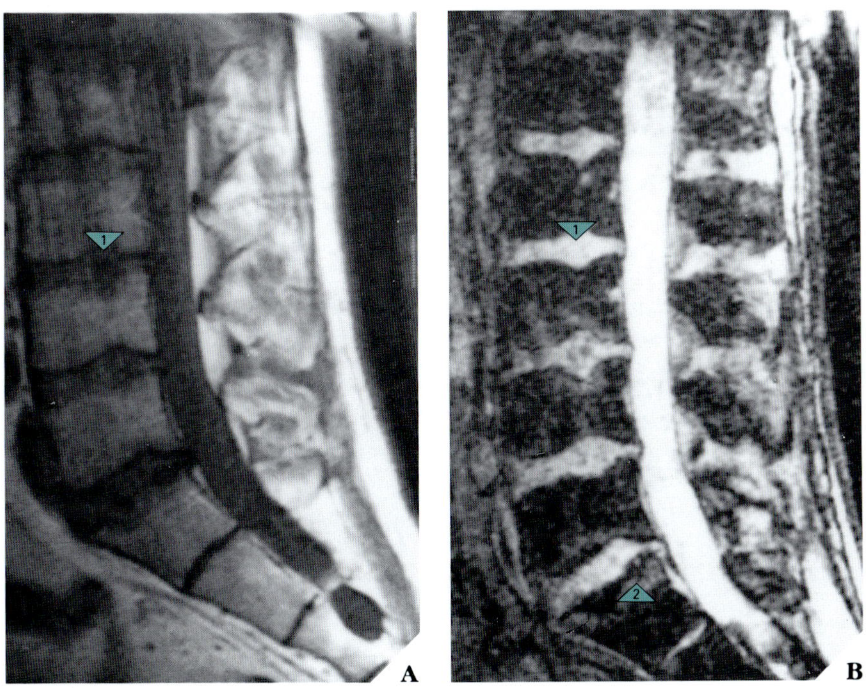

A B

FIGURE 9.26. Lumbar Scheuermann's Disease. A Sagittal section [SE 800/20]. On this T1WI of a teenager all of the disc interspaces are narrowed, irregular, and slightly bulging. Small protrusions of the discs into the endplates are related to small hernations of disc material called Schmorl nodes. **B** Sagittal section [GRASS 1000/12, flip angle 22.5 degrees]. On this PDWI, the discs demonstrate variable SI at each level. The lower the signal intensity the greater the disc degeneration, which is more severe than expected for a young patient. The endplate irregularities are due to Schmorl's nodes. The high SI changes of the superior endplate of the first sacral vertebral body, are secondary to reactive changes. Arrows **1** Schmorl's node. Arrow **2** Endplate reactive changes.

Foraminal Stenosis

Foraminal stenosis can be secondary to many anatomic changes, including facet or ligamentous hypertrophy, endplate osteophytes, disc protrusion, subluxation, and trauma (Fig. 9.18). Sagittal images are well suited to the identification of foraminal stenosis. The lumbar nerve roots should be visible in the superior portion of the foramen with the radicular vein in the inferior segment. Since the foramina in the cervical region are smaller than those in the lumbar region, axial images are generally better for nerve root evaluation.

Acquired Spinal Stenosis

A combination of many hypertrophic findings frequently leads to acquired spinal stenosis (Fig. 9.18). In this condition the anterior to posterior dimension of the spinal canal measures less than 10 mm. The shape of the canal is also frequently deformed and may be triangle- or kidney-bean-shaped in configuration, on axial images. A single finding may be present, such as a large disc protrusion. MRI can help identify which components are involved and to what degree.

Postoperative Findings

Postoperative evaluation of the spine is always difficult. Distinguishing the changes due to surgery from changes due to primary degeneration is very challenging. Most scars are seen as low signal regions on MRI. Postdiscectomy interspaces usually have low SI as well. Small bulging contours of the posterior dural sac, and pseudomeningoceles are common postoperative changes. Occasionally, large symptomatic postoperative pseudomeningoceles are encountered. Their configuration is different from congenital meningoceles since the large CSF space insinuates into the posterior muscle planes.

MRI holds promise for the differentiation of recurrent disc from scar. Scar tissue is very vascular

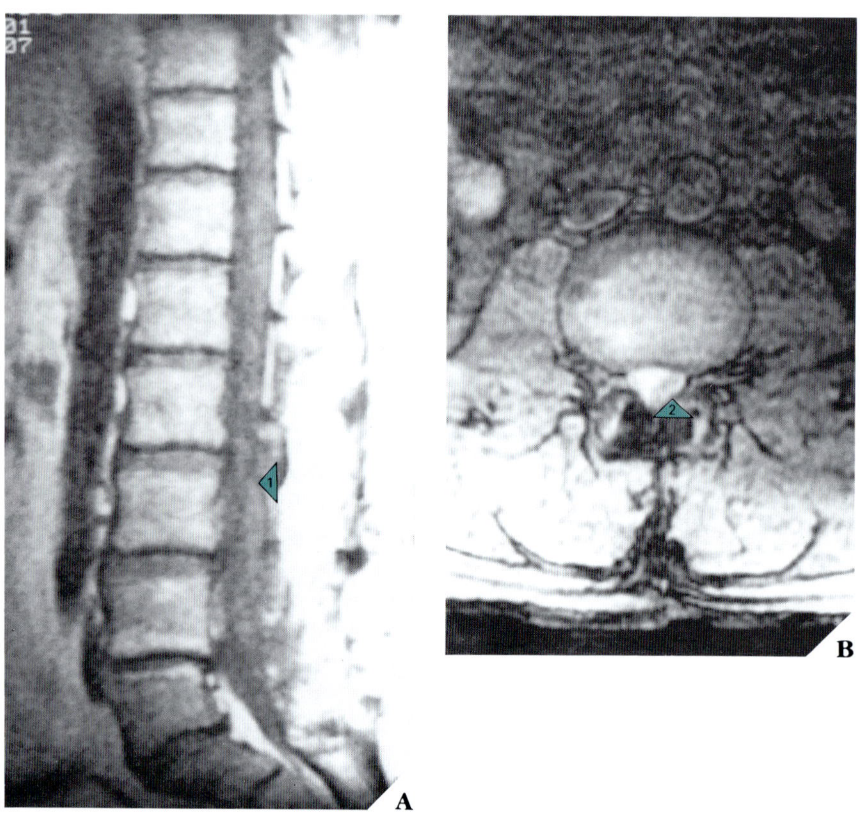

FIGURE 9.27. Lumbar Arachnoiditis. **A** Sagittal section [SE 800/20]. On this T1WI, a long laminectomy is seen posteriorly in the lower lumbosacral area. There is loss of the normal posterior epidural fat and an ill-defined posterior margin of the thecal sac, secondary to postsurgical changes. The conus medullaris and cauda equina are also ill-defined. The findings are subtle but suggest distortion, which could be secondary to arachnoiditis. **B** Axial section [GRASS 1000/12 flip angle 22.5 degrees]. This axial PDWI is made above the surgical site. Distortion of the roots is highly suggestive of arachnoiditis. The individual nerve roots are clumped together along the left posterior margin of the dural sac due to arachnoiditis. Arrow **1** Ill-defined cauda equina. Arrow **2** Clumped nerve roots.

in contrast to disc material. Gd-DTPA has been shown to be helpful (11). On contrast-enhanced T1WI, scar increases in SI in comparison to disc protrusions. However, differentiation is still difficult when both disc and scar are present at the same time.

Arachnoiditis

Arachnoiditis (Fig. 9.27) usually occurs secondary to surgery or myelography (22). Less commonly, it occurs secondary to infection, trauma, hemorrhage, or rupture into the CSF by such tumors as epidermoids or cholesteatomas. Myelography is still the most sensitive diagnostic exam, but MRI frequently shows suggestive findings. Clumping or traction of the nerve roots to the margins of the dural sac are characteristic. Occasionally, the roots are so clumped that they mimic a tethered cord. In rare instances, there is severe distortion of the spinal cord, multiple intradural cysts, and spinal cord syrinx. Cerebral spinal fluid flow artifacts may also be absent, suggesting abnormal CSF movement.

Posterior Longitudinal Ligament Ossification

Cervical posterior longitudinal ligament ossification (Fig. 9.28) is a relatively rare finding but its identification is very important (17). It occurs most commonly in the Asian population but it can be seen in any race. The patients usually present with a slowly progressing myelopathy due to spinal stenosis and cervical cord compression. There is marked thickening and ossification of the posterior longitudinal ligament. It may mimic a large osteophyte but is separate from the endplates. Other degenerative findings may be absent. A large low SI mass in the anterior position of the spinal canal, which deforms the spinal cord, is characteristic.

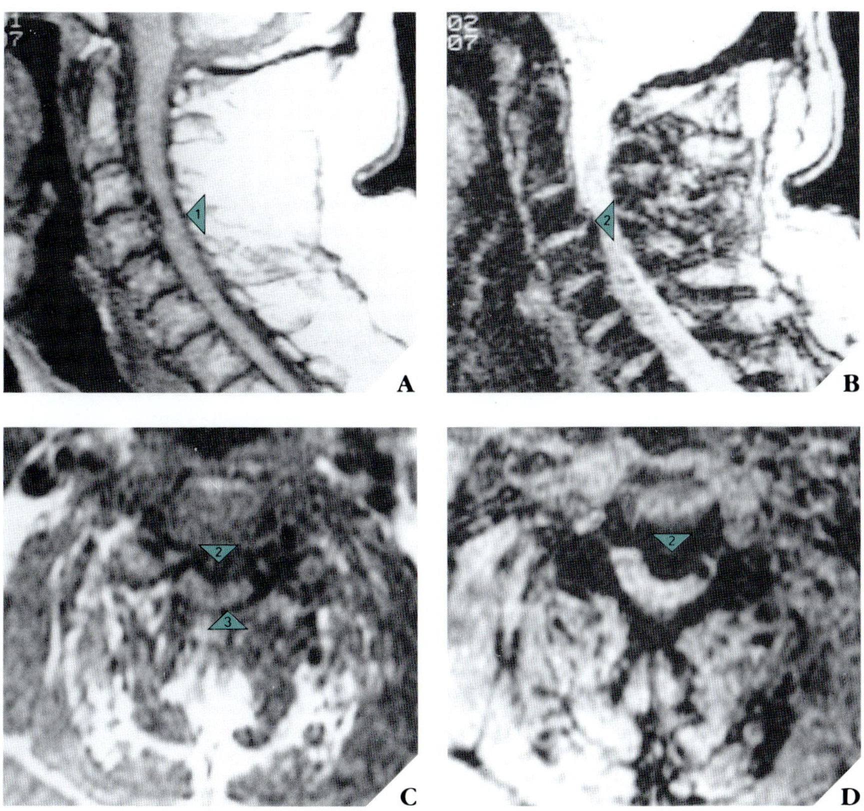

FIGURE 9.28. Ossification of the Cervical Posterior Longitudinal Ligament. A Sagittal section [SE 800/20]. A band of signal void secondary to ossification of the posterior longitudinal ligament bridges at least two posterior portions of the mid-cervical vertebrae. The dense thickened ligament causes anterior impression on the cervical cord. The calcification is not seen clearly since the CSF and cortical bone are all low SI on this T1WI. The disc spaces are well-maintained and the cord is otherwise normal. **B** Sagittal section [GRASS 1000/12, flip angle 22.5 degrees]. The high gray-scale contrast of this PDWI more accurately defines the severity of the canal narrowing and the signal void of the calcified ligament. Note that the affected disc is not degenerated as would be expected with hypertrophic degenerative disease. **C** Axial section [SE 800/20]. The degree of spinal cord compression is dramatically shown on the axial T1WI. The internal anatomy of the cord is barely visible. The cord has a flattened kidney-bean shape, secondary to compression from a signal void anteriorly located mass. **D** Axial section [GRASS 1000/12, flip angle 22.5 degrees]. The hypertrophied ossified ligament is dark on all pulse sequences because of the calcified matrix. The cord is not well-seen within the markedly deformed spinal canal. Arrow **1** Distortion of cervical cord. Arrows **2** Ossified ligament. Arrow **3** Cord compression.

ARTHROPATHY

Rheumatoid Arthritis

The high cervical spine changes of rheumatoid arthritis are well evaluated by MRI (Fig. 9.29) (1). The odontoid is usually eroded or thinned by pannus. Anterior subluxation of the first cervical vertebra on the second due to an incompetent atlantoaxial joint is common. The subluxation can be accentuated by flexion. MR images in flexion and extension can be helpful. The first cervical vertebra may also settle inferiorly over the odontoid, so that the anterior arch of the first vertebra articulates with the anterior body of the second. Both of these deformities lead to compression of the spinal cord and possible neurologic injury. Since these patients are very fragile, MR presents a great advantage over myelography and CT.

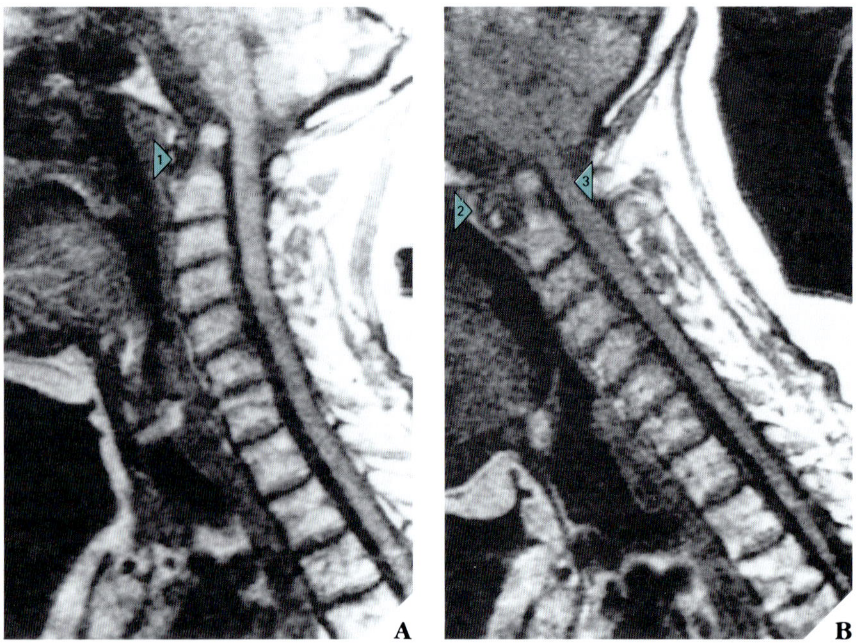

FIGURE 9.29. Atlantoaxial Rheumatoid Arthritis. A Sagittal section [SE 800/20]. In patients with rheumatoid arthritis, some dynamic information can be obtained from flexion and extension MRI studies of the upper cervical spine. In the neutral to slightly extended position, the anterior arch of C1 has "settled" slightly inferiorly on the odontoid. There is a widened predental space and a mild impression on the cervicomedullary junction by the odontoid, which extends into the posterior fossa. **B** Sagittal section [SE 800/20]. In flexion, there is further widening of the predental space as the atlas subluxes on the axis. The area of intermediate SI, between the anterior arch of the first cervical vertebrae and the odontoid, is pannus formation. The odontoid protrudes deeply into the foramen magnum. This is due to craniovertebral settling and is not secondary to invagination of the skull base. Special care must be taken not to injure the patient in these maneuvers. Arrow **1** Settling of anterior arch of C1. Arrow **2** Widened predental space. Arrow **3** Cervicomedullary compression.

Ankylosing Spondylitis

The findings of ankylosing spondylitis, such as syndesmophyte formation, square vertebral bodies, and fusion of the vertebral bodies, can be visualized by MRI (Fig. 9.30). Fusion of the facet joints is more difficult to evaluate. Enlargement of the CSF spaces in the lumbar region can also be seen. Fusion of the sacroiliac joints produces loss of the joint space. Subtle enthesopathy is better seen on plain films. Similar findings can be seen in Reiter's syndrome and psoriatic arthritis though in these disorders the findings are usually less symmetric and diffuse.

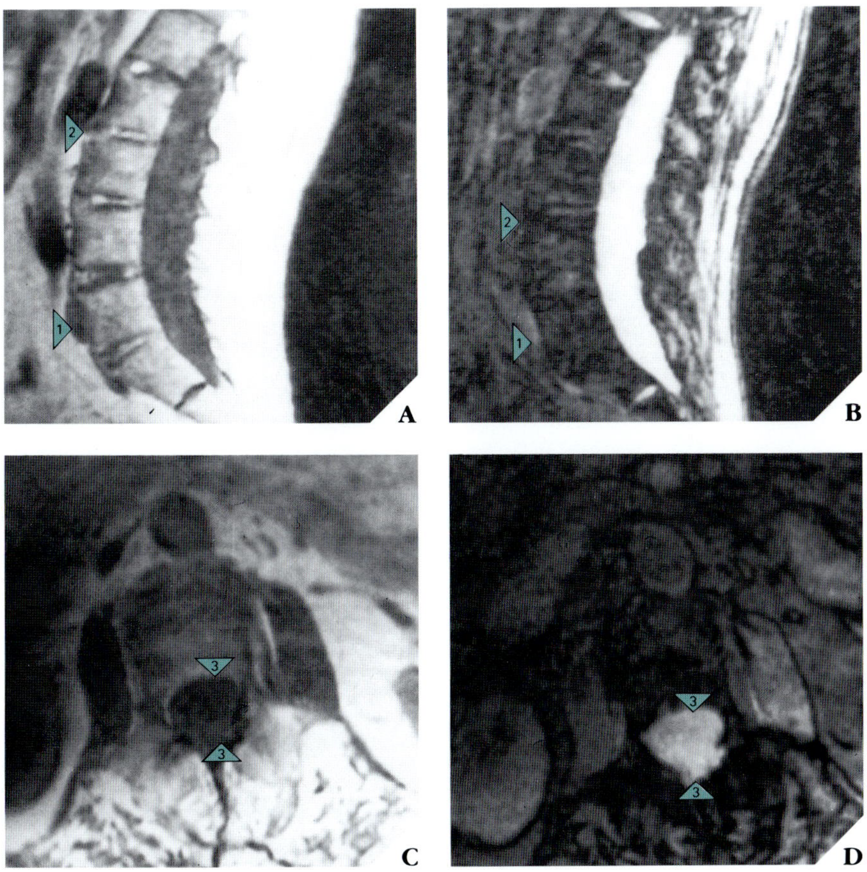

FIGURE 9.30. Lumbar Ankylosing Spondylitis. A Sagittal section [SE 800/20]. The squaring of all vertebral bodies is a characteristic finding of ankylosing spondylitis. This produces a "bamboo spine" appearance. The spinal canal is broad in its anterior to posterior dimension. **B** Sagittal section [GRASS 1000/12, flip angle 22.5 degrees]. The disc spaces are narrowed somewhat and the discs themselves demonstrate some loss of SI. There are no major hypertrophic changes to suggest that common degenerative disease of the discs is the etiology. **C** Axial section [SE 800/20]. The spinal canal is large and normal on this T1WI. **D** Axial section [GRASS 1000/12, flip angle 22.5 degrees]. The spinal canal is expanded with a scalloped posterior margin. This is another characteristic finding of ankylosing spondylitis. Arrows **1** Square vertebral bodies. Arrows **2** Bamboo-shaped spine. Arrows **3** Enlarged canal diameter.

TRAUMA

MRI has had a surprisingly important impact on the evaluation of spinal trauma (4,26). It generates unique information about the spinal cord and ligaments which is unavailable through any other technique. Plain radiographs and CT, however, are still the mainstay for defining the level of injury and subtle bony fractures. MRI is particularly helpful in situations where patients have a known myelopathy, or where more information is needed to decide on the most appropriate surgical intervention. Patients are imaged early, before the use of traction or other intervention makes the exam difficult or impossible to complete. Titanium or graphite tongs, and other nonmagnetic traction devices, are available to image the acute or chronic cord injury patient.

Compression Fractures

Compression fractures are seen as wedge-shaped deformities of the vertebral bodies (Figs. 9.31, 9.32). The SI is usually low, due to the compacted calcium components and the altered magnetic susceptibility of acute hemorrhage. Bright SI from methemoglobin and other paramagnetic blood breakdown products of subacute hemorrhage are not

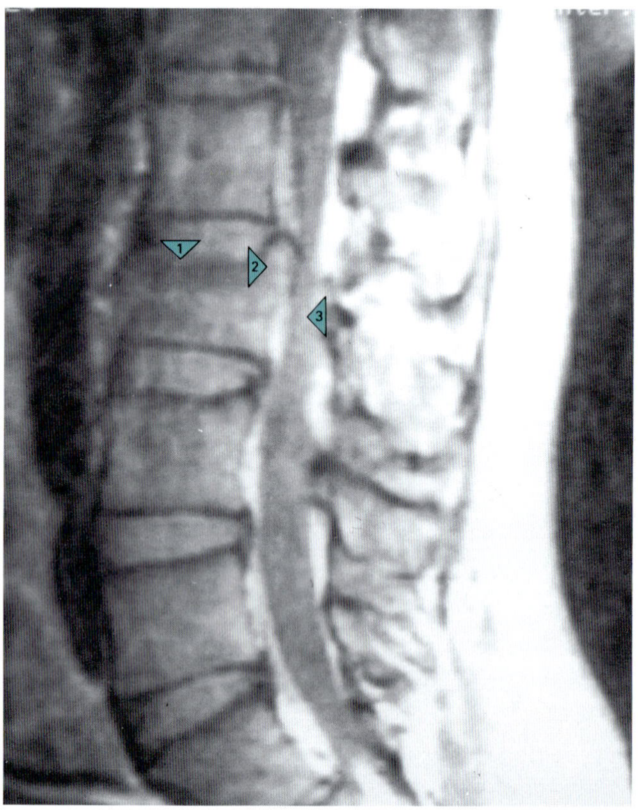

FIGURE 9.31. Compression Fracture. Sagittal section [SE 800/20]. This T1WI demonstrates a burst fracture involving the superior aspect of the third lumbar vertebra. A large fragment has been displaced posteriorly, resulting in canal compromise. Note that the posterior longitudinal ligament does not appear to be ruptured. The lower signal of the vertebral body may be due to compressed calcium components and acute hemorrhage. Arrow **1** Fracture of vertebral body superior aspect. Arrow **2** Retropulsed fragment. Arrow **3** Narrowed canal.

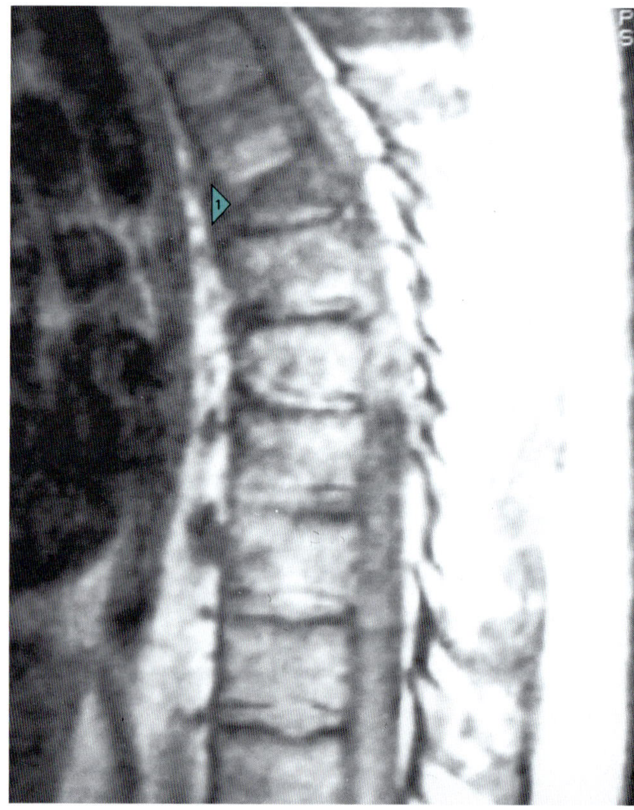

FIGURE 9.32. Compression Fracture. Sagittal section [SE 800/20]. In this patient, an anterior wedge-compression fracture has occurred acutely in the upper thoracic spine. The vertebral body demonstrates relatively low SI due to compression of bony vertebral elements and acute hemorrhage. At high magnetic fields, there may be low SI from acute hemorrhage, due to the magnetic susceptability changes of deoxyhemoglobin. Arrow **1** Wedge compression fracture.

seen in the marrow spaces. Posterior displacement of fracture fragments into the spinal canal can be seen which deform the spinal canal and cord. MRI can also define subluxations and other alignment abnormalities on direct sagittal images. CT, however, is a better diagnostic tool to define the exact number and configuration of the fractures.

Disc Protrusions

Occasionally, disc protrusions may occur with trauma and have similar findings to those seen in degenerative disease. In these cases however, the discs may not be degenerated (Fig. 9.33). Those patients who have pre-existing spinal stenosis of any etiology may have serious cord injuries without the associated findings of fractures or instability.

Epidural Effusion and Hematoma

Subacute epidural hematomas can be high SI on T1WI MR images. The normal epidural fat may be obliterated or deformed. Subacute hemorrhage has a high signal both on T1WI and on PDWI and T2WI. Displacement of the low SI dura away from the margins of the spinal canal with associated cord or root compression may be evident (Fig. 9.33).

Cord Injury

Spinal cord compression, contusion, hemorrhage, and transection are frequently visible on MRI. Displacement or distortion of the surface of the cord can be seen on T1WI and can be secondary to disc protrusion, bony fragment, severe subluxation, or epidural hematoma. Mild cord contusions are seen as regions of nonvisualization of the normally evident, internal cord architecture. Usually, visualization of the gray and white matter structures is possible, though deformed by imaging artifacts (5). More severe edematous cord contusions lead to fusiform cord enlargement and high signal on PDWI and T2WI (Fig. 9.33). Intraparenchymal hemorrhage will vary in appearance, depending upon the timing of the exam. The magnetic properties of the degradation products of hemorrhage change rapidly with time (similar to intraparenchymal brain hemorrhages). The SI of immediate cord hemorrhage is low or isointense with normal cord. Subacute blood will be high SI on essentially all SE sequences. Long-term follow-up, even years later, may be necessary to exclude a syrinx from prior hemorrhagic necrosis of the central cord. Delayed progressive ascending neurologic deterioration may be present.

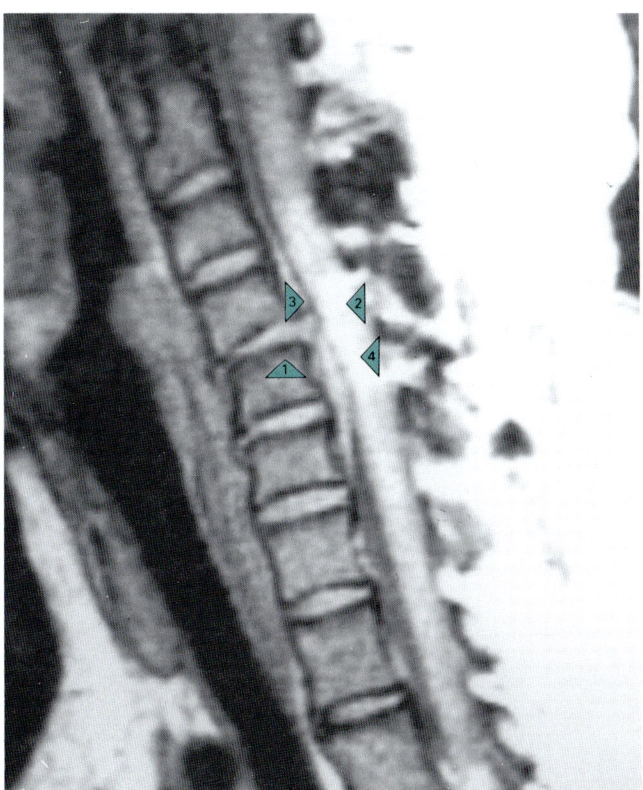

FIGURE 9.33. Acute Cervical Hyperflexion Injury. Sagittal section [SE 1800/20]. On this T1WI there is focal kyphosis, centered at the fourth–fifth cervical interspace. The interspace is narrowed anteriorly and widened posteriorly. Since there are no major fractures, this suggests a ligamentous injury. There is thinning, though not complete rupture of the posterior longitudinal ligament, where the posterior low SI margin of the low SI disc outline is thinned. The cervical disc is slightly protruding. There is an anterior epidural hemorrhagic fluid collection that lifts the anterior margin of the dura of the posterior margin of the vertebral bodies. The high SI from the cervical cord suggests a contusion. Arrow **1** Widened posterior aspect of disc spaces. Arrow **2** Disc protrusion. Arrow **3** Anterior epidural hemorrhage. Arrow **4** High SI cord.

Ligamentous Injury

MRI gives unique information about ligamentous injuries (Figs. 9.33 and 9.34). Flexion and extension radiographs and the configuration of the spinal curvature can be used to infer ligamentous injuries by demonstration of acute subluxation or abnormal motion. MRI can directly image the anterior and posterior longitudinal ligaments which bridge from one vertebral endplate to the next, over the disc interspaces. They are usually not visible at the vertebral body level (since they are low SI like cortical bone) unless there is an effusion or hemorrhage which peels the ligaments off the bone. At the level of the interspaces, disruption of the normally thin signal void of the ligaments is evidence of rupture of the ligaments. The presence of a torn ligament may alter the choice of surgical management. If the ligament is intact, it is more likely that correcting the abnormal spine curvature will remove free bony fragments from the spinal canal. With hyperextension injuries, the anterior longitudinal ligament will be ruptured; with flexion injuries, the posterior longitudinal ligament may be injured.

METABOLIC DISORDERS

Osteoporosis

Osteoporosis is only indirectly evaluated on MRI (Fig. 9.35) (27). Since MRI cannot accurately assess calcium, CT or plain radiographs are more accurate in demonstrating calcium loss. However, the secondary manifestations or complications of osteoporosis are well seen. Compression fractures, scoliosis, and kyphosis can be seen in severe cases.

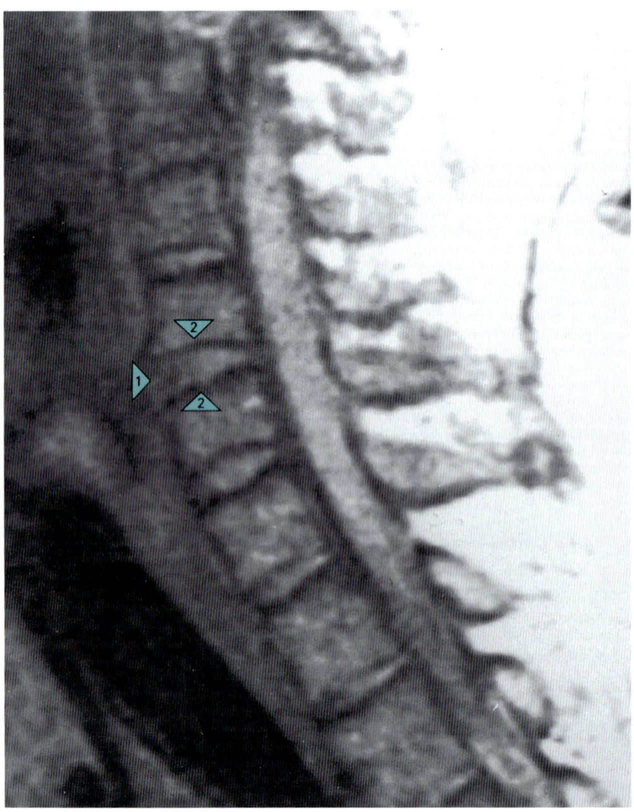

FIGURE 9.34. Rupture of the Cervical Anterior Longitudinal Ligament. Sagittal [SE 800/20]. This patient suffered an acute hyperextension injury that resulted in a tear of the anterior longitudinal ligament at the fourth–fifth cervical interspace. The intervertebral disc space is also anteriorly widened. Note that the normal, thin, low SI line of the ligament (overlying the discs and bridging from one vertebral body to another) is not seen at the affected level. Arrow **1** Ruptured ligament. Arrows **2** Widened anterior disc space.

There are no characteristic SI changes of the marrow spaces. However, MRI can help exclude metastases or other diffuse marrow processes which may have the identical appearance on radiography.

Paget's Disease

Paget's disease of bone or osteitis deformans is a relatively common disorder, usually affecting individuals over the age of fifty (Fig. 9.36). Any bone may be affected but most common is the lumbar spine followed by skull, pelvis, and femur. The disease is due to an accelerated, disorganized formation and resorption of bone, that may have a viral etiology. The balance between formation of immature woven bone and accelerated osteoclastic activity results in differing appearances of Paget's disease: totally lytic (osteoclasts predominate), sclerotic (new woven bone), or a mixture of both.

The MRI appearance of Paget's disease also depends on the phase of the disease process. When sclerotic, the cortical bone of the vertebral body is thickened, demonstrating low SI on all pulse sequences. In the vertebral body this can result in the "picture frame" vertebra, caused by the cortex thickening to outline the vertebral body. MRI is used to show encroachment on the neural foramen or spinal canal by thickened bone. The marrow spaces have a fatty high SI on short TR/TE images, which may be due to either increased fatty marrow or slow blood flow through the highly vascular area. Lytic areas are well-delineated on CT and radiographs, but are poorly detected by MRI due to a variable SI, either high or intermediately low, depending on the nature of the matrix (fibrous or myeloid) produced.

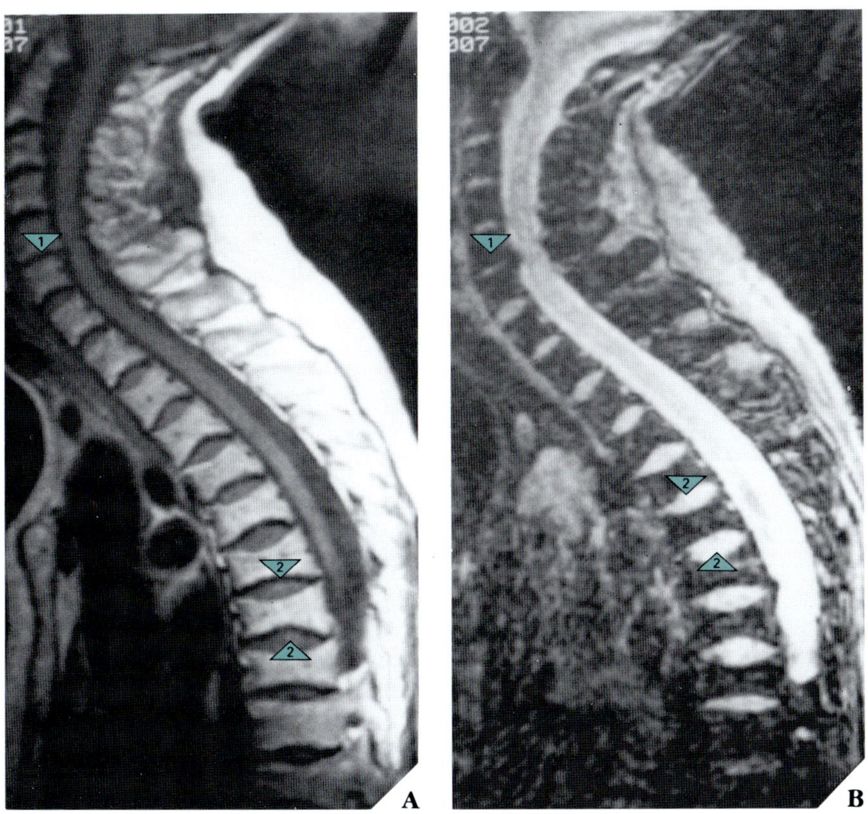

FIGURE 9.35. Thoracic Osteoporotic Compression Fractures. A Sagittal section [SE 800/20]. There is bi-concave compression of the thoracic spine and some exaggeration of spinal lordotic curvature. The disc spaces in the thoracic region are more elliptical than normal secondary to distortion of the endplate. Note that the signal of the marrow spaces is normal. **B** Sagittal section [GRASS 1000/12, flip angle 22.5 degrees]. The biconcave distortion of the endplates is well-demonstrated. There is some slight mottling of signal from the vertebral bodies but there is no overall suggestion of high SI tumor infiltration. Arrows **1** Degenerated disc with osteophyte. Arrows **2** Biconcave vertebral bodies.

Sarcomatous degeneration occurs in 2–10 per cent of Paget's patients and is a grave complication most often representing osteosarcoma or secondly fibrosarcoma. Giant cell sarcomas occur more commonly in the skull. Unless a large destructive area has occurred, it is difficult to distinguish a sarcomatous area in the midst of the heterogeneous signal properties of Paget's diseased bone. For this purpose, CT remains a better tool for evaluation of bone integrity.

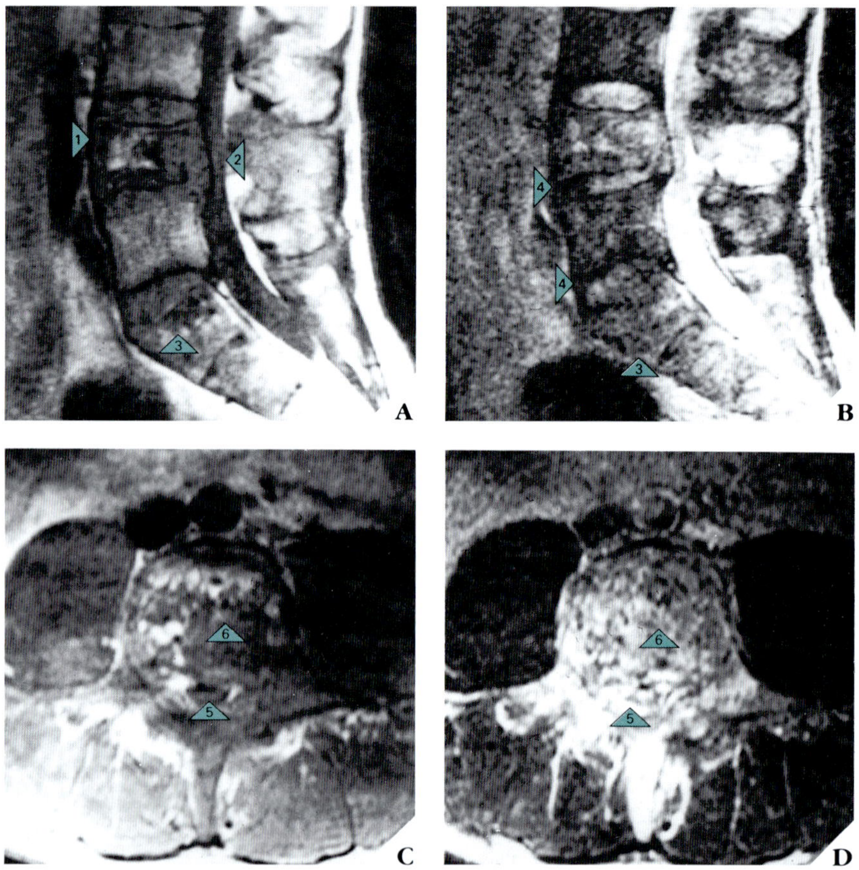

FIGURE 9.36. Lumbar Paget's Disease. A Sagittal midline section [SE 800/20]. The fourth lumbar and the first sacral vertebrae demonstrate patchy high and low SI regions on this T1WI. The fourth lumbar vertebra is slightly expanded. There is secondary encroachment on the spinal canal. Note that there is a replacement by tissue of the normal low SI ring of the cortical bone, at the periphery of these vertebrae. This suggests destruction or replacement of the cortical bone. **B** Sagittal midline section [SE 2000/80]. On this T2WI there is loss of signal from the two lower lumbar interspaces, consistent with degeneration of the discs. There is increased SI from the affected vertebrae. The nerve roots are seen as a low SI band in the bright CSF because they are compacted by the expanded vertebral body. Distinction of these findings from a tumor would be impossible. **C** Axial section [SE 800/20]. This T1WI through the fourth lumbar vertebrae shows only small pockets of higher SI which may be small islands of normal fatty marrow. The low SI lesion extends into the left pedicle with marked expansion. The dural sac is markedly compressed by the expanded vertebral body. **D** Axial section [SE 2000/80]. The fourth lumbar vertebral body is inhomogeneous with multiple, small, low SI regions. This low SI suggests sclerosis of the vertebral body. A computed tomogram would better evaluate the dense bony changes. Arrows **1** Expanded vertebral body. Arrow **2** Dura. Arrows **3** Involvement of sacrum. Arrows **4** Decreased signal from degenerated discs. Arrows **5** Narrow thecal sac. Arrows **6** Pagetic bone.

MARROW DISORDERS

MRI has allowed a whole new assessment of the marrow spaces not obtainable in the past (18,21,27). Since the fatty and erythropoietic components of the marrow can be evaluated without interference from the calcium-containing components, MRI has some advantages. With special pulse-sequence selective fat- or water-imaging techniques, it is actually possible to image the two components separately. The normal marrow spaces have balanced components of fat and water constituents. Alteration of this balance can be recognized by MRI. Well into adulthood the spine is an important site for hematopoiesis while the long bone marrow spaces usually are replaced with fatty marrow. Therefore, most of the marrow spaces of the spine are not as bright on T1WI as the long bones.

If the hematopoietic components are selectively destroyed, as in aplastic anemia, radiation, or chemotherapy, fatty marrow will predominate. These marrow spaces demonstrate very high SI on T1WI and low SI on PDWI and T2WI (Fig. 9.37) (20). If the fatty spaces are displaced by hematopoietic tissue (normal or hypertrophic), by tumor, or by any high water-containing tissues, this marrow has low SI on the T1WI and high SI on PD/T2WI (see Chapters 6 and 10). If the marrow spaces are universally abnormal, the relative signal changes may be difficult to recognize because the patient acts as his or her own control for normal.

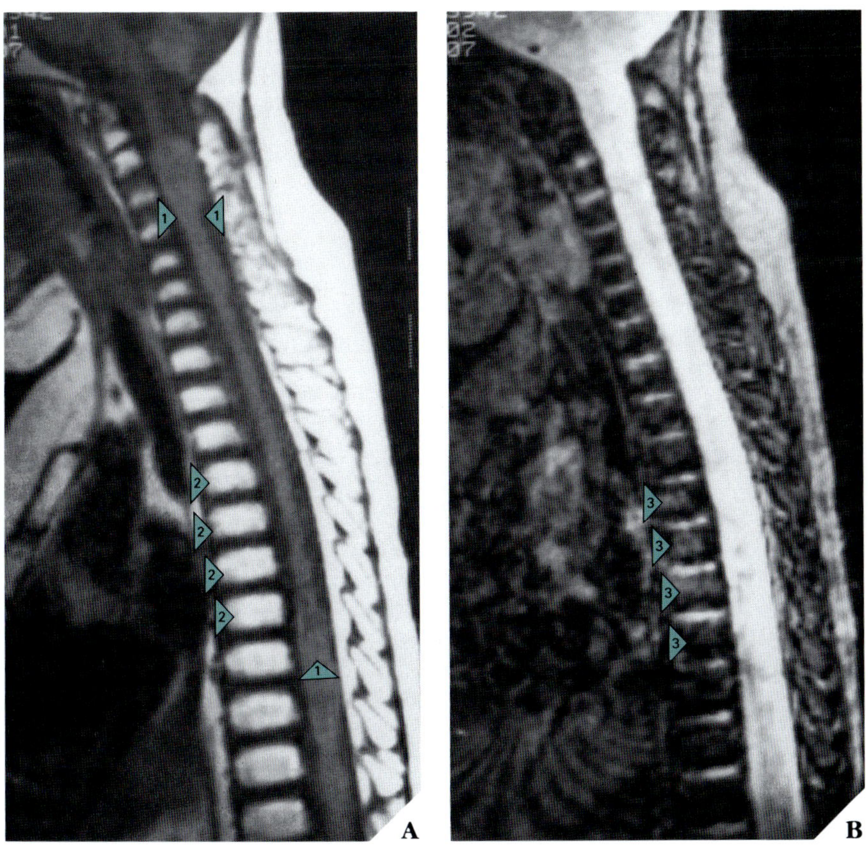

FIGURE 9.37. Postradiation Fatty Marrow Replacement. A Sagittal spaces of the cervical and upper thoracic region have high signal on this T1WI. This child had received whole axis radiation to the spine for spinal metastatic medulloblastoma. Note that the vertebrae are very bright and homogeneous due to replacement of the hemopoietic marrow with fat. The spinal cord is demonstrated by an irregular fusiform configuration secondary to drop metastases implanted on the surface of the cord. **B** Sagittal midline section [GRASS 1000/12, flip angle 22.5 degrees]. The marrow spaces are slightly inhomogeneous, but there are no regions that suggest tumor deposition. The tumor implants on the surface of the spinal cord are not as well seen on this PD pulse sequence. Arrows **1** Medulloblastoma. Arrows **2** Fatty conversion of marrow. Arrows **3** Patchy signal.

For example, in myelofibrosis all of the marrow spaces may be diminished, leading to patchy low SI changes (Fig. 9.38).

If there is complete replacement of all of the marrow spaces by fibrous or calcified tissue, there will be no signal from the marrow. Such low SI can be seen with any sclerotic disorder including metastases, infection, trauma, bone islands, discitis, and osteopetrosis. Multiple blood transfusions in patients with anemia can lead to iron deposition in the marrow spaces. The iron can cause magnetic susceptibility variations, producing low SI from the marrow and liver structures, due to a markedly decreased T2 (Fig. 9.39).

Sickle Cell Disease

Hemoglobinopathies such as sickle cell disease result in a hyperplasia of the hematopoietic bone marrow in response to chronic anemia. As part of the normal aging process, bone marrow changes from hematopoietic red to fatty yellow appearing as high signal on T1WI. In chronic anemias, such

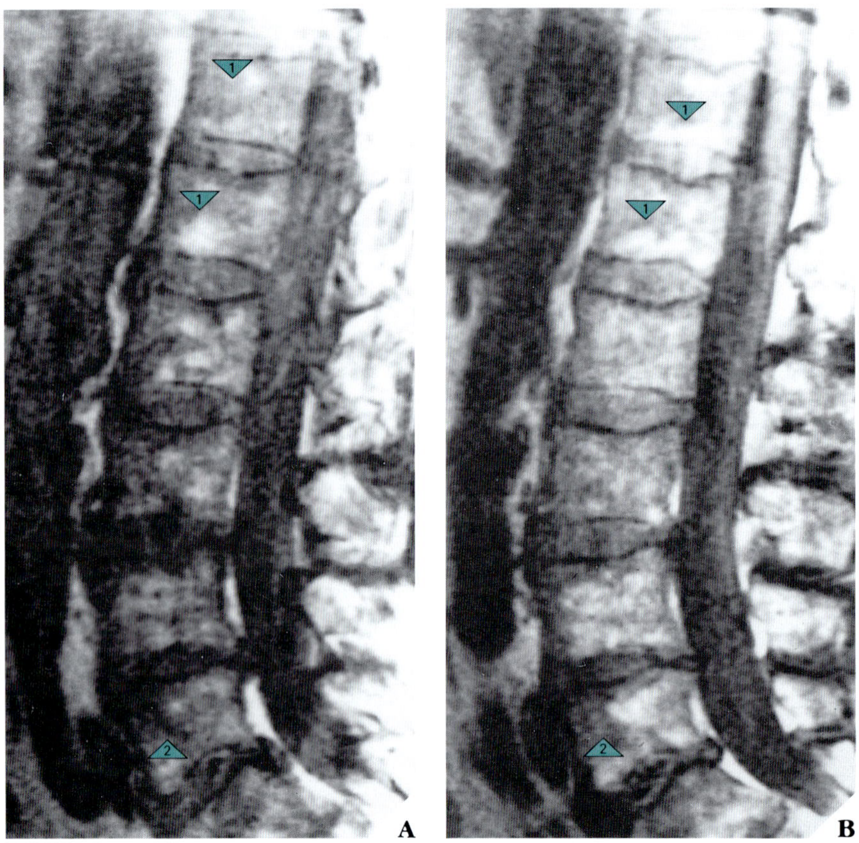

FIGURE 9.38. Lumbar Myelofibrosis. **A** Sagittal midline section [SE 800/20]. This sagittal T1WI demonstrates irregular high SI regions related to inhomogeneous distribution of the fat within the bone marrow. This patient had chronic leukemia with extensive prior chemotherapy. In these patients high SI marrow spaces on PDWI or T2WI may not be present. The marrow changes are not due to tumor infiltration but rather to fibrosis of the marrow spaces, which leads to low signal changes of both the fatty and hemopoietic marrow components. Irregular signal changes of the marrow spaces can also be seen in patients with renal failure or those on steroid therapy. **B** Sagittal midline section [SE 800/20]. Focal regions of irregular fatty marrow deposits are also seen on this T1WI. Arrows **1** Fatty marrow. Arrows **2** Fibrotic areas of myeloid narrow spaces.

as sickle cell disease, the marrow is reconverted to myeloid elements (18,21). This reconversion results in a decreased SI in comparison to the normally bright vertebral bodies on T1WI. This process may be focal, but is usually more global in the vertebral bodies. The appearance on T2WI is variable depending upon cellularity, presence of infarcts, and the amount of tissue water present (see Chap. 6).

Sludging of the sickled cells can lead to anoxia, ischemia, and infarction of the bone. In the spine osteonecrosis characteristically occurs at the central portion of the vertebral growth plate. An endplate with a central cleft, typically termed an "H"-shaped vertebra, results. When completely infarcted, this cleft in the endplate should have a low SI on all pulse sequences. Bone marrow infarction has a variable appearance on MRI depending on the stage of the ischemia and infarction. Low SI areas on short TR/TE images that become higher SI areas on T2WI are believed to represent acute infarctions. Chronic infarctions are more characteristically a focus of low SI in the midst of higher SI fatty areas on short TR/TE images.

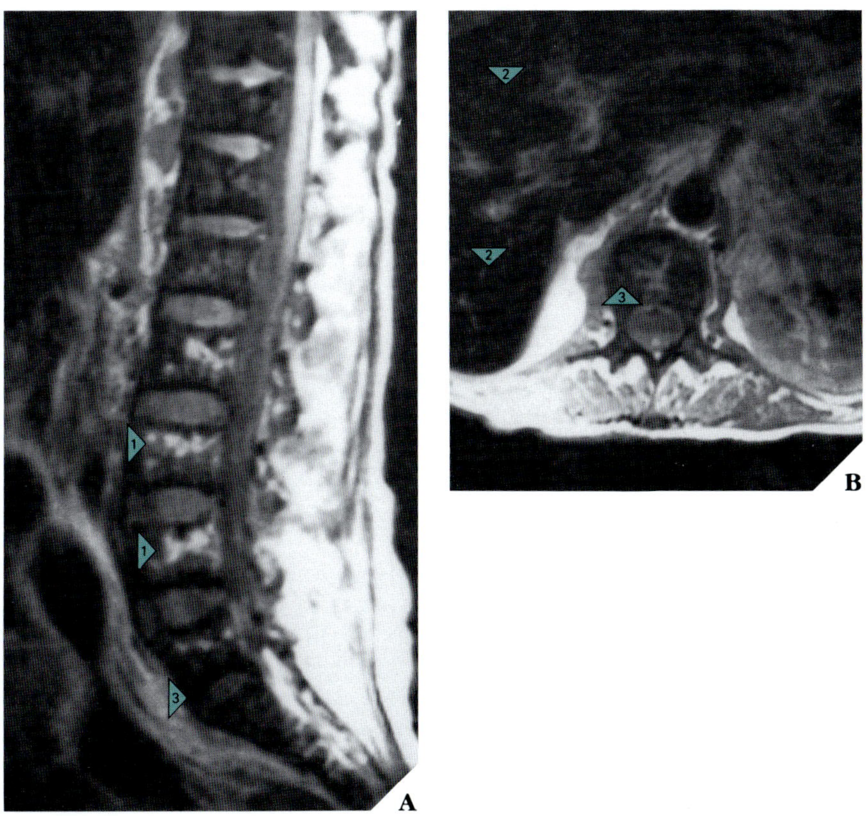

FIGURE 9.39. Iron Depostion Within the Bone Marrow and Liver. A Sagittal section [SE 800/20]. This patient with leukemia had been treated with multiple blood transfusions. This sagittal midline T1WI of the lumbar spine demonstrates very low SI from all of the bone marrow spaces. Only a small amount of normal high SI fatty marrow is seen. This finding could be due to bony sclerosis or fibrosis of the marrow spaces. In this case however, iron deposition into the marrow spaces and chemotherapy fibrosis account for the low signal intensity. **B** Axial section [SE 800/20]. This axial T1WI also demonstrates low signal changes of the bones. Note the very low SI from the liver which is normally intense. This confirms the deposition of iron, causing magnetic susceptibility changes leading to very low SI. Arrows **1** Relatively normal fatty marrow. Arrows **2** Signal void from liver. Arrows **3** Dark signal from iron deposition and fibrosis of marrow.

REFERENCES

1. Aisen A, Martel W, Ellis J, McCune WJ: Cervical spine involvement in rheumatoid arthritis: MR imaging. Radiology 1987;165:159–163.
2. Altman N, Altman D: MR imaging of spinal dysraphism. AJNR 1987;8:533–538.
3. Chakeres D, Allen D, Christoforidis AJ: Spine, In Christoforidis AJ: Atlas of Cross-Sectional Anatomy, with Nuclear Magnetic Resonance and Computed Tomography, Philadelphia, WB Saunders, 1987:130–159.
4. Chakeres D, Flickinger F, Stokes B, et al: Magnetic resonance imaging of acute canine spinal cord trauma model. AJNR 1987;8:5–10.
5. Curtin AJ, Chakeres DW, Buelas R, et al: MR imaging artifacts of the axial internal anatomy of the cervical spinal cord. AJNR (1989;10:19–26).
6. Edelman RR, Stark DD, Sanja S, et al: Oblique planes of section in MR imaging. Radiology 1986;159:807–810.
7. Ehman RL: MR imaging with surface coils. Radiology 1985;157:549–550.
8. Geremia GK, Russel EJ, Clasen RA: MR imaging characteristics of a neurenteric cyst. AJNR 1988;9:978–980.
9. Grenier N, Kressel H, Schiebler M, et al: Normal and degenerative posterior spinal structures: MR imaging. Radiology 1987;165:517–525.
10. Han J, Benson J, Kaufman B, et al: Demonstration of diastematomyelia and associated abnormalities with MR imaging. AJNR 1985;6:215–219.
11. Hueftle M, Modic MT, Ross JS, et al: Lumbar spine: Postoperative MR imaging with Gd-DTPA. Radiology 1988;167:817–824.
12. Laakman RW, Kaufman B, Han JS, et al: MR imaging in patients with metallic implants. Radiology 1985;157:711–714.
13. Mamourian A, Briggs R: Appearance of Pantopaque on MR images. Radiology 1986;158:457–460.
14. Modic M, Feiglin D, Piraino D, et al: Vertebral osteomyelitis: assessment using MR. Radiology 1985;137:157–166.
15. Modic M, Masaryk T, Ross J, Carter J: Imaging of degenerative disk disease. Radiology 1988;168:177–186.
16. Modic M, Pavlicek W, Weinstein M, et al: Magnetic resonance imaging of intervertebral disk disease. Radiology 1984;152;103–111.
17. Nichols JN, Tehranzadeh J, Vaccarino FP: Posterior longitudinal ligament ossification myelopathy and hypothyroidism. Comput Tomogr 1987;11:61.
18. Papavasiliou C, Trakadas S, Gouliamos A, et al: Magnetic resonance imaging of marrow heterotopia in haemoglobinopathy. Eur J Radiol 1988;8:50–53.
19. Porter BA, Hastrup W, Richardson ML, et al: Classification and investigation of artifacts in magnetic resonance imaging. Radiographics 1987;7:271–2874.
20. Ramsey R, Zacharias C: MR imaging of the spine after radiation therapy: easily recognizable effects. AJNR 1985;6:247–251.
21. Rao V, Fishman M, Mitchell D, et al: Painful sickle cell crisis: bone marrow patterns observed with MR imaging. Radiology 1986;161:211–215.
22. Ross J, Masaryk T, Modic M, et al: MR imaging of lumbar arachnoiditis. AJNR 1987;8:885–892.
23. Ross J, Masaryk T, Modic M, et al: Vertebral hemangiomas: MR imaging. Radiology 1987;165:165–169.
24. Rubin J, Enzmann D: Imaging of spinal CSF pulsation by 2DFT MR: significance during clinical imaging. AJNR 1987;8:297–306.
25. Sze G, Krol G, Zimmerman R, Deck M: Malignant extradural spinal tumors: MR imaging with Gd-DTPA. Radiology 1988;167:217–223.
26. Tarr R, Drolshagen L, Kerner T, et al: MR imaging of recent spinal trauma. J Comput Assist Tomogr 1987;11(3):412–417.
27. Vogler JB, Murphy WA: Bone marrow imaging. Radiology 1988;168:679–693.
28. Wang H, Rosenbaum A, Reid C: Pediatric patients with achondroplasia: CT evaluation of the craniocervical junction. Radiology 1987;164:515–519.
29. Wehrli FW, MacFall JR, Glover GH, et al: The dependence of nuclear magnetic resonance (NMR) image contrast on intrinsic and pulse sequence timing parameters. Magn Reson Imaging 1984;2:3–16.

CHAPTER

TEN

Javier Beltran

Musculoskeletal Tumors

TECHNICAL CONSIDERATIONS

▼▼▼ *IMAGING PLANES, FIELD OF VIEW, AND PULSE SEQUENCES*

Musculoskeletal tumors are diagnosed because they produce clinical symptoms, palpable masses, and/or radiographic abnormalities. Primary bone tumors in general are well seen on plain films, but soft tissue tumors are elusive to radiographic detection. Depending on the location of the tumor and its extension as determined by palpation or by x-rays, the MRI examination can be tailored to each lesion (9). Most often axial and coronal or sagittal sections are sufficient for complete coverage of the lesion. In some anatomically complex areas, such as the pelvis, scapula, or spine, all three orthogonal planes may be necessary. When imaging tumors involving the long bones, it is extremely useful to use oblique planes of section along the longitudinal axis of the bone, for this view will display the intramedullary or soft tissue extension of the tumor in one single section.

A large FOV with the body coil can be used to image large tumors, but whenever possible, a smaller FOV with surface coils should be used since it provides better contrast and spatial resolution.

As in other types of pathology, T1- and T2-weighted pulse sequences are necessary for tumor imaging. As is discussed below, the SI changes displayed by some tumors can provide information regarding tissue characterization and tumor extension. In order to keep imaging time to a minimum, T1WI with short TR/TE can be obtained in one plane, and long TR with short and long TE can be obtained in another orthogonal or oblique plane. A third plane can be included if necessary, using proton density and T2-weighted pulse sequences.

Fast imaging techniques with short flip angles (GRASS, FLASH, FISP) can also be used instead of the time-consuming, long TR, SE pulse sequences. T2WI can be obtained using short TR (in the range of 40–60 msec), short TE (15–25 msec), and a large flip angle (50–90 degrees) (42). One of the disadvantages of fast scanning techniques is the increased sensitivity to magnetic field inhomogeneities using long TE (24).

Proton chemical shift imaging can generate selective water and fat images by virtue of the different resonance frequencies of the aliphatic and water protons in tissues (40,43). Chemical shift images can be obtained using multiple methods. Dixon's technique is one of the most frequently used (16). Contrast between fatty and hematopoietic marrow and between normal and abnormal marrow is accentuated using these techniques.

Contrast between pathologic processes such as tumor or infection and normal fatty marrow can also be enhanced using inversion recovery techniques with short inversion time (STIR). Using this pulse sequence, the normal yellow marrow signal is nulled. This provides high contrast with the red marrow, which appears bright due to the additive effects of T1 and T2 contrast using STIR. Pathologic processes also show high SI using STIR (10,40).

▼▼▼ *ACQUISITION MATRIX AND SLICE THICKNESS*

A matrix of 128 × 256 with 5-mm slice thickness and 1-mm interslice gaps is normally sufficient for tumor imaging. An occasional very small tumor may require thinner slices at the expense of decreased S/N, which has to be compensated for by increasing the NEX.

▼▼▼ *SURFACE COILS*

The use of surface coils is recommended to obtain consistently high resolution images with high S/N. Imaging of tumors involving the extremities with surface coils can be performed with any of the commercially available coils. Occasionally, it may be necessary to use large FOV (for example, to image large tumors, or to compare the normal extremity with the abnormal one). In these cases, the body coil can be used.

▼▼▼ *CONTRAST AGENTS*

Paramagnetic contrast agents have been used successfully for MR imaging of CNS tumors, but its use in musculoskeletal tumors has not yet been clearly defined. Preliminary experience indicates that gadolinium DTPA produces enhancement of viable and vascularized parts of tumors on T1WI, providing an improved S/N, better differentiation between necrotic tumor, edema, and viable tumor, and decreased imaging time (if T2WI are excluded from the imaging protocol). The clinical usefulness of this information has yet to be determined (4,20,33,41).

TUMOR STAGING

Amputation was the standard surgical treatment for musculoskeletal neoplasms until advances in chemotherapy, radiation therapy, reconstructive surgery, and diagnostic techniques fostered new interest in limb-salvaging procedures (18,19). Depending on the site, extension, histological type of the tumor, and presence of metastasis, different limb-salvaging procedures can be performed, including intracapsular excision, marginal excision, wide excision, and radical resection (19). Therefore, the accurate assessment of tumor extension using different diagnostic techniques becomes very important in making surgical decisions.

Multiple imaging modalities are available for tumor diagnosis and staging, including plain films, bone scintigraphy, arteriography, arthrography, CT, and MRI. Experience accumulated during the last few years has helped to define the role of these techniques in specific clinical situations.

Six parameters should be evaluated for tumor staging: (a) tumor detectability; (b) intramedullary extension; (c) soft tissue extension; (d) intraarticular extension; (e) cortical disruption; and (f) neurovascular involvement.

TUMOR DETECTABILITY

Primary or secondary bone neoplasms are accurately detected using combinations of plain films, conventional tomography, bone scintigraphy, and, more recently, CT. The location in the bone, pattern of destruction, type of margin, internal matrix, presence or type of periosteal reaction, and condition of the cortex are well-established plain film criteria for diagnosis and assessment of tumor activity (32). The only two exceptions are cases of subtle or early bone permeation that may escape plain film detection, and myeloma. Bone scintigraphy is positive in most cases of permeative bone destruction, but myeloma often shows negative plain films and negative bone scans, unless there is a pathologic fracture. In these cases, MRI can add significant information regarding tumor detection (Fig. 10.1).

Soft tissue tumors, on the other hand, have consistently eluded plain film detection. When it was introduced over a decade ago, CT represented a significant improvement in tumor detection and staging. However, recent reports indicate that, by virtue of its striking soft tissue contrast resolution,

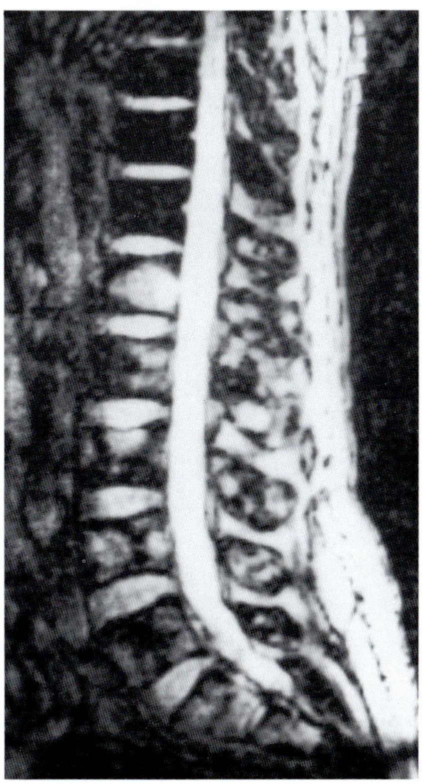

FIGURE 10.1. Multiple Myeloma. Midline sagittal plane [GRASS 1000/20, flip angle 12 degrees]. There are multiple high SI lesions in all lumbar vertebral bodies and spinous processes. Plain films and bone scintigraphy in this patient were normal.

MRI is superior to CT, not only in the evaluation of the extent of the tumor but also in tumor detection (15) (Fig. 10.2).

INTRAMEDULLARY EXTENSION

Well-conducted comparative studies between different imaging modalities, with pathologic correlation (4,21) indicate that MRI is significantly superior to bone scintigraphy and CT for the evaluation of the intraosseous extent of the tumor. The reasons for the improved visualization of intramedullary tumor involvement are related to the existing high contrast between the normal bone marrow and the tumor on practically all MRI pulse sequences (Fig. 10.3), and the possibility of obtaining sections along the longitudinal axis of the long bones using oblique planes. Bone scintigraphy often demonstrates extended uptake due to increased bone turnover in the vicinity of the tumor. CT is limited to the axial plane. In addition, false positive tumor extension can be interpreted in the metaphysis where reinforced cancellous bone produces increased density on CT.

SOFT TISSUE EXTENSION

Clinically useful information regarding soft tissue extension of tumors arising in the bone and extending to the surrounding soft tissues or primary soft tissue tumors includes tumor dimensions, relationship to fascial planes, compartmental involvement, and individual muscle involvement, in addition to neurovascular and articular involvement (discussed in the following sections) (5,15,18,19). Published series indicate that CT and MRI are equally accurate in evaluating both tumor dimension and the relationship with neurovascular structures and joints. MRI, however, has been found to be superior to CT in the detection of individual muscle involvement and in the delineation of the anatomic compartment (15) (Fig. 10.4). These data are important to the surgeon, since marginal resection is considered adequate only when the tumor specimen boundaries are histologically negative. Another surgical principle emphasizes the necessity of resecting the entire muscle involved by tumor but preserving uninvolved muscles in the same compartment (13,29).

Many malignant tumors demonstrate a halo of high SI on T2WI which can represent muscular in-

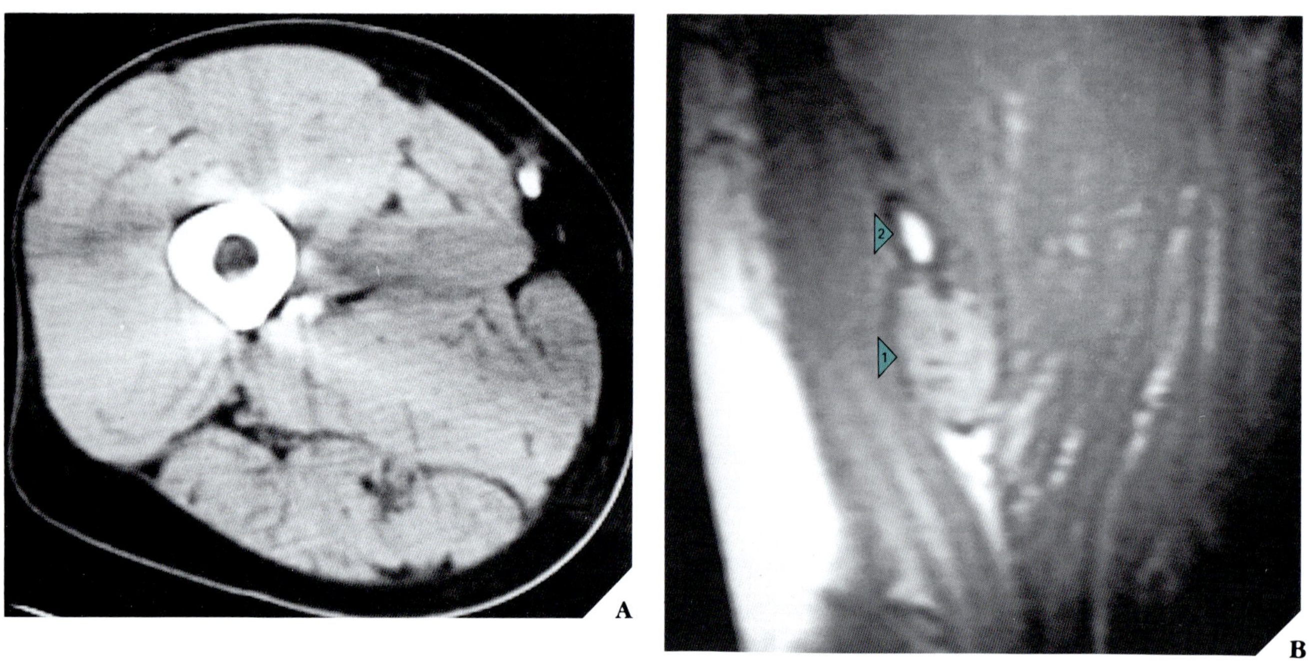

FIGURE 10.2. Synovial Sarcoma. A Axial contrast-enhanced CT through the midthigh region. This patient presented clinically with a palpable mass in the inner aspect of the thigh. Percutaneous biopsy demonstrated synovial sarcoma. CT was performed to assess size and extension of the tumor. Beam-hardening artifacts and low contrast between tumor and surrounding muscle prevent visualization of the lesion. **B** Sagittal MRI [SE 2000/80] through the midthigh in the same patient demonstrates a well-demarcated 3-cm tumor. The high SI area above the tumor probably represents hemorrhage from a previous biopsy. Arrow **1** synovial sarcoma. Arrow **2** hemorrhage.

filtration or reactive edema (2) (Fig. 10.5). In most cases the distinction between edema or tumor infiltration is not significant, since the margin of the resection must include the high SI halo. In some instances, however, it may be necessary to make this distinction, in which case gadolinium DTPA may provide improved delineation of the tumor (33).

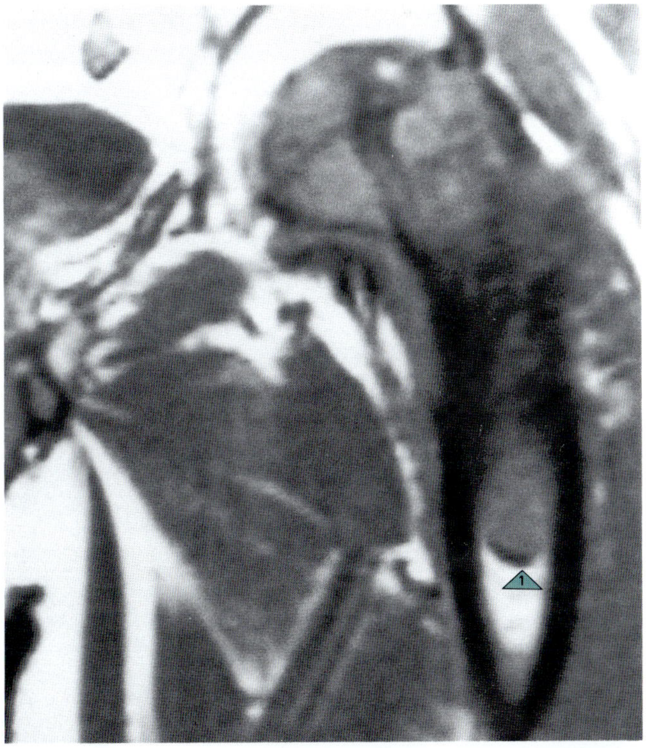

FIGURE 10.3. Non-Hodgkin Lymphoma of the Proximal Left Femur with Intramedullary Extension. Coronal plane [SE 800/20]. There is replacement of the normal high SI fatty marrow by homogeneous low SI lymphoma on this T1WI. The distal margin of the tumor contrasts sharply with adjacent high SI fat (arrow **1**).

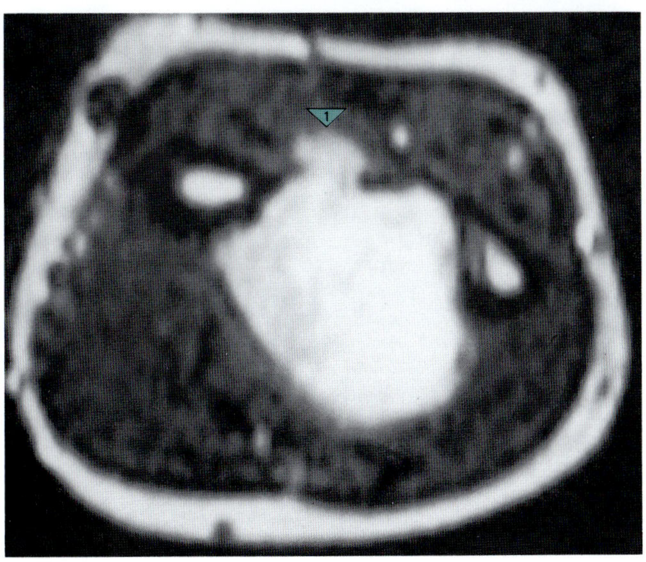

FIGURE 10.4. Synovial Sarcoma with Transcompartmental Extension. Axial section through the forearm [SE 2000/80]. There is a well-demarcated, high SI tumor involving the posterior compartment of the forearm, extending to the anterior compartment through the interosseous membrane (arrow **1**).

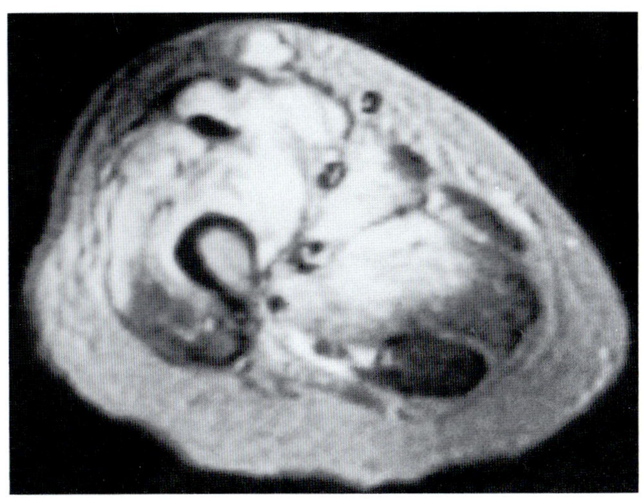

FIGURE 10.5. Malignant Undifferentiated Sarcoma of the Thigh. Axial section [SE 2000/80]. There is an extensive area of ill-defined high SI involving the anterior and posterior compartments of the thigh. The distinction between reactive edema and tumor infiltration cannot be made on the basis of this MRI, without gadolinium DTPA enhancement.

INTRAARTICULAR EXTENSION

The necessity of predicting preoperatively the intraarticular extension of the tumor is as important as determining the soft tissue extension in order to optimize resection boundaries. Most reported series indicate that MRI is comparable to CT in assessing intraarticular extension (15) (Fig. 10.6).

CORTICAL DISRUPTION AND PERIOSTEAL REACTION

In one published series (34) plain films and CT were found to be superior to MRI in assessing cortical bone abnormalities, including destruction, expansion, and permeation. In a more recent report, however, the accuracy of all three techniques was found to be similar (15). Using high-field surface coil imaging, small areas of partial or complete bone destruction can be well assessed with MRI (Figs. 10.6 and 10.7). Diffuse cortical thinning and expansion may be difficult to differentiate from tumor pseudocapsule, since both show low SI in all pulse sequences (see Fig. 10.21).

Periosteal reaction is well detected with plain films and CT, but state-of-the-art MRI can also depict this abnormality (Figs. 10.6 and 10.8). Axial sections can demonstrate cortical irregularity with concentric low SI rings representing layers of reactive periostitis (Fig. 10.9).

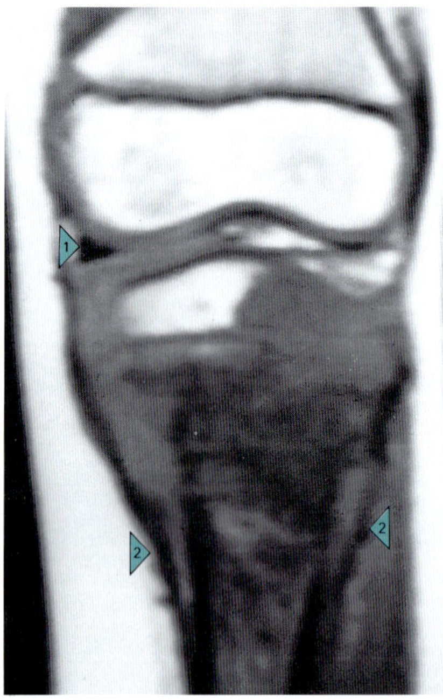

FIGURE 10.6. Osteosarcoma with Intraarticular Extension. Coronal section [SE 800/20]. There is a low SI tumor involving the proximal tibia, extending proximally to the tibial epiphysis and into the joint space. The meniscus is in contact with the tumor. Note the cortical bone destruction and periosteal elevation produced by the tumor. Arrow **1** meniscus. Arrows **2** periosteal elevation.

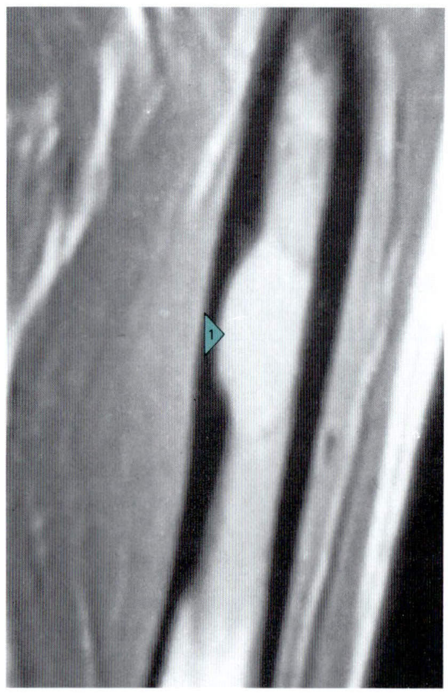

FIGURE 10.7. Metastatic Tumor to the Femur. Surface coil coronal section [SE 2000/20]. There is an intermediate SI lesion in the midshaft of the femur producing localized endosteal scalloping of the medial cortex (arrow **1**).

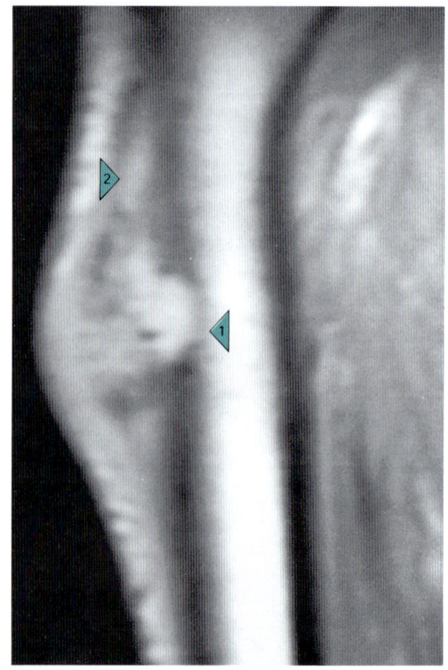

FIGURE 10.8. Osteosarcoma with Soft Tissue Extension. Sagittal section [SE 2000/20]. There is discontinuity of the anterior cortex of the tibia, with soft tissue extension of the tumor, and periosteal elevation. The intramedullary extent of the tumor was seen more clearly on T2WI (not shown). Arrow **1** cortical disruption. Arrow **2** periosteal elevation.

NEUROVASCULAR INVOLVEMENT

Neurovascular structures can be adherent to the tumor, invaded by the tumor, displaced by the tumor but not adherent to it, or have no relationship to the tumor. Tumor invasion of the neurovascular bundle can be predicted only if the tumor completely encases these structures (Fig. 10.10). If the tumor is adherent to the neurovascular structures, one has to assume that they are involved, unless a

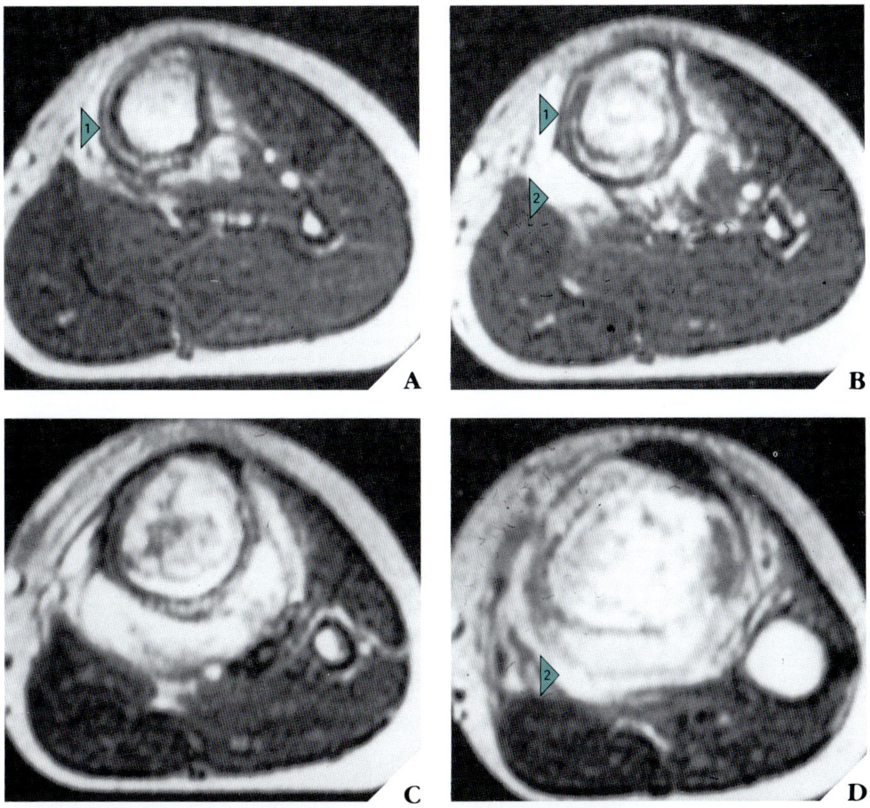

FIGURE 10.9. Ewing's Sarcoma with Periosteal Reaction. Four consecutive axial sections through the proximal tibia [SE 2000/80] showing high SI of the bone marrow space on this T1WI. There is thinning of the cortex and several concentric rings around the cortex, which represent different layers of periosteal reaction ("onion skin"). The tumor extends into the soft tissues around the tibia. Arrows **1** periosteal reaction. Arrows **2** soft tissue tumor extension.

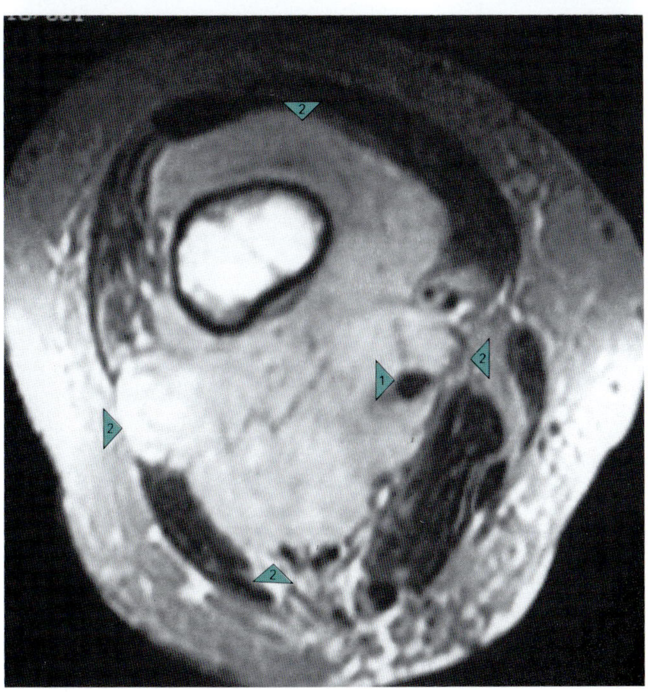

FIGURE 10.10. Malignant Fibrous Histiocytoma; Vascular Encasement with Occlusion of the Popliteal Vein. Axial section through the distal thigh [SE 2000/80]. The popliteal artery is surrounded by tumor. The popliteal vein is not identified. An angiogram demonstrated popliteal vein occlusion. The tumor involves the medullary space of the distal femur and extends into the soft tissues surrounding entirely the distal femur. Arrow **1** popliteal artery. Arrows **2** tumor.

soft tissue plane (for example, fat) is present between the tumor and the neurovascular structures (Figs. 10.11 and 10.12). Intravenously enhanced CT and MRI are equally accurate in assessing neurovascular involvement.

Tumors of the osseous spine often extend to the spinal canal. In these cases, MRI is superior to CT in assessing spinal canal invasion, unless intrathecal contrast material is injected prior to CT examination (1) (Fig. 10.13).

TUMOR RECURRENCE POSTTREATMENT

Distinction of active tumor from treatment-induced changes (radiation therapy and/or chemotherapy) may be difficult. Available data (38,39) indicate that if low SI on T1WI and T2WI is present at the site of previous surgery or radiotherapy, the probability of tumor recurrence is small. However, if high SI is seen on T2WI and the patient has had only surgery, the probability of recurrent tumor is high. If

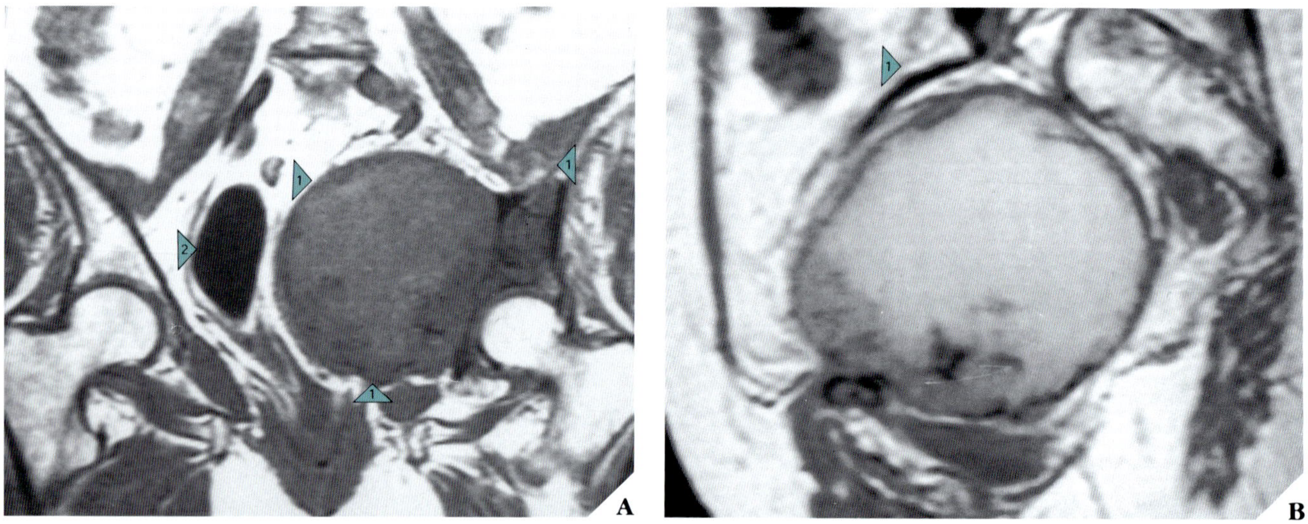

FIGURE 10.11. Large, Low Grade Chondrosarcoma Adherent to the Iliac Artery. A Coronal section [SE 800/20]. There is a large, low SI tumor arising from the left acetabulum and extending into the soft tissues of the pelvis, displacing the bladder. Arrows **1** chondrosarcoma. Arrow **2** bladder. **B** Sagittal section [SE 2000/80]. The internal iliac artery is adherent and anteriorly displaced by the tumor and adherent to it. Arrow **1** internal iliac artery.

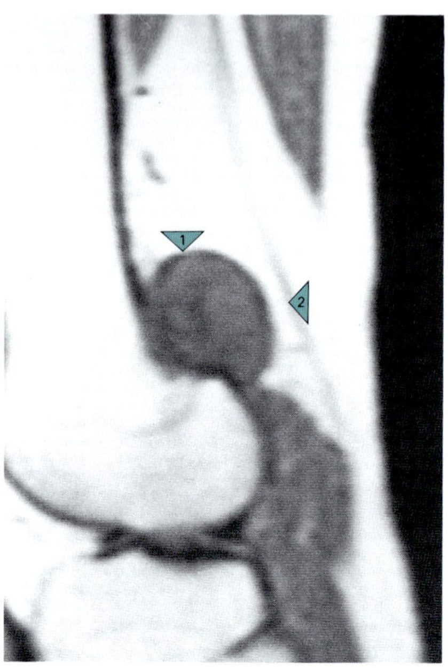

FIGURE 10.12. Parosteal Osteosarcoma. Sagittal section [SE 800/20]. There is a low SI tumor in the posterior aspect of the distal femur, well separated from the popliteal artery by a fat plane. There is no evidence of intramedullary extension. Arrow **1** tumor. Arrow **2** popliteal artery.

the patient has had radiation therapy and high SI is seen on T2WI, differentiation between radiation-induced changes and tumor recurrence is not possible using noncontrast-enhanced MRI techniques (Fig. 10.14). Lack of enhancement postgadolinium-DTPA injection is a reliable indicator of absence of tumor recurrence in these cases.

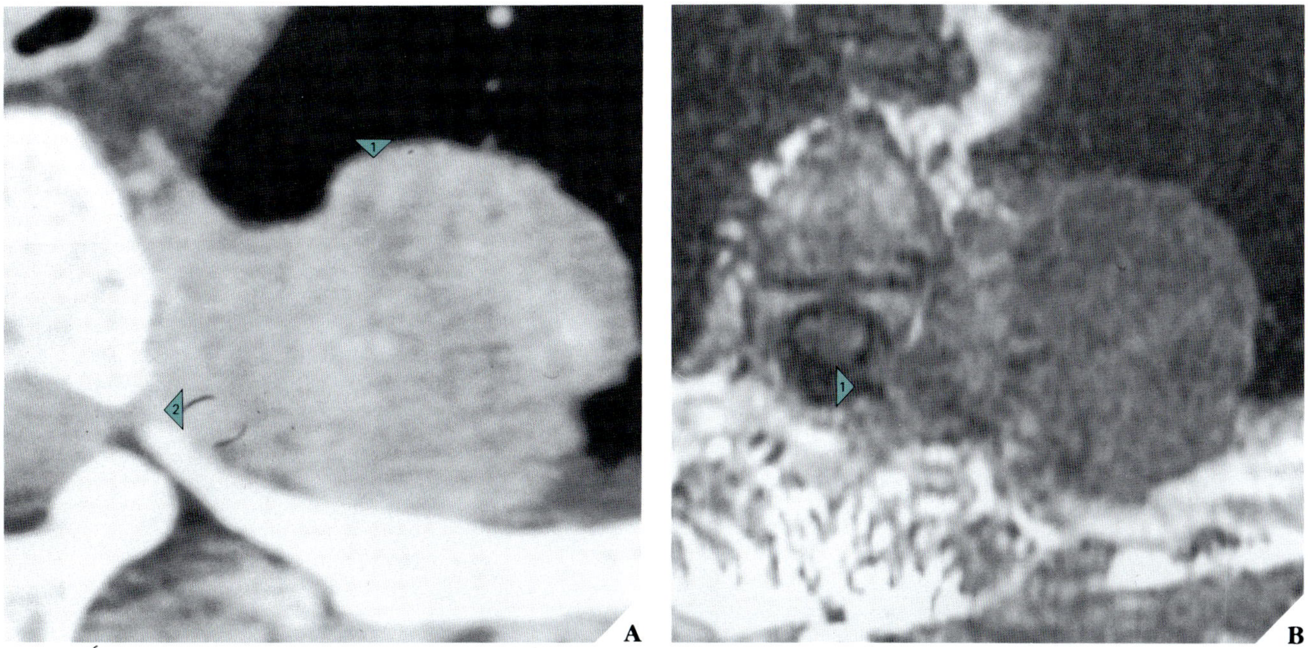

FIGURE 10.13. Paraspinal Chondrosarcoma Extending into the Spinal Canal. A Axial CT. The tumor is well seen but its extension through the neuroforamen cannot be assessed without intrathecal contrast material. Arrow **1** tumor. Arrow **2** neuroforamen. **B** Axial MRI [SE 800/20] in the same patient demonstrates more clearly the tumor extension through the neuroforamen into the spinal canal (arrow **1**).

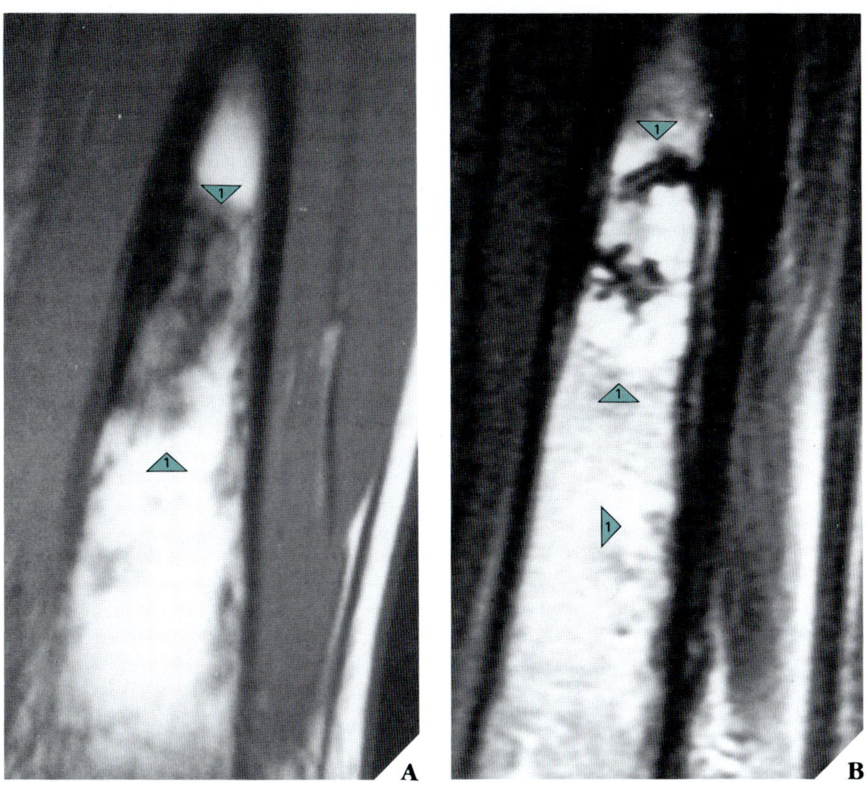

FIGURE 10.14. Non-Hodgkin Lymphoma (Postradiation Therapy Changes). Coronal sections through the femur [**A** SE 2000/20, **B** SE 2000/80]. There is an ill-defined area of low SI on PDWI **(A)** in the midshaft of the femur, with a corresponding area of high SI on T2WI **(B)**, extending distally. These images were obtained approximately 6 months postradiation therapy. Recurrent lymphoma could not be excluded. Percutaneous biopsy demonstrated postradiation changes. Arrows **1** postradiation therapy SI changes.

TUMOR CHARACTERIZATION

The early efforts directed toward calculating T1 and T2 relaxation times in order to histologically characterize tumoral tissue (14) have not been rewarding. Based upon T1 and T2 calculations, malignant tumors cannot be differentiated from benign tumors (35). Definite relaxation characteristics have been found in some tissues, including tumors containing fat, areas of necrosis and liquefaction, hemorrhage, and in some tumors such as aneurysmal bone cyst and giant cell tumor (35). However, the clinical usefulness of these data remains unknown.

Morphologic characteristics of the tumor as determined by MRI may add some information. Most malignant tumors of the bones and soft tissue demonstrate a halo of high SI on T2WI, representing either reactive edema of the surrounding tissues or tumor infiltration (Fig. 10.5). This finding is unusual in benign lesions (2).

Intravenous injection of gadolinium DTPA combined with dynamic imaging with fast pulse sequences has shown significantly higher signal increase in malignant tumors than in benign tumors or peritumoral edema (20).

Following is a discussion of the MRI characteristics seen in some frequent tumors and tumor-like conditions of the musculoskeletal system, which may assist in making a specific diagnosis.

BONE TUMORS AND TUMOR-LIKE CONDITIONS

Osteoid Osteomas

Osteoid osteomas are well evaluated with plain films, bone scintigraphy, and CT when looking for the presence and location of the characteristic radiolucent nidus. Occasionally, MRI may be requested to exclude other conditions. High SI bone marrow edema on T2WI has been described in few reported cases, with or without demonstration of the nidus (22,44) (Fig. 10.15). The high SI probably

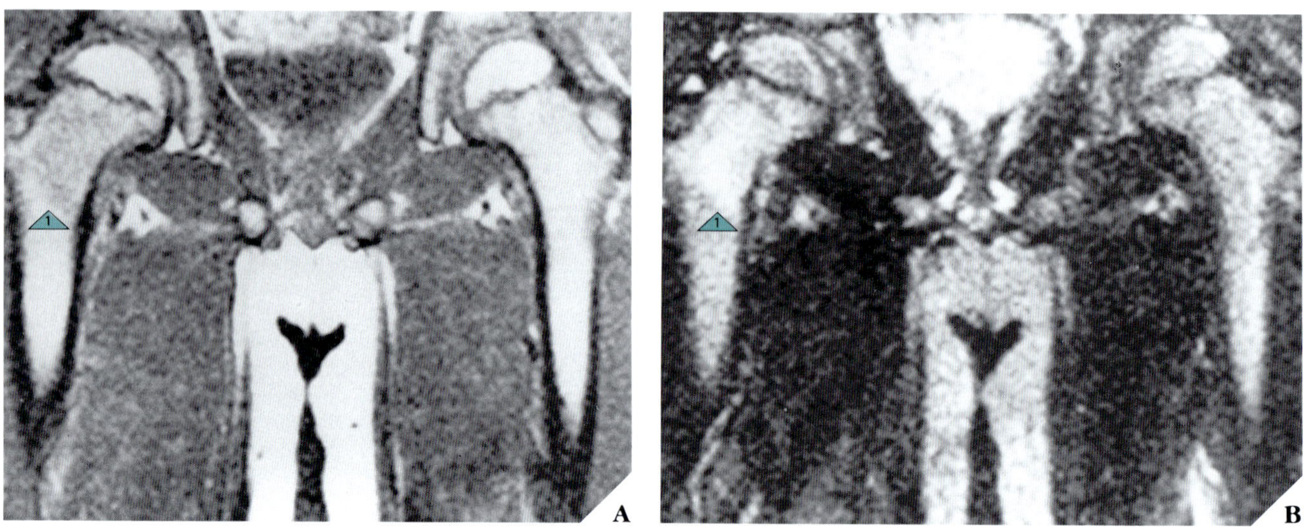

FIGURE 10.15. Osteoid Osteoma. Coronal sections [**A** SE 800/20, **B** SE 2000/80]. There is an area of low SI on T1WI **(A)** and high SI on T2WI **(B)** in the region of the femoral neck. This 14-year-old patient complained of right hip pain for several months. Initial plain films were normal. This MRI was obtained 6 months after the onset of symptoms. Follow-up radiographs obtained one year after the MRI demonstrated a classical nidus in the medial aspect of the femoral neck. Arrows **1** bone marrow edema.

reflects hyperemia and bone marrow edema. Differential diagnosis in these cases includes transient osteoporosis, infection, and trauma (see Chapter 6).

Fibrous Dysplasia

As in many other tumor and tumor-like conditions of the bone, the SI changes seen in fibrous dysplasia are nonspecific, but in some cases these changes may suggest the diagnosis. Pathologically, the main component of this entity is a fibrous collagenous matrix, with irregular areas of woven bone which may be well mineralized. It has been shown that hypocellular fibrous tissue has a short T2, thus providing little or no signal on regular SE T2-weighted pulse sequences (37). In addition, calcified tissue has long T1 and very short T2 relaxation times because of its relatively fixed molecular state. Therefore, both major components of fibrous dysplasia would complement each other to produce low SI on T1WI and T2WI (Fig. 10.16).

Cartilaginous islands or nodules have been described in some cases of fibrous dysplasia. Cartilaginous tissue has more water content than fibrous tissue, thus the high SI on T2WI is increased. The presence of cartilaginous tissue would explain the relatively hyperintense signal seen in some cases of fibrous dysplasia. Fluid-filled areas have been described in fibrous dysplasia, probably representing degeneration and cellular necrosis. These cystic spaces would be responsible for the most hyperintense lesions seen occasionally on T2WI (Fig. 10.16).

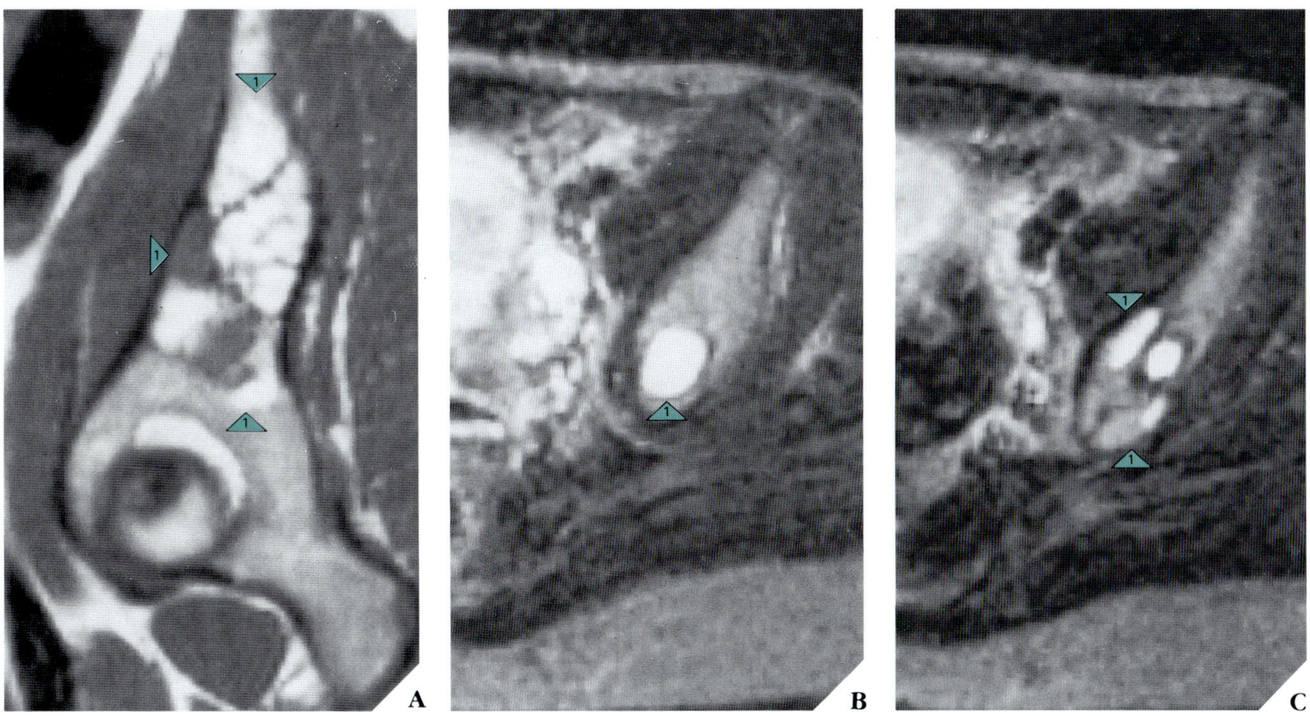

FIGURE 10.16. Fibrous Dysplasia. A Sagittal section [SE 550/22], **B** and **C** axial sections [SE 2500/80], through the left hip. There are well-demarcated areas of low, intermediate, and high SI in the supraacetabular region, extending into the iliac crest, probably representing a combination of fibrous tissue, cystic areas, and cartilaginous islands (arrows **1**).

Nonossifying Fibroma

Nonossifying fibroma is a very common asymptomatic lesion seen in the long bones when performing MRI for other reasons. Healed nonossifying fibromas show well-demarcated areas of low SI on any pulse sequences, reflecting the sclerosis seen on plain films. Active nonossifying fibromas demonstrate well-delineated, cortical-based areas of low SI on T1WI and high SI on T2WI surrounded by a low SI rim (Fig. 10.17).

Cartilage-Forming Tumors

Hyaline, mature cartilage is composed of a small number of cells and a large, water-rich matrix of collagen and mucopolysaccharides. The high water content of the hyaline cartilage contributes to the very high SI seen on T2WI. Hyaline cartilage is present in articular surfaces, growth plates of long bones, and in benign or low grade malignant tumors, including enchondromas (Fig. 10.18), osteochondromas (Fig. 10.19), and some chondrosarcomas (see Fig. 10.11). The "lobules within lobules" appearance has been described as a rather characteristic finding in hyaline cartilage-matrix tumors (11).

The high SI on T2WI produced by the hyaline cartilage allows assessment of the thickness of the cartilaginous cap present in osteochondromas (30) (Fig. 10.19). A thickness greater than 3 mm suggests malignant transformation (11,26). The presence of a bursa adjacent to the tip of an osteochondroma can be readily identified with MRI if there is bursitis.

Giant cell tumors are studied well with conventional techniques. Limited experience with MRI suggests that most giant cell tumors demonstrate low SI on T1WI and uniform high SI on T2WI. Inhomogeneity of the SI on T2WI has been found in the more aggressive type of giant cell tumors (25) (Fig. 10.20).

Aneurysmal bone cysts often demonstrate septations and fluid-fluid levels on CT. MRI can depict similar changes. Other MRI findings described in aneurysmal bone cysts include different SI of the cystic parts of the tumor, probably due to different

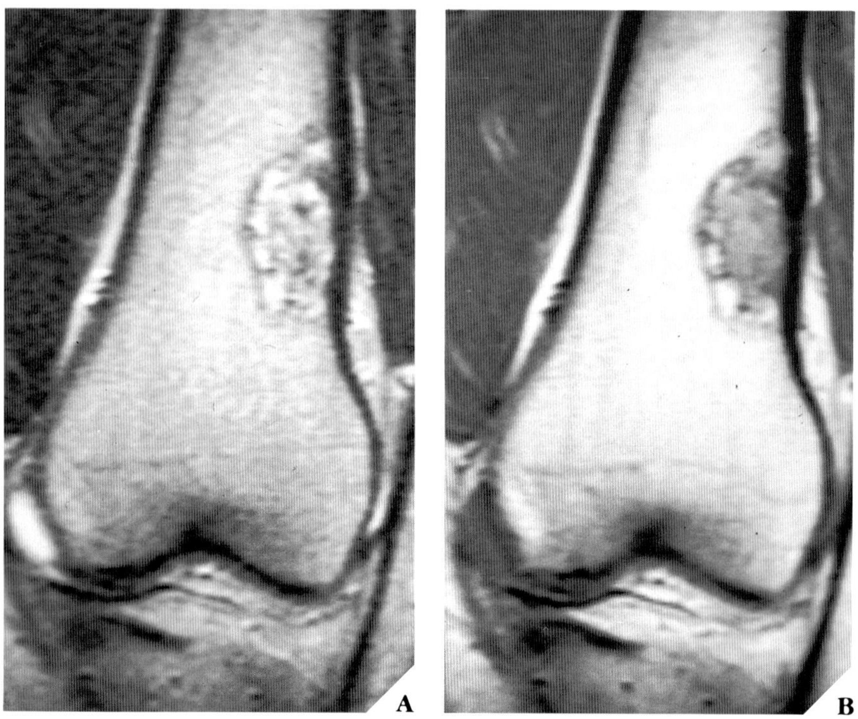

FIGURE 10.17. Nonossifying Fibroma. Coronal sections [**A** SE 2000/20; **B** SE 2000/80] through the distal femur showing a well-circumscribed tumor with inhomogeneous areas of low and high SI.

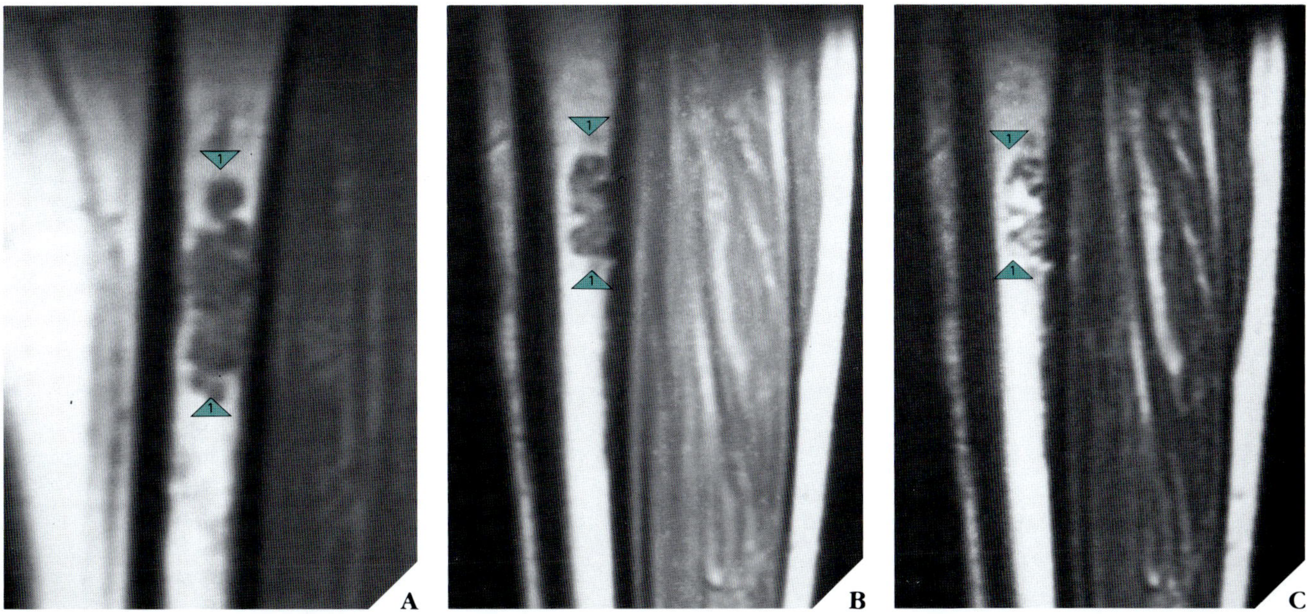

FIGURE 10.18. Enchondroma. A Coronal section [SE 800/20], **B** and **C** sagittal sections [SE 2000/20 and SE 2000/80], through the femur. There is a polylobulated intramedullary lesion demonstrating low SI on T1WI and PDWI (**A** and **B**), and a mixed pattern with areas of high and low SI on T2WI (**C**). High SI regions on T2WI represent cartilaginous matrix, and the low SI regions on PDWI and T2WI (**B** and **C**) are due to calcifications, as seen on plain films. Arrows **1** enchondroma.

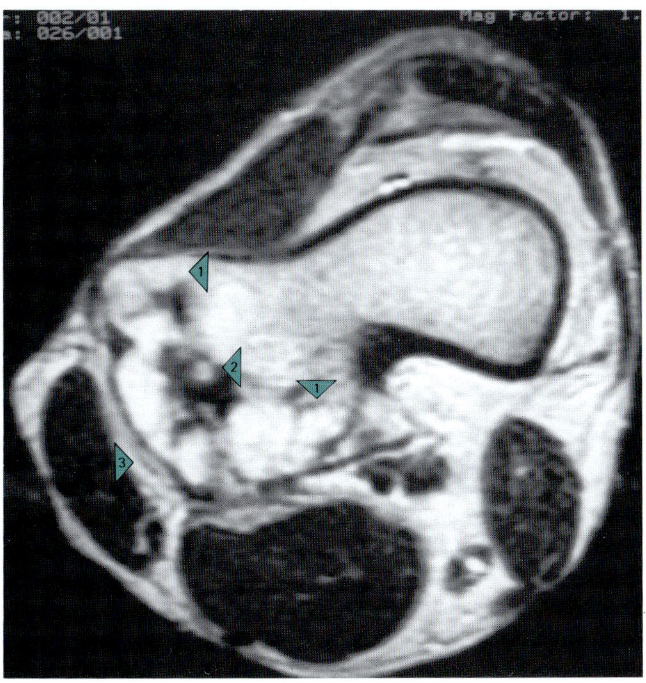

FIGURE 10.19. Osteochondroma. Axial section through the distal femur [SE 2000/80]. There is a large osteochondroma with high SI islands of hyaline cartilage mixed with low SI calcifications. Note the "lobules within lobules" appearance. A thin cartilaginous cap is identified in the surface of the tumor. The bone marrow of the humerus extends into the base of the osteochondroma. Arrows **1** hyaline cartilage lobules. Arrow **2** matrix calcifications. Arrow **3** cartilaginous cap.

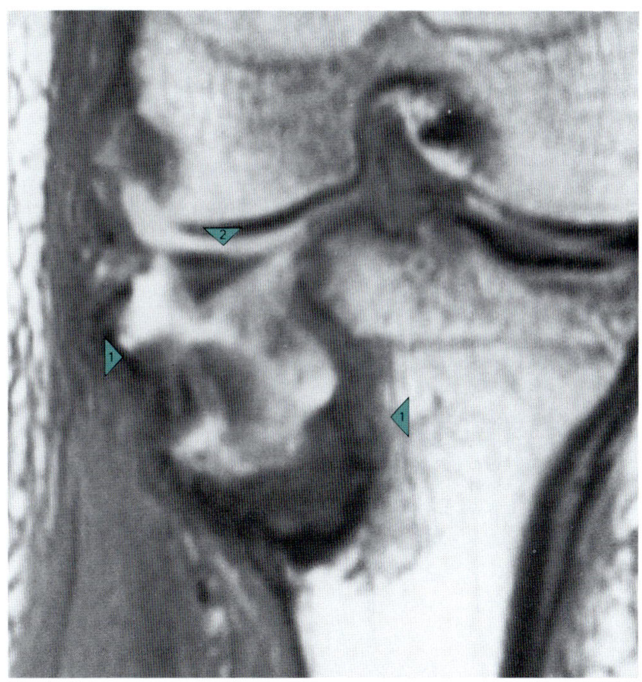

FIGURE 10.20. Giant Cell Tumor. Coronal section [SE 800/20] through the knee. There is a tumor expanding the lateral aspect of the proximal tibial epiphysis, extending into the articular surface, associated with a pathological fracture. The lateral meniscus is in contact with the tumor. Arrows **1** giant cell tumor. Arrow **2** lateral meniscus.

stages of hemoglobin breakdown in the hemorrhagic content of the tumor, and small diverticular-like outpouches in the periphery of the tumor (3,27) (Fig. 10.21).

Lymphoma, Metastasis, and Multiple Myeloma

No specific MRI characteristics are seen in frequent tumors such as lymphoma, metastasis, or multiple myeloma (Figs. 10.1 and 10.3). Most of these lesions are well detected with MRI by virtue of their lower SI on T1WI than the surrounding bone marrow, and high SI on T2WI, as it occurs in the vast majority of tumors. When these tumors involve the spine they often produce collapse of the vertebral bodies. In these cases differentiation from osteoporotic collapse may be difficult on the basis of plain films. In the acute stage of osteoporotic fracture the SI of the collapsed vertebral body is low on T1WI and high on T2WI. The same changes occur in pathologic collapse. Early experience indicates that gadolinium DTPA produces enhancement

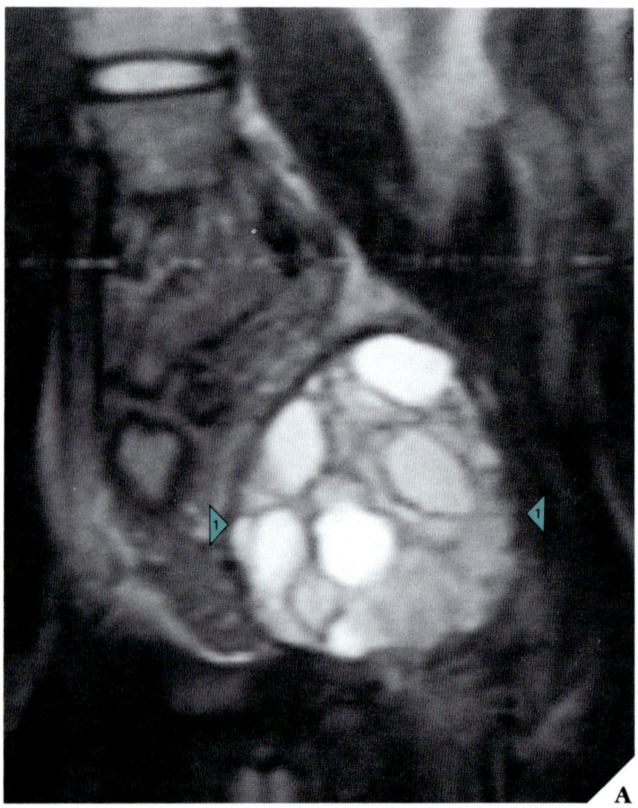

FIGURE 10.21. Aneurysmal Bone Cyst. A Coronal section through the pelvis [SE 2000/80]. **B** Sagittal section [SE 500/20]. There is a large expansile tumor arising from the left acetabulum extending into the pelvis. There is a low SI rim around the tumor which could represent pseudocapsule or expanded cortex. There are multiple cystic spaces inside the tumor, with fluid-fluid levels **(B)**, and different SI on T1WI **(B)** and T2WI **(A)**, probably due to different stages of the hemoglobin breakdown products. Arrows **1** aneurysmal bone cyst. Arrow **2** fluid-fluid level inside one of the cystic spaces.

of the tumoral vertebrae on T1WI, but no enhancement is seen in the osteoporotic vertebrae (17). Old osteoporotic fractures show high SI on T1WI and low SI on T2WI, due to fatty bone marrow infiltration. This pattern differs from tumoral collapse, allowing for an easy distinction (Fig. 10.22).

Osteosarcomas and Ewing's tumors are frequently studied with MRI to assess their soft tissue and intramedullary extension (7) (Figs. 10.6, 10.8, 10.9, and 10.12). The SI characteristics of these tumors are not different from other lesions, with the exception of the rare case of telangiectatic osteosarcoma, which can demonstrate fluid levels on MRI (Fig. 10.23). Plain film findings and biopsy provide the necessary information for histologic diagnosis. Phosphorus 31 and proton MR spectroscopy have shown some promise in evaluating tumor response to chemotherapy (6).

▼▼▼ *SOFT TISSUE TUMORS*

With some exceptions, most soft tissue tumors show low SI on T1WI and high SI on T2WI.

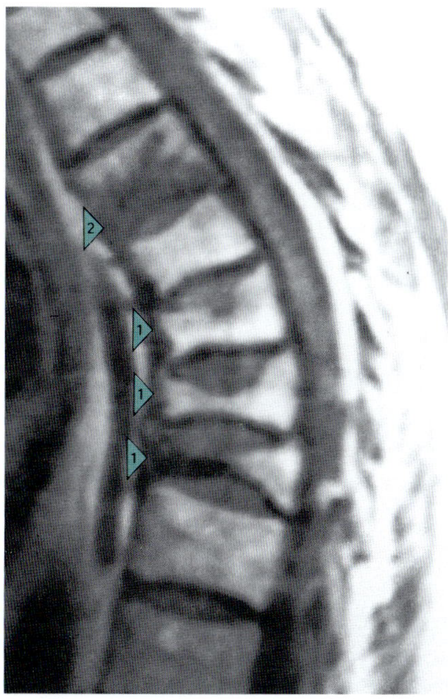

FIGURE 10.22. Chronic Osteoporotic Collapse and Metastasis. Sagittal dorsal spine [SE 800/20]. There are chronic osteoporotic collapses of T10, T11, and T12 vertebral bodies, with high SI fatty marrow and kyphosis. The vertebral body of T8 is not collapsed but shows low SI on this T1WI, due to metastasis. Arrows **1** osteoporotic collapse. Arrow **2** metastasis.

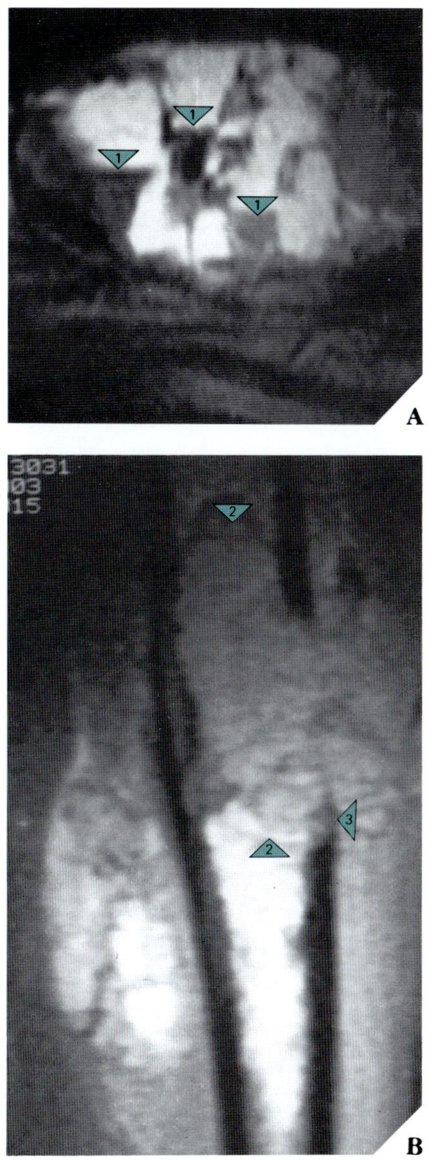

FIGURE 10.23. Telangiectatic Osteosarcoma. Coronal plane [SE 2000/20.] Two coronal consecutive sections through the arm. **A** There are multiple fluid levels within the large soft tissue component of the tumor. The high SI of the fluid levels in the upper portion of the cystic spaces is due to hemorrhagic events. **B** The intramedullary extension of the tumor is better demonstrated in this section. Note the cortical breakthrough posteriorly. Arrows **1** fluid levels. Arrows **2** intramedullary tumor. Arrow **3** cortical disruption.

Lipomas

Lipomas, fatty components of liposarcomas, and tumors with areas of hemorrhage may demonstrate high SI on T1WI (36) (Figs. 10.24, 10.25). Some soft tissue tumors may exhibit low SI on T2WI. Apparently this is related to the presence of abundant collagen and hypocellularity of the tumor but it does not reflect the biological behavior of the lesion (37).

Fibromas

Neurofibroma, malignant fibrous histiocytoma, and aggressive fibromatosis are among the tumors that may show diffuse or segmental low SI on T2WI (Fig. 10.10). Plexiform neurofibromas are characterized by their lobulated undulating contour. A typical pattern of well-demarcated low SI center within the lobules is best seen on T2WI. This low SI center is due to increased fibrous tissue (8) (Fig. 10.26).

Hemangiomas

Hemangiomas involving the soft tissues or bone also demonstrate low SI on T1WI and are markedly hyperintense on T2WI. In addition, a characteristic striated or mottled pattern probably corresponding to loose fibrous connective tissue separating endothelial-lined, dilated vascular channels can be seen in large cavernous hemangiomas (12,28,31,45) (Fig. 10.27).

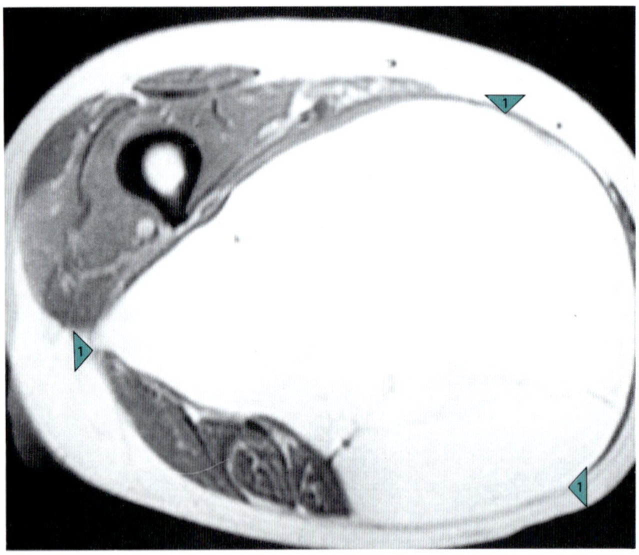

FIGURE 10.24. Lipoma. Axial section through the thigh [SE 800/20]. There is a large high SI mass in the posterior compartment of the thigh, producing compression of the adjacent muscles. Homomgeneous high SI on T1WI and low SI on T2WI (not shown) are typical of lipomas. Arrows **1** lipoma.

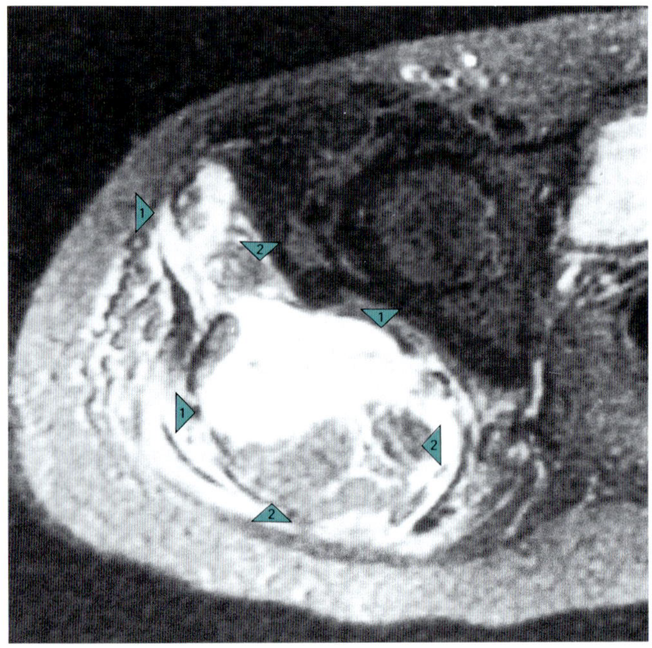

FIGURE 10.25. Liposarcoma. Axial section through the pelvis [SE 2000/80]. There is a large tumor with mixed SI characteristics. The high SI areas on this T2WI correspond to the solid component of the liposarcoma. The low SI areas, similar to the subcutaneous fat, correspond to the fatty component of the tumor. Arrows **1** solid component. Arrows **2** fatty component.

NONTUMORAL SOFT TISSUE MASSES

Because of the above-described advantages of MRI over other imaging techniques for the evaluation of soft tissue masses, nontumoral lesions are frequently referred for MRI assessment. Careful analysis of the MRI characteristics should help in differentiating nontumoral masses from soft tissue neoplasms and prevent unnecessary or potentially harmful biopsy procedures.

Cysts

Synovial and ganglion cysts frequently present as a palpable mass in the vicinity of a joint. Thin needle aspiration rather than large needle biopsy can be performed if there is clinical concern of a neoplasm. MRI of cysts and solid tumors may be similar, but cysts tend to exhibit a more uniform distribution of the signal than neoplasms, and they show higher SI on T2WI than solid tumors (see Synovial Cysts in Chapter 7).

Pseudoaneurysms

Pseudoaneurysms are evaluated by palpation, auscultation, ultrasound, and/or angiography. Occasionally, however, they may be confused with a soft tissue solid neoplasm and referred for MRI examination. For obvious reasons, biopsy should be

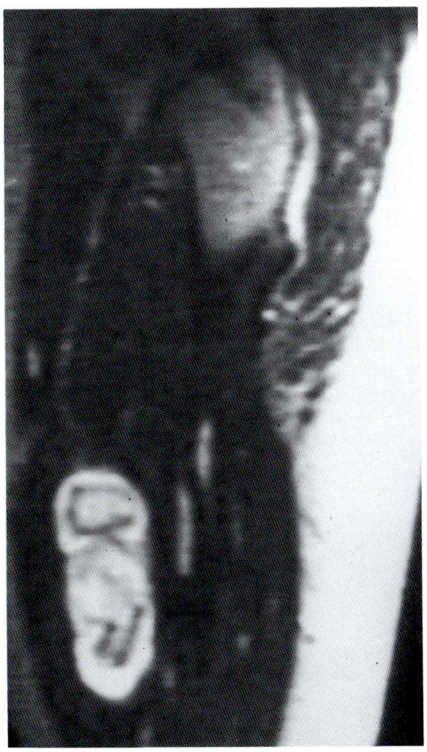

FIGURE 10.26. Plexiform Neurofibroma. Coronal section [SE 2000/80] through the anterior thigh, demonstrating a high SI lobulated tumor with a central low SI area on this T2WI.

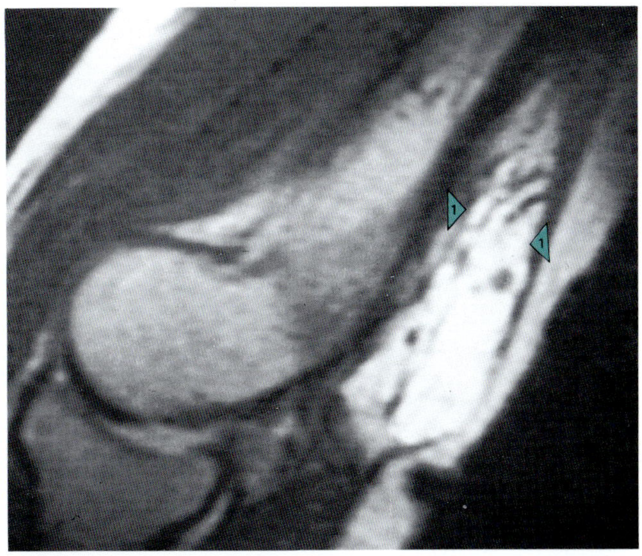

Figure 10.27. Hemangioma. Sagittal section through the elbow [SE 2000/80] demonstrating an area of high SI with the characteristic striated pattern (arrows **1**).

avoided. MRI in pseudoaneurysms demonstrates a characteristic flow-related artifact in the direction of the phase-encoding gradient (Fig. 10.28).

Lymphadenopathy

Lymphadenopathy may be difficult to differentiate from solid tumor on MRI. Enlarged lymph nodes have a tendency to enlarge in clusters, in

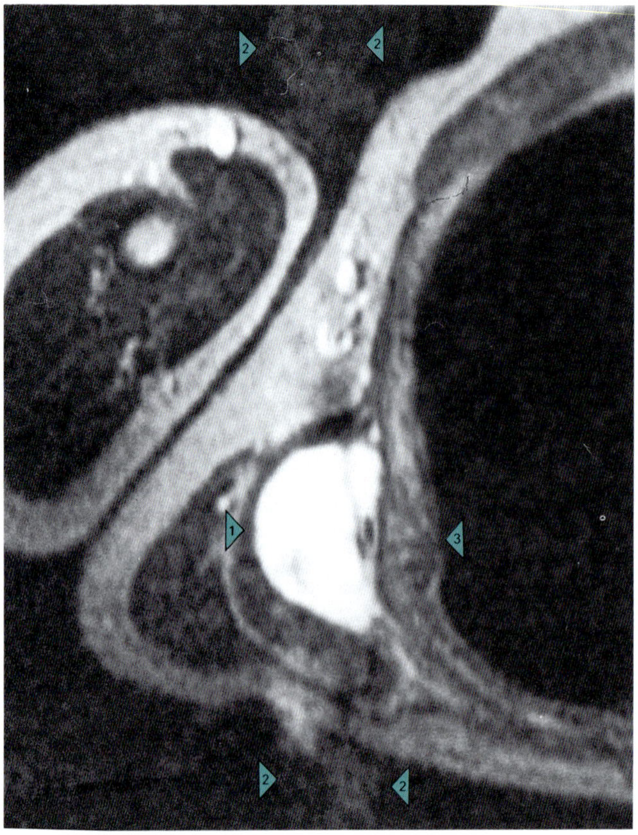

FIGURE 10.28. Pseudoaneurysm of the Intercostal Artery. Axial section through the midchest wall [SE 2000/80] demonstrating a high SI mass with characteristic artifact in the direction of the phase-encoding gradient (superior to inferior) produced by flowing blood inside of the aneurysm. A healing rib fracture is noted adjacent to the mass. Arrow **1** pseudoaneurysm. Arrows **2** flow artifact. Arrow **3** rib fracture.

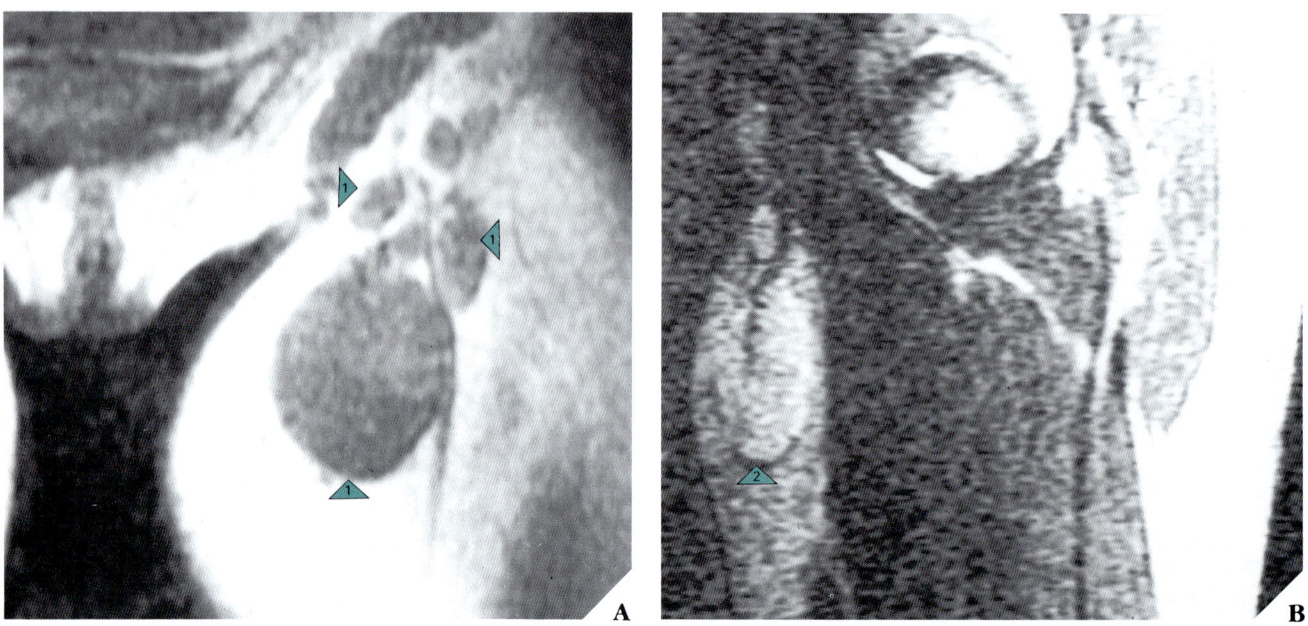

FIGURE 10.29. Lymphadenopathy. A Coronal section [SE 800/20] through the inguinal region demonstrating a cluster of low SI lymph nodes (arrows **1**). **B** Sagittal section [SE 2000/80] through the largest lymphadenopathy in the same patient, showing SI similar to that of the subcutaneous fat. Arrow **2** lymphadenopathy.

specific areas, such as the inguinal region and axilla. On MRI they exhibit more uniform SI than other solid neoplasms, and their SI on T2WI is similar to that of the subcutaneous fat (Fig. 10.29).

Abscesses

Abscesses can demonstrate SI characteristics similar to those of fluid-filled cysts (see Chapter 11).

Hematomas

Hematomas may simulate a fluid collection during their early stages, but as the products of the breakdown of the hemoglobin are released, the SI on T1WI tends to increase (23) (Fig. 10.30).

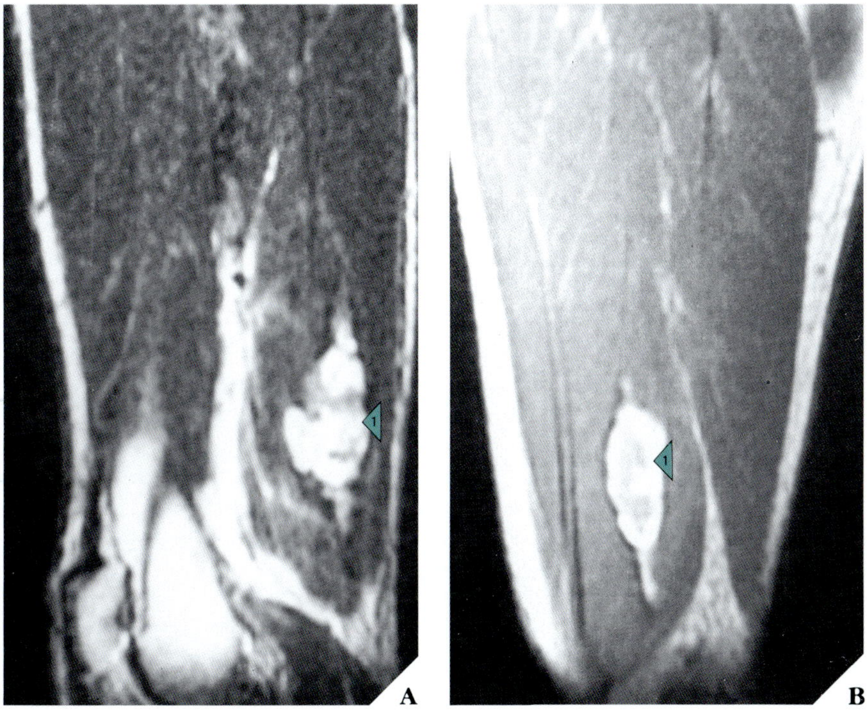

FIGURE 10.30. Subacute Hematoma. **A** Surface coil sagittal section [SE 2000/20] demonstrating a high SI lobulated hematoma in the posterior aspect of the thigh on this T2WI (arrows **1**). **B** Coronal section [SE 800/20] also showing high SI on T1WI (arrow **1**).

REFERENCES

1. Beltran J, Noto AM, Chakeres DW, et al. Tumors of the osseous spine: staging with MR imaging versus CT. Radiology 1987;162:565–569.
2. Beltran J, Simon DC, Katz W, et al. Increased MR signal intensity in skeletal muscle adjacent to malignant tumors: pathologic correlation and clinical relevance. Radiology 1987;62:251–255.
3. Beltran J, Simon DC, Levy M, et al. Aneurysmal bone cysts: MR imaging at 1.5T. Radiology 1986;58:689–690.
4. Bloem JL, Doornbos J, Taminiau AHM, et al. Gd-DTPA-enhanced MR imaging of bone tumors. Scientific exhibit presented at the 74th meeting of the RSNA, Chicago, December 1988.
5. Bloem JL, Taminiau AHM, Euldernik F, et al. Radiologic staging of primary bone sarcoma: MR imaging, scintigraphy, angiography, and CT correlated with pathologic examination. Radiology 1988;169:805–810.
6. Bottomley PA. Human *in vivo* NMR spectroscopy in diagnostic medicine: clinical tool or research probe? Radiology 1989;170:1–15.
7. Boyco OB, Ciry DA, Cohen MD, et al. MR imaging of osteogenic sarcoma and Ewing's sarcoma. AJR 1987;148:317–322.
8. Burk DL, Brunberg JA, Kanal E, et al. Spinal and paraspinal neurofibromatosis: surface coil MR imaging at 1.5T. Radiology 1987;162:797–801.

9. Burk DL, Dalinka MK, Schiebler ML, et al. Strategies for musculoskeletal magnetic resonance imaging. Radiol Clin North Am 1988;26:653–672.
10. Byder GM, Young IR. MR imaging: clinical use of the inversion recovery sequence. J Comput Assist Tomogr 1985;9:659–675.
11. Cohen EK, Kressel HY, Frank TS, et al. Hyaline cartilage-origin bone and soft-tissue neoplasms: MR appearance and histologic correlation. Radiology 1988;167:477–481.
12. Cohen EK, Kressel HY, Perosio T, et al. MR imaging of soft-tissue hemangiomas: correlation with pathologic findings. AJR 1988;150:1079–1081.
13. Collin C, Hajdu SI, Goldbold J, et al. Localized, operable soft tissue sarcoma of the lower extremity. Arch Surg 1986;121:1425–1433.
14. Damadian R. Tumor detection by nuclear magnetic resonance. Science 1971;171:1151–1153.
15. Demas BE, Heelan RT, Lane J, et al. Soft-tissue sarcomas of the extremities: comparison of MR and CT in determining the extent of disease. AJR 1988;150:615–620.
16. Dixon WT. Simple photon spectroscopic imaging. Radiology 1984;153:189–194.
17. Dooms G, Maldague B, Malhem J, et al. MR differential diagnosis between benign and malignant vertebral collapses. Presented at the 74th meeting of the RSNA. Chicago, December 1988.
18. Enneking WF, Spanier SS, Goodman MA. A system for the surgical staging of musculoskeletal sarcoma. Clin Orthop 1980;153:106–119.
19. Enneking WF. Staging musculoskeletal neoplasms. Skeletal Radiol 1985;13:183–194.
20. Erlemann R, Reiser M, Stoeber U, et al. MR imaging with static and dynamic Gd-DTPA studies in neoplasms of the musculoskeletal system. Presented at the 74th meeting of the RSNA. Chicago, December 1988.
21. Gillespie T, Manfrini M, Ruggieri P, et al. Staging of intraosseous extent of osteosarcoma: correlation of preoperative CT and MR imaging with pathologic macroslides. Radiology 1988; 167:765–767.
22. Glass RB, Poznanski AK, Fisher MR, et al. MR imaging of osteoid osteoma. J Comput Assist Tomogr 1986;10:1065–1067.
23. Gomori JM, Grossman RI, Goldberg HI, et al. Intracranial hematomas: imaging by high-field MR. Radiology 1985;157:87–93.
24. Hendrik RE, Kneeland JB, Stark DD. Maximizing signal-to-noise and contrast-to-noise ratios in FLASH imaging. Magn Reson Imaging 1987;5:117–187.
25. Herman SD, Mesgarzadeh M, Bonakdarpour A, et al. The role of magnetic resonance imaging in giant cell tumor of bone. Skeletal Radiol 1987;16:635–643.
26. Hudson TM, Hamlin DJ, Enneking WF, et al. Magnetic resonance imaging of bone and soft tissue tumors: early experience in 31 patients compared with computed tomography. Skeletal Radiol 1985;13:134–146.
27. Hudson TM, Hamlin DJ, Fitzsimons JR. Magnetic resonance imaging of fluid levels in aneurysmal bone cysts and in anticoagulated human blood. Skeletal Radiol 1985;13:267–270.
28. Kaplan PA, Williams SM. Musculocutaneous and peripheral soft-tissue hemangiomas: MR imaging. Radiology 1987;163:163–166.
29. Karakousis CP, Emrich LJ, Rao U, et al. Feasibility of limb salvage and survival in soft tissue sarcomas. Cancer 1986;57:484–491.
30. Lee JK, Yao L, Wirth CR. MR imaging of solitary osteochondromas: report of eight cases. AJR 1987;149:557–560.
31. Levine E, Wetzel LH, Neff JR. MR imaging and CT of extrahepatic cavernous hemangiomas. AJR 1986:147:1299–1304.
32. Lodwick GS, Wilson AJ, Farrell C, et al. Determining growth rate of focal lesions of bone from radiographs. Radiology 1980;134:577–583.
33. Petterson H, Eliasson J, Egund N, et al. Gadolinium-DTPA enhancement of soft tissue tumors in magnetic resonance imaging—preliminary clinical experience in five patients. Skeletal Radiol 1988;17:319–323.
34. Petterson H, Gillespie T, Hamlin DJ, et al. Primary musculoskeletal tumors: examination with MR imaging compared with conventional modalities. Radiology 1987;164:237–241.
35. Petterson H, Slone RM, Spanier S, et al. Musculoskeletal tumors: T1 and T2 relaxation times. Radiology 1988;167:783–785.
36. Sundaram M, McGuire MH, Herbold DR, et al. High signal intensity of soft tissue masses on T1 weighted pulse sequences. Skeletal Radiol 1987;16:30–36.

37. Sundaram M, McGuire MH, Schajowicz F. Soft-tissue masses: histologic basis for decreased signal (short T2) on T2-weighted MR images. AJR 1987;148:1247–1250.
38. Vanel D, Coffre C, Couanet D, et al. Potpourri of limitations and pitfalls of MRI imaging in primary musculoskeletal tumors. Presented at the 74th meeting of the RSNA. Chicago, December 1988.
39. Vanel D, Lacombe MJ, Couanet D, et al. Musculoskeletal tumors: follow-up with MR imaging after treatment with surgery and radiation therapy. Radiology 1987;164:243–245.
40. Vogler JB, Murphy WA. Bone marrow imaging. Radiology 1988;168:679–693.
41. von Schulthess GK, Kuoni W, Wuthrich R, et al. Soft-tissue and bone lesions examined with 1.5-T MR imaging and Gd-DTPA. Presented at the 74th meeting of the RSNA. Chicago, December 1988.
42. Wehrli FW: Introduction to fast-scan magnetic resonance. General Electric Co. 1986.
43. Wismer GL, Rosen BR, Buxton R, et al. Chemical shift imaging of bone marrow: preliminary experience. AJR 1985;145:1031–1037.
44. Yeager BA, Schiebler ML, Schmidt RG, et al. MR imaging of osteoid osteoma of the talus. J Comput Assist Tomogr 1987;11:916–917.
45. Yuh WTC, Kathol MH, Sein MA, et al. Hemangiomas of skeletal muscle: MR findings in five patients. AJR 1987;149:765–768.

CHAPTER ELEVEN

Vijay Chandnani
Javier Beltran

Musculoskeletal Infections

GENERAL CONSIDERATIONS

Musculoskeletal infections are common disorders which may present in an acute course, or in a slower, more insidious manner, and which may pose a diagnostic and therapeutic challenge for many specialists in medicine. Early diagnosis and treatment is essential to prevent growth disorders and limb deformity (14,18,19). In the preantibiotic era, mortality was common and, even today, morbidity remains significant.

The four major types of musculoskeletal infections are osteomyelitis, abscess, septic arthritis, and tenosynovitis. These infections may result either from the hematogenous spread of a distant septic focus or from the extension of a contiguous septic source. Another route of infection is direct implantation of septic organisms, which may occur from animal or human bites, surgery, or the implantation of prostheses. In any given individual, several of these factors may be responsible for promoting a musculoskeletal infection.

Many radiologic modalities can be used to diagnose musculoskeletal infections, each having its own inherent advantages and limitations. These include plain films, nuclear medicine, CT, and MRI.

Plain Film Radiography

Plain film radiography with or without tomography is a sensitive technique for detecting bony abnormalities, such as periosteal reaction, permeative lucencies, sequestrum, involucrum formation, and the presence of cloacae. However, these changes are not typically demonstrated until at least 10–14 days after the onset of infection, or until at least 35–40 per cent of bone destruction has occurred (5).

Abnormalities such as healing comminuted fractures and diabetic neuroarthropathy, which are difficult to differentiate from other associated skeletal lesions, are best demonstrated by radiography. Soft tissue changes such as the obliteration of fat planes and joint effusions can be demonstrated as early as 3 days, but these changes are relatively nonspecific (4).

Computerized Tomography

Computerized tomographic scanning has greater sensitivity in diagnosing musculoskeletal infections than roentgenography (11), and is also better at detecting intraosseous or soft tissue gas. Computerized tomography is also useful in differentiating soft tissue from bony infections. It can differentiate an organized, drainable fluid collection from cellulitis, and is helpful in surgical planning (1,2,11). Its application, however, is limited in patients with metallic implants, and in those lesions involving the small bones of the hand and feet. Beam hardening artifacts also can be problematic in accurately depicting the soft tissues around the bones. Clinical experience as well as experimental studies of musculoskeletal infections in laboratory animals, have shown CT to be less sensitive than MR imaging in detecting osteomyelitis and abscesses (4,5).

Nuclear Medicine

Varying combinations of scintigraphic techniques, including technetium 99m-MDP, gallium 67, and indium 111-leukocyte-labeled scintigraphy, are considered to be quite sensitive for demonstrating bone and soft tissue infections. However, the perceived anatomic extent of disease is sometimes inaccurate, and it is not always possible to differentiate bone from soft tissue infection or an abscess from cellulitis (4). Moreover, the accuracy of scintigraphic imaging is diminished in cases of chronic osteomyelitis (2) or in osteomyelitis associated with healing fractures or postsurgical healing. The labeling procedure for In-111 scans is long (3–4 h), and a question has been raised regarding the long-term radiation damage to the lymphocytes (15). Moreover, there have been reports of osteomyelitis associated with negative scintigraphic scans (12).

Magnetic Resonance Imaging

Magnetic resonance imaging has also been described as being efficacious in the evaluation of musculoskeletal infections (3–6,10,16). Its advantages include improved spatial resolution over scintigraphy, and good differentiation of osteomyelitis from soft tissues infections, and cellulitis from abscesses. The sensitivity of MRI in detecting infections is greater than that of plain film radiography and CT, and is equal to that of scintigraphy. Magnetic resonance imaging can accurately stage the extent of infection and has the capability to perform images in multiple planes. Moreover, there is little image degradation in the presence of nonferromagnetic prostheses, even in imaging the tissue surrounding a prosthetic hip. A major advantage of MRI is that it does not use ionizing radiation (5,6,16).

TECHNICAL CONSIDERATIONS

IMAGING PLANES AND PULSE SEQUENCES

As in cases of other suspected musculoskeletal diseases, evaluation of musculoskeletal infections requires both T1WI with short TR/TE, and PDWI/T2WI with long TR/short and long TE. T1WI and PDWI provide excellent depiction of the anatomy, including vascular structures, muscles, cartilage, and bone. T2WI are reliable in depicting the changes associated with hyperemia and infection in the bone marrow and surrounding soft tissues. In some cases, short thau inversion recovery pulse sequences (STIR) can help in delineating the extent of the inflammatory changes.

At least two imaging planes should be obtained for accurate depiction of the infection (17). In anatomically complex areas such as the pelvis, spine, foot, and hand, three planes may be necessary.

OTHER TECHNICAL PARAMETERS

Slice thickness of 5 mm with 1-mm gaps in a 256 × 128 matrix with 2 NEX is sufficient in most cases.

Surface coil imaging is strongly recommended whenever possible, especially when dealing with infections in the small bones of the hands and feet. Different types of surface coils have been discussed in previous chapters.

ACUTE OSTEOMYELITIS

Acute osteomyelitis is mainly a disease of childhood, although it can occur in adult patients and in neonates. Fever, pain, swelling, redness, and limited motion are the classical clinical features. Males are affected more commonly than females in the childhood and adult forms, but males and females are equally affected in the neonatal form. One or more bones may be affected; the polyostotic form is common in infants. In infants and children the long tubular bones are the most commonly affected area, whereas in adults, the small bones of the hands and feet, as well as the spine are most frequently involved (18). Cases of osteomyelitis are usually secondary to infection with *Staphylococcus aureus* (18).

Pathologic changes of inflammation within the bone include vascular engorgement, edema, cellular response, and abscess formation. Penetration of the infection into the cortex is facilitated by the osteoporosis resulting from the inflammatory response. With further extension of the infection, periosteal elevation can take place. The elevation of the periosteum may cause interruption of blood supply to the outer surface of the cortex, and subsequent sequestrum formation. If the infected fluid penetrates the periosteum, it may lead to cellulitis or abscess formation in the surrounding soft tissues (18).

The earliest evidence of acute osteomyelitis that is discernible on MRI is areas of increased SI in the medulla seen on T2WI (4–6,10,16) (Fig. 11.1). The

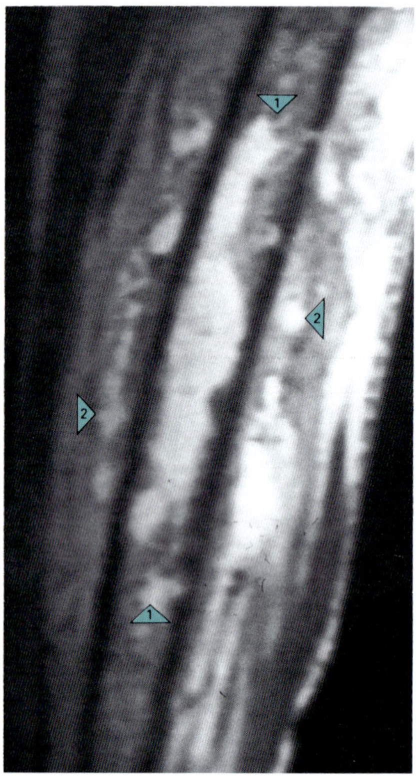

FIGURE 11.1. Hematogenous Acute Osteomyelitis of the Femur. Sagittal plane [SE 2000/80]. There is a well-defined area of increased SI in the medullary space of the midshaft of the femur representing osteomyelitis in this intravenous drug abuser. There are soft tissue inflammatory changes with multiple small abscesses adjacent to the femur. Arrows **1** Osteomyelitis. Arrows **2** Small soft tissue abscesses.

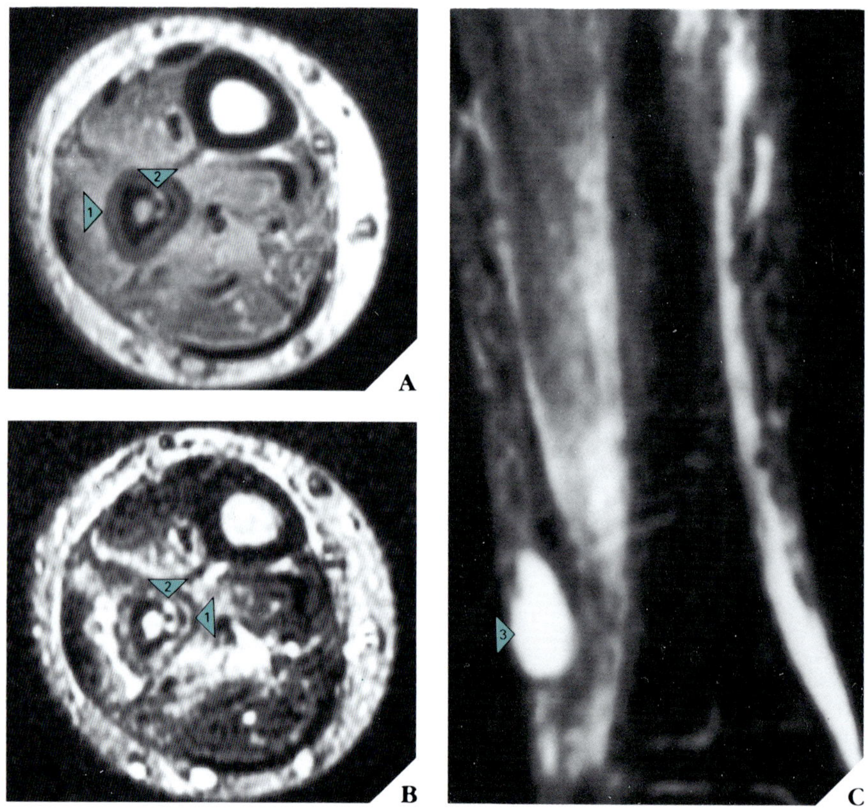

A

B

C

FIGURE 11.2. Acute Osteomyelitis with Periosteal Reaction. Axial plane **(A)** [SE 2000/20] and **(B)** (SE 2000/80), and coronal plane **(C)** [STIR 2000/160/30] through the distal leg in a patient with inflammatory symptoms. Differential diagnostic considerations were Ewing's sarcoma and acute osteomyelitis. MR images demonstrate high SI of the bone marrow space of the fibula on T2WI **(B)**, with a "double ring" around the fibula reflecting a periosteal reaction. There is also high SI on T2WI in a small segment of the cortex due to cortical bone infiltration of the infection. Soft tissue edema around the fibula is noted on T2WI **(B)**. Coronal STIR image **(C)** demonstrates a fluid collection adjacent to the fibula, representing a soft tissue abscess. Arrows **1** Periosteal reaction. Arrows **2** Cortical infection. Arrow **3** Abscess.

changes that result from inflammation within the cortex are more elusive, but may be seen as increased SI on T2WI (Fig. 11.2). The presence of inflammation within the subperiosteal space may be seen also as high SI areas on T2WI surrounding the cortex. The elevated periosteum itself may be imaged as a black line within the high signal of the subperiosteal inflammation surrounding the cortex (Figs. 11.2 and 11.3).

These changes of osteomyelitis are relatively nonspecific on MRI and may be seen in a variety of other pathological conditions, such as tumors, bone marrow edema, and intramedullary hemorrhage. However, correlation with the clinical findings often clarifies the specific diagnosis. Associated soft tissue fluid collections representing an abscess favor the diagnosis of infection (see Fig. 11.2).

Cortical infection alone, without intramedullary infection (osteitis), can occur by direct spread of the infection from a source adjacent to the cortical bone. In these cases, periosteal elevation occurs first, followed by cortical infection, and finally bone marrow infection. Magnetic resonance imaging can help in defining the soft tissue and assessing the presence or absence of bone infection (see Fig. 11.3).

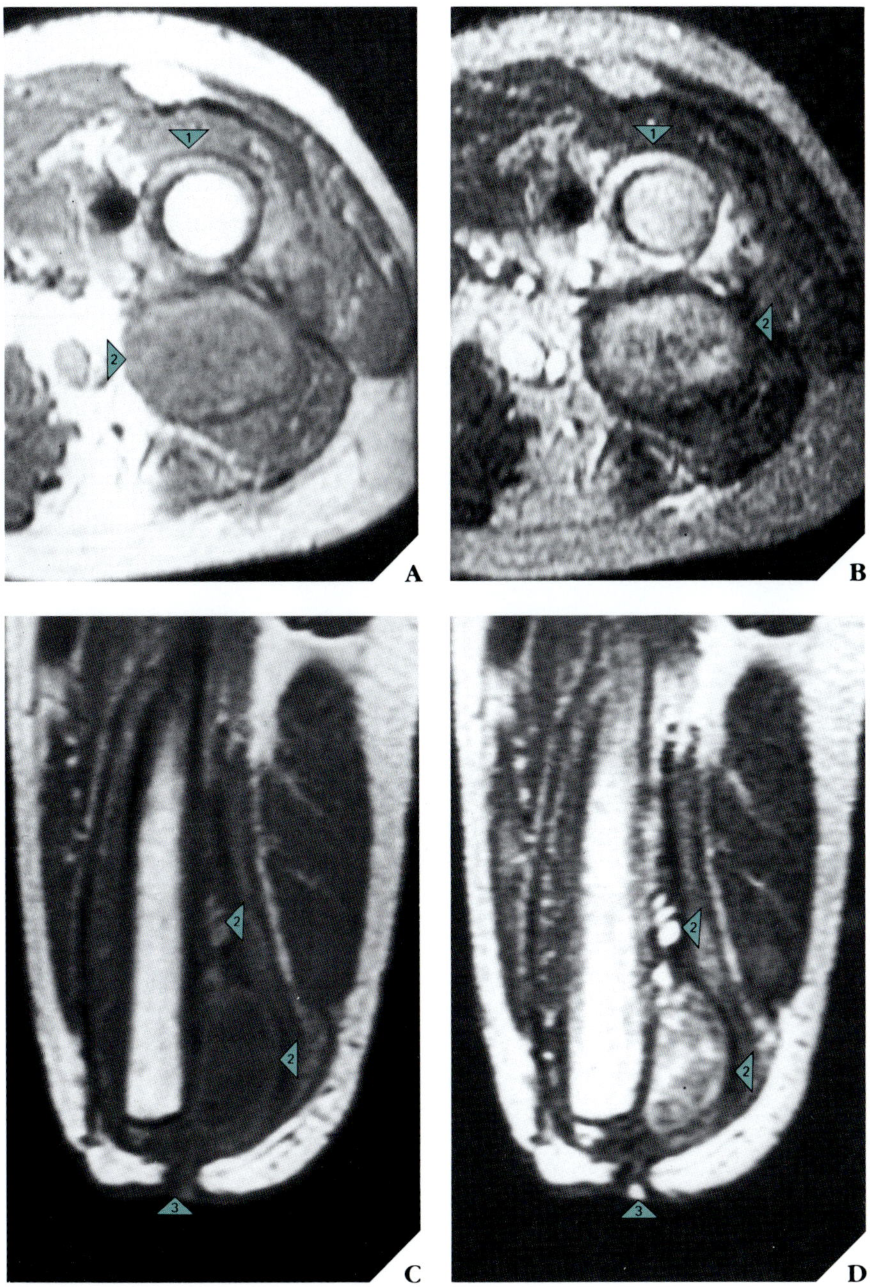

FIGURE 11.3. Bone Infection by Contiguity. Axial plane **(A)** [SE 2000/20] and **(B)** [SE 2000/80], and sagittal plane **(C)** [SE 2000/20] and **(D)** [SE 2000/80] through the thigh. This patient had an above-knee amputation and, following surgery, clinical signs of infection became evident with a draining sinus in the stump. MRI in the coronal plane (**A** and **B**) demonstrates a low SI ring around the femoral cortex representing periosteal reaction. The periosteal reaction is separated from the cortical bone by a high SI ring, which most likely represents inflammatory tissue. The SI of the bone marrow space is normal. An abscess with low SI at the center is seen adjacent to the femur. Sagittal images (**C** and **D**) demonstrate the draining sinus, several soft tissue abscesses, and normal SI of the bone marrow. This constellation of findings is consistent with infection of the bony surface by contiguity, but without bone marrow infection. Arrows **1** Periosteal reaction. Arrows **2** Soft tissue abscesses. Arrows **3** Draining sinus tract.

SUBACUTE AND CHRONIC OSTEOMYELITIS

Acute osteomyelitis can proceed to its subacute or chronic forms in specific clinical circumstances. Partial or inadequate antibiotic therapy, for example, can result in an infection produced by organisms with decreased virulence, or in the increased resistance of the host to treatment (18).

Interruption of the blood supply to the cortex by fluid in the subperiosteal space can lead to cortical necrosis and sequestrum formation. Sites of active infection may be indicated by the presence of one or more intraosseous abscesses (Brodie's abscesses). These abscesses are circumscribed lesions that tend to be located in the ends of a long tubular bone. They are especially common in male children, and typically occur in either the proximal or distal metaphyseal regions of the tibia. They are characteristically 1–4 cm in size and their walls are lined by inflammatory granulation tissue (18). There may be cloacae draining to the skin surface.

While roentgenographic criteria for detection of subacute and chronic osteomyelitis include the presence of cloaca, involucrum, sequestrum, and Brodie's abcess, in clinical practice the radiograph often shows sclerotic and lucent regions scattered among regions of bony thickening and deformity. Thus, accurate differentiation of healing from infected areas is not always possible.

Computerized tomographic imaging, as well, depends upon demonstrating the presence of cloacae, involucrum, and sequestrum in detecting chronic osteomyelitis, rather than on direct imaging of the infectious process (11,17).

Brodie's Abscess

Brodie's abscess appears on MRI as a well-demarcated area that shows low SI on T1WI and high SI on T2WI; the surrounding regions of thickened sclerotic bone show low SI on T1WI and T2WI (Fig. 11.4). Bone marrow edema around the intraosseous abscess can often be seen as ill-defined areas of high SI on T2WI (Fig. 11.5). The granulation tissue lining the inner wall of the abscess may be seen as a high SI band on T2WI producing a "double-line" sign similar to that seen in avascular necrosis (Fig. 11.6). The smooth, rounded configuration of the abscess is, however, different from the undulating pattern of avascular necrosis. In the vast majority of cases, clinical findings are instrumental in distinguishing these entities.

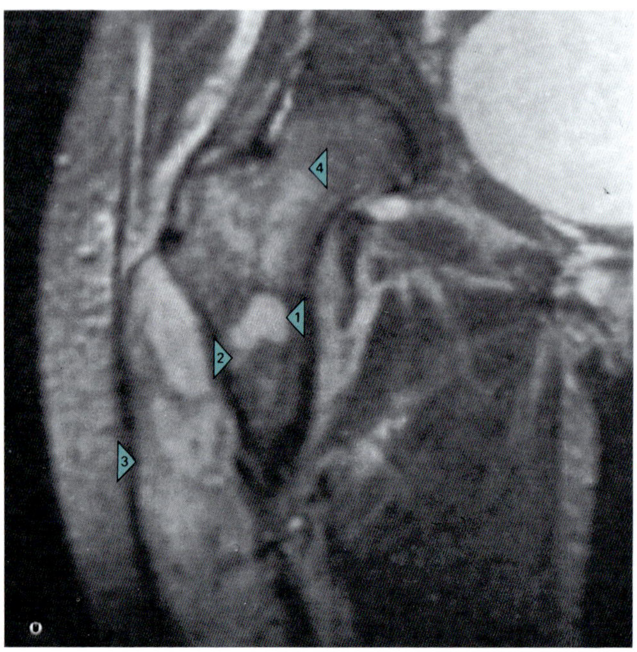

FIGURE 11.4. Subacute Osteomyelitis (Brodie's Abscess) with Soft Tissue Abscess. Coronal plane [SE 2000/80]. There is an intraosseous abscess in the proximal femur. The low SI rim corresponds to a sclerotic margin seen on plain films. The Brodie's abscess drains to the adjacent vastus lateralis muscle through a sinus tract. At surgery, the high SI area lateral to the femur was a large soft tissue abscess involving the vastus lateralis muscle. There is high SI bone marrow edema of the femoral neck. Arrow **1** Brodie's abscess. Arrow **2** Draining sinus. Arrow **3** Soft tissue abscess. Arrow **4** Bone marrow edema.

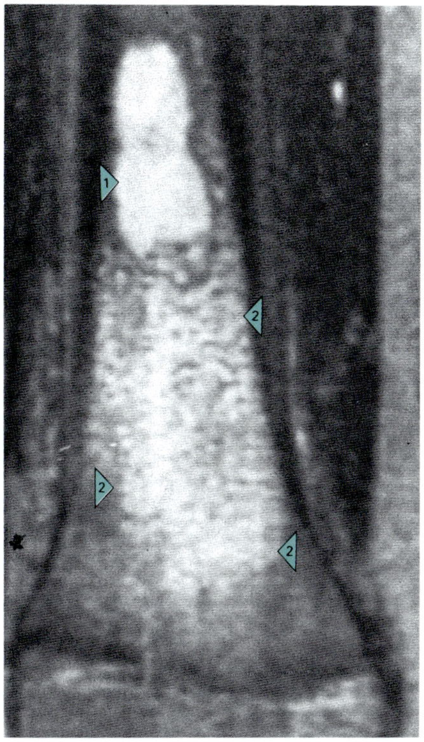

FIGURE 11.5. Brodie's Abscess with Bone Marrow Edema. Coronal plane [SE 2000/100] through the distal femur. There is a well-demarcated, high SI intraosseous abscess. Distally, there is an ill-defined area of high SI corresponding to bone marrow edema extending to the metaphysis. Arrow **1** Brodie's abscess. Arrows **2** Bone marrow edema.

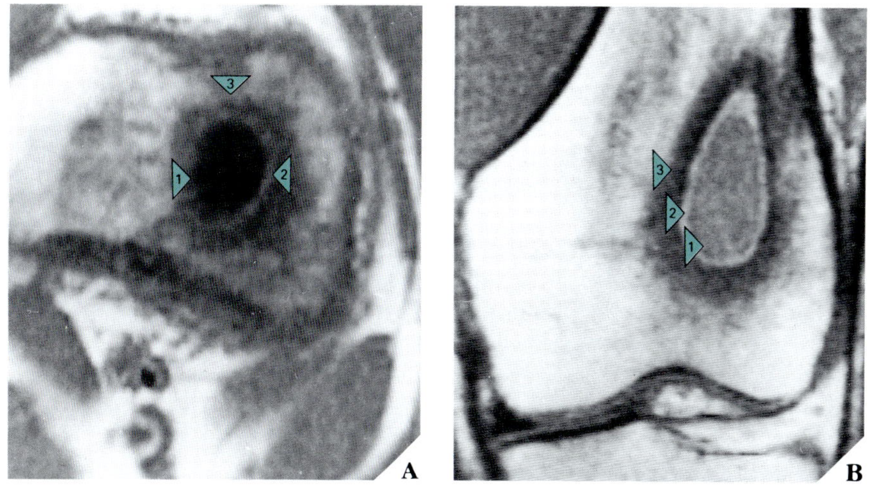

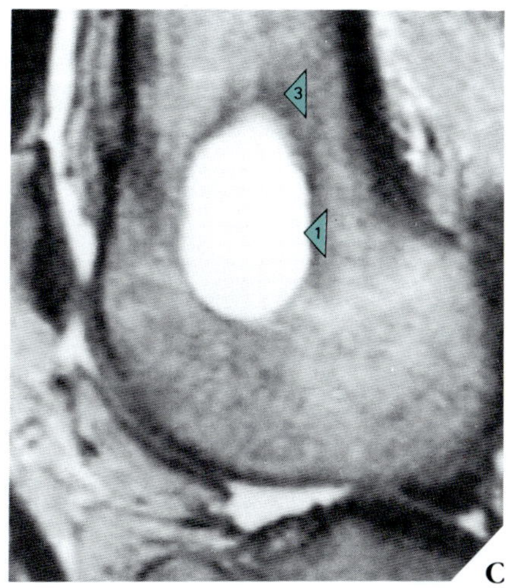

FIGURE 11.6. Brodie's Abscess with Double-Line Sign. Axial plane **(A)** [IR 1317/428/59], coronal plane **(B)** [SE 575/30], and sagittal plane **(C)** [SE 1800/100] through the distal femur. On T1WI (**A** and **B**), the central fluid portion of the abscess shows low SI and is surrounded by a higher SI rim which, at surgery, represented granulation tissue. The outer low SI band fading peripherally represents bony sclerosis seen on plain films. On T2WI **(C)**, the fluid material inside the abscess and the layer of granulation tissue have similar high SI and cannot be discriminated. Arrows **1** Purulent fluid material inside of the abscess cavity. Arrows **2** Layer of granulation tissue. Arrows **3** Peripheral rim of sclerosis.

Sequestrum

Sequestrum formation in chronic osteomyelitis is similar to cortical bone on MRI appearing as a low SI fragment on T1WI and T2WI. Surrounding this low SI fragment is an area of high SI on T2WI, which represents inflammatory or granulation tissue. The tissue is well-demarcated from the adjacent normal medullary intermediate SI (Fig. 11.7). The sequestra may also be seen protruding through the cortex.

The specificity of MR imaging is limited, and similar images may be obtained with complex, noninfected healing fractures. The specificity of MR images is improved when there is demonstration of a draining sinus tract (Fig. 11.8). The ability to image in multiple planes improves the sensitivity in detecting these findings of chronic osteomyelitis (4,17).

SOFT TISSUE INFECTIONS

Infections in the soft tissues may occur from transperiosteal spread of infected fluid from osteomyelitis, from direct spread from another contiguous source, from direct implantation, or from hematogenous spread. Cellulitis implies a diffuse inflammation of the soft tissues. Abscess implies a well-organized fluid collection, which may be surrounded by a fibrous capsule.

Cellulitis

Magnetic resonance imaging depicts areas of cellulitis as poorly marginated regions of increased SI diffusely infiltrating the soft tissues (Fig. 11.9).

Abscess

An abscess is demonstrated on MRI as a rounded

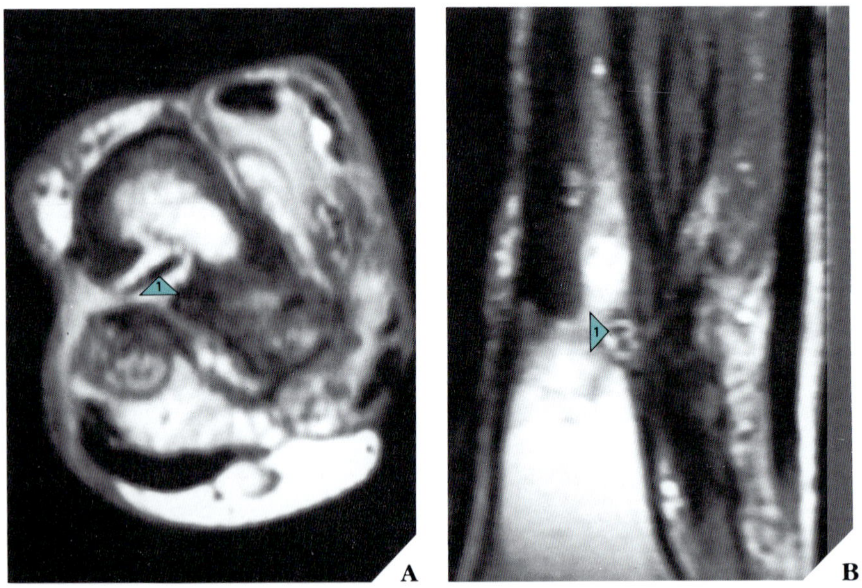

FIGURE 11.7. Chronic Osteomyelitis with Sequestrum. Axial **(A)** and sagittal **(B)** planes [SE 2000/80] through distal tibia and fibula. This patient sustained a fracture of the tibia which healed, but subsequently developed chronic osteomyelitis with a skin draining sinus. MRI was performed to determine the extent of the infection in the medullary space. Axial and sagittal T2WI demonstrate a small sequestrum surrounded by a well-demarcated area of low SI. The SI of the medullary space is otherwise normal. At surgery a sequestrum surrounded by a small amount of granulation tissue was found, without other signs of bone marrow infection. Arrows **1** Sequestrum.

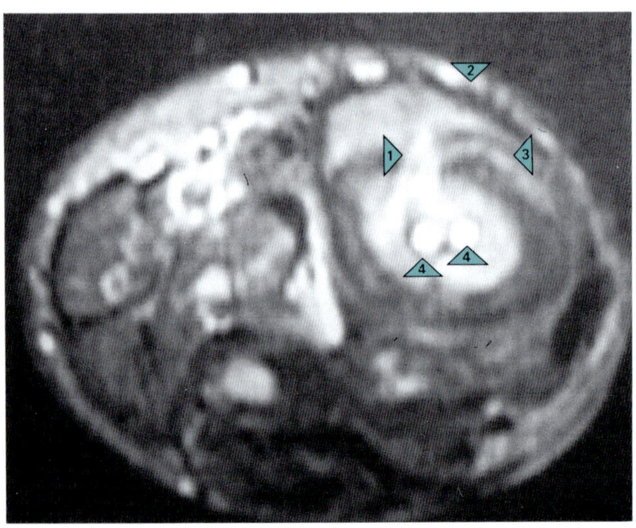

FIGURE 11.8. Chronic Osteomyelitis with Draining Sinus and Subperiosteal Abscess. Axial plane [SE 2000/80] through the distal tibia. This patient sustained a distal tibial fracture treated with two intramedullary nails that were subsequently removed because of the development of chronic osteomyelitis. The SI of the bone marrow of the tibia is high compared with that of the fibula. The tibial cortex is broken anteriorly and there is a high SI fluid collection surrounded by a low SI thickened periosteum. Anteriorly, the cortex is split by a high SI line representing intracortical infection. Note the two rounded high SI areas in the center of the medullary cavity; these represent the tracts of the removed intramedullary nails. Arrow **1** Draining sinus. Arrow **2** Thickened periosteum over subperiosteal abscess. Arrow **3** Intracortical infection. Arrows **4** Intramedullary tracts.

collection of increased SI on T2WI which may be surrounded by a low SI rim (3,4). This low signal intensity rim represents the fibrous capsule about the abscess (Fig. 11.10). Many abscesses show a relatively lower signal intensity within their center, as compared with their periphery. The reason for this is not clear, but it may represent a lower proteinaceous content of the central portion of the abscess (Fig. 11.11).

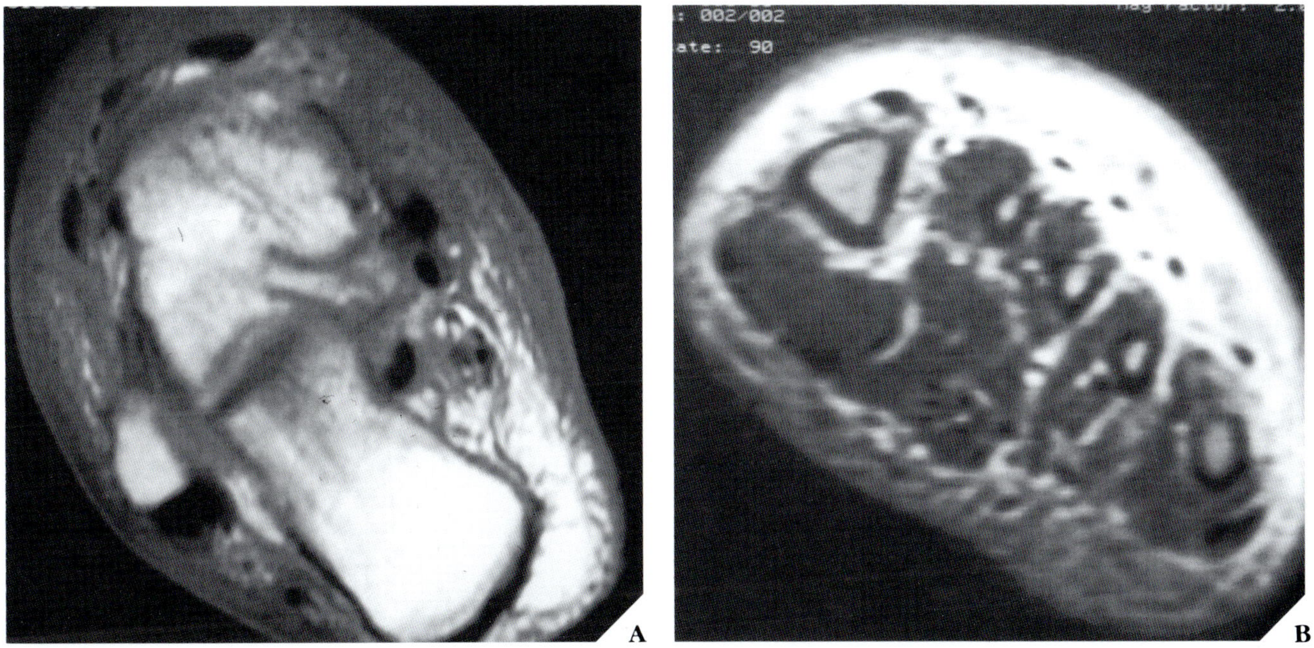

FIGURE 11.9. Cellulitis Involving the Dorsum of the Foot. Oblique axial, long axis **(A)** [SE 800/20] and coronal, short axis **(B)** [SE 2000/80] planes through the foot in a patient with inflammatory signs in the dorsum of the foot. There is cellulitis involving the dorsal and lateral surfaces of the foot with low SI on T1WI **(A)** and high SI on T2WI **(B).** No drainable fluid collection is present.

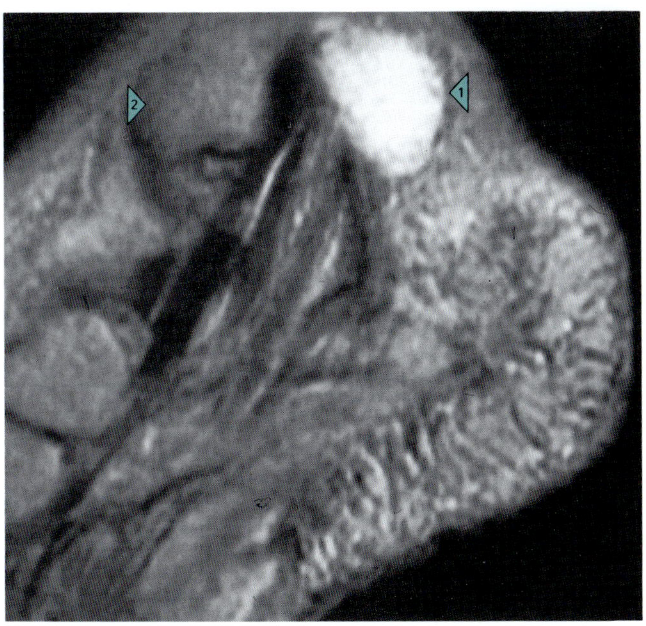

FIGURE 11.10. Soft Tissue Abscess. Sagittal plane [SE 2000/80]. There is a high SI fluid collection adjacent of the medial malleolus in this patient with a diabetic foot. Arrow **1** Abscess. Arrow **2** Medial malleolus.

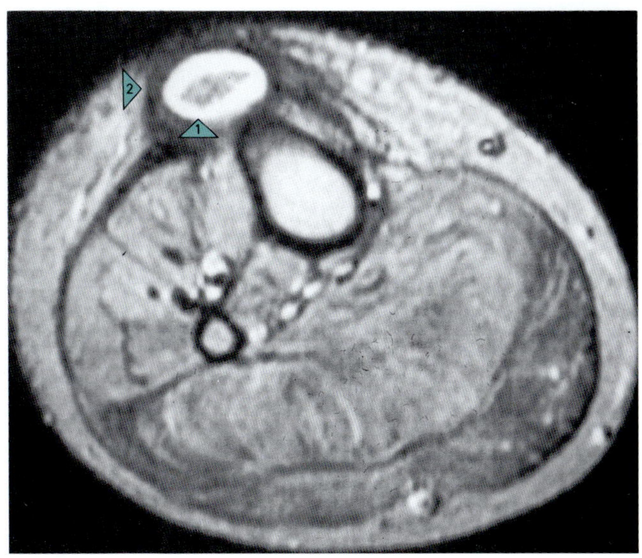

FIGURE 11.11. Soft Tissue Abscess with Low SI Center. Axial plane [SE 2000/80]. There is a high SI soft tissue fluid collection anterior to the tibia, with a lower SI central area in this patient following above-knee amputation. The abscess is surrounded by a thick, low SI capsule. Note the high SI edema involving diffusely the subcutaneous fat and muscles of the stump. Arrow **1** Abscess. Arrow **2** Capsule.

SEPTIC ARTHRITIS AND TENOSYNOVITIS

Infectious arthritis and tenosynovitis may also result from the mechanisms of hematogenous spread, direct extension from a contiguous source, or direct implantation. Although septic arthritis and tenosynovitis may present in any age group or distribution, they most commonly present in children, in a monoarticular distribution in the lower extremity. In response to an infectious agent, usually *Staphylococcus aureus* or *Neisseria gonorrhea,* synovial membrane becomes edematous, swollen, and hypertrophied. These characteristics are accompanied by increased production of synovial fluid (18).

The role of MR imaging in detecting infectious arthritis or tenosynovitis remains to be defined. It shows the presence of fluid within the joint capsule or within a tendon sheath as increased SI on T2WI distending the capsule (Fig. 11.12) or tendon sheath (Fig. 11.13). However, based on SI changes alone, it is not possible to distinguish between infected and uninfected articular and tendon sheath fluid. The presence of edema, osteomyelitis, or abscess surrounding the joint or tendon sheath indicates that the fluid has an infectious etiology. Joint or tendon fluid aspiration may be necessary in cases such as diabetic foot where precise determination of the presence or absence of infection is important for patient management.

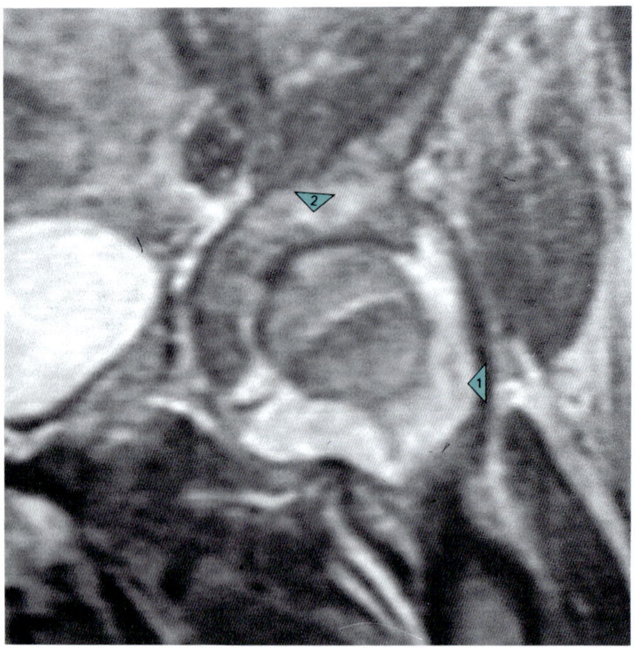

FIGURE 11.12. Septic Arthritis of the Hip with Osteomyelitis of the Acetabulum. Coronal plane [SE 2000/80]. There is a high SI joint effusion on the left hip in this young patient. In addition, there is an ill-defined area of high SI in the acetabulum, corresponding to an area of osteomyelitis. Arrow **1** Joint effusion. Arrow **2** Focus of osteomyelitis.

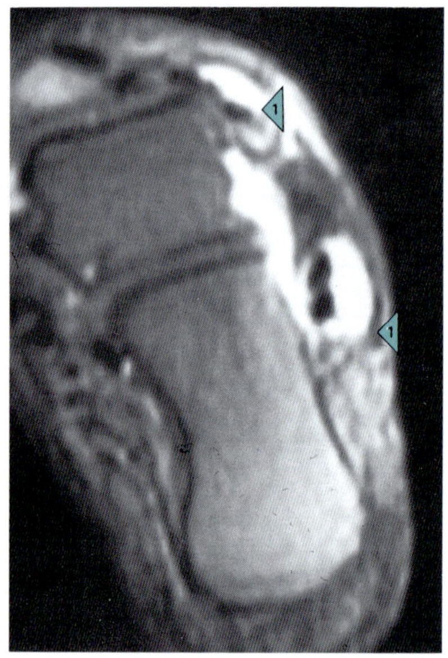

FIGURE 11.13. Infected Synovitis of the Foot. Oblique coronal plane [SE 2000/80]. There is high SI fluid distending the synovial sheaths of the peroneal tendons and extensor tendons of the foot in this patient with a diabetic foot. On aspiration, the fluid was found to be infected. Arrows **1** Distended synovial sheaths.

DIABETIC FOOT

Diabetic foot can be defined as a group of pathologic abnormalities which occur in the feet of patients with diabetes mellitus. These abnormalities include ischemia, sensory neuropathy, and infection, all of which may lead to tissue necrosis. Amputation may be required in up to 40 per cent of patients with diabetic foot. Early diagnosis, accurate staging, aggressive treatment of infection, and, in some cases, revascularization procedures including vascular reconstruction and angioplasty, are essential for limb salvage (10,15).

Magnetic resonance imaging is more sensitive than scintigraphy in evaluating cases of diabetic foot with osteomyelitis, and has the added advantage of improved spatial resolution (21). Osteomyelitis of the small bones of the foot can be accurately depicted (Fig. 11.14). Cellulitis can be discriminated from drainable abscess (see Figs. 11.9 and 11.10). Synovitis in the tendon sheaths (see Fig. 11.13) and joint effusion can be detected, although differentiation between infected and noninfected synovitis may require fluid aspiration and culture. Neuroarthropathy can be distinguished from osteomyelitis by virtue of the low SI changes seen on T1WI and T2WI within the bone marrow space of the neuropathic joint versus high SI on T2WI seen in osteomyelitis (see Chapter 8).

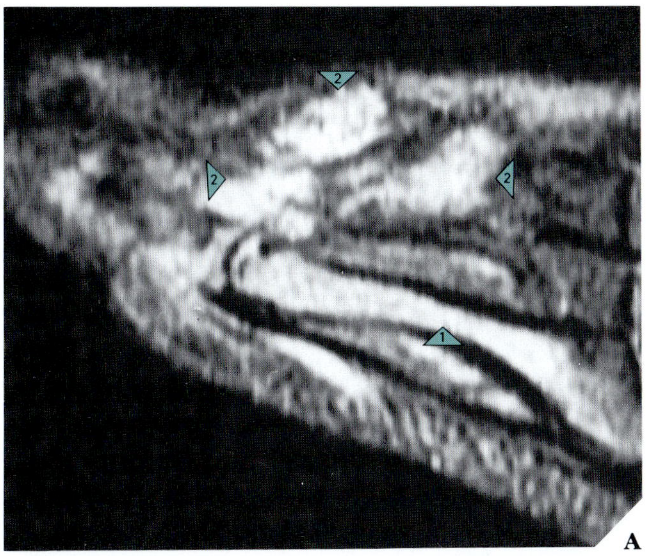

A

FIGURE 11.14. Diabetic Foot. Osteomyelitis and Soft Tissue Abscesses. Sagittal **(A)** and coronal **(B)** planes [SE 2000/80] through the foot in a young diabetic patient with clinical evidence of infection. There is high SI in the bone marrow space of the fifth metatarsal, indicating acute osteomyelitis. Several soft tissue abscesses are evident in the dorsal aspect of the foot. Arrow **1** Osteomyelitis. Arrows **2** Abscesses.

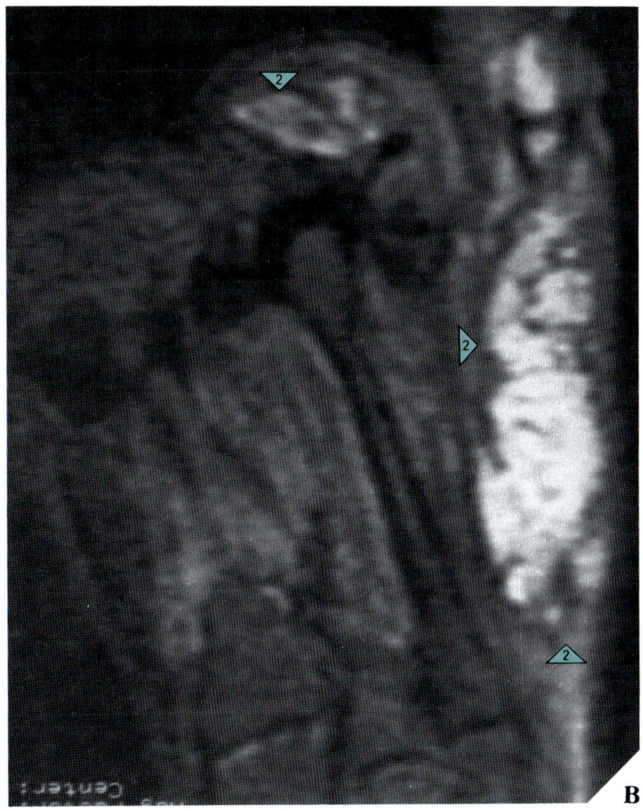

B

SPINAL INFECTIONS

Discitis is inflammation of the intervertebral disc. When caused by an infectious agent, usually *Staphylococcus,* the term used is *septic discitis.* Infection is often hematogenously spread, but may arise from a penetrating injury, surgery, or paraspinal nidus. Septic discitis is a serious medical problem and can lead to life-threatening complications if not appropriately treated. If the discitis is sterile, as might occur with lumbar disc degeneration, arthritis, chemonucleolysis, or trauma, then it is called *aseptic discitis.* Most patients with aseptic discitis have a self-limited medical course and do not require specific medical therapy.

Disc space narrowing and juxtavertebral endplate changes are characteristic plain film findings that generally suggest discitis. Magnetic resonance imaging provides additional, unique information about the signal within the disc itself, as well as within the bones and surrounding soft tissues (Fig. 11.15). It is more sensitive than plain films or CT, and is also more specific than radionuclide bone scans (16). As a result, MRI has become the modality of choice for evaluating spinal infections.

Septic Discitis

Septic discitis demonstrates disc space narrowing on MRI. The vertebral endplates are low SI and indistinct on T1WI. On T2WI and PDWI the vertebral endplates may be of increased signal, but the major finding in the disc space is inappropriately high signal (Fig. 11.16).

Pyogenic Septic Discitis

Pyogenic septic discitis usually derives from *Staphylococcus* infection. Characteristic findings are paraspinal soft tissue thickening and edematous changes in the paraspinal musculature. The earliest MRI finding is obliteration of the fat planes between the paraspinal musculature and the outer margin of the disc. Later, a mass may extend anteriorly to involve the paraspinal musculature, or posteriorly to infect the epidural space.

Psoas Muscle Abscess

Psoas muscle abscess formation is relatively common and has parallel signal properties of low signal on T1WI that become high signal on T2WI and PDWI (Fig. 11.17) (20).

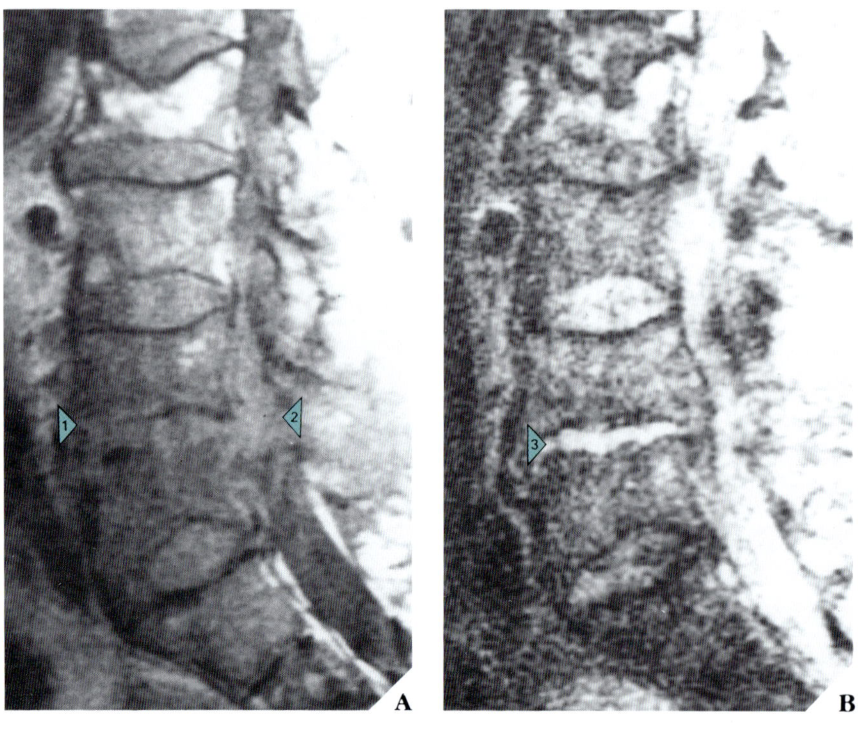

FIGURE 11.15. Septic Discitis. Sagittal plane **(A)** [SE 2000/20] and **(B)** [SE 2000/80]. Two months following surgery for spinal stenosis and disc herniation, this patient returned with leg spasms, pain, and fever. The disc space between the fourth and fifth lumbar vertebra is narrowed and there is a low signal on PDWI **(A).** The normal higher signal of fatty marrow in the vertebral bodies is diminished. The vertebral endplates are partially indistinct, and a lenticular infected paraspinal mass extends posteriorly into the epidural space. On T2WI **(B),** the infected disc is inappropriately high SI. Epidural extension of the infection shows intermediate low SI. Arrow **1** Narrowed disc space. Arrow **2** Epidural spread. Arrow **3** Persistent high SI of the infected disc.

Aseptic Discitis

In contrast, aseptic discitis characteristically will have a disc space that is narrowed and relatively low signal on all pulse sequences, as expected in disc degeneration. There are variable vertebral endplate changes depending on the stage of inflammation in aseptic discitis. High signal on T1WI due to fatty infiltration may occur. Low signal vertebral endplates on all pulse sequences suggest sclerosis. The acute stages of aseptic discitis are indicated by a low signal on T1WI that becomes bright on T2WI and PDWI. However, unlike pyogenic septic discitis, there is no paraspinal mass. Metastatic disease to the spine has similar signal properties, but metastasis very rarely occurs at a disc space.

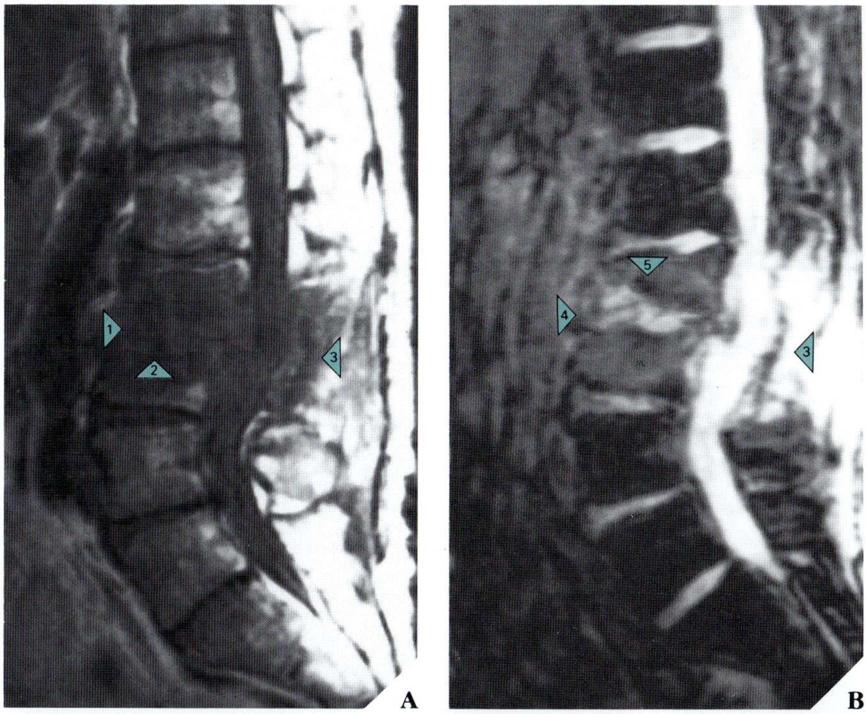

FIGURE 11.16. Postoperative Infection. Sagittal plane **(A)** [SE 800/20] and **(B)** [GRASS 1000/12]. Laminectomy of the second and third lumbar vertebra had been performed previously. The patient returned with fevers and back pain. The L2-L3 intervertebral disc is low SI and the vertebral endplates are no longer distinct on T1WI **(A).** Reversal of the normal lordotic curve and retrolithesis is present. Low SI of the adjacent vertebral bodies is present. T2WI **(B)** shows wedge compression of the second lumbar vertebra and partial collapse of the third. The relatively high SI in the disc space and in the vertebral bodies suggest osteomyelitis. Moderate canal stenosis is present. Arrow **1** Low SI disc space on T1WI. Arrow **2** Indistinct vertebral endplate. Arrows **3** Laminectomy site. Arrow **4** High SI of the infected disc on T2WI. Arrow **5** Collapsed high SI vertebral body.

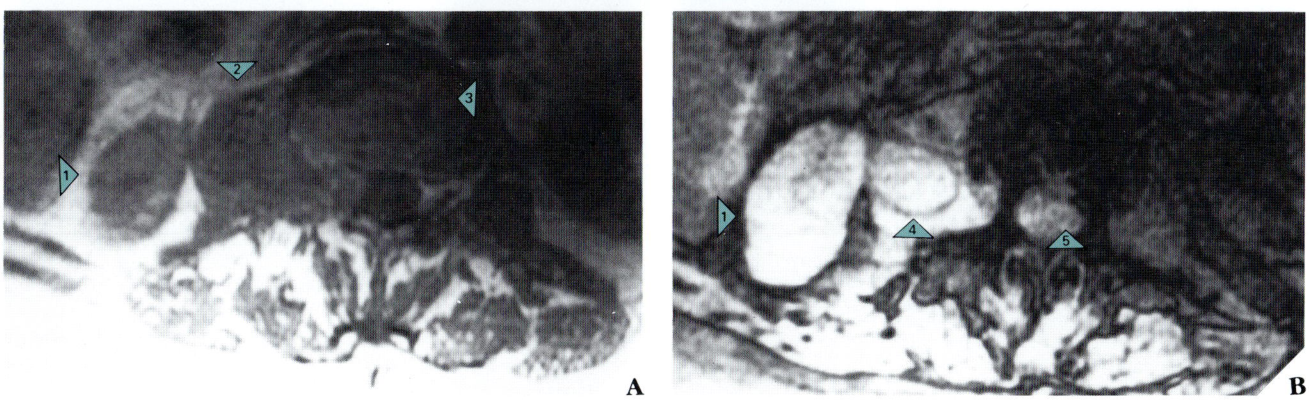

FIGURE 11.17. Paraspinal Abscess. Axial plane **(A)** [SE 800/20] and **(B)** [GRASS 1000/12]. The right psoas muscle is swollen and enlarged, with low SI on T1WI **(A)** and high SI on T2WI **(B).** An encapsulated abscess cavity is present posteriorly. CT-guided drainage found gram positive cocci. There is high SI near the skin surface due to surface coil proximity. Arrows **1** Right kidney. Arrow **2** Enlarged psoas muscle. Arrow **3** Intervertebral disc space. Arrow **4** Psoas abscess. Arrow **5** Thecal sac.

Fungal Infections

Fungal infections of the disc usually have less paraspinal soft tissue thickening and may be difficult to distinguish from aseptic discitis (Fig. 11.18). However, signal changes in the vertebral bodies are often quite extensive. Indolent fungal infections and tuberculosis cause fewer disc space reactive changes and may not have increased signal on T2WI. Large paraspinal masses have been reported in tuberculous spondylitis (9).

At times it may be difficult to distinguish septic from aseptic discitis, and biopsy may be necessary. However, the lack of a paraspinal mass argues strongly against a pyogenic septic discitis. The excellent depiction of paraspinal soft tissues showing the full extent of infection is a distinct advantage of MRI over other modalities.

Vertebral Osteomyelitis

Osteomyelitis confined to the vertebral body alone is a relatively uncommon finding. When evident, however, it is usually in immunocompromised patients. Signal properties in MRI may be impossible to distinguish from spinal metastasis, and biopsy may be necessary. The infected marrow space of the vertebral body will have a low signal on T1WI and a higher signal on T2WI or PDWI. Extensive paraspinal soft tissue swelling or abscess formation may also be present. When vertebral osteomyelitis extends from an infected disc space, the diagnosis may be more obvious. Bone destruction and vertebral body collapse may result in canal compromise (Fig. 11.19).

Epidural Abscess

Epidural abscess is an infected fluid collection between the spinal dura and the bony spinal canal. Mass effect within this confined space may result in cord or nerve root compression and have disastrous neurologic complications. The classic history is back pain, fever, and neurologic compromise, usually in a diabetic or septic patient. However, since not all of these clinical signs may be present, this diagnosis may be difficult to make. Confusion with osteomyelitis, disc protrusion, hemorrhage, or spinal metastasis may occur. Magnetic resonance imaging provides an excellent noninvasive means of diagnosis, since the epidural space can be directly visualized.

The epidural abscess may have low-to-intermediate signal intensity on T1WI, depending on whether the abscess is predominantly liquified or,

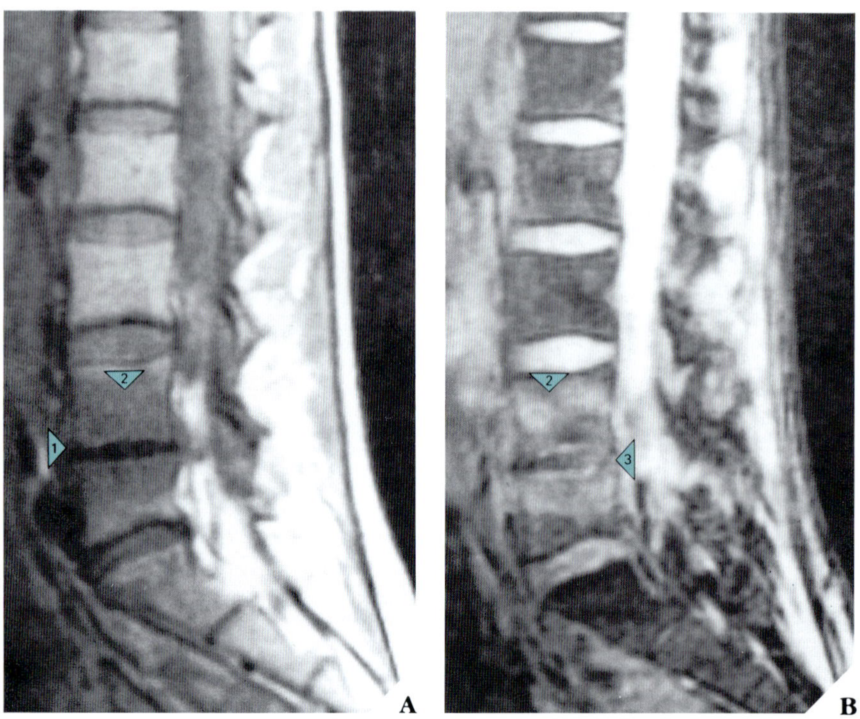

FIGURE 11.18. Fungal Discitis. Sagittal plane **(A)** [SE 800/20] and **(B)** [GRASS 1000/12]. The disc space between the fourth and fifth lumbar vertebra is narrowed. The disc space is low SI on T1WI **(A).** There is low SI in the adjacent vertebral bodies suggesting either sclerosis or marrow replacement. On T2WI **(B),** the vertebral bodies are high SI, but unlike pyogenic infection, the disc space is of only moderately increased signal. The vertebral endplates are still visualized and there is very little paraspinal soft tissue thickening. Arrow **1** Narrow disc space. Arrow **2** Secondary changes in vertebral bodies. Arrow **3** Some signal remains in infected disc.

as often occurs, composed of thick, fibrous inflammatory tissue. The abscess usually has a high signal on T2WI or PDWI, but, again depending on the contents, may be lower.

Gd-DTPA may be useful in cases where the abscess is isointense to cord on T1WI. The abscess capsule and, to some extent, its contents are enhanced by Gd-DTPA (Fig. 11.20), which provides an excellent means of demonstrating the full extent of the process. However, this does not confirm the diagnosis of abscess, since almost any epidural process, such as meningioma or metastasis, would have a similar appearance.

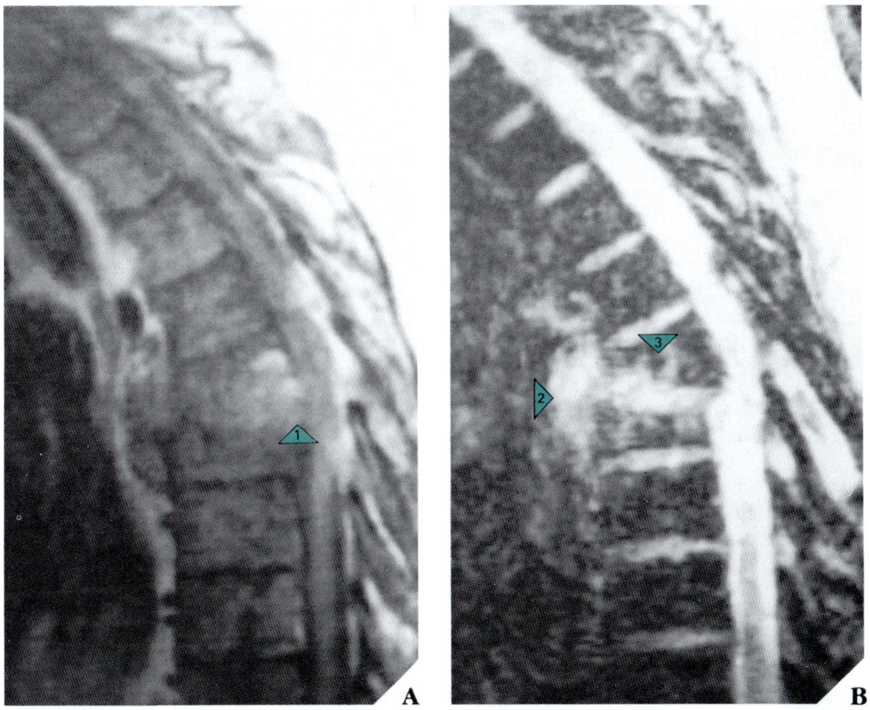

FIGURE 11.19. Discitis and Osteomyelitis. Sagittal plane **(A)** [SE 800/20] and **(B)** [GRASS 1000/12]. In this diabetic patient, a midthoracic vertebral body is wedge-shaped secondary to partial collapse of the inferior margin, mimicking metastasis. Biopsy was needed to confirm the diagnosis. There is associated kyphosis and spinal stenosis. The vertebral endplates are absent. The disc space is obliterated and there is an anterior paraspinal mass. The outline of the thoracic cord by spinal fluid is no longer visualized. Paraspinal abscess has also formed anteriorly. Arrow **1** Epidural spread with cord compression. Arrow **2** Paraspinal abscess. Arrow **3** Vertebral body destruction.

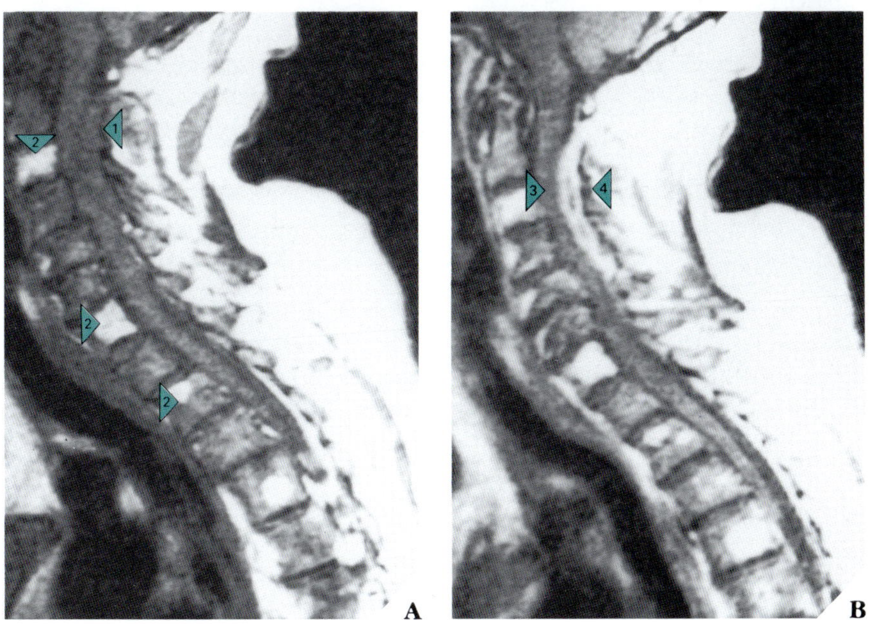

FIGURE 11.20. Epidural Abscess. Sagittal plane **(A)** [SE 800/20] and **(B)** [GRASS 1000/12, post-GD-DTPA]. This elderly female presented with neck pain and fever. The high signal foci scattered through the vertebral bodies are incidental vertebral hemangiomas. On T1WI **(A)**, the cervical cord appears to thicken from the second cervical vertebra to the fifth. After enhancement with Gd-DTPA **(B)**, the high SI lenticular posterior epidural abscess extending from the second cervical vertebra to the fifth, is seen compressing the spinal cord anteriorly. A thin cleft located centrally was the only liquid portion of the abscess; the remainder was thick, fibrous inflammatory exudate. Arrow **1** Epidural soft tissue mass. Arrows **2** Vertebral hemangiomas. Arrow **3** Cord compression. Arrow **4** GD-DTPA-enhanced epidural abscess with low SI central fluid cavity.

REFERENCES

1. Altman N, Harwood-Nash DC, Fitz C, et al: Evaluation of the infant spine by direct sagittal computed tomography. AJNR 1985;6:65–69.
2. Arger Ph, Coleman BG, Dalinka MK: Computed tomography in orthopedics. Orthop Clin North Am 1983;14:217–232.
3. Beltran J, et al: Experimental infections of the musculoskeletal system: evaluation with MR imaging and Tc-99m and Ga-67 scintigraphy. Radiology 1988;167:167–172.
4. Beltran J, Noto AM, McGhee RB, et al: Infections of the musculoskeletal system: high field strength MR imaging. Radiology 1987;164: 448–454.
5. Berquist TH: Musculoskeletal infection. In: Berquist TH et al, Eds. Magnetic resonance of the musculoskeletal system. Raven Press, 1987, pp. 109–126.
6. Berquist TH, Brown ML, Fitzgerald RH, May GR: Magnetic resonance imaging: application in musculoskeletal infection. Magn Reson Imaging 1985;3:219–230.
7. Datz FL, Thorne DA. Effect of chronicity of infection on the sensitivity of the In-111 labeled leukocyte scan. AJR 1986;147:809–812.
8. Edmons ME. The diabetic foot: pathophysiology and treatment. Clin Endocrinol Metab 1986; 15:889–916.
9. deRoos A, Van Meerten EL, Bloem JL, et al. MRI of tuberculous spondylitis. AJR 1986:146:79–82.
10. Fletcher BD, Scoles PV, Nelson AD: Osteomyelitis in children: detection by magnetic resonance. Radiology 1984;150:57–60.
11. Golimbu C, Firooznia H, Ralii M: CT of osteomyelitis of the spine. AJR 1984;42:159–163.
12. Handmaker H: Acute hematogenous osteomyelitis: has the bone scan betrayed us? Radiology 1980;135:787–789.
13. Kaufman J, Breeding L, Rosenberg N: Anatomic location of acute diabetic foot infection. Its influence on the outcome of treatment. Am Surg 1987;53:109–112.
14. Langenskiöld A: Growth disturbance after osteomyelitis of femoral condyles in infants. Acta Orthop Scand 1984;55:1.
15. Maurer AH, Millmond SH, Knight LC, et al. Infection in diabetic osteoarthropathy: use of indium labeled leukocytes for diagnosis. Radiology 1986;161:221–223.
16. Modic MT, Feiglin DH, Piraino DW, et al. Vertebral osteomyelitis: assessment using MR. Radiology 1985;157:157–166.
17. Quinn SF, Murray W, Clark RA, Cochran C. MR imaging of chronic osteomyelitis. J Comput Assist Tomog 1988;12:113–117.
18. Resnick D, Niwayama G: Diagnosis of bone and joint disorders, pp. 2523–2754. Philadelphia, WB Saunders, 1988.
19. Trueta J, Amato VP: The vascular contributions to osteogenesis. III. Changes in the growth cartilage caused by experimentally induced ischemia. J Bone Joint Surg (Br) 1960;42:571.
20. Weinreb JC, Cohen JM, Maravilla K. Iliopsoas muscles: MR study of normal anatomy and disease. Radiology 1985;156:435–440.
21. Yuh WTC, Corson JD, Baraniewski HM, et al. Osteomyelitis of the foot in diabetic patients: evaluation with plain films, 99mTc-MDP bone scintigraphy, and MR imaging. AJR 1989;152: 795–800.

FIGURE CREDITS

Figures 11.1, 11.4, and 11.14. From Beltran J, Noto AM, McGhee RB, et al: Infections of the musculoskeletal system: high field strength MR imaging. Radiology 1987; 164:448–454.

Figure 11.5. Courtesy of Dr. Francisco Aparisi, Hospital "La Fe," Valencia, Spain.

Figure 11.6. Courtesy of Dr. Jose Vilar, Hospital Peset, Valencia, Spain.

INDEX

Numbers in **boldface** refer to Fig. numbers

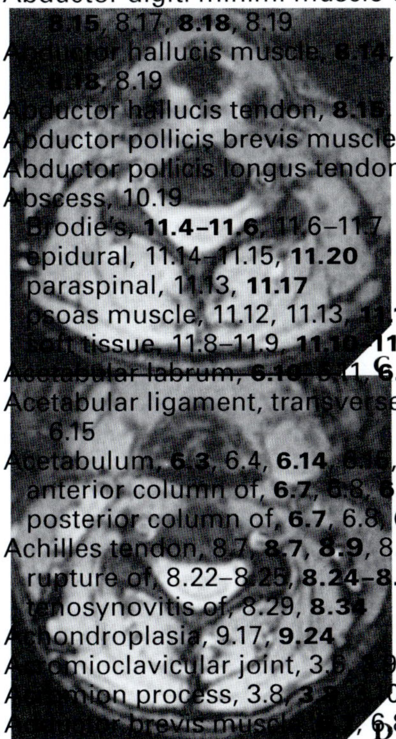

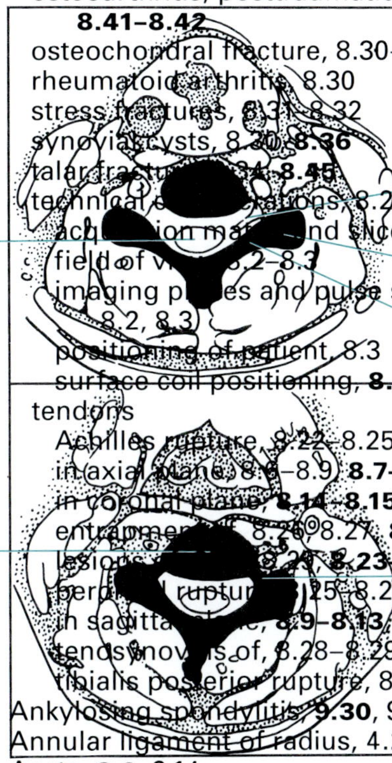

small low signal foci are evident due to dense cortical bone. There is no bony destruction or expansion.

Perineural Cysts

Tarlov or perineural subarachnoid cysts are simple regions of fluid dilatation of the axillary sheaths of the nerve roots in or near the neural foramen (Fig. 9.8). They may become quite large (> 1.5 cm) and mimic tumors, with enlargement of their respective foramina. Preservation of adjacent epidural fat, multiple cysts, and lack of symptoms are common findings for Tarlov cysts. Differentiation of a Tarlov cyst from a tumor, particularly a cystic neurofibroma, may be impossible on MRI and myelography may thus be required.

Conjoint Nerve Root

Conjoint nerve roots, which occur when two roots exit through one enlarged root sheath, are common normal variations (Fig. 9.9). They are by far most common in the lumbar region. Characteristically, the dural sac and axillary root sheaths are bulbous at the lateral recess. The visualization of double roots within the sheath and the preservation of the epidural fat are the most helpful findings to exclude the presence of disc protrusion.

Residual Pantopaque

Since Pantopaque (iophendylate, myelographic contrast) has a fatty, oil base component, it has an SI similar to subcutaneous fat, with a high signal on T1WI (Fig. 9.10) (13). The distribution is characteristic as well, with visualization of a fluid-fluid level or multiple small droplets. Chemical shift artifacts can help confirm the presence of fatty Pantopaque in the dural sac. Trapped Pantopaque which does not fall to the dependent portion of the canal suggests arachnoiditis.

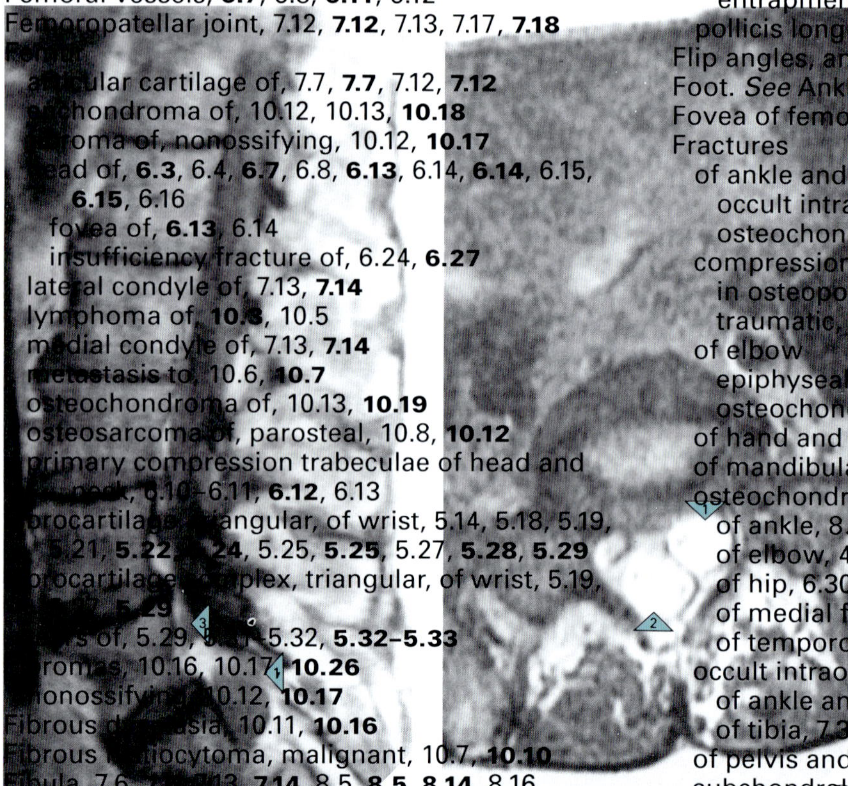

FIGURE 9.8 Sacral Tarlov Root Cyst. A Parasagittal [SE 1500/20]. This T1WI of the lumbar spine shows a large Tarlov cyst of CSF signal surrounded by epidural fat. The nerve root directly superior is normal in size. **B** Axial section [SE 1500/80]. On this axial T2-weighted image, the perineural Tarlov cyst is of high water signal similar to that of the caudal thecal sac. Also note that the chemical shift artifacts in the left-to-right frequency-encoded direction produce asymmetric signal margins. Arrows **1** Tarlov cyst. Arrow **2** Dural sac. Arrow **3** Normal nerve root.

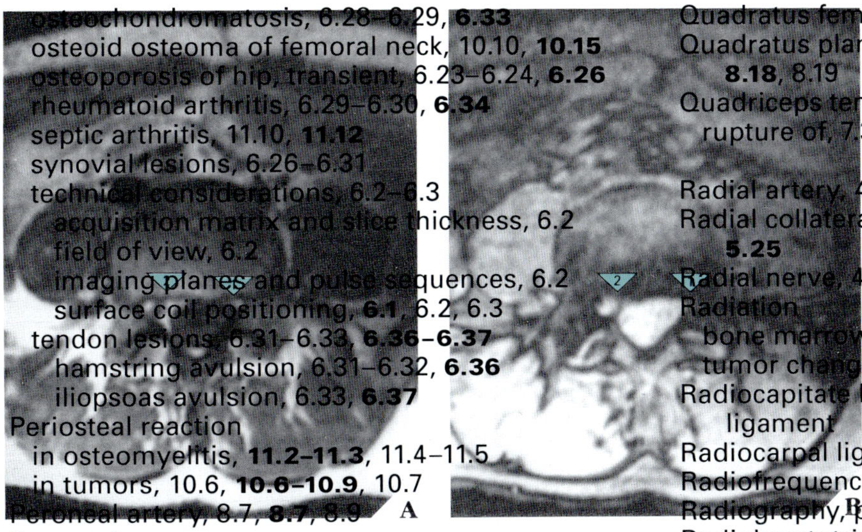

FIGURE 9.9. Lumbar Conjoint Nerve Root Sheath. **A** Axial section [SE 800/20]. A conjoint nerve-root sheath is seen (on the left) on this T1WI. The conjoint root sheath is bulbous in comparison to the normal sheath (on the right). The absence of fat surrounding the nerve root on the left should not be confused with a disc herniation. **B** Axial section [GRASS 1000/12, flip angle 22.5 degrees]. This PDWI confirms the CSF nature of the asymmetric dural sac and nerve-root sheaths secondary to a conjoint root (on the left). Arrows **1** Conjoint nerve root. Arrows **2** Normal nerve root.

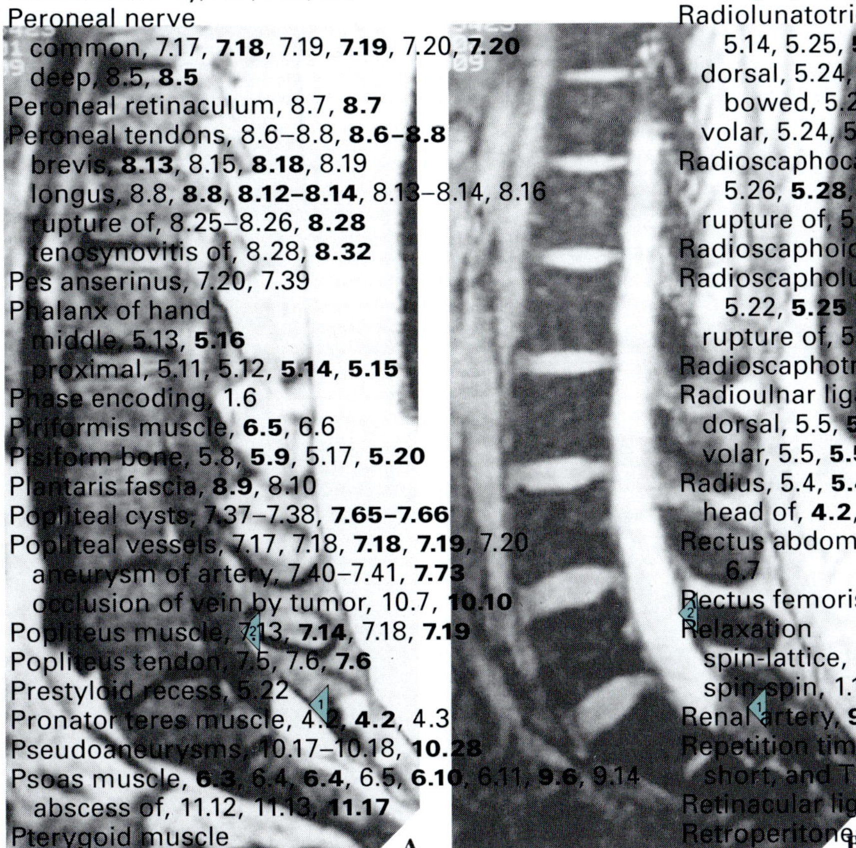

FIGURE 9.10. Pantopaque in Lumbar Spine. **A** Sagittal section [SE 800/20]. In this supine patient, Pantopaque (iophendylate) myelographic contrast layers are seen in the caudal sac. The fatty oil-based contrast has a high SI similar to that of the subcutaneous fat on this T1WI. The CSF is dark. Notice that a chemical shift artifact is present, with a low SI superior margin as seen in the vertebral bodies; this finding confirms that this is not hemorrhage but fat. **B** Sagittal section [GRASS 1000/12, flip angle 22.5 degrees]. On this PDWI the CSF is high SI whereas the oil-based contrast is of lower SI, similar to that of the retroperitoneal fat. The signal of the subcutaneous fat is higher than the Pantopaque because of its closer proximity to the coil. Arrows **1** Pantopaque. Arrows **2** CSF.

the patients have sigmoid scoliosis of the thoracic spine with a correcting curve so that the alignment of the head is not affected. MRI can help exclude the presence some unexpected disorders such as syrinx, but in general MRI has not made a significant impact on the evaluation of this disorder.

Spondylolisthesis

Spondylolisthesis (Fig. 9.15) occurs when a superior vertebral body is shifted anteriorly or posteriorly (reverse spondylolisthesis) in relation to the adjacent inferior body. This subluxation can be secondary to acquired causes such as pars interarticularis defects, trauma, fractures, destructive tumors, or severe degeneration of the facets or discs. The congenital cases are frequently quite severe with marked subluxation. These congenital cases occur most commonly at the lumbosacral junction. Absent or dysplastic facets allow severe anterior subluxation, to the point that the lumbar vertebrae slide into the pelvis.

In patients with bilateral pars interarticularis defects and spondylolisthesis, there will be elongation of the anterior to posterior dimension of the spinal canal. This occurs because the posterior elements remain fixed while the vertebral body slides forward and opens the canal. Spondylolisthesis secondary to degeneration of the discs and the facets results in spinal stenosis. In this case the posterior elements slide forward with the vertebral body and the hypertrophic facet changes. This results in dorsal as well as ventral narrowing of the spinal canal.

Facet Anomalies

Significant isolated facet anomalies are uncommon. There are usually multiple associated findings including unilateral hypoplasia of one facet and pedicle. The contralateral facet may show rotation, hypertrophy, and deformity. Those patients with severe vertebral body anomalies frequently have associated posterior element anomalies as well.

Spinal Dysraphism

Spinal dysraphism is due to nonclosure of the primary neural tube and most often affects the posterior elements of the spine (2). The exact etiology for this failure of closure is not known. MRI, US, and serum testing for elevated fetal alpha-protein in the mother can be used as prenatal screening exams. At this time, the FDA has not approved MRI as a routine exam during pregnancy. The mildest form of spinal dysraphism is an incidental common variation in the lumbosacral region occurring when there is spina bifida without any other abnormality.

Meningoceles are extensions of the dural sac through a defect in the spinal canal (Fig. 9.14). They may be covered by skin, subcutaneous fat, or mus-

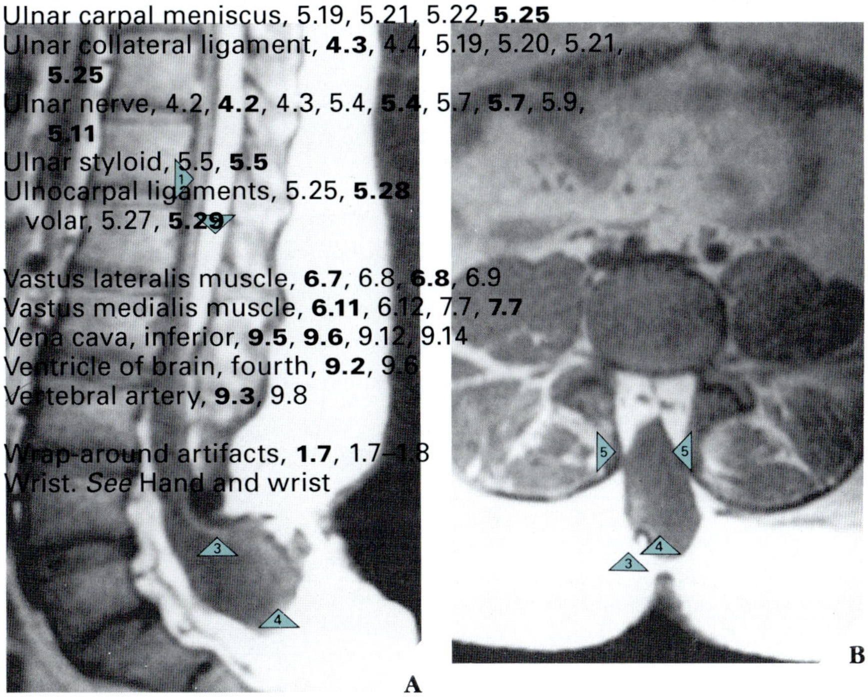

FIGURE 9.14. Lumbosacral Spinal Dysraphism. A Sagittal section [SE 800/20]. On this T1WI from a pediatric patient, a meningomyelocele is seen with a large low SI CSF sac containing a thin segment of tethered cord, which extends into the posterior spine. The dysraphic segment, or posterior element defect, extends approximately from the fourth lumbar vertebrae through the sacrum. The widened lumbar canal and the large amount of epidural fat in the posterior thoracic canal are also part of this congenital anomaly. Note that the lower thoracic and lumbar spinal cord is dysplastic and small. **B** Axial section [SE 800/20]. The defect in the posterior elements is well-identified in the axial plane. The spinal canal is large with anterior epidural fat. The meningocele extends posteriorly. A small lipoma in the tethered cord appears as a small high SI region within the posterior aspect of the sac. Arrow **1** Thin dysplastic cord. Arrow **2** Intracanicular lipoma. Arrows **3** Tethered cord to dysraphic segment. Arrows **4** Meningomyelocele. Arrows **5** Spinal dysraphism.